THE ATOMIC SCIENTISTS
A Biographical History

The Wiley Science Editions

THE ATOMIC SCIENTISTS
A Biographical History

Henry A. Boorse,
Lloyd Motz,
and
Jefferson Hane Weaver

WILEY

Wiley Science Editions
John Wiley & Sons, Inc.
New York • Chichester • Brisbane • Toronto • Singapore

Publisher: Stephen Kippur
Editor: David Sobel
Managing Editor: Frank Grazioli
Editing, Design, and Production: Publication Services

Library of Congress Cataloging-in-Publication Data

Boorse, Henry A. (Henry Abraham). 1904–
    The atomic scientists: a biographical history / Henry A. Boorse,
Lloyd Motz, and Jefferson Hane Weaver.
       p.   cm. — (Wiley science editions)
    Bibliography: p.
    ISBN 0-471-50455-6
    1. Nuclear physics—History. 2. Atomic theory—History.
I. Motz, Lloyd. 1910–   . II. Weaver, Jefferson Hane. III. Title.
IV. Series
QC773.B66 1989
539′.09–dc19   88-35181
                                              CIP

Printed in the United States of America

89 90 10 9 8 7 6 5 4 3 2 1

Preface

In a sense scientists, particularly physicists, have been outsiders at every stage of civilization. Before the modern scientific era—that is, before the beginning of Newtonian physics—scientists were outsiders because of their iconoclastic role in society. Driven by an irrepressible curiosity about the world around them and the need to know "how things work," they were the eternal doubters, the annoying questioners, and, above all, the heretical skeptics, who rejected theologic dogma, mysticism, and superstition as sources of belief. As such they were, of course, outcasts, constantly under the threat of persecution. With the development of Newtonian physics, not only did scientists emerge from the class of outcasts, but they also became esteemed, revered, and almost enshrined. However, scientists were still outsiders because they worked in mysterious ways to obtain incomprehensible results and to develop "strange" concepts and theories that were understood, if at all, by only a few other scientists. Since nonphysicists had no way of verifying the pronouncements of the leading physicists of the day, they had to accept such statements on faith. With the discoveries of the two great theories that guide us today in our search for an understanding of the universe—the theory of relativity and the quantum theory—scientists have become more isolated from the lay public than ever, for these theories deal with concepts and lead to conclusions that are extremely subtle and not explainable in terms of our "common sense" ideas.

The lives, the activities, and the drive of the great physicists, as presented in this book, reveal that although they were outsiders from the point of view of their professions, they were and are very much in the mainstream from the point of view of their daily lives and their relationships to those around them. They had their great successes, of course, and the incredible exhilarations that come with the discoveries of natural truths. But they had their disappointments, too, for the path of the scientist is tortuous, beset with false leads, and often comes to a dead end. The artist, the poet, the musician, the novelist, and the playwright create their own truths, in a sense, over and over again, but scientists cannot do this; they are severely restricted by nature, and each of nature's truths can be discovered only once. That, of course, implies an exclusion principle in that the names of two different scientists are rarely associated with the same great discovery. The competition among scientists "to be the first" is therefore extremely keen, sometimes traumatic, and often acrimonious.

Unfortunately, most people learn science, if they learn it at all, as a series of dull, abstract principles that sprang full-blown from the minds of dull abstract people. We believe that this book will disabuse people of this notion and reveal that the lives of scientists are as interesting and exciting as the science they pursue. Here we have integrated the biographies of the scientists with their science and presented their discoveries in a way that makes them more relevant to our own lives.

Henry Boorse
Lloyd Motz
Jefferson Weaver

Figure Credits

All photographs in this book are provided courtesy of the American Institute of Physics. In addition, grateful acknowledgment is made to the following:

Bainbridge Collection: page 209

Bancroft Library, John Hagemeyer: page 196

E. Scott Barr Collection: page 368

Berlin University, Westphal Collection: page 197

Margrethe Bohr Collection: page 183

Niels Bohr Library: pages 4, 7, 64, 74, 81, 119, 139, 176, 213, 235, 268, 275, 286, 307, 341, 353, 383, 395, 413, 439

Brookhaven National Laboratory: page 242

Burndy Library: pages 18, 24, 162

Case Western Reserve University: page 99

Cornell University, Sol Goldberg: page 397

Elliott & Fry, Ltd.:   page 310

Harvey of Pasadena: page 402

Herzfeld Collection: page 422

A. B. Lagrelius & Westphal: page 172

Lawrence Radiation Laboratory, Robert L. Weber: page 379

Los Alamos Scientific Laboratory: page 322

Massachusetts Institute of Technology: page 360

W. F. Meggers Collection: pages 77, 91, 96, 104, 109, 116, 124, 170, 224, 234, 255, 331

Meggers Gallery of Nobel Laureates: pages 98, 115, 133, 218, 293, 311, 337, 354, 363, 375, 378, 391, 400, 411, 438

Lotte Meitner-Graf: page 423

Moffett Studios: page 230

William G. Myers Collection: pages 48, 108

National Bureau of Standards Archives: page 45

Physics Today Collection: pages 57, 106, 165, 246, 303, 359

Francis Simon: page 280

Stanford University: page 443

Studio France Presse: page 362

Times WideWorld Photo: page 357

Uhlenbeck Collection: page 243

Weber Collection: page 217

Contents

1 The Foundations of Atomic Theory

For wheresoe'er I turn my ravished eyes,
Gay gilded scenes and shining prospects rise,
Poetic field encompass me around,
And still I seem to tread on classic ground.

—Joseph Addison, *Letter from Italy*

Atomism in Antiquity: Lucretius (99–55 B.C.)

To understand the ancient origin of the concept of atoms, we must recognize that science as we now understand it was not pursued in the classical world. The Greek intellectual was preoccupied with questions of philosophy. In his attempts to solve philosophical problems, the idea of the smallest particle of matter, an atom, came into being. Democritus, for example, who inherited the concept of the atomic theory from Leucippus, utilized it to solve a specific philosophical problem. Democritus's ideas were, in turn, adopted by Epicurus as a basis for a materialistic philosophy. This philosophy was preserved and widely disseminated through the Western world by the celebrated work *De Rerum Natura* of the Roman poet Titus Lucretius Carus, better known as Lucretius.

De Rerum Natura had the most profound effect on the thought of Western scholars some 17 centuries later, when experimental science was reborn. In England the atomic idea taught by *De Rerum Natura* was especially fruitful; it greatly influenced Robert Boyle, the founder of modern chemistry, and Isaac Newton, the father of mathematical physics. Their influence, in turn, was immensely powerful in affecting the views of the able empiricists whose experiments grounded and then extended atomic theory.

But to understand the work and its setting, its background needs a brief explanation. In antiquity, philosophy first concerned itself with efforts to explain the natural world. The oldest bits of systematic thought of this kind are attributed to Thales of Miletus, who lived about 600 B.C. He was concerned with reducing the "manifold of phenomena" to some kind of unity. To do so, he assumed that a "primary matter," modified in various ways, produced all the matter that we recognize about us. This idea of primary matter has survived with extraordinary persistence, and it will reappear from time to time in the following pages. "Prout's hypothesis," offered in 1815 by William Prout, is perhaps its most famous reincarnation. But Thales did not stop with a "philosophical" solution. Although his supporting arguments are lost, we know that he identified primary matter as water.

Heraclitus (ca. 480 B.C.), who followed Thales, was concerned with a very different problem, that of change. His experience convinced him of the thesis that

1

to be was to change. Therefore, he reasoned, the primary matter from which all things are made must exhibit this principle. Fire fulfills it well, for as it exists it is continually composed of different burning matter. Because it is not the same from one instant to the next, fire was identified with primary matter.

Parmenides, whose influence prevailed at about the same time as that of Heraclitus, was also concerned with the problem of permanency and change. But his conclusion, interestingly enough, was in direct contrast: permanency is real and change is only an illusion. Parmenides's argument, a metaphysical one, runs as follows:

> All that *is*, together forms being . . . that which is must be *one*, that is, it must possess unity, for if it were manifold, there would have to be something which divides it. But outside being, nothing is. Therefore, there is nothing that can divide being, and, therefore, being is *one*. This one being is also unchangeable. For what could be the meaning of change? It could mean either the transition from one kind of being to another . . . or the transition from non-being to being . . . . The transition from one kind of being to another kind of being . . . really amounts to being remaining what it is . . . . Because non-being is not, . . . it cannot become anything. (Van Melsen, p. 15)

One may be justified in having serious reservations about these arguments, which may seem tautological but, however we view them today, Parmenides made a profound impression on his contemporaries and on Democritus, whose opinions gained wide acceptance some 70 years later. Democritus saw that, if change is impossible, then a science of nature is not possible. To solve the problem, he contradicted Parmenides with the assertion that being is not *one*, but is divided in many "beings," each permanent and indivisible. These "beings" he called atoms. Democritus also asserted that *nothing* does exist—*nothing* implying what we would call a *void*. Thus, *to be*, for him, meant to be *atom* or to be *void*. According to this view, change consists only in the rearrangement of eternal and unchanging atoms. This part of his argument agreed with Parmenides. The permanence of atoms is real and the change one sees is an illusion since the permanent atoms always remain. To make change intelligible, he postulated that atoms are infinite in number and ceaselessly in motion in infinite space. All atoms consist of the same primary material. The only differences in atoms are their size, shape, and weight. These characteristics manifest themselves in differences visible in the different objects around us.

The ideas of Democritus, as noted above, were adopted by Epicurus as the basis for a materialistic philosophy. The purpose of this philosophy was to abolish superstitious fears of intervention by the gods in the world of humans and of the punishment of the soul in life after death. Epicurus attempted to do this by showing that the world is rational, that is, governed by laws of nature. All that exists is corporeal. The intangible is nonexistent or empty space. In this way Democritus's ideas were incorporated in Epicurus's philosophical system. As for punishment in the afterlife, he asserted that the soul perishes with the body. All of these beliefs were later faithfully expounded by Lucretius.

A careful reading of Lucretius's work *De Rerum Natura* offers some surprises. In the third paragraph following the opening of Book 2, Movements and Shapes of Atoms, he discussed the motion of dust motes visible in a beam of light. We have all seen such motions and perhaps dismissed them as arising from convection currents in the air (which, in fact, they are on this large scale). Lucretius, however, suggested something different—"their dancing is an actual indication of underlying movements of matter that are hidden from our sight . . . you . . . see many particles under the impact of invisible blows changing their course . . . this way and that, in all directions. You must understand that they all derive this restlessness from the atoms. It originates with the atoms, which move of themselves. . . . So the movement mounts up from the atoms and gradually emerges to the level of our senses . . ."

These words are indeed arresting, not only because they assert a kinetic theory of matter but also because they tell us that the observed motion of very small (actually microscopic) particles in a gas is a consequence of atomic agitation. Just this kind of motion was found for microscopic particles in a liquid by the British botanist Robert Brown in 1827. Ever since it has been known as "Brownian motion." At the time of its discovery its origin was a complete mystery, despite the fact that a kinetic hypothesis of matter had already been suggested by Robert Boyle, Robert Hooke, Daniel Bernoulli, and others. In fact, it was not until after Rudolf Clausius and James Clerk Maxwell had firmly established the kinetic theory of gases in 1860 that suggestions were even made as to the origin of the Brownian motion. It is interesting to speculate how much faster the progress of science might have been if the Victorian scientists had been as aware of Lucretius as their illustrious predecessors, Boyle, Newton, and Hooke, had been.

Another interesting statement which Lucretius makes a few paragraphs later in Book 2 relates to falling bodies: " . . . through undisturbed vacuum all bodies must travel at equal speed though impelled by unequal weights." This statement is, of course, completely at variance with Aristotelian ideas. The establishment of its validity, however, had to await the experimental methods in mechanics introduced by Galileo.

Of Lucretius, the author of *De Rerum Natura*, relatively little is known. He seems to have been of Roman origin and to have been born sometime between the years 98 to 95 B.C., dying in 55 B.C. Greek teachers of the Epicurean sect settled in Rome during Lucretius's youth and soon formed close ties with the governing class, which had developed a new taste for philosophy. The inference that Lucretius belonged to this class is found in the manner in which he addresses Memmium, a Roman aristocrat and friend of the author, at the beginning of *De Rerum Natura*. According to one account, Lucretius became mad as a result of taking a love philter, composed *De Rerum Natura* in lucid intervals, and finally died by suicide at the age of 44. This story is doubtful. A work characterized by such imagery and continuity of thought hardly seems the effort of an unsound mind. The internal evidence is that the poem was left in an unfinished state because of the death of the author.

Vortices and Particles: René Descartes (1596–1650)

The concept of the atomic theory had its origin in the philosophies of Democritus and Epicurus, whose ideas were preserved by Lucretius. But 17 centuries elapsed before those of learning again concerned themselves seriously with these ideas. With the rebirth of experimental science at the height of the Renaissance, the scholars of western Europe were caught in the grip of a germinating scientific revolution.

While Johannes Kepler was immersed in the researches that gave validity and new meaning to the Copernican heliocentricity hypothesis and Galileo Galilei was laying the foundation for Newtonian mechanics, there occurred at La Haye in the province of Touraine, France, an event of the greatest consequence to the subsequent development of mathematics and to the scientific outlook for more than a century—the birth of René Descartes on March 31, 1596.

René's father, Joachim, was a successful lawyer and counselor to the local parliament of Brittany. His mother, Jeanne Brochard, was, unfortunately, destined to live less than a year after his birth. Like many others who achieve intellectual fame, René was a delicate child; throughout his early years it was feared that he would not survive to attain maturity. But delicate health may not always be a detriment to intellectual development. In René's case, his afflictions led to a novel way of life. When he entered the newly founded Jesuit College at La Flèche in Anjou at the age of eight, his health was such that the rector accorded him the unusual privilege of pursuing his studies abed until midday. This practice, conducive to reflection in his active mind, was continued by Descartes throughout life. The eight years he spent at La Flèche he remembered not only with affection but with gratitude for the superior instruction. In later years he spoke of the school as one of the best in Europe.

But an increasing dissatisfaction with formal study led Descartes at 16 to leave La Flèche and to settle for a time in Paris. Instead of leading a social life in the capital, he spent a year in almost solitary study before turning again to formal instruction, this time at the University of Poitiers. His desire to understand people and society and to observe life firsthand led him to a new, and for a scholar, unusual act. In May 1617, at the age of 21, he enlisted as a volunteer in the military school at Breda, Holland. His two years at Breda are noteworthy in only one respect; he met there, quite by chance, the rector of the college at Dorchecht,

René Descartes (1596–1650)

Isaac Beekman, an accomplished mathematician, who became his good friend. This friendship continued through Descartes's later life.

After two years at Breda, Descartes enlisted in the army of the Duke of Bavaria. In November 1619, he was stationed at Neuberg on the Danube. Because of the severe weather he shut himself in a warm room (not in a stove as is sometimes fancifully stated) and gave himself completely to deep meditation. Here he conceived the basis for his own system of philosophy—the idea that what is seen clearly and distinctly must be true. To amass further experience, he continued his wanderings all over Europe for the next eight years, sometimes in military service and sometimes out.

In 1628, at the age of 32, Descartes had had his fill of wandering and decided to find a haven in which he could at last devote himself to philosophical reform and contemplation. Holland had won its independence from Spain and was enjoying a peaceful period of prosperity. It was a country in which new academies and universities were being founded in an atmosphere of religious toleration. Therefore, it was in Holland that Descartes chose to carry out the work to which he had dedicated his life. Here he lived for the next 20 years, as he described it, "in thirteen different places and in twenty-four different abodes." Despite his desire for domestic privacy, it must not be imagined that Descartes always led a solitary life. He had many friends, including Isaac Beekman, the mathematician, and Constantin Huygens, secretary to the Prince of Orange and father of the natural philosopher Christian Huygens, founder of the wave theory of light. Descartes was not hesitant in engaging in learned disputes. "Indeed," writes one of his biographers, "he had a special talent for dividing scholars among themselves, and for prolonging the disputes he had aroused."

Despite the absence of scholarly publication during his years of wandering, Descartes now began developing those ideas that had been maturing since his youth. His first work was a great treatise on the physical system of the world, entitled *Le Monde, un Traité de la Lumière*, which he finished in 1633. On the eve of its publication the news reached Descartes of Galileo's censure by the Inquisition. Always sensitive to Catholic authority and opinion, and realizing that his views, like Galileo's, agreed with a Copernican system of the world, he immediately withdrew the work. Although parts of *Le Monde* appeared subsequently in the *Discours de la Méthode* in 1637 and the accompanying essays *La Dioptrique, Les Météores*, and *La Géométrie*, together with further developments in the *Principia philosophiae* in 1644, it was never published in its entirety. His *Meditationes de prima philosophia* was published in 1641 and, finally, the *Traité des passions de l'aime* in 1649.

In the same year, 1649, M. Channut, French ambassador to the Swedish court and an admirer of Descartes, suggested to Queen Christina that, as a patroness of learning, she would do well to have this learned philosopher as an ornament in the regal circle. In due time Descartes received the royal command and, with reluctance, departed for Stockholm in October. But the northern climate and the necessity of interviews with the Queen at five in the morning were too much for his frail constitution. On February 1, 1650, he contracted an illness that developed into pneumonia. Ten days later he was dead. Not yet 54 years old, he had exerted a profound influence that continued to dominate European learning for more than another half century.

Because of Descartes's wide influence he is included among those who have shaped atomistic thought, even though his views were contrary to the atomic hypothesis. His system of the world as a group of vortices was widely accepted and was in fashion at Cambridge years after Newton had propounded the theory of universal gravitation. Descartes was the first to attempt a mechanical explanation of the solar system. His explanation did not endorse the notion of action at a distance but, instead, proposed a material method for the transmission of motion to heavenly bodies. Descartes proposed that celestial matter was everywhere in motion in the form of vortices, a single mechanical principle that explained the movements of the planetary system. To many contemporary scientists, Newton's idea of planets suspended in empty space and retained in orbit by an invisible force seemed the essence of occult doctrine. Nevertheless, Descartes's vortex theory, on closer inspection, proved to be mechanically unsound. It was not susceptible to calculation and precise prediction, characteristics that won for Newtonian ideas the adherence of 18th-century astronomers and brought science from the medieval world into the modern.

Descartes could not accept the theory of atoms and the void of Democritus. To a large extent, Descartes's system followed from the basic tenet expressed in Principle XVI of his *Principia*: ". . . it is absolutely inconceivable that nothing [i.e., emptiness] should possess extension." To exist, a void must have extension; but, says Descartes, emptiness cannot have extension. Thus, space, which is all around us, having extension, can have no void and, consequently, must be a continuum of matter.

The way in which Descartes worked out a world picture on this basis has interest because it produced ideas that affected the shape of physics in later years. Newton and Boyle were both strongly influenced by Descartes, Newton by his mathematical and inertial ideas and Boyle by his system of smallest particles. For, strange as it may appear, Descartes was forced to develop a theory of particles in constructing his scheme of the universe. The way in which a particulate scheme can exist in a material continuum was developed in his vortex theory. Descartes did this by imagining a group of vortices in which a particular vortex represents the vortex of the solar system. In the beginning, this central vortex, like the others, consisted of a number of small parts, each rotating around its own center as well as around the central start of the vortex. According to Descartes, this kind of motion ground the corners of the celestial matter so that the parts became rounded. The resulting globules were supposed, as the result of inertia, to move to the outside of the vortex; the parings of the original particles, ground to fine dust, tended to collect at the center of the vortex. This "finer matter" is the so-called *first* matter of Descartes and constitutes the sun and stars. The spherical particles are the *second* matter; they envelop the central accumulation of the first matter and constitute the heavens.

There is also a *third* matter produced from the original particles. Some of the filings produced by rubbing pass through the revolving spherical particles, are caught, twisted, and channeled. This coarsest material was assigned the property of *mass*. A body possessed more or less mass depending on the amount of coarse matter it contained. This material could also form a dense crust around a central nucleus. Such a "degraded" star could be caught up by another vortex and settle into the latter where its velocity was the same as the velocity of that part of the vortex.

Such phenomena accounted for the presence of the planets. One may compare these ideas of Descartes with the mode in turbulent theories of the origin of the solar system.

While these considerations sound very speculative and contrived to the present analytical mind, the whole system in its day was highly regarded by contemporary philosophers. It must not be overlooked that Descartes's ideas contained some very notable advances over previous systems and so paved the way for the modern astronomical ideas instituted by Newton. Thus, we recognize that, according to Descartes, matter in the solar system is not different from that on earth. By contrast, the Aristotelian system had taught that the sphere encompassed by the radius to the moon contained "corruptible" material; that is, it was a space in which change was possible. But the sphere beyond the moon was "incorruptible" and in it no change occurred. Descartes's system also banished spirits and genii, which even Kepler had considered as guardians of the planetary motions. Finally, it presented a view of the cosmos in which events occurred on a rational, physical basis as they do on earth. This rational view, despite its many shortcomings, marks Descartes's scheme as a substantial advance toward modern ideas.

Particles in the Atmosphere: Robert Boyle (1627–1691)

From the millions that constitute humanity, there occasionally emerge persons of exceptional character and ability who consciously devote themselves to the larger interests of all people. One such figure was Robert Boyle. Most of our information about his life comes from an abbreviated autobiography, *Philaretus*, written in his twenties, and a supplementary life by Thomas Birch. Both fascinating documents are included in the *Collected Works of the Honourable Robert Boyle* published by Birch in 1744. No one can read these histories without being deeply impressed by Boyle's Anglican conviction, his incorruptible integrity, and his high-minded view of the worth and dignity of all humanity. He conceived of himself as an instrument

Robert Boyle (1627–1691)

of the Divine plan for the betterment of humanity, a result to be achieved through charity and the application of Baconian philosophy—the improvement of the human condition through science.

In a day when science hardly existed and had little prestige, Boyle's social position and attractive, courteous personality exerted a powerful influence in raising its social status. He was one of the founding Fellows of the Royal Society and a member of its first council. Throughout his residence in London he regularly attended the society's meetings and remained one of its most influential members. His peers elected him president in 1680, but he refused the honor on the ground that the society's charter required the taking of an oath, an act that his conscience did not allow.

Boyle was deeply religious. It was said that he never mentioned God without a preliminary pause for veneration, a custom that his friends vouchsafed was unfailing over a period of 40 years. He spent much time in meditation, wrote many tracts on religion and theology, and was able to quote passages of Scripture as freely in Greek as in English.

Boyle was a director for 16 years (1661–1677) of the famed East India Company. He took the post because of his interest in propagating Christianity among native populations throughout the world, and he was influential in persuading his fellow directors to do more for the improvement of the social condition of the natives and less toward their exploitation. To the same end he personally supported translations of the Bible into Turkish, Arabic, and Malayan languages and into American Indian dialects for the Massachusetts Colony; he saw to the distribution of these translations to the natives. About a third of his income was spent for charitable purposes and for science.

Boyle never married, perhaps because of his religous conviction, but probably because of poor health. During the last 23 years of his life, when he lived in London, he lived with his elder sister Katherine, Viscountess Ranelagh. Despite his station and his wealth, Boyle lived very simply. He was sincere, exceptionally courteous, and considerate of the feelings of others. As a part of his self-discipline he schooled himself to listen carefully to the opinions of others. His interest in people made him much sought after in London, and he received a constant stream of visitors from abroad, although he complained privately that many of them wasted his time. He freely published the details of his experiments and his scientific ideas, never attempting to use them for gain or for self-aggrandizement.

The details of Boyle's life are useful for an understanding of his outlook, his milieu, and his influence on science. His ancestors were a landed Herefordshire family who could trace their lineage to pre-Conquest times. His father, after a Cambridge education and training in law at the Middle Temple, London, saw little future for himself in England and emigrated to Ireland in 1588, where he eventually accumulated a huge fortune. He built a great country estate, Lismore, in the province of Munster, and, in 1620, was created Earl of Cork. His fourteenth child and seventh son, Robert, was born at Lismore on January 25, 1627.

The boy's schooling was begun at home by tutors, but at the age of eight he was sent to Eton. There he showed a great interest in learning and apparently found reading more stimulating than play. He had been at Eton almost four years when his father settled him briefly at Stalbridge, a new estate acquired in Dorsetshire. In October of that year (1638) at the age of 12, he was sent with his brother and a Swiss tutor to Geneva to complete his education.

Unlikely as it may seem, the most determinative event for his later life was a violent electrical storm that occurred during his stay in Geneva. It made a profound and lasting impression on young Boyle. He viewed the storm as a demonstration of Divine power and the awe he felt made him resolve to devote his life to those activities that promote God's work on earth. But the channel through which to perform his self-imposed task was revealed to him only as the result of a chance acquaintance which he made after his return to England. Because of pecuniary difficulties resulting from the rebellion in Ireland in 1642, and the temporary loss of the family estates, his return had been delayed until 1644, when he was 17. During a brief sojourn in London with his sister, Lady Ranelagh, he met a Polish Protestant refugee, Samuel Hartlib, who supplied the direction he was seeking.

Hartlib was a utopian, interested in founding a "philosophical" college in which experiment would be the method of exploring nature, and useful knowledge the aim of inquiry. Boyle was immensely impressed with these ideas; in a few months, when he retired to Stalbridge, which he had inherited on his father's death, he began to study science in earnest: anatomy, medicine, natural philosophy, astronomy, chemistry, and physics, together with theology. Except for two years in Ireland, which were required for the supervision of his property, Boyle spent most of the next nine years at his Dorset manor. All the while he maintained his contacts with science in London through Hartlib, and became especially interested in chemistry.

On his return from Ireland in 1654, Boyle was persuaded by John Wilkins, warden of Wadham College, to take up residence in Oxford. Wilkins was a devoted advocate of the new experimental philosophy. Before going to Oxford he had been the moving spirit of a small but distinguished scientific group in London. He had a remarkable gift for inspiring friendship and respect, and an unusual talent for bringing people together to exchange ideas and for serious discussion of experimental natural science. The group that met regularly in his rooms at Wadham included John Wallis and Seth Ward, respectively the Savilian professors of geometry and astronomy, Johnathan Goddard, warden of Merton College, Ralph Bathurst, fellow and subsequently president of Trinity College, Dr. Thomas Willis, physician, later Sedleian professor of natural philosophy, and Christopher Wren, fellow of All Souls College, destined to be the celebrated architect who would rebuild London after the great fire of 1666. Some of these men had been part of Wilkins's group in London; later all became founding fathers of the Royal Society.

Wilkins's invitation must have been very attractive to Boyle because it offered a congenial and stimulating atmosphere in which to carry on his theological studies as well as his experimental science. But Boyle always felt that he was not a "Professor of Philosophy nor a gown-man." This conviction, coupled with the state of his health and the desire for sufficient space to experiment, caused him to reside in lodgings in the High Street next to University College. His stay in Oxford began in June of 1654, when he was 27 years old, and continued for 14 years.

As far as physical science was concerned, these were the great productive years of his life. During this time he conducted his famous experiment on the properties of the air and published the observations on which he had reflected since he began taking a serious interest in science. Boyle's research on the properties of the atmosphere was inspired by von Guericke's sensational experiments at Magdeburg, Germany, described in a book entitled *Mechanica hydraulica-pneumatica*, published in 1657 by a Jesuit, Gaspar Schott, at Frankfurt. Boyle obtained a copy and was fascinated with all he read. Von Guericke had shown that if two hollow metal hemispheres were placed base to base and air was removed from the spherical interior thus formed, the pressure of the atmosphere was sufficient to hold the halves together against the force exerted by two teams of six horses pulling in opposite directions. But the air pump used to evacuate the hemispheres seemed to Boyle capable of improvement and the experiments susceptible to wider inquiry. Boyle needed an assistant who knew what the pump was meant to do and who was inventive enough to make a better one. Willis, a member of the Wilkins group, recommended a Christ Church student named Robert Hooke, who had been his assistant in chemistry. Hooke, then in his early twenties, accepted Boyle's offer, went to live in his lodgings, and very quickly produced a successful pump for the pneumatic experiments. This effort began a fruitful association between the two men that lasted throughout Boyle's lifetime and gave Hooke his entrée into the Royal Society and the opportunity to achieve the fame he later enjoyed. Without Boyle's friendship, it is doubtful if his genius would have been realized. Hooke was not prepossessing either in appearance or manner. He lacked tact and had little skill in personal relations; moreover, in 17th-century England, he suffered the even greater handicap of not being a "gentleman."

In 1660, the first results of the pneumatic experiments were published in *New Experiments, Physico-Mechanical, Touching the Spring of the Air and Its Effects*. Two basic results were recorded: (1) Air, like other material substances has weight, and (2) it contains a "vital Quintessence" essential to animal life, although much of it serves no such purpose. Boyle thus recognized the nonhomogeneity of air, a fact not proved until the discovery of oxygen by Joseph Priestly. From the discovery that air has weight Boyle concluded, after further experiments, that the atmosphere near the earth is compressed and that its pressure is sufficient to support a mercury column about 29 inches high. *New Experiments* was followed by a series of related essays, many of which had been written or partially written during the preceding 15 years. In the *Certain Physiological Essays* (1661) and the *Origins of Forms and Qualities According to the Corpuscular Philosophy* (1666), as well as in the *Sceptical Chymist* (1661), Boyle set forth his unqualified adherence to the corpuscular (atomic) hypothesis. It is difficult, however, to find succinct statements of his views in these papers because his style is prolix and he talks around his subject, often without coming directly to grips with it.

The Wilkins group, which had been so stimulating to Boyle when he first moved to Oxford, gradually broke up as a result of the changes accompanying the end of the Commonwealth and the advent of the Restoration. Wilkins himself left Oxford in 1659 to become, very briefly, the master of Trinity College, Cambridge. Goddard had left the year before to take up an appointment as professor of physics in Gresham College, London. The Royal Society was founded there in November of 1660. Thus, the center of science in England was transferred to London. In 1662 Boyle recommended his assistant Hooke as the Society's curator, ending for a period their close association. Boyle's experiments, however, continued with other assistants. Despite periodic trips to London, he did not move there permanently until 1668 when he was 41 years old. From then until his death 23 years later, he made his home with his sister Lady Ranelagh in Pall Mall.

In 1672 Boyle published an "Essay about the Origin and Virtue of Gems" in which he noted the regular structure of crystals as evidence for an atomic theory of matter. This essay was followed in 1675 by an account of experiments (made more than 20 years earlier) on "Quicksilver Growing Hot with Gold." Boyle believed that he had evidence for transmutation. Newton, then immersed in his own secret alchemical experiments, wrote to Henry Oldenberg, the secretary of the Royal Society, that Boyle's result "may possibly be an inlet to something more noble, not to be communicated without immense damage to the world . . ." In his Cambridge years, Newton, despite his great and penetrating insight into the mechanism of nature, nevertheless had a strange streak of magic in his makeup, a quality that has been illuminated by his biographers.

In reflecting on Boyle's research, it is interesting to observe that his methods were rather different from those of his contemporaries and very much like those of the present day. Rather than attempting much of the work with his own hands (although he did not abstain by any means from actual experimentation himself), he employed assistants to construct equipment and carry out manipulations while he supplied the ideas and supervision. His voluminous publications were also produced in much the same way; he did not write his tracts but dictated his thoughts to secretaries.

These deviations from the usual procedures of his day are quite understandable in view of his special circumstances. Because of his social position and affluence, it would have been considered improper for him to do much of the laboratory work himself. In fact, it was considered socially disgraceful for a man of title to be familiar with workmen and the processing arts. But such familiarity was necessary if science was to progress. This is one of the reasons that, throughout his life, Boyle resisted all attempts on the part of others to obtain a title or high public office for him. He commented in his autobiography on his fortunate situation, remarking that a lower birth would have exposed him to the inconveniences of a "lower descent." On the other hand, had he been the eldest of a great family, he would have been forced into entanglements in public affairs to further family interests.

Another reason for his research and writing methods was his health. Shortly after he took up residence at Stalbridge, and before he was 20, Boyle began to suffer from severe attacks of kidney stones. Throughout his adult years he was almost excessively anxious about his health and became so addicted to dosing himself with all kinds of nostrums that his physician warned him that his cures were sometimes worse than his disease. His regimen also consisted in observing a rigid diet, so that he never ate for pleasure but only to keep himself alive. In physique he was tall and thin, and in later years his strict diet made him look pale and emaciated. Gilbert Burnet, Bishop of Salisbury and his longtime friend, said in Boyle's funeral sermon that "he escaped the smallpox, but for forty years laboured under such a feebleness of body and such a lowness of strength and spirits that it was astonishing how he could read, meditate, try experiments and write as he did."

Boyle did write all his life. In his later years, during the 1680s, his papers were mostly of a religious nature. In the latter part of the decade he began to fail, but he continued writing industriously to the end. Death overtook him on December 30, 1691, as he lay quietly in bed correcting proofs of his *Essay on the General History of the Air*, a subject on which he had spent a lifetime of reflection. As Birch, his biographer, remarks: "His life was spent in the pursuit of nature . . . and in the most rational as well as devout adoration of its divine Author."

Both in his *New Experiments* and in his *General History of the Air*, Boyle hypothesized that air is composed of several kinds of particles that have different functions. In the former paper he considered that the elasticity (compressibility) of the air arises from static elastic atoms in contact or, alternatively, from atoms being caught up in a whirling motion, an idea evidently borrowed from Descartes's vortex theory. In his *General History of the Air*, which reflects a more mature period, Boyle was more explicit and talked of three kinds of particles composing the atmosphere: first, vapors from various sources; second, smaller or "subtile" particles that carry magnetic effects and light; and, third, particles that produce the elastic behavior of the air. (The idea that elementary particles are of different sizes is also in accord with Cartesian notions.) But the elastic particles, which might be similar to coiled springs, he seemed to imply, may be *mixed* with others that beat off their neighbors solely because of their whirling motion.

Boyle's outlook was strongly conditioned by Descartes. His experiments and the questions they raised went straight to the heart of the Cartesian physics, that is, that *extension* is synonymous with matter, and that a true vacuum is impossible because it can have no dimensions. If it could be shown that a true vacuum existed, then Descartes's position would have to be abandoned. Boyle could not do this because removing the air from a receiver was insufficient to produce a vacuum. What about the more "subtile" material, the ether that the Cartesian theory envisioned? Boyle tried to demonstrate the existence of this subtle material in several experiments that he described in a later publication, but his attempts yielded no positive results. What he has to say about the problem, however, is basic not only to atomic theory but to any physical theory as well.

Matter and Motion: Robert Hooke (1635–1703)

If by some miraculous machine we could be transported back through three centuries of time to the city of London, then being rebuilt after the great fire of 1666, it is quite likely that we would soon notice an unusual-looking figure as it scurried along on its business through the city streets. This figure would be none other than Robert Hooke, for, if we are to believe Richard Waller, secretary of the Royal Society, "he went stooping and very fast . . . having but a light body to carry, and a great deal of spirits and activity . . . "

Waller published Hooke's *Posthumous Works* in 1705 with a dedication to Newton, who was then president of the Society. Such a dedication is very difficult to understand, unless it was made as a gesture to Newton's eminence. Hooke had had the misfortune to cross Newton's path with respect to priorities for discoveries in optics and celestial mechanics. Newton had a pathological aversion to engagement in any kind of controversy. As time passed, his involvement with Hooke filled him with such a loathing that he would neither publish his book, *Opticks*, nor accept the presidency of the Royal Society while Hooke was alive. It must be apparent, therefore, that a posthumous publication with a dedication approved by Newton must make Hooke appear as unfavorably as possible, for Newton could be both petty and mean and seldom forgot or forgave any slight or argument. With this background it is interesting to see what Waller has to say about Hooke.

As to his person he was but despicable, being very crooked, though I have heard from himself and others that he was straight till about sixteen years of age when he first grew awry, by frequent practice turning with a turnlathe and the like incurvating exercises being but of a thin weak habit of body, which increased as he grew older so as to be very remarkable at last: this made him but low of stature though by his limbs he should have been moderately tall. He was always very pale and lean, and latterly, nothing but skin and bone . . . his eyes grey and full with a sharp, ingenious look while young . . . his chin sharp and forehead large, his head of a middle size. He wore his own hair of a dark brown color very long and hanging neglected over his face, uncut and lank, which about three years before his death, he cut off, and wore a periwig. He went stooping and very fast (till his weakness a few years before his death hindered him), having but a light body to carry, and a great deal of spirits and activity, especially in his youth.

He was of an active, restless, indefatigable genius almost to the last, and always slept little to his death, seldom going to sleep till two, three or four o'clock in the morning and seldomer to bed, often continuing his studies all night, and taking a short nap in the day. His temper was melancholy, mistrustful and jealous, which increased upon him with his years. (Hooke, pp. xxvi–xxvii)

Most of this description is probably correct, but Waller goes on to say that in his younger years Hooke freely communicated his scientific ideas "till some accidents made him a crime close and reserved," and that he always led a "monastic life . . . so like a hermit or cynic too penuriously." That this estimate is wrong in important respects has become clear as the result of the publication of Hooke's diaries in 1935, the tercentenary of his birth. Hooke never meant the diaries for publication, and their contents were of such a candid nature that for more than two centuries it was considered unsuitable to put them in print. It seems clear, therefore, that the impression we get from reading the diaries is entirely trustworthy: that he led a very active, full life; that he possessed warm, human feelings; and that, unlike Newton, he harbored no lasting grudges or ill will. A brief review of his life will help to give some perspective of his place in science, his effect on his contemporaries, and the achievements that he bequeathed to those who came after him.

Robert Hooke was born the son of a curate in a small parish at Freshwater in the Isle of Wight on July 18, 1635. He was a sickly infant afflicted with a weak stomach and, as a young boy, was troubled almost continuously with headaches. Despite these painful handicaps he was lively and quick to learn, but as regular lessons under his father made his headaches worse (perhaps with reason!) the arrangement was given up. For several years thereafter he had no regular schooling but spent his time making mechanical contrivances and sketching, for which he showed such unusual aptitude in the latter of these endeavors that, on his father's death in 1648, he was sent to London at the age of 13 with the idea that he would be an apprentice to the great portrait painter, Sir Peter Lely.

But Robert soon tired of art and entered the Westminster School instead where he made excellent progress in mathematics, especially geometry, and in mechanics, Latin, and Greek. In 1653, at 18, he entered Christ Church, Oxford, as a chorister; but because church music had been suspended during the Commonwealth, he was in effect on a small scholarship. After two years at Oxford, he was engaged as an assistant in chemistry by Dr. Thomas Willis, one of the illustrious group dedicated to science whose focus was John Wilkins, then master of Wadham College. Wilkins was the moving spirit behind the new experimental philosophy that was growing up in England, and perhaps more than any one man he was later responsible for the founding of the Royal Society. Through active support of science Boyle had just been lured to Oxford, and because of similar interests Wren, then a fellow of All Souls, became a member of the club. When Boyle was casting about for an intelligent assistant to improve his air pump, Willis suggested Hooke. The resulting engagement determined the whole of Hooke's life, for he was ever afterward associated with Boyle and Wren and, through them, with the Royal Society and the work that it sponsored.

There is little need to comment on the great success of his pump. The results established Boyle's reputation as a scientist and Hooke's as an able experimenter and mathematician. He continued with Boyle until 1662 when the Royal Society, founded two years earlier at Gresham College in London, felt that it needed a curator. The curator was required to take charge of the Society's apparatus and to "furnish three or four considerable experiments" every day they met. Since they met once a week, this was a formidable assignment, but Hooke accepted it, after having been recommended by Boyle. It is a monument to Hooke's fertile imagination that he carried out his duties with outstanding skill for many years. Certainly his demonstrations and discussions contributed in no small measure to the success of the society in its early years.

The Royal Society's meetings at Gresham soon led to an academic appointment for Hooke. The distinguished mathematician Isaac Barrow, who came to Gresham at the same time as Hooke, resigned as professor of geometry in 1664 to become the first Lucasian professor of mathematics (and the predecessor of Newton) at Cambridge. Hooke was chosen to succeed him, but owing to an irregularity in the election he was not confirmed in the post until the next year. At about the same time, he also accepted the obligation of delivering the Cutlerian lectures on industrial processes at Gresham. Despite ill health and his later duties as surveyor for the city of London and secretary of the Royal Society, he carried on the curator's duties and the two lectureships for over thirty years. It is not difficult to believe Waller's assertion that he seldom went to sleep "till two, three or four o'clock in the morning."

Shortly after becoming a Gresham professor in 1665, Hooke published his *Micrographia* under the auspices of the Royal Society. The work represents not only the results of his original research and observations using the microscope he had constructed, but it contains a number of beautifully drawn illustrations that further emphasize his unusual gifts as a draftman. Unlike most of the scientific work of that time, Hooke's work was written in English and, thus, was readily accessible to a wide audience. It was immediately popular and earned the praise of the learned world. Samuel Pepys, the indefatigable busybody, said of it: "Before I went to bed, I sat up till two o'clock in my chamber reading Mr. Hooke's *Microscopical Observations*, the most ingenious book that I ever read in my life." We know also that Newton read it with care and made copious notes on it.

In September 1666, the great fire of London destroyed approximately 13,000 dwelling houses, more than 80 parish churches, St. Paul's Cathedral, 44 city companies' halls, and most of the other public and semipublic buildings, including the Guildhall and the Royal Exchange. The fire was hardly cold before Wren, Evelyn, and Hooke presented plans for rebuilding. Hooke's plan, submitted through the Royal Society, does not survive, but it appears to have been one with streets at right angles to one another, as in modern American cities. In the end none of the plans was adopted, but Wren and Hooke became the dominant members of a rebuilding committee for the city.

The reconstruction was at its height between 1668 and 1674; these were years of furious activity for Hooke. He spent countless hours in conferences, committee work, inspections on site, conferring and transacting business with Wren. From these meetings he would rush away to give his Cutlerian or Gresham lectures, to conduct an experiment at the Royal Society, to a book auction, or to his favorite diversion—discussing some aspect of science with a friend in the coffee houses, the taverns, or at a private home. His closest friend and associate was Wren, but his relationship with Boyle also remained close and he often dined with him and his sister, Lady Ranelagh. He had many friends in the Royal Society and his diary abounds with their names.

In 1667, Henry Oldenberg, who had been the active secretary of the Royal Society (two were originally appointed, the famous and able John Wilkins being the other), died and Hooke was elected in his place. Oldenberg's passing must have been a great relief to Hooke for there was cordial dislike between the two from the start. Hooke had serious disputes both with Oldenberg, about credit for the invention of the spring-driven timepieces, and with Newton, about discoveries in optics. It seems clear that Oldenberg did his best to keep Newton's dislike of Hooke active. Oldenberg, despite his important service to the Royal Society, including his founding and editorship of the *Philosophical Transactions*, the earliest periodical in the world devoted exclusively to natural science, appears to have been something of a rascal. It is possible that he passed the details of Hooke's spring-driven clock (which Hooke had tried to patent about 1660) to Christian Huygens and then, in turn, tried to patent Huygens's watch inventions for which Hooke claimed precedence. Whatever the merits of this dispute, it is clear that Hooke, with his fertile mind and hundreds of inventions, was too busy thinking up new ones to perfect any more than a few of the old ones. Oldenberg, as secretary, corresponded widely with academies and persons on the Continent and made a trade of intelligence. Whether any of this was treasonable is also questionable. In June 1667, at the height of the Dutch War, Oldenberg was arrested and locked up in the tower. He was released two months later without any formal charge ever having been placed against him.

Hooke's term as secretary was not notable and ended in 1682 when he was 47. He continued to be active as curator, in his lectureships, and to report on a variety of subjects to the Royal Society. Much of this activity is described in his *Posthumous Works*. In 1687 his niece, Grace Hooke, to whom he was deeply attached, died at the age of 28. He never seemed to recover fully from this blow, and the death in 1691 of Boyle, with whom he had always been intimate, must have increased his unhappiness.

Despite his prodigious energy, Hooke suffered from poor health throughout his life, being continually affected with dizziness, noises in the head, an uneasy digestion, and insomnia. He continually experimented on himself with all kinds of medicines and purges and endured periodic bouts of melancholy. Hooke also suffered emotionally all his adult life because, by 17th-century standards, he was not born a gentleman; his life, however, was spent largely in the company of men who were so regarded. Because of this anomalous status, his directness was considered by many as uncouth and his lack of social graces must have prejudiced many who might have admired his genius. It is hardly any wonder that in a life so afflicted by poor health and a sensitivity to his social standing, he should have been accounted by many as ill-tempered and morose.

In 1696 his health, always precarious, became seriously impaired and his condition grew steadily worse in the few years that remained to him. He gradually lost his sight and his strength. In the last year of his life he was bedridden. He died on March 3, 1703, at the age of 67.

In this abbreviated biography much of the detail of Hooke's scientific contribution has been passed over. His best-known contribution is the law of the elastic behavior of solids, known as Hooke's law. In its simplest form it may be stated as follows: "within the elastic limit in a solid, extension is proportional to the applied stress." This law is of the most fundamental importance in the design of structures, instruments, and mechanical devices. Hooke had an acute physical insight and was the first to state clearly that the motions of bodies in the solar system must be regarded as a mechanical problem. His astronomical views and his presentment of a universal gravitation were sufficiently profound to bring him into sharp conflict with Newton, but he was unable to develop his ideas for lack of mathematical skill. He demonstrated the rotation of Jupiter and reasoned that the orientation of comets' tails must be due to a repelling action originating in the sun. His efforts to discover motion of the "fixed" stars later led to Bradley's discovery of the aberration of light.

In the field of optics he developed an imperfect wave theory of light with the vibrations transverse to the direction of propagation. He anticipated the idea of interference, and he independently observed the diffraction of light. As a result he was able to discover and explain the property of "resolving power," the limitation on the magnification of an optical system. His interest in the improvement of microscopes and their use resulted in his *Micrographia*. He showed how to improve the illumination of objects under examination and pointed out the advantages of liquid immersion of the object and the objective lens of the microscope.

Coupled with his gifted insight into the structure of nature were Hooke's great mechanical gifts. His development of the air pump for Boyle only began a long train of inventions that greatly improved scientific devices. In the field of astronomical instruments he invented telescopic sights, increased the accuracy of measurement by the addition of fine screw adjustments, and designed a clock-driven telescope mount so that a given area of the sky could be kept under constant observation without the necessity for human adjustment. In the field of timekeeping, he developed spring-driven clocks and invented the first machine for cutting clock wheels. He invented what we now call the "universal" joint (used in automobiles to transmit power from the engine to the rear driving wheels) and the spirit level with the moving bubble, which is essential to all building construction and in many scientific instruments. Returning to optics, Hooke invented the iris diaphragm, so necessary for controlling the sharpness of images and now used in all cameras. This list of innovations is by no means all that came from his fertile brain, but it shows the great range of his mind, his genius for invention, and his outstanding ability to construct the visions of his imagination.

One of the principal reasons for including Hooke as an eminent atomist lies in his insistence on the kinetic view of matter. In this subject he seems to have been more penetrating than Newton. If we are to infer anything from Proposition 23, Theorem 18, of Book II of the *Principia*, then we must conclude that, in addition to being an atomist, Newton had specific views about the nature of a gas. He assumed that gas atoms were static but mutually repulsive with a force that varied as the distance. In this respect, however, there appears to be a curious dichotomy in Newton's views, for he says in Query 5 in his *Opticks*, "Does not . . . light . . . " act upon bodies "heating them, and putting their parts into a vibrating motion wherein heat consists?" The same idea that there is internal motion in bodies occurs in Query 31 when he speaks about the heat produced in chemical mixing. But, further on in the same query, he observes that this "vast contraction and expansion seems unintelligible, by feigning the particles of air to be springy and ramous, or rolled up like hoops [a model suggested by Boyle], or by any other means than a repulsive power." Whatever he precisely intended, John Dalton understood him to hypothesize static atoms and adopted his hypothesis without question.

While Hooke's kinetic hypothesis differs from the current view, nevertheless his advocacy of such ideas was much in advance of his time. He considered the particles of all bodies to be in incessant motion, those of differing mass having different speeds. A heavy particle has a low vibration frequency and a light one, a high frequency. Energy exchanges take place only between particles having the same frequency. He also thought that all particles everywhere are immersed in a subtile fluid (we meet again the idea of an all-pervading ether) that communicates motion. Hooke's explanation of the manner in which the earth holds its atmosphere is surprisingly modern, although his explanation of gas expansion is not.

A *Wave Theory of Light: Christian Huygens (1629–1695)*

In the physical sciences one finds few investigators born into families of power and influence who could also write their names in large letters on the history of their chosen subject. Despite their general rarity, we find two such men not only in the same century but as contemporaries with close birth dates and almost identical life spans. Robert Boyle, son of the first Earl of Cork, the man who founded modern chemistry and who, for years, was one of the mainstays of the Royal Society, was born on January 25, 1627, and died December 30, 1691. Christian Huygens, son of the outstanding intellectual, poet, and diplomat, Constantin Huygens, grandson of the widely influential secretary of William the Silent, was born April 14, 1629 and died June 8, 1695.

Constantin Huygens, Christian's father, has been considered by some critics as the most brilliant figure in Dutch literary history. He not only displayed extraordinary gifts in music, art, and poetry, but he also possessed unusual physical beauty and vigor and was one of the most accomplished athletes of his time. As a young man he went to England several times in various diplomatic capacities, studied at Oxford, and, in 1622, when only 26, was given an English knighthood by King James I. This paragon, for no other adjective seems to fit him, married his cousin Susanna van Baerle in 1627 and settled in The Hague. Four sons and a daughter were born to this marriage, Christian being the second son. Both Christian and his elder brother Constantin early showed signs of intellectual brilliance and they were carefully educated at home under their father's watchful supervision. When Christian was eight years old his mother died, and after a short time the family moved to a large new house at Voorburg, not far from The Hague. Here the elder Huygens, a member of the Council of State, entertained many visitors from France, including René Descartes who was then composing the philosophical and mathematical works that had such a profound influence on the thought of Europe for many decades to come. It can easily be imagined how strong Descartes's influence was on the Huygens family and how much it affected the outlook of Christian who was already beginning to show signs of unusual ability in mathematics. In 1645 both Christian and Constantin were enrolled at Leiden University. Descartes's *Prin-*

Christian Huygens (1629–1695)

cipia had just been published the year before; as a consequence, during the next few years the university became a battleground between the ideas of Aristotle and Descartes. In this atmosphere of intellectual ferment Christian continued his studies in mathematics and commenced others in mechanics. After two years he left Leiden and continued for two more at the College of Breda where he also studied law. But mathematics soon commanded his complete attention and in 1651, at the age of 21, his first published work, *Cyclometriae*, a treatise written to expose the mathematical fallacies of Gregory de St. Vincent, appeared. This was followed in 1654 by a larger mathematical work in which many geometrical problems were solved algebraically, an accomplishment that established him among the leading mathematicians of the time. In the same period he began working on the improvement of the telescope, especially the reduction of chromatic and spherical aberration errors. An improved objective lens, made with the help of his brother Constantin, led to the discovery in 1655 of Titan, the sixth satellite of Saturn, although another year elapsed before Huygens announced the discovery of Saturn's rings. Previous telescopes had not afforded sufficient resolution to determine the exact nature of the puzzling image of this planet.

In July 1655, Huygens, then 26, made his first trip to Paris to meet the leading astronomers and mathematicians of France. This visit not only encouraged his work in astronomy but impressed on him the need for more accurate timekeeping in such measurements. The solution of this problem led him to investigate the pendulum as an isochronous element and to apply it to the movement of a practicable clock. It must be remembered that the clocks of this period were notoriously inexact timekeepers. Although Galileo had noted the approximately isochronous nature of pendulum swings as early as 1581, only a few crude applications of this observation to clocks had been made. Huygens's first pendulum clock was completed in 1657, and the *Horologium* describing his work on the clock appeared the following year.

The necessity for accurate timekeeping was not only a scientific astronomical problem but a practical one as well. The problem of the determination of longitude at sea still remained to be solved and an accurate clock was needed for this solution. The States-General of the Netherlands had offered a prize of 10,000 florins for a method of longitude determination. In 1641 Galileo had submitted a method based on occultations of the moons of Jupiter together with an impractically crude pendulum clock, but this proposal was not accepted. To develop a suitable marine clock Huygens examined the amplitude-time dependence of pendulum motion and found that isochronous vibrations required a cycloidal motion of the pendulum bob. To achieve such a motion in an actual model, he hung the lower rigid part of the pendulum from a thin flexible metal strip that could fit itself, as the pendulum was drawn aside, to the contour of a cycloidal-shaped metal guide. One of these was located on each side of the pendulum support. Huygens patented this device and made some profit from it, but its usefulness was short-lived.

The pendulum clock was never practicable at sea owing to the ship's motion; moreover, in this particular model, the friction of the flexible metal strip on its cycloidal guide introduced errors large enough to compensate for the introduction of the guide itself. For small oscillations Huygens showed that the simple pendulum was also highly isochronous. The use of such oscillations by Clement in London in 1680, combined with his anchor escapement, made the cycloidal pendulum obsolete. Nevertheless, attempts to improve the pendulum clock as a timekeeping instrument at sea occupied Huygens throughout his life.

In November 1660, Huygens was again in Paris discussing Saturn and lens-grinding with astronomers, pendulums with mathematicians, and timekeepers with the clockmakers. He went to meetings of leading scientists called the "Montmor" group (after their host) where he met many of the savants who later constituted the French Academy of Sciences. In March 1661, he departed for London to meet the men who had organized the Royal Society and who were meeting regularly at Gresham College—in particular, Brounker, Wallis, Goddard, Boyle, and Oldenberg. The latter, as first secretary of the Society, remained a faithful correspondent of Huygens and kept him informed of the Society's activities. Even after assuming leadership in the French Academy, Huygens always regarded the Royal Society with the greatest admiration and considered it the preeminent scientific society in existence. Boyle's work on the properties of the air impressed him especially. On his return to Holland he had a copy of Boyle's pump made so that he could repeat the experiments himself.

In 1663 Huygens was again briefly in London to study the organization of the Royal Society. Subsequently, he went to Paris to receive an award from Louis XIV for his development of the pendulum clock. He returned to Holland after wintering in the French capital, but it was clear that he would soon be recalled to a position in the new Academy of Sciences that Colbert, the King's minister, was arranging. Two years later, in 1666, the Academy came into being, and Huygens was installed in the Bibliothèque du Roi to guide its activities.

His own work entered a new phase as he turned from astronomy and clock-making to pursue wider problems in mechanics. Most elementary textbooks in physics dealing with what we now call Newton's three laws of motion seem to imply that these concepts emerged full-blown from Newton's mind without any previous understanding of them by his predecessors or contemporaries. Despite Newton's genius, it is clear that what we call the first law was known to Galileo, Descartes, and Huygens; with respect to the second law, Newton himself says that "what Mr. Huygens has published since about centrifugal force I suppose he had before me." In fact, Huygens must have used the concepts behind the second law in 1659 to arrive at his theorems on centrifugal force. The proper appreciation of the second law, which relates force, mass, and acceleration, also requires that a distinction be made between mass and weight. But Huygens did this as well in his treatment of centrifugal force 1659.

The results of Huygens's study appeared in 1673 in his great work, *Horologium oscillatorium*. It contained, in addition to his studies relating to the pendulum and its period of oscillation, the solution of the general problem of the center of oscillation, the theory of evolutes, the cycloid as its own evolute, theorems on the composition of forces in circular motion, and the general idea of the conservation of energy. We know that Newton read the work and admired it and that he considered Huygens the "most elegant writer of modern times." At about this time Gottfried Wilhelm von Leibnitz, the great German mathematician, and Huygens were intimately associated, Leibnitz having come to Paris in 1672. In 1674 Huygens presented to the French Academy Leibnitz's first paper on differential calculus, a study in which Huygens never became proficient.

It is appropriate at this time to say something about Huygens's health, which was never robust. Ever since his youth he had been subject to a kind of debility accompanied with violent headaches and insomnia. In 1670, little more than three years after his Paris appointment, he had an attack of such severity that he made preparations for his own death, including arrangements for the transfer of his papers to the Royal Society. However, he recovered sufficiently in a few months to travel to his home in Holland where he spent the winter, returning in the spring of 1671. In March 1676, he again became ill, but before being completely incapacitated he returned to The Hague. This time with Colbert's permission he stayed for two years, not returning until June 1678. During this period Olé Roemer discovered the velocity of light by measurements of the intervals between the eclipses of Jupiter's moons. This discovery undoubtedly must have drawn Huygens's thoughts to his earlier studies in optics, for during the two years in the quiet of his country home he developed the wave theory of light for which he is particularly remembered in physics. All of this work was set forth in his *Traité de la lumière*, which was not published until 1690, although he reported it in detail to the Academy on his return from Holland in 1678.

It may seem somewhat farfetched that in a development of atomic theory one should discuss the nature of light. Nevertheless, any detailed picture of modern physics is impossible without an understanding of the interaction of light both visible and invisible with atoms themselves. From the standpoint of the development and acceptance of physical theories it is interesting that Huygens's wave theory of light was developed almost coincidentally with Newton's corpuscular theory. Both published the results of their researches years after the concepts were formed, Huygens in 1690 and Newton in his *Opticks*, which first appeared in 1704. Both theories were incomplete, and acceptance of either one involved uncertainties that varied with the phenomena chosen for explanation. Newton's corpuscular theory easily explained the straight-line propagation of light, whereas a wave theory required a propagation like that of sound—a spreading of the wave around corners. But the wavelength of light is so short that this effect is obscured. Even though Grimaldi had observed the effect before 1665, it was not firmly established for approximately another century. By applying the laws of mechanics to his corpuscles of light, Newton accounted for the manner in which light bends when passing from one medium to a more dense one, such as from air to water. His explanation required, however, that light travel faster in water than in air. With the wave theory just the opposite is true. Thus an *experimentum crucis* was presented to determine which of the two theories was correct, but the means to perform the experiment were not available at that time.

Largely because of Newton's overwhelming scientific prestige, the corpuscular theory gained almost universal acceptance and retained it until the opening of the 19th century. At that time Thomas Young demonstrated clearly the interference phenomena required by the wave theory; Fresnel used the theory to explain a wide range of diffraction effects shortly thereafter. If this was not enough to discredit the corpuscular theory, the *coup de grâce* was dealt by Fizeau in 1849 when he performed the crucial experiment that measured the velocity of light in water and found it to be smaller than in air. Nevertheless, the ensuing half century disclosed experiments whose explanation demanded a corpuscular hypothesis.

But we must return to Huygens to follow the later events of his life. After his report on the wave theory in 1678, only another year elapsed before he again fell ill; fortunately, this attack was not severe enough to force him to give up work for any extended period. However, early in 1681 he suffered a violent relapse and, when able to travel, departed once more for Holland. Although the French Academy was now thriving, Huygens's position was by no means secure. There were internal rivalries in the society of pro- and anti-Cartesian factions. Huygens was not completely in one or the other. He accepted Descartes's vortices as the scheme of the world and, even after the appearance of the *Principia* in 1687, never could be persuaded that Newton's universal gravitation was anything more than a fiction. He could not accept the fact that matter could act on distant matter without some understandable mechanism intervening.

Huygens was not ready to return to Paris until early in 1683. At this time the Dutch East India Company again became interested in marine clocks and Huygens spent the summer working on this project. Colbert, his patron, died in September 1683, and the opposition to Huygens's return was sufficient to prevent it. In the meantime France became increasingly anti-Protestant and the feeling of the Academy against foreigners became stronger than ever. Despite all attempts in the next few years he was never able to resume his position.

During the period 1685 to 1690 Huygens tried to perfect a marine pendulum clock, but these efforts were unsuccessful. After this time he returned to spring-driven movements but his efforts reached no conclusive outcome. In 1687 his father died at the advanced age of 91 and, in the following year, his brother went to England with William III, who, with Mary, succeeded to the English throne following the deposed James II. Huygens now looked forward to a position in England and, in 1689, spent the summer months there. He attended a meeting of the Royal Society at Gresham College, met Boyle on several occasions, and Newton for the first time. Later he went to Cambridge and it is recorded that early on a bright July morning he and Newton rode together in the coach from Cambridge to London. Huygens was then 60 years old and Newton 47. The *Principia* had been published two years before and, although Newton's fame was everywhere growing, he was then in one of his moods in which he was weary of science. Indeed this journey was prompted by his application for the post of provost of King's College, a post which, incidentally, he did not receive. There is unfortunately no record of what the two men talked about on this journey, however interesting it would be to know it. It is possible that they said very little, for Huygens's English was never fluent and Newton was often remote and taciturn. Huygens must have felt that his fortunes were ebbing and Newton, in turn, must have been concerned with efforts to change his own.

After returning from England, Huygens spent his remaining years in Holland. He never married, and the decline of his scientific work and his health, coupled with the fact that he was now cut off from close contact with others who were active in science, led to an increasing sense of loneliness. In March of 1695 he again fell ill and his condition became progressively worse. He died on June 8, 1695.

Some idea of his powerful scientific imagination, as well as the basis of the wave theory of light, can be gained from a review of his *Traité de la lumière*. After a charming preface in which he explained why he was publishing his discoveries and why he waited so long to "bring this work to the light," he presented a systematic, almost axiomatic, development of the principles that guided him in his analysis of optical phenomena. In the first few pages of his treatise Huygens boldly outlined the principal lines along which he proposed to conduct his investigations and stated the essential problem that must be solved to account for the propagation of light in straight lines and to explain why rays of light emitted by different bodies do not interfere when they cross. Huygens stated the problem as follows: "For I do not find that any one has yet given a probable explanation of the first and most notable instrument of light, namely why it is not propagated except in straight lines, and how visible rays, coming from an infinitude of diverse places, cross one another without hindering one another in any way" (Huygens, p. 2). To account for some of the observed properties of light Huygens immediately introduced the idea that light "consists in the motion of some sort of matter," for it was clear to him that only motion can account for such things as "fire and flame" from which light stems. He went on to introduce the idea that light is transmitted from a source to the observer by the "movement of matter which exists between us and the luminous body." Here we have again the Cartesian concept of an ether filling all space which later was so useful in the hands of Michael Faraday and James Clerk Maxwell even though it finally had to be discarded with the introduction of the theory of relativity.

With the introduction of these ideas of motion as the cause of light, Huygens implicitly accepted a kinetic (that is, a molecular) theory of matter, for he remarked that " . . . in the true Philosophy, . . . one conceives the causes of all natural effects in terms of mechanical motions." He further emphasized the kinetic basis of his theory of light by stating that " . . . all of those bodies that are liquid, such as flames, and apparently the sun and the stars, are composed of particles which float in a much more subtle medium which agitates them with great rapidity and makes them strike against the particles of the ether which surrounds them and which are much smaller than they" (Huygens, p. 10). He compared the emission of light with the emission of sound and pointed out that the cause in both cases is motion but that the motion responsible for light must be exceedingly more rapid than that causing sound.

Having introduced an ether with kinetic properties, Huygens first developed the idea that light is propagated through the ether by the continual collision of the particles of the ether, and from this he went on to the concept of spherical waves. These waves, he argued, are emitted when the particles in a source of light, such as the flame of a candle, are set vibrating. Each point in the source then emits concentric spherical waves, with the waves from separate points moving independently of each other.

To account for the propagation of light in a straight line and to explain the various observed optical phenomena, Huygens introduced his famous concept of partial waves. Here he explained how the successive wave fronts emanating from a point source of light can be constructed from previous wave fronts by picturing each point on a previous wave front as generating partial wavelets. The new wave front is then the surface that simultaneously touches one point of each of the partial wavelets that originate from all the points on a previous wave front. This concept led to an extremely powerful technique for following the propagation of light through various media and for explaining such phenomena as reflection, refraction, and diffraction.

Although Huygens had no idea of the modern conception of the nature of light, which is the electromagnetic theory originated by Maxwell, his wave theory for the propagation of light is still as correct and useful today as it was when he first introduced it. Indeed, Huygens's concept of partial waves as contributing to the final wave front has been carried over by Richard Feynman into the modern wave concept of matter to explain more clearly the quantum behavior of such elementary particles as electrons.

Newton on Particles and Kinetics: Isaac Newton (1642–1727)

Although Isaac Newton is best known for his theory of universal gravitation and his three laws of motion, he also wrote extensively on atomism; many of his ideas about atomic theory appeared in his *Opticks* which was published in 1704. Whatever Newton proposed had enormous influence on his contemporaries and those who followed, especially, as we shall see, on John Dalton, the founder of experimental atomic chemistry. Newton's station in the history of atomism is secured primarily through this indirect influence, the effect of his vision on the ideas of others.

The celebrated French mathematician, the Marquis de l'Hôpital, writing of Newton, said, "I picture him to myself as a celestial genius." These words state, perhaps most aptly, the common reverence that Newton's name has inspired in the generations since his death. More than a century later, the great English poet William Wordsworth, contemplating the heroic marble statue in the antechapel at Trinity College, Cambridge, was moved to write

> . . . Newton, with his prism and silent face
> The marble index of a mind
> Voyaging through strange seas of thought alone.

The Victorians especially built his image to almost godlike proportions. Certainly Sir David Brewster's biography published in 1832 did much to launch this attitude. But Brewster knew more than he wished to write, for despite Newton's greatness—and, indeed, no one has ever questioned how sublimely gifted he was—some facets of his character were distinctly strange and unpleasant.

In 1946, the Royal Society's tercentenary celebration of Newton's birth, which should have been commemorated in 1942 but had to be postponed because of World War II, brought to public notice new evaluations of Newton as a man. One of these, by the late Lord John Maynard Keynes, revealed a strange and hitherto

Isaac Newton (1642–1727)

unknown side of this formidable man. But, however strange Newton actually may have been, Keynes also reinforced the reverence Newton has received as that owed, as l'Hôpital says, to a "celestial genius."

Isaac Newton was born on Christmas Day, 1642, at the manor of Woolsthorpe, a small farm close to the village of Colsterworth in Lincolnshire, England. The nearest town of any size to appear on an ordinary map is Grantham, which lies about eight miles to the north. Newton's birthplace still stands today in an excellent state of preservation; the visitor may see the room in which he was born and the room that he used as a study during the years when Cambridge University was closed, when he laid the foundation of his future greatness. Several apple trees grow on the lawn, perhaps as reminders of the popular tale regarding the origin of the theory of universal gravitation.

The Newton family were inconspicuous country farmers and Newton's father, Isaac, was apparently a quite ordinary man; his mother, Hannah Ayscough, seems to have been a capable but average country woman. Nowhere in his forebears can one find evidence of outstanding ability; his emergence as one of the greatest minds of all time seems, therefore, all the more remarkable. Newton's father died at the early age of thirty-six, a few months after his marriage. As if to compound this misfortune, Newton was so small at birth that it was said he could have been put in a quart mug, and he was not expected to live. But in this instance, as in many other aspects of his life, Newton confounded the expectations of those about him.

When Newton was only two years old his mother remarried, this time the Rev. Barnabas Smith, rector of the small parish of nearby North Witham. The boy was spared the trials of a stepfather, being left throughout his youth at Woolsthorpe with his maternal grandmother. In 1656 the Rev. Smith died and Newton's mother, with her three children by this marriage, a son and two daughters, returned to the Newton farm. At this time young Isaac was fourteen years old and had been for several years a student in the grammar school at Grantham.

It might be expected that he would have shown evidence of the peerless mind that was destined to revolutionize human thought, but we find instead a rather solitary boy, interested in constructing mechanical devices and in books, but indifferent to formal study. A schoolboy fight picked by a student who did better at his lessons than Newton awakened a desire to surpass his rival and gradually his schoolwork became outstanding.

On his mother's return to Woolsthorpe, Newton was taken out of school, with the hope that he would devote himself to working the family lands. But farmwork was alien to his awakened and active mind. On trips to the Grantham market with the family servant he would speed off to seek out the books in the apothecary's home where he had boarded when at school. His continuing absorption in intellectual pursuits and his dislike of farm life led his maternal uncle, rector of a neighboring parish and graduate of Trinity College, Cambridge, to persuade the boy's mother to prepare him for the same training.

Fortunately, reason prevailed and young Newton was reentered in the Grantham school. And thus it was that in June of 1661 there appeared in Cambridge a withdrawn and intense 18-year-old of country background and manners who, following the star of his destiny, had arrived at the gates of that ancient university. In the early years at Trinity, as at Grantham, Newton attracted little attention. Having little money he had to earn his way as a "subsizar," a student assigned to odd jobs and to assisting his tutor. Because of the combination of slender financial resources, a sober background, solitary habits, and the gift of intense concentration, it appears likely that he mixed little with his fellow students. In fact, Newton was so inconspicuous that we know practically nothing of his undergraduate life except what he was willing to write in later years. It is known that he applied himself assiduously to mathematics and that he read Euclid (which he considered at first a "trifling book") and then Descartes's geometry. Also, references in his *Opticks*, published many years later, show that he had begun to make careful observations of natural phenomena, for he describes in particular the measurements he made on halos about the moon seen on the night of February 19, 1664.

Newton completed his B.A. degree in January 1665. Because the seniority list for the students of that year is missing, there is no official report of his standing, but from his own account it is clear that he had made phenomenal progress in mathematics and in "natural philosophy," as physics was then called. He had made a strong impression on Isaac Barrow, Lucasian professor of mathematics in the university, but, as far as we know, on no one else.

Barrow recognized the potential in his intense young student and encouraged Newton to remain at Trinity and continue his studies. Sometime during the summer of 1665 the plague, which had broken out earlier in London, assumed epidemic proportions and began to spread through England. Trinity College was officially closed early in August, but from the records it appears that Newton had already returned to Woolsthorpe. Between this date and the spring of 1667, when the university reopened, he laid the foundation of all the great discoveries that have immortalized his name. As Andrade remarks, "In those two great years he had to the full two priceless gifts which no one enjoys today, full leisure and quiet." What he did with it we can hear from Newton himself:

> In the beginning of the year 1665, I found the method for approximating series and the rule for reducing any [power] of any binomial to such a series [i.e., binomial theorem]. The same year in May [probably while still at Cambridge] I found the method of tangents of Gregory and Slusius, and in November had the direct method of fluxions [differential

calculus] and the next year in January had the theory of colours and in the May following I had the entrance into the inverse method of fluxions [integral calculus] and in the same year I began to think of gravity extending to the orb of the moon . . . and having thereby compared the force requisite to keep the moon in her orb with the force of gravity at the surface of the earth, and found them to answer pretty nearly. All this was in the two years 1665 to 1666, for in those years I was in the prime of my age for invention and minded mathematics and philosophy more than anytime since. (Newton, I, x, 41)

The magnitude of this accomplishment is difficult to comprehend, and it is more astonishing when we think that it was all done when Newton was in his 23rd and 24th years. Although Newton tells us specifically just when he developed the concept of differential and integral calculus, the theory of colors, and when he had the idea of "gravity extending to the moon," it is curious that he does not mention one of his most fundamental and celebrated achievements—an understanding of the laws of motion. To a large extent this must have been accomplished before he could say that he had found the comparison of the force requisite "to keep the moon in her orb" with "the force of gravity at the surface of the earth" . . . "to answer pretty nearly." It meant that he had not only generalized the idea of inertial mass proposed by Descartes and Galileo and had taken the step to equate it to gravitational mass but had gone a great step further and incorporated the notions of force, mass, and acceleration, including centripetal acceleration, into the universal relationship that lies at the heart of dynamics. This result, his second law, states that force equals mass times acceleration.

It is worthwhile to point out here that Newton's world scheme, based on the idea of gravitation, presents an entirely different approach to the understanding of the solar system and the cosmos from that stemming from the Cartesian views, which were widely held at that time. Instead of mechanical vortices that were supposed to keep the planets in their paths about the sun, Newton substituted the mysterious and unseen inertial and gravitational forces that, together, guide these orbs in their revolution through the depths of an empty space. It is hardly any wonder that, even in Cambridge, Cartesian ideas were accepted long after the publication of Newton's great work, the *Principia*, in which all these topics were explained.

Newton returned to Cambridge in the spring of 1667. The following October he was elected a fellow of Trinity and, early the next year, in March, took his M.A. In these years he was extending his study of optics and the development of calculus, but he had also bought chemicals and a furnace, apparently for chemical experiments. The nature of these experiments has long been obscure, but, as Lord Keynes's article "Newton the Man" pointed out, Newton had a streak of magic in his makeup. He had a passionate and driving interest in alchemy and spent years in its study and experimentation. Because Newton was excessively secretive, this mystical side of his nature was not revealed until the "Portsmouth Papers" disclosed his preoccupation with occult ideas and with certain esoteric aspects of Biblical and theological inquiry, concerning which he left a huge collection of manuscripts, many of which have not been investigated to this day.

Despite Newton's reluctance to reveal his activities and his discoveries, he presented to Barrow in 1669 a paper summarizing his work on the binomial theorem and calculus. This work and, perhaps, Barrow's knowledge of Newton's optical researches, which included the construction of a small reflecting telescope (quite an achievement at that time), produced an unexpected change in Newton's fortunes. Barrow resigned his chair in the latter part of the summer; on October 29, 1669, Newton, who had published nothing and was not yet 27, became the Lucasian professor of mathematics.

Newton's optical researches had convinced him that lenses free from chromatic aberrations can not be constructed; accordingly, he turned to a reflecting mirror to produce a telescope in which sharp images could be obtained. The choice of metal for the mirror, its casting, and the subsequent polishing were all done by Newton's own hand. After his death, the composition of the alloy that he used was found to be very close to what is now known as speculum metal and the polishing was done with techniques that with slight changes would be acceptable in modern practice.

This uncanny knack of devising correct procedures and finding correct solutions makes it seem almost as though Newton were at times endowed with the kind of superhuman vision and with the magic he so earnestly sought. His telescope had sufficient resolving and magnifying power to show the moons of Jupiter and to observe the phases of Venus. He then made a second and somewhat improved instrument; an account of it was read to the Royal Society in January 1672, when he was elected a fellow. A month later he forwarded his first paper describing some of his optical studies on refraction and reflection and his proof that white light consists of colors of different refrangibility. A second paper, delivered the following year, contained a corpuscular theory of light and the supposition of an all-pervading ether. In this paper, Newton hypothesized that the vibrations of a denser ether in a substance of higher index of refraction produced "fits" of easy reflection and easy refraction, depending on the phase of the vibration with respect to that of the incident corpuscle. In this way Newton accounted for interference and diffraction and color effects in thin films. All of this inevitably aroused disagreement and controversy, some with Hooke and some with scientists on the Continent. Newton was so painfully sensitive to this intrusion on his thoughts and his time that he said his "affection to philosophy" was worn out and he was as little concerned with it "as a countryman about learning."

Unfortunately for Hooke and for Newton, this was not to be their only clash. Another dispute occurred in 1679 regarding the path of a falling body, taking into account the earth's rotation. Still another clash occurred in 1684 as a result of a discussion that took place among Hooke, the astronomer Edmund Halley, and Sir Christopher Wren. These three men, members of the Royal Society, agreed that the gravitational force decreases inversely as the square of the distance, despite the lack of a general proof. For Halley an even more interesting question was a solution to the astronomical problem of the path taken by a body moving around a center of force which exerts an attraction that varies as the inverse square of the distance. Hooke professed to have solved this problem but did not produce his result. After waiting for several months Halley decided in August 1684 to go to Cambridge and ask Newton. He was astonished to get the immediate reply "an ellipse." Newton thereupon offered to show Halley the solution but was unable to locate it among his papers. Halley returned to London with Newton's assurances that he would reproduce the calculation. It was forwarded in November and, after reading it, Halley again went to Cambridge for further discussions with Newton. The result of these conversations was a report in February 1685 to the Royal Society of Newton's work and the appearance of a paper "De Motu Corporum," which was the beginning of the *Principia*. It is entirely characteristic that Newton should have completed the solution of the problem of universal gravitation, which he had started in 1666, as well as the motions of the satellites in a central field of force without revealing it to anyone.

It was clear to Halley that the great body of Newton's work had to be made available to the scientific world, and he set about coaxing Newton to prepare it for publication. To expedite it, Halley himself assumed the cost of its publication, placated Newton in the controversies stirred up in its preparation, saw the book through the press, and arranged for its sale. The whole work, which opened a new world to science and philosophy, was completed in about 18 months and appeared for sale in the summer of 1687. Newton was then 44 years old.

The task of preparing the *Principia* seems to have exhausted Newton and put him again in the frame of mind in which science no longer held his interest. During the spring of 1687, when he was completing the *Principia*, he took part in the legal proceedings by which Cambridge University defied King James II in his efforts to change the religious basis for the conferring of the M.A. degree. The following year James II was deposed and Newton was elected to represent Cambridge University in the so-called Convention Parliament. Newton's residence in London and the formation of new friendships called attention to his modest way of life. It was suggested that a man of his eminence should be placed in a government position as a tangible mark of national gratitude. Newton seems to have acquiesced to this idea, but it was not fulfilled for another five years.

In 1689 Newton's mother died. Despite his residence at Cambridge, the record shows that Newton over the years returned to Woolsthorpe very frequently, sometimes spending long periods there. His relationship with his mother seems to have been close, and a strong bond of affection existed between them. In her last illness it is said that he sat up all night with her. Whether it was his mother's death, his changing interests, or the delay of his friends in finding him a high-salaried administrative post, Newton became dissatisfied and depressed. By 1693 he had fallen into periods of deep melancholy in which he imagined that his friends had deserted him or were attempting to involve him in unpleasant situations. He suffered long periods of sleeplessness, and his usually suspicious nature became greatly exaggerated. For instance, he wrote Pepys, whom he did not know well, saying that they must never see each other again. He also wrote his friends that Charles Montague, longtime fellow of Trinity and later Earl of Halifax, who was working in his behalf, "bore him a grudge and was false to him." This strange behavior was certainly symptomatic of a nervous breakdown but Newton did not, as some accounts claim, become insane. Toward the close of 1693 he seems to have recovered and it is interesting that Montague, whose friendship he had questioned, should have been the person to whom he eventually owed his appointment in London. As Chancellor of the Exchequer, Montague arranged matters so that Newton became Warden of the Mint. This appointment was useful to both men; Montague was effecting the recoinage of the currency, and Newton gave the problem his complete attention. In 1699 Newton was advanced to Master of the Mint.

In the years that followed, long-deferred honors began to accrue. In 1703 Newton was elected president of the Royal Society, a post to which he was reelected annually for the rest of his life. The first edition of his *Opticks* appeared in the next year. In 1705 Queen Anne and a royal retinue journeyed to Cambridge to confer knighthood on Newton at the scene of his great achievements. In 1713, with the assistance of Roger Cotes, Plumian professor of astronomy at Cambridge, a second edition of the *Principia* was published. A third edition was prepared in Newton's lifetime by Henry Pemberton. His health remained good until 1725 when he suffered an attack of inflammation of the lungs and was advised to move to the purer air of Kensington. In February 1727 he felt unusually well and took the opportunity to drive in his coach to London to preside at a meeting of the Royal Society. But the rigors of the journey proved too much for him and he suffered a relapse. Returning to Kensington, he gradually grew weaker and died on March 20 at the age of 84.

Newton was buried in Westminster Abbey with the greatest honors England could bestow. On his monument is written: "Let mortals rejoice that such and so great an ornament of the human race has existed." As great a man as Einstein said of him more than two centuries later, "Nature to him was an open book whose letters he could read without effort." Perhaps the most thought-provoking comment was made by Andrade, speaking at the Newton Tercentenary, when he said: "I feel that Newton derived his knowledge by something more like a direct contact with the unknown sources that surrounded us, with the world of mystery, than has been vouchsafed to any other man of science."

A Kinetic Theory of Gases: Daniel Bernoulli (1700–1782)

In politics, in music, in finance, and other fields of human endeavor, history occasionally furnishes us with the story of an illustrious family whose members, over a series of generations, continue the contributions that founded their fame. The Churchills, the Bachs, the Rothschilds are among them. But in science we find fewer outstanding examples—the Bernoulli family being perhaps the foremost. "No fewer than 120 of the descendants of the Mathematical Bernoullis have been traced genealogically and of this considerable posterity the majority achieved distinction—sometimes amounting to eminence—in the law, scholarship, science, literature, the learned professions, administration and the arts. None were failures" (Bell, p. 131).

In the earliest records, the Bernoulli family resided in Antwerp but, because of their adherence to Protestantism, they moved briefly to Frankfurt in 1583 and later to Basel, Switzerland. There the brothers Jacques (Jacob) (1654–1705) and Jean (John) (1667–1748), the father of Daniel, first demonstrated the mathematical genius for which the family is noted. In 1676, although only 24, Jacques traveled to England where it is said he was "admitted to the meetings of Robert Boyle, Hooke, and other learned and scientific men," evidently the Royal Society. After six years of travel in Europe he returned to Basel in 1682 where he opened a public seminary for experimental physics. Five years later he was appointed to the chair of mathematics in the University of Basel, a position he held until his death.

Daniel's father, Jean, though an able mathematician, would be characterized today as a disturbed person. As Newman remarks, "He was violent, abusive, jealous, and when necessary, dishonest" (p. 772). Jean's first academic post was the chair of mathematics at Groningen, Holland. His difficulties there were many; nevertheless, he continued in this position for ten years. During his sojourn at Groningen Daniel was born. In 1705 the family returned to Basel where Jean accepted the chair of mathematics left vacant by his brother's death. He had already been accused of dishonesty at Groningen, and he was then suspected of cheating in a prize competition for the solution to the isoperimetrical problem (the shortest perimeter enclosing a given area), first proposed by his brother Jacques nine years earlier. His relations with Daniel were even more reprehensible. He not only disliked his son and discouraged his efforts to study mathematics but he also adopted a repressive severity toward the boy calculated to undermine his self-confidence. Despite this hostility, Daniel began studying mathematics at 11 under the tutelage of his older brother Nicholas and later continued his studies in Italy. Before surrendering to the family passion, Daniel thought he preferred medicine; after completing this study, he was for a time a physician.

But medicine was only a passing fancy and, when Daniel was 25, both he and Nicholas accepted appointments as professors of mathematics at St. Petersburg, Russia. The collaboration was short-lived; eight months later, in July 1726, Nicholas lay dead of a fever. Daniel stayed on another seven years, enduring as well as he could the Russian manners and climate. He returned to Basel in 1733 as professor of anatomy and botany, an appointment that was changed later to professor of experimental and speculative philosophy. He spent almost fifty years in these posts, during which he won or shared no fewer than ten prizes awarded by the Paris Academy of Sciences, a record equaled only by his friend, the great mathematician, Leonhard Euler. He was elected to almost every learned society in Europe, and he succeeded his father as Foreign Associate of the Academy of Paris. His work in his later years was concerned mostly with the development of the theory of probability. Although he suffered from asthma, he remained vigorous and active until almost 80. He lived only a few years after his retirement, passing away in Basel on March 17, 1782.

Bernoulli's discussion of gas dynamics in the tenth section of his *Hydrodynamics* (1738) was the first mathematical treatment of that branch of physics now known as the kinetic theory of gases. It is remarkable on many counts. First, it presented his picture of a gas as composed of an enormous number of particles in very rapid chaotic motion. This description came at a time when the physical nature of gases was quite unknown, the atomic nature of matter highly speculative, and the kinetic hypothesis equally so. Second, his theory shows mathematically that gas pressure can be accounted for as a result of direct particle impacts on the walls of the container. Third, he clearly recognized the distinction between heat and temperature, which was understood only after the work of Joseph Black years later. Fourth, he assumed that the addition of heat increases the speed of the gas particles—a statement of the conversion of heat to mechanical energy that was established by James Joule's work over a century later. Fifth, he recognized that the pressure-volume relation may not hold at higher densities than those of normal air. Sixth, he saw that the validity of the pressure-volume relation holds only if the temperature is kept constant. Finally, he inferred that the particle velocity is not dependent on the pressure. It is amazing that all these ideas, without exception, are valid.

The Atom as a Center of Force: Roger Joseph Boscovich (1711–1787)

The early atomists seem, with few exceptions, to have pictured the atom as a very small, massive, and hard particle. Newton in his *Opticks* is explicit when he observes that "it seems probable to me that God in the beginning formed matter in solid, massy, hard, impenetrable, moveable particles." Certainly many of the atomists who lived after him adopted this picture, and the "little billiard ball atom," as one might call it, seemed prevalent in the minds of the early chemists as well as those who developed the kinetic theory of gases. But this idea was not universally held, for the picture originated by Roger Joseph Boscovich and presented in his *Theory of Natural Philosophy*, first published in 1758, is entirely different.

Boscovich was one of the early supporters of Newtonian ideas in Italy. His interest in atomic ideas seems to have followed from his study of Newton's mechanics, specifically those of collision. If one pictures atoms as hard spheres, which are homogenous (being without parts) and incompressible, then a dilemma arises if one imagines collisions between them. Since they cannot deform, they must in general suffer an instantaneous change of velocity in collision. But, said Boscovich, this conclusion violates the law of continuity which must be observed. Accordingly, atoms cannot be hard, rigid, massive spheres, and some other model must be sought.

On the basis that atoms cannot come into contact Boscovich built up a picture consisting of point centers having position and mass but no extension. To account for the physical behavior of bodies composed of atoms he assumed that at large distances atoms attract according to the gravitational forces postulated by Newton. As one atom approaches another at distances smaller than those appropriate to Newtonian attraction, the attractive force at first increases, then decreases to zero, becomes a repulsion which increases and then decreases to zero, then an attraction, and so on. These alternating attractions and repulsions exist as spheres of force around an atom, and only when a very close approach is involved does the force becomes asymptotically repulsive.

The number of these spheres of attraction and repulsion was not specified. The necessity for this alternation of force arose from the need to explain certain physical actions, such as cohesion and vaporization, as well as chemical actions. In the instance of the vaporization of liquids, let us picture the atoms in the liquid as being very close together. Then an attempt to compress the liquid brings into play the asymptotic repulsion of the atoms and the volume cannot be much reduced even with very large forces of compression. On the other hand, if actions (such as heating) occur to increase the distance between the atoms, cohesion or attraction comes into play and acts until the distance becomes too great, when it falls to zero. To account for atoms of vapor, a repulsion is assumed that takes place when the distance between the atoms is sufficiently great. The atoms can thus exist as vapor, in which phase they exert a pressure. This pressure, according to Boscovich, is a consequence of the mutual repulsion of the vapor particles. Finally, if the distance grows larger, the force of universal gravitation exerts an attraction.

Using the alternating attraction and repulsion forces, it was possible to account in a qualitative way for the melting and boiling points in a given substance, as well as for the differences in these temperatures for various materials. The explanation of crystalline forms also followed. With the hypothesis of a spherical, hard atom, in an atmosphere of repelling caloric, how could a given substance be assumed to display a given crystalline form rather than an amorphous one? Or, if such structures were possible, why did a substance always show the same crystalline form? Indeed, how did chemical reactions take place at all in this picture of hard, spherical atoms? Evidently other *ad hoc* assumptions were needed. But with the Boscovichian atom, reactions occurred when the atoms' force fields meshed so as to produce a resultant attraction. That some compounds were more stable than others depended on the strength of the force fields involved. These few examples illustrate the qualitative attractiveness of Boscovich's theory—a theory that is now hardly known but one that in the 19th century enjoyed a considerable vogue. Both Humphrey Davy and Michael Faraday, especially the latter, were more sympathetic to the atomic theory in this form—the atom as a seat of forces—than to the Daltonian picture.

Boscovich himself had an interesting history. He was born in Dubrovnik, Yugoslavia (then Ragusa), on May 18, 1711. His father and grandfather were merchants; his mother, Pava Bettera, who came from a cultivated Italian merchant family, had the hardihood to live to the astonishing (in that day at least) age of 103 years. The family were serious and pious Roman Catholics. Roger was the eighth of nine children and the youngest of six sons. At about the age of eight, he was enrolled in Ragusa's Jesuit College. Then, in September 1725, when just over 14, he was sent to Rome to begin studies for the priesthood. After two years as a novitiate, he entered the Collegium Romanum. He distinguished himself in mathematics, physics, and astronomy, and was one of the first to advocate Newtonian ideas in Italy. He began teaching in 1732 while doing research in astronomy. He published his first paper in 1736 on the determination of the sun's equator and on the period of the sun's rotation. Other papers followed on various subjects, including the shape of the earth and the aberration of the fixed stars. In 1740, at the age of 29, he was appointed to the chair of mathematics at the Collegium Romanum. Not until 1744, however, did he take his vows as a priest and become a full member of the Society of Jesus. Following his ordination at 33, he spent 14 years as a teacher and scholar. His scientific papers, the mapping of the Papal States, the figure of the earth, the confirmation of Bradley's theory of the aberration of light, and his own reputation as a brilliant intellect earned him admission to scientific societies all over Europe. But Boscovich had other than scholarly sides to his character; his ability as a poet as well as a scientist gave him entrée to an active social life in the highest ecclesiastical, academic, and diplomatic circles. He became, it appears, something of a social climber and a snob. He was sent to Vienna in 1757 in the capacity of a scholar-diplomat. There, between April and the following February of 1758, Boscovich wrote his *Theoria philosophiae naturalis*, a work that had been maturing during the previous 12 years. It developed atomic theory, as described above, from a completely new point of view, focusing attention on the fields of force around the atom rather than on the atom itself.

After returning to Rome, Boscovich was again sent on a long tour of Europe beginning in 1759. The reason for this journey is not clear but he was away from Rome for over four years. The first part of this tour was spent in Paris and London, where he arrived in May 1760. There he was introduced to many of the outstanding personages of the era, attended several meetings of the Royal Society, met Benjamin Franklin, and saw him demonstrate electrical experiments. He also visited Oxford and Cambridge. His impression on the members of the Royal Society was such that he was elected a foreign fellow, but not until after his departure in December 1760. After leaving England he traveled on the Continent and journeyed as far as Constantinople, returning to Rome in 1763. He was then 52 years old, and he was soon to realize that the star of his success had passed its zenith. Whether in the meantime his *Theoria* had caused his superiors to view him with suspicion or whether his absence had been so long as to demand a replacement is not clear; at any rate his professorship was no longer open to him. However, in the Seminarium Romanum he was free to continue his studies and his research. In 1764 he accepted the chair of mathematics at the University of Pavia, which under the Austrian government was then in a very backward state. He remained at Pavia until 1769 when he was transferred to the Scuola Palatina in Milan. Also situated in this city was the Jesuit College of Santa Maria de Brera for which Boscovich had designed an observatory during his stay at Pavia and at which he became, in addition to his duties at the Scuola Palatina, a lecturer in astronomy. Boscovich's prominence as an emissary of the Vatican, coupled with his sense of importance derived from his academic honors all over Europe, his social connections with the nobility, and his proprietary feeling concerning the Brera Observatory soon brought him into impossible relations with his colleagues. In August 1772, during an absence on vacation, he was dismissed from his post at the observatory. He then went to Venice from where he sent his resignation the following year. Much of his own money had been spent for instruments at the observatory, and his financial situation was now precarious. Added to this misfortune, pressure against the whole Jesuit order had been growing for years all over Europe and was then coming to a head. On June 8, 1773, Pope Clement XIV signed an order suppressing the society. Boscovich was then 62 years old, and his future must have appeared exceedingly dark. Fortunately, powerful friends in France persuaded King Louis XV to create a special post for him in the French Navy as Director of Naval Optics. Despite Louis's death the following year, the post was confirmed by his successor, Louis XVI. Protected by the great and powerful whom he so successfully cultivated, Boscovich remained in Paris for nine years until July 1782. Then, for reasons of health and his desire to prepare his collected works for publication, he left his post and returned to Italy. There, in various places in northern Italy, he completed his editing in the next three years. In the fall of 1786, he began to show signs of mental disturbance that developed into pathological melancholia. Saved from an attempted suicide he lived on in mental confusion until February 13, 1787, when a series of medical complications ended his life.

References

E. N. daC. Andrade, *Sir Isaac Newton*. New York: Doubleday & Co. , 1958.

E. T. Bell, *Men of Mathematics*. New York: Simon & Schuster, 1937.

Encyclopædia Britannica. London, 11th ed., 1910.

Robert Hooke, *Posthumous Works of Robert Hooke*. London: Richard Waller, 1705.

Christian Huygens, *Treatise on Light*. Trans. S. P. Thompson. New York: Dover, 1912.

James R. Newman, *The World of Mathematics*. New York: Simon & Schuster, 1956.

Isaac Newton, *Portsmouth Collection*.

Andrew G. Van Melsen, *From Atomos to Atom*. Trans. Henry J. Koren. Pittsburgh: Duquesne University Press, 1952.

2 *The Foundations of Atomic Chemistry*

To see a World in a grain of sand,
And a Heaven in a wild flower,
Hold Infinity in the palm of your hand,
And Eternity in an hour.

—William Blake, *Auguries of Innocence*

The Birth of Atomic Theory: John Dalton (1766–1844)

It is a sobering thought that 2,000 years of atomic speculation produced no mind able to formulate atomic theory in questions simple enough for direct experimental answers. Even Isaac Newton's derivation of Robert Boyle's law on the assumption of gas particles that repelled each other with a force decreasing inversely as the distance and, some half century later, Daniel Bernoulli's brilliant deduction of the same law on a kinetic particle hypothesis opened no doors to the world of the atom. It would be logical to expect that if the greatest minds of two millennia had found no way to question nature in atomic terms, only a demigod might be expected to see where mortals were blind. But no demigod appeared. Instead, the fates sent an unprepossessing country schoolmaster of silent mien and uncouth manners who, in a chance flash of revelation, caught a glimpse of an open door and passed through.

John Dalton came into this world on September 5, 1766, the son of a poverty-stricken weaver in Eaglesfield, Cumberland County, a shire in the north of England. When it came time, the boy was sent to what was known as the Pardshaw Hall School, near his village, where his formal education continued through his 11th year. At that time the master, a Mr. John Fletcher, resigned, and Dalton himself reopened the school first in his own home and then in the local Friends' Meeting House. How anyone could entrust a boy 12 years old with the regular teaching of his contemporaries boggles the imagination. Had he been thrust into this position because of outstanding brilliance, we would wonder somewhat less, but Dalton was never characterized as such. We can only surmise that the dogged perseverance, insensitive self-reliance, and patient concentration that marked his adult years had already attracted his elders' attention.

After two years as a schoolmaster and one of farming, Dalton, then only 15, left home to join his brother Jonathan and a cousin in Kendal, about 40 miles to the south, where they conducted a school for the children of the Friends. During the next 12 years (1781–1793) he taught and studied. In the last eight years of this period he laid the foundation for his future discoveries through a fortunate friendship with John Gough, an impressive and remarkable young man, some ten years Dalton's senior. Although handicapped by blindness from infancy, Gough possessed those qualities that Dalton lacked: brilliance of mind, comfortable means, grace of person, and a broad intellectual training. Together with these assets, he harbored a taste for science and mathematics, and these were the subjects that the two studied most often, with Gough as tutor and Dalton as student. Meteorology was one of their mutual pursuits, and, in March 1787, Gough persuaded Dalton to start a journal of weather observations. So devoted did he become to this pursuit that, with but a few lapses, he continued it to the day preceding his death. Unlikely as this avenue might appear, it was through meteorology that Dalton was eventually led to his explanation of chemical reactions and to the undying renown that he acquired as the founder of modern atomic theory.

The formative years at Kendal came to an end in 1793. In the spring of that year, largely through Gough's recommendation, Dalton received the offer of an appointment as tutor in mathematics and natural philosophy at the New College in Manchester. He accepted the offer and took up his abode in the college buildings. Dalton was ready for a larger sphere of activity, and his acceptance of this post at the age of twenty-six marks the beginning of his intellectually mature years.

Shortly after his arrival, Dalton published his first scientific treatise, "Meteorological Observations and Essays," the data for which had been accumulated at Kendal. In it we find most of the ideas with which he was concerned during the next ten years, ideas on problems associated with "mixed gases" found in the atmosphere. Less than a year later, he joined the Manchester Literary and Philosophical Society. It is interesting that his first communication to the Society was not on the properties of gases but on the subject of color vision. In this paper he discussed his red-green color blindness, a defect that has since become known as "Daltonism." Since in the next five years he presented but one more scientific communication, it is reasonable to conclude that his duties at New College absorbed all his effort except for his meteorological observations. Had his connection with the college continued, he probably would not have had the leisure for original investigation. But, in 1799, the college closed and moved to York. Deprived of his comfortable, though demanding, position, Dalton reverted to private teaching, supporting himself in this capacity for most of the remainder of his life.

Fortunately for science, the Philosophical Society permitted him to occupy one of the lower rooms of its house on George Street, which he used as a schoolroom, study, and laboratory through the succeeding years. He had never before carried on systematic experimental research, but, once this was begun, he brought to fruition in a few years all the ideas for which he is remembered: the diffusion of gases, the law of partial pressures, the phenomena of water vapor in the atmosphere, including the dew point, and the chemical atomic theory. From this impressive list of contributions it is reasonable to assume that Dalton must have brought to his studies a high order of experimental skill. The judgment of his contemporaries was, curiously enough, quite the opposite.

At the close of the 18th century, the characteristics of gases were still poorly understood. The kinetic hypothesis, the fundamental basis for understanding the behavior of gases, although already proposed by Robert Hooke and treated mathematically by Bernoulli, was hardly known despite the stature of these champions. The scientific world, dazzled by Antoine Lavoisier's genius, made chemistry the vogue of contemporary investigation. Mysterious phenomena were made plausible by "chemical" assumptions. The puzzling fact that oxygen and nitrogen in the air remained thoroughly mixed, and did not separate into distinct strata despite their different specific gravities, was accounted for by saying that the two gases had entered into a "loose" chemical union. Just what this was, in comparison with examples of regular chemical union, could only be characterized as "loose" thinking. Another example of this same kind of "chemical" action was exemplified by the movement of water molecules into the air and their continued presence as water vapor.

Dalton, being innocent of chemical notions when he formed his ideas on these subjects, approached them from a physical point of view. Being influenced by Newton and his particle conception of gases, he advanced a physical theory of diffusion, of partial pressures, and of the behavior of water vapor in air. All these ideas were developed in a long paper entitled "Experimental essays on the constitution of mixed gases; on the force of steam or vapour from water and other liquids in different temperatures, both in a Torricellian vacuum and in air; on evaporation; and on the expansion of gases by heat," and read before the Literary and Philosophical Society in October 1801. The physical theory of diffusion depended upon Dalton's assumption of Newton's mutually repelling particles. Attacking the problem experimentally, he showed that, at any given temperature, there was a maximum pressure of water vapor in the air, the saturation pressure, which had a constant value regardless of whether air or any other gas were present. In fact, the result was the same if no gas was present, and the water evaporated into a vacuum. This situation led naturally to the law of partial pressure—that each gas of a mixture exerts a partial pressure that equals the pressure it would exert if it filled the container alone.

Analyzing the saturation-pressure phenomenon, he showed that any attempt to increase the saturation pressure resulted only in condensation. Furthermore, if the partial pressure of water vapor in the air was less than the pressure at which saturation took place, a sufficient reduction in temperature produced saturation, that is, the dew point was reached. Some of this material, e.g., his explanation of diffusion and his law of partial pressures, were included in his *New System of Chemical Philosophy*, Part I, published in 1808.

His theory of diffusion, which he assumed accounts for the inevitable and complete mixing of two dissimilar gases in a confined space, led him to the prediction that high enough up in the atmosphere, the separation of the gases takes place. To test his theory he began analyzing samples of air. His method of analysis consisted of mixing air with nitrous oxide (NO) to produce a reaction; he found that the proportions in which the two gases united chemically depended on the manner in which the reaction occurred. The investigation of this variability extended from the latter part of 1802 through the summer of 1803. From his notebooks it seems clear that not until August 4 did the reason for the variability of the reacting proportions become clear. The conclusion he reached may be given in his own words.

> If 100 measures of common air be put to 36 of pure nitrous gas in a tube 3/10ths of an inch wide and 5 inches long, after a few minutes the whole will be reduced to 79 or 80 measures, and exhibit no signs of either oxygenous or nitrous gas.

> If 100 measures of common air be admitted to 72 of nitrous gas in a wide vessel over water, such as to form a thin stratum of air, and an immediate momentary agitation be used, there will as before be found 79 or 80 measures of pure azotic gas for a residuum.

If in the last experiment, *less* than 72 measures of nitrous gas be used, there will be a residuum containing oxygenous gas; if *more* then some residuary nitrous gas will be found. . . . *these facts clearly point out the theory of the process: the elements of oxygen may combine with a certain portion of nitrous gas, or with twice that portion but with no intermediate quantity.* In the former case nitric acid is the result; in the latter nitrous acid. . . . But as both of these may be formed at the same time, one part of the oxygen going to one of the nitrous gas, and another to two, the quantity of nitrous gas absorbed should be variable. (Roscoe & Harden, pp. 32–33)

Here it is worth emphasizing the way in which science actually develops as compared with the way popular accounts lead the reader to believe it does. No great deductive discovery ever springs suddenly with noonday clarity into an investigator's head. Usually experimental evidence hints at a generalization whose scope is hazy; a great deal of thought must be spent on the idea to link it with the known relevant facts. In this way the concept begins to take a definite shape whose validity and limits finally come into some sort of focus. The ultimate value and importance of a new discovery can rarely, if ever, be fully apprehended at the time of its enunciation; it is in what follows that its true value is established.

With the facts of definite proportion in a chemical reaction finally unmasked, what was its meaning? Imbued with the physical, particle picture of gases, Dalton was led to the assumption that the free particles of a given gas—atoms, as he soon called them—are all alike, and a chemical reaction is only the combination of an atom of one substance with that of another. Thus was the atomic theory at last placed on an experimental basis. But there was more to it. By the following month, Dalton saw that from the masses of two simple substances that react completely and an assumption regarding the formula of the compound, one can infer the relative weights of the constituent atoms. He immediately constructed a table of relative atomic weights together with a set of symbols for the elements, and at the October meeting of the Literary and Philosophical Society he presented a resumé of this work. We have no information about the members' reaction to this communication; probably only a handful were present at the meeting, and it is doubtful that any of those had a glimmering of its real importance.

At about this time or shortly after, Dalton accepted an invitation to deliver the 1803 Christmas lectures at the Royal Institution. The Institution, destined to play a great role in British science, was then only newly founded under the direction of the vigorous and colorful Count Rumford, alias Benjamin Thompson, the Massachusetts Tory who, on the eve of the Revolutionary War, fled a wife considerably his senior, several children, and the mounting ire of his colonial neighbors. For the scientific direction of the institute, Rumford, with characteristic foresight, had already hired a brilliant and personable young Cornishman, then in his early twenties, one Humphry Davy. Davy arranged Dalton's stay and the details connected with the presentation of his lectures. Certainly Dalton must have communicated his atomic theory to Davy, but all we know for certain is that the theory was presented to his audiences at that time. Although Davy finally acknowledged the power of the atomic theory, he never accepted its basic tenets or the reality of atoms.

Dalton's papers of 1801–1803 propounding the atomic hypothesis were not published until 1805 when they appeared in the latest volume of the society's memoirs. In the meantime, Dr. Thomas Thomson of Glasgow, a chemist of repute and author of a widely used treatise on chemistry, visited Dalton in 1804 to discuss obscurities in his diffusion theory. Although this discussion was not too fruitful, Thomson listened eagerly to Dalton's chemical experiments and his atomic theory. When the next edition of Thomson's textbook appeared in 1807, it contained the first generally available description of the atomic theory. Dalton's own account was not published until the following year in his *New System of Chemical Philosophy*, Part I.

The diffusion hypothesis that had led him to his atomic theory was never straightened out. At first, Dalton assumed that the atoms of a gas were all the same size, but Thomson's objections led him to change his view and to assume that, although all the atoms of a given substance are identical in all respects, the atoms of different substances are of different size. Still later he reverted to his original belief that all atoms are of the same size. The diffusion theory was also complicated by the assumption that, while atoms of gas A repel each other and those of gas B do likewise, atoms of A and B exert no mutual force on each other. This fit well enough with his law of partial pressures. But if the respulsion were due to "the atmosphere of heat," that is, caloric (an idea borrowed from Lavoisier), that surrounded all atoms, it was hard to see why atoms of A and B did not act on each other. This point also was never resolved. Dalton spent the years after 1804 elaborating on the ideas he had already expounded and in further chemical experiments. None of these experiments produced any memorable results. After he had enunciated the atomic theory, it seemed that he became hopelessly confused in a maze of rapidly accumulating, but seemingly contradictory, chemical data.

The acceptance of the atomic theory was slow, and among Englishmen it found few champions. On the continent, and especially in France, it fared better. Joseph Louis Gay-Lussac's experiments, published in 1809, demonstrated that gases combine in equal volumes or in small volume ratios. These experiments were explained in 1811 in terms of Dalton's theory by the Italian physicist Amedeo Avogadro; explained, that is, if one were willing to believe in atoms and "half atoms." Dalton, in fact, had assumed that the free particles in any elemental gas were atoms and he never distinguished between what we now call atoms and molecules. The distinction between atoms and molecules of an element, and the tangle that developed in determining atomic weights as the result of Dalton's arbitrary rules for chemical combination, had to wait another half century for the analyzing genius of Stanislao Cannizzaro. Even though many chemists, like Davy, did not accept the reality of atoms, nevertheless the atomic theory became widely accepted as a potent tool in understanding chemical reactions.

In the year following his discovery of the law of atomic combination Dalton became a boarder in the home of the Rev. William Johns on George Street close to the society's building, an arrangement that continued with complete amity for 26 years until the Johnses moved to the suburbs of Manchester. In these years, Dalton's life was much like that of the hero of Jules Verne's *Around the World in Eighty Days*—he was "un homme mathématiquement précis." His days in the laboratory never varied, and his neighbors knew the time to the minute from his observations of his outdoor thermometers. In the evenings, he finished his day at nine, had dinner, and then sat with the family quietly puffing his pipe. He took little part in the conversation but remained interested and listening, occasionally interjecting a remark or a laconic witticism. Every Thursday afternoon, "shop" was closed for his weekly recreation. After his noonday meal, he always set out for the bowling green adjoining the Dog and Partridge Tavern where he entered into the game with a zest and enthusiasm that amused all the spectators. His usual staid and austere manner was dropped, and he pursued the ball and waved his arms as if to direct it in its course. A few modest bets, always carefully noted, were part of the game, which was always followed by tea and a smoke on his long pipe. Then he walked home in time for a final reading of his barometer and thermometers. On Sundays, dressed in his Quaker knee breeches, gray stockings, and buckled shoes, he always attended public worship twice. He appeared to feel it his duty to express his religious observance in this way, although he adopted little of the Quaker phraseology and never discussed his views on religion. It was as though he felt an obligation to observe the outward form of his faith to maintain the morale of others or to escape their censure, whereas his own emotional life was not vitally concerned with it. In most company he was withdrawn, a condition that stemmed partly from his physical makeup and partly from his country background and lack of youthful social training. Apparently, many people were repelled by his physical awkwardness and lack of social grace. Dr. Davy, the brother of Sir Humphrey Davy, said that "his aspect and manner were repulsive . . . his voice harsh and brawling, his gait stiff and awkward." The testimony of a fellow member of the Literary and Philosophical Society was much the same: "His voice was deep and gruff, and his articulation thick, indistinct and mumbling; so much as to give a stranger almost the idea of uncouthness. . . ." From these descriptions, it is at once clear why Dalton was never very successful as a lecturer, although this social awkwardness alone does not seem to be the cause. As a demonstrator he was not skillful and many of his lecture experiments failed. This lack of experimental skill even extended to his researches, for his biographer, Dr. W. C. Henry, writes that: "His instruments of research, chiefly made by his own hands, were incapable of affording accurate results, and his manner of experimenting was loose, if not slovenly." He did not possess "the manual dexterity or the mental habits or temperament essential to rigorous experimental determinations." Nevertheless, those close to him, especially in his scientific work, recognized his ability and his devotion. The Manchester Literary and Philosophical Society voted him secretary in 1800, vice president in 1808, and president in 1817. He was continuously reelected to the latter office throughout his life. But here again his lack of social understanding and the limitations of his outlook prevented his incumbency from being distinguished or inspiring. He was not a reader, often boasting that he "could carry his library on his back and yet had not read half the books." This point of view characterized his extreme reluctance to increase the small store of scientific books of the society, even though it was then the only such library in Manchester. It must be remembered also that the society had professed literary as well as philosophical pursuits. As chairman, however, Dalton was inclined to discourage the reading of literary essays. It must have been amusing for those sufficiently detached, but very annoying to the authors, to hear him pass judgment that such readings "contributed no positive facts to the stock of knowledge and proved nothing."

As the value of Dalton's theory became recognized he began to receive appropriate recognition. The French Academy elected him a corresponding member in 1816 and the Royal Society, with Davy as president, elected him a regular member in 1822. In 1826 the Royal Society again honored him with the award of a Royal Medal and one of the two George IV prizes of 50 guineas. In 1830, the French

Academy awarded him its highest honor by electing him one of its eight foreign associates, a place previously held by Davy, who had died the year before. When the British Association for the Advancement of Science held its meeting in Oxford two years later, the university conferred on him the degree of Doctor of Laws along with the illustrious Michael Faraday, Robert Brown, and Sir David Brewster. The following year at the Cambridge meeting it was announced that the government had awarded him an annual pension of £150. Three years later this pension was doubled. In 1834 he received the LL.D. degree from Edinburgh University and in the same year he was presented to King William IV at a levee in St. James Palace. For this ceremony, it was necessary to wear a military sword, a requirement that was not only out of keeping with his character but also with his Quaker beliefs. The problem was solved by dressing him in his Oxford Doctor of Science gown. It was reported that the King's greeting was "Well, Mr. Dalton, how are you getting on in Manchester, all quiet I suppose?" This remark was not quite as vacuous as it sounds, since the King had reference to the riots that had occurred in Manchester some years before. Dalton, in characteristic fashion, answered: "Well, I don't know—just middlin', I think." When asked by a friend if he did not consider his manner of address a breach of court etiquette, he replied with a lapse into his native Cumberland dialect: "Mebbe sae, but what can yan say to sic like fowk?"

It is fortunate that his honors should have come in time, for in the spring of 1837 Dalton was afflicted with two paralytic strokes from which he never recovered. Though able to continue his work at a slower pace, his mental vigor began to diminish, so that by 1840 a paper contributed to the Royal Society was declined to save his reputation. In June 1842, the British Association met in Manchester, and it was hoped that he might be made president of the meeting. But by this time his articulation, never clear, was so bad as to be unintelligible and his health so infirm as to render the appointment impossible. Nevertheless he was made vice president—the last public recognition during his life of the eminence he had achieved in the scientific world. He lingered for two more years, conscientiously taking his meteorological observations until a stroke on July 27, 1844, freed him forever from his self-imposed obligation.

Not everyone familiar with atomic science knows that the chemical atomic theory, that is, that elementary substances in a chemical reaction combine atom with atom, was proposed by others before Dalton. The first proponent of such a theory appears to have been an Irish chemist and teacher of chemistry working in London, Bryan Higgins (1737–1820) and the second was his nephew William Higgins (1769–1825). Because it is sometimes claimed that Dalton has received credit that rightfully should go to the Higginses, it is necessary to summarize their contributions so that readers may be able to judge for themselves.

Bryan Higgins's ideas were formulated in four publications: "Syllabus of Chemical and Philosophical Inquiries" (1775), "A Philosophical Essay Concerning Light" (1776), "Experiments and Observations Relating to Acetous Acid, etc." (1786), and "Minutes of a Society for Philosophical Experiments and Conversations" (1795). The atomic theory developed in these works has been summed up as follows:

> 1. There are seven elements—earth, air, water [the elements of the ancients to which Higgins added] acid, alkali, phlogiston and light. Fire [the fourth element of the ancients] Higgins assumed to be a compound of phlogiston and light. 2. Atoms of the elements are

all hard, and completely spherical or nearly so. 3. Atoms attract one another by forces depending on distance and polarity. The forces between atoms are dependent on the inverse nth power of the distance between them, i.e., force $= f(1/r^n)$. 4. Fire pervades all bodies and gives effects opposed to attraction. 5. Gases are produced by the addition of fire to solids and liquids. 6. Combination in definite proportions ("saturation") is an effect of attractive and repulsive forces. 7. Molecules of binary compounds always contain one atom of each constituent. 8. Particles in gases are surrounded by repelling atmospheres of fire. In chemical combination these atmospheres must be broken or blended (Partington, p. 269).

A comparison with Dalton's atomic ideas shows some points of similarity, such as spherical atoms, gaseous particles separated by a repelling atmosphere of fire (heat), and combination in definite proportions. The first two, spherical atoms and gaseous particles separated by a repelling atmosphere of fire, represent more or less the general thinking of atomists of the time. The third is a consequence of the assumption of atomic combination. As Dalton read very little of the work of others, there is little reason to suppose that he knew about these ideas.

The details of chemical reaction imagined by Higgins will not be elaborated upon but his theory shows considerable ingenuity and his work reflects the influence of Newton's ideas. "[T]he explanation of combination in definitive proportions ('saturation') in terms of attractive and especially repulsive forces between the atoms is the first example of the application of the Newtonian theory to chemical changes" (Partington, p. 272).

Although Bryan Higgins was not taken very seriously by the scientific community at that time, the same cannot be said of his nephew William Higgins who published a book called *A Comparative View of the Phlogistic and Antiphlogistic Theories* in 1789. He attacked the then very fashionable phlogistic doctrine (the name given to the so-called "burning principle" according to which, when a substance burned, phlogiston was assumed to be given off into the air). Because atomistic speculations were introduced as merely another weapon in the intellectual arsenal that William Higgins turned on the proponents of the phlogistic doctrine, Higgins's book did not attract the attention that it might have had it concentrated on atomic theory alone. In any event, Higgins's atomistic speculations were muddled because he never specified the origin or nature of the forces causing atoms to combine. If they were seen as central forces, that is, gravitational forces, they would vary inversely as the square of the distance between atomic centers. But there is no indication that Higgins embraced this view. If the atoms were hard spheres (in conformity with the expressed ideas of Bryan Higgins), then we might assume that chemical combination would put the atomic spheres in contact. Assuming contact and atoms of equal size, the distance between the centers of all the atoms would be alike, at least until combinations were considered involving more than six atoms with the one under consideration. None of these ideas occurs in Dalton's atomic theory so it seems improbable that Dalton knew of Higgins's theory.

The parallels between Higgins's and Dalton's atomic theory were first brought to the attention of the scientific world in November of 1810, when Davy discussed them in his Bakerian lecture of that year. Davy's standing was such that it at once publicized Higgins throughout Europe and set off a long controversy between the two rival theorists; the inevitable suggestion arose that Dalton was guilty of plagiarism. Dalton's prior claims were strongly supported by Dr. Thomas Thomson, chemist and editor of the *Annals of Philosophy*, by Dr. W. C. Henry, and by a succession of Manchester chemists, including Angus Smith, Roscoe, and Schorlemmer.

Considering that Dalton read very little, that he had a limited acquaintance with chemists, and that his theory began as a *physical* theory connected with the diffusion of gases and the mutual interaction of mixed gases in the atmosphere, we find it unlikely that he knew of Higgins's work before Davy's announcement in 1810. This was affirmed later by many of Dalton's friends. While it must be admitted that Higgins's idea of atomic combination in chemical reactions predated Dalton's, the credit for a scientific discovery goes to the investigator who gives it a solid foundation and brings it to the scientific world on this basis. In the opinion of his contemporaries, William Higgins failed to do this work. Consequently, Davy, who called him to the attention of scientists in 1810, passed this severe judgment on Higgins at his death in 1825: "It is impossible not to regret that he did not establish principles which belong to the highest department of chemistry, and that he suffered so fertile and promising a field to be entirely cultivated by others; for though possessed of great means of improving chemistry, he did little or nothing during the last thirty years of his life."

The Volume Combination of Gases: Joseph Louis Gay-Lussac (1778–1850)

Dalton's atomic theory, which envisioned chemical combination as an atom to atom union, was immediately most useful in settling the question of combination in definite proportions. But like most newly born theories, it included basic assumptions that needed experimental confirmation. The one assumption that, from the standpoint of the subsequent development of chemistry, produced a morass of confusion was the supposition that the ultimate naturally occurring unit of any element was the single atom itself. This "monatomic" hypothesis created difficulties in understanding chemical reactions that required almost half a century to dissipate.

What is most remarkable is that this confusion need hardly have arisen. A set of experimental results published by Joseph Louis Gay-Lussac in 1809, the year after the appearance of Dalton's first volume of *A New System of Chemical Philosophy*, contained the data that might have avoided these difficulties, although Gay-Lussac's data alone would not have been sufficient. Even more remarkable is that the proper interpretation of Gay-Lussac's experiments (demanding a deep understanding of the possibilities of the atomic hypothesis—deeper certainly than Dalton himself possessed) was supplied just two years later by an obscure Italian scholar, Amedeo Avogadro.

Gay-Lussac showed that if gases enter into chemical reaction, they do so in numerically simple volume ratios, and the volume of the products, if gaseous, may be expressed by simple integral numerical ratios to the volumes of the original reactants. For example, if one unit volume of nitrogen reacts with one of oxygen, two volumes of nitrous oxide are produced. If one volume of nitrogen reacts with three of hydrogen, two volumes of ammonia result. These simple relationships indicated some very fundamental aspect of chemical combination, but what was it?

At that time, the explanation was obscured by several factors. First, there was no general agreement as to whether or not chemical combination occurred in definite proportion; second, no distinction between the atom and the molecule of a given substance had yet been made. Once Dalton's hypothesis was accepted, the first difficulty was settled. But the second distinction had to be made along with a far-reaching supposition as to the number of molecules in equal volumes of gases. That was Avogadro's contribution. However, before we examine Avogadro's work we review what Gay-Lussac accomplished.

In his personal and professional life Joseph Louis Gay-Lussac may be described as the always-successful man. He was born on December 6, 1778, in St. Léonard in the department of Haute-Vienne. His father, Antoine Gay, was a judge who, after establishing his residence at St. Léonard, acquired nearby property named Lussac. He then added this name to his surname to distinguish himself from others having the same family name. Young Joseph Louis grew up in the troubled times preceding and during the French Revolution; as a consequence, he received his early education at home. His father was imprisoned briefly as a suspected "aristocrat" after the outbreak of the Revolution in 1789 and did not consider it safe to send his son to Paris for schooling until 1795. After two years of preparatory study in Paris, Joseph Louis was admitted at the age of nineteen to the École Polytechnique after writing a brilliant examination.

In 1800, he transferred to the École des Ponts et Chaussées and became research assistant to the great chemist Berthollet who had just returned from service on Napoleon's personal staff with the French army in Egypt. Berthollet recognized young Gay-Lussac's outstanding abilities, adopted him as a "son-in-science," and gave him a place in his private laboratory at Arcueil. This fortunate association greatly assisted Gay-Lussac's rapid professional advancement. In 1802, he was appointed demonstrator for Professor Antoine F. Fourcroy at the École Polytechnique while continuing his work with Berthollet. As the latter's pupil he began work in the same year on the expansion of gases by heat and discovered the very important principle that we now know as Charles's law, namely, that the ratio of volume to absolute temperature of an enclosed gas is constant if the pressure is held constant. Before publishing his results he was chagrined to find that the same conclusion had been reached 15 years earlier by Charles, who had not considered it sufficiently important to publish.

In 1804, as a result of balloon flights in Germany and Russia, it was reported that the strength of the earth's magnetic field decreased with altitude. The French Academy, wishing to verify the result, selected Gay-Lussac and Jean B. Biot to make the test. On August 24 they ascended from the garden of the Conservatoire des Arts et Métier in a hydrogen-filled balloon and reached an altitude of 13,120 feet. They found, despite some difficulties, no change in the magnetic field. Dissatisfied with the results, but well aware of the danger involved, Gay-Lussac volunteered to make a second ascent alone to improve the accuracy of the measurements and to obtain meteorological data not then available. He accomplished this feat on September 16, rising to a record height of 23,040 feet. Despite an extreme temperature of $-9.5°C$, he carried out a series of magnetic measurements over a period of several hours, measured the temperature and humidity of the air, and obtained air samples to find out if the composition of the atmosphere changed with altitude. The results showed no diminution of the magnetic field and no change in the composition of the atmosphere within the height attained. Despite the negative results, the whole undertaking cannot help but excite admiration for so daring an investigation, as well as respect for the high competence evident in managing such a flight alone and in securing all the scientific information for which the venture was organized. It is worth noting that the possibility of aerial navigation had been demonstrated by the Montgolfier brothers only 20 years previously and that the hazards of using hydrogen as a lifting gas had been shown by fatal accidents in the intervening years.

Two weeks after this flight Gay-Lussac, in conjuction with Alexander von Humboldt (later to win fame as a naturalist), read a paper that publicized the first results on the volume combination of gases. Further research on this subject was delayed by a long trip through Europe which Gay-Lussac took with Von Humboldt for the purpose of making magnetic measurements. While this work was still in progress, he was elected in 1806, at the age of 28, to the French Academy. His investigation on the volume combination of gases was finished in 1808 and the results read before the Sociéte Philomathique on the last day of that year.

Gay-Lussac married in 1808 in a manner that offers an interesting sidelight on his character. While in a linen draper's shop in Paris he saw a young woman clerk reading a book on chemistry. He promptly paid court to the lady and married her, after which he proceeded to have her educated as he thought proper, in English and Italian. Probably as a consequence of assuming family obligations, Gay-Lussac's attention, following the completion of his study of the combination of gases, turned increasingly to applied chemistry. His reputation was such that his services as a technical consultant were in wide demand. He was already active as a member of the Consultative Committee on the Arts and Manufactures, and this interest increased in the years ahead.

In 1808 he was also appointed professor of physics in the Sorbonne, a post he held until 1832; on January 1, 1810, he succeeded Fourcroy as professor of chemistry at the École Polytechnique. His last outstanding piece of pure research, in the period 1811 to 1815, was on prussic acid. In the succeeding years, he became superintendent of the government gunpowder factory in 1818 and, in 1829, assayer to the mint. However, he continued to maintain his laboratory work in chemistry and to lecture at the École. A report quoted by J. R. Partington on his lectures in 1820 states that "he had a slender and handsome figure, his voice was gentle but firm and clear, his diction terse and choice and the lecture was a superlative specimen of continuous unassailable experimental reasoning. " In 1831, when 52, he was elected to the Chamber of Deputies to represent his native Department of Haute-Vienne and in 1839 he entered the House of Peers, having been created a peer by King Louis Philippe. He was editor with Arago of the *Annales de chimie et physique*, a task which he continued almost to the end of his days.

In summing up his personal life we quote Sir Humphry Davy. He was "quick, lively, ingenious and profound with great activity of mind and great facility of manipulation." But his early outstanding success in both academic and public life made him increasingly aware of his own merit, a feeling manifested by an aloofness and chilly reserve that increased as he grew older. He became, in fact, somewhat of a prig, but this should not blind us to the fact that he was a bold, energetic, and courageous man whose work, characterized by high competence, perseverance, and a consuming passion for accuracy, greatly enriched science and humanity. He died in Paris on May 9, 1850, at the age of 71.

Atoms and Molecules—Avogadro's Law: Amedeo Avogadro (1776–1856)

The clarification of atomic weights, the formulas of chemical compounds, the development of the periodic table of elements, and, indeed, a rational science of chemistry, all have their roots in the application of Avogadro's principle that equal volumes of different gases at the same temperature and pressure contain the same number of molecules.

The deduction of this law and its application to various chemical combinations on the basis of Gay-Lussac's measurements is a model of presentation and clear-headed logic based on experimental evidence. By his principle Avogadro correctly deduced the chemical formula for water, ammonia, nitrous oxide, nitric oxide, and nitrogen dioxide. This principle stemmed from a new and revolutionary hypothesis, namely that the ultimate particles of many of the elementary gases such as hydrogen and nitrogen are molecules, that is, combinations of two, or sometimes more, atoms. Ideas on combination then current made this suggestion too implausible for the chemists of that day to accept and the science of chemistry fell into a morass of "systems" that greatly impeded progress. Recovery finally came through the genius of Stanislao Cannizzaro who dispelled the confusion by the thoroughgoing application of Avogadro's principle.

Amedeo Avogadro—or Lorenzo Romano Amedeo Carlo Avogadro di Quaregna e di Cerreto, to use his full name—was born in Turin, Italy, on August 9, 1776, the son of Count Filippo and Anna Vercellone Avogadro. Of his early life and interests, and his ability as a student, scant information exists. All that is known is that his family had been prominent since the 11th century in the political life of the nearby province of Vercelli. His forebears for generations had been lawyers for the Roman church. As a consequence, "Avogadro," denoting this occupation, became the family name, and the original family name was lost.

Amedeo, as might be expected, was destined for the legal profession and his education was directed to that end. In 1792, at the age of 16, he received his bachelor of laws degree and, three years later, that of doctor of ecclesiastical law. During the next four years, he served as a lawyer in several government offices. But at the end of this time he had apparently made up his mind about his life's work, for he abandoned his legal career and devoted his full energies to the study of physics and mathematics for the next five years. During this period, at the age of 27, he

Amedeo Avogadro (1776–1856)

published, with his brother Felice, his first original paper entitled "Analytical Essay on Electricity." This was followed shortly by other papers on the same subject. As a result of these publications he was elected to membership in the Turin Royal Academy of Sciences in 1804. Shortly after his election he was appointed tutor in the Royal College of Vercelli, and, in 1809, professor, a post he held until 1820. His famous principle, published in 1811 when he was 35, was composed there.

The 14 years at Vercelli were followed by an appointment as professor of mathematics and physics at Turin. However, owing to political unrest and the attendant reactionary policies imposed on the university, his professorship was canceled in 1822. As a sop, he was made professor emeritus and given a small pension. In 1832 a change of government restored the canceled professorship.

Avogadro's former chair was not immediately offered to him but was given instead to the celebrated mathematician Augustin-Louis Cauchy. After only a year at Turin, Cauchy accepted an appointment in Prague; a year later Avogadro was reinstated in his former professorship. He held this post until his retirement at the age of 74 in 1850. He lived six more years, passing away in Turin in 1856, at the age of 80.

Factual details of his life give very little insight into the kind of man Avogadro was but his portrait suggests a gentle person of slight build. His features show a prominent hooked nose that gave him a rather birdlike appearance. Except for his initial collaboration with his brother Felice, all of his mature work was done alone. He attended no meetings and sought no colleagues with whom to discuss scientific ideas. He was modest to the point of self-effacement, apparently because of a complete indifference to the world rather than any fear of it. Although he did not live to see the recognition of the true value of his principle, nevertheless he attained a measure of scientific acclaim in his lifetime. To this acclaim he seemed completely oblivious and never utilized his recognition to push himself or to attain scientific or political position. He cultivated no influential friendships and seemed completely devoid of personal ambition. All of this might lead to the conclusion that he was a recluse, but that would be quite wrong. He married and reared a family of six children, two sons and four daughters. He formed warm personal relationships, and he not only took his duties as a teacher seriously, but was successful also in imparting his knowledge to students. He was thoughtful of others and it was even stated that he resigned his professorship so that his protégé, Felice Chio, might be advanced. Although he published little, he was a prodigious writer, leaving at his death 75 volumes of manuscript, each containing approximately 700 pages of handwritten notes consisting of résumés on astronomy, organic and inorganic chemistry, and physics. He was, in short, a perpetual student, happy with his family and a small circle of friends, oblivious to most of the external world, and devoted to the quiet life of contemplation and scholarship.

It is important to recognize that Avogadro's principle is a twofold one. It states first that equal volumes of gases (at the same temperature and pressure) contain the same number of molecules. Second, the elemental particles of a gas may consist of more than one atom, that is, a nitrogen molecule is diatomic, N_2. This is so universally accepted today that it is scarcely noted, but, in Avogadro's time and later, this hypothesis was not accepted. Its rejection was based on a widely held theory of chemical combination, advanced by Jakob Berzelius, that assumed an electric charge on individual atoms. Like atoms were assumed to hold like charges and hence to repel each other. Berzelius was one of the most influential chemists of his era. Avogadro, on the other hand, was unknown and his ideas appear to have been equally so until Cannizzaro's time. The result was that chemistry fell into confusion, a situation that emphasizes the necessity in science of the reiteration of opposing ideas until a secure solution is reached.

In a gaseous reaction, such as hydrogen with chlorine to produce hydrochloric acid, we find that one volume of hydrogen combining with one of chlorine produces two volumes of hydrochloric acid gas. We assume that the gases are polyatomic — specifically, in this instance, diatomic. Diatomic molecules consist of two atoms joined together by a force of attraction. It is clear that in the production of HCl, two volumes should result. In the formation of water vapor from a reaction of gaseous hydrogen and oxygen, experiment shows that if the reaction is complete, that is, with none of the original constituents remaining, two volumes of hydrogen must combine with one of oxygen. We see that if hydrogen is and assume oxygen to be diatomic, the production of two volumes of water vapor indicates that the formula for water is H_2O.

Next we need to see whether this simple view is consistent with other facts obtained by weighing. Suppose we adopt a standard volume such that when it is filled with hydrogen at some standard temperature and one atmosphere pressure, the weight of the enclosed hydrogen is 2 grams. On the same basis, the weight of chlorine in the initial experiment above is found to be 71 grams. Experiment shows that each container of the product, HCl, weighs 36.5 grams. If a table of atomic weights is based on hydrogen as the unit weight, then on this basis, $H = 1$, $Cl = 35.5$, and HCl weighs 36.5 grams. A volume of oxygen equal to the "standard" volume chosen for hydrogen weighs 32 grams; accordingly H_2O should thus have a molecular weight 18, in agreement with experiment. If further experiments of this kind had been made on the synthesis of the oxides of nitrogen, all of the results would have been found to be consistent with the starting hypotheses. Furthermore, the atomic weight of nitrogen from the volumes involved would have been found to be approximately 14.

The Search for a Primordial Matter: William Prout (1785–1850)

The notion that all matter is composed of the same primary substance and that when organized in different ways produces the various elements occurs far back in antiquity. Empedocles, about 440 B.C., suggested that only four elements exist: earth, air, fire, and water. But Aristotle, the giant intellect of the ancient world, could not accept immutable elements. Instead, he proposed a world in which visible matter was the result of the incorporation of "qualities," such as hot, dry, cold, and wet, with the great substratum of formless primary matter. This doctrine of the four elements led easily to the idea of transmutation, for were not all the elements composed of the same material and only modified in their qualities? Thus, the mere substitution of wet for dry should be sufficient to change fire into air. However naïve these ideas seem to us now, we should not forget that they were accepted by scientists up to the time of Boyle and Newton, a mere 350 years ago.

An idea long established in the minds of humanity dies very hard. Although the revolution introduced into chemistry by Boyle swept away the four elements, it pronounced no such fate for the concept of a "first material." How then did this idea fare in the newer world of Daltonian theory, where atoms were distinguished by their different masses? As the first halting experiments in the new chemistry

William Prout (1785–1850)

measured these masses, many turned out to be closely integral, if the value for the atomic weight of hydrogen is taken as the basic unit and common denominator. How natural, therefore, for anyone steeped in the ancient lore to perpetuate the idea of the "first material" by identifying it as hydrogen.

The suggestion that the atoms of all elements are simply combinations of hydrogen first appeared in two anonymous papers published successively in the *Annals of Philosophy* for the years 1815 and 1816. Later in 1816, Thomas Thomson, the editor of *Annals of Philosophy*, disclosed that the author was a London physician named William Prout, a man with a deep interest and great ability in chemistry. Prout's suggestion soon became known as Prout's hypothesis. For more than a century it stood as a guidepost in the geography of atomic research. In the end, the direction to which it pointed proved erroneous but not until many explorers had thoroughly mapped the wilderness of atomic weights. It was important because it pointed the way to regions that had to be explored before atomic theory could safely widen its horizons.

It is worth reviewing briefly the vicissitudes that Prout's hypothesis endured in order to see how pervasive its influence has been. Thomas Thomson, himself a chemist of note and the first to publicize Dalton's atomic theory, had been measuring atomic weights for several years when Prout's papers appeared. Thomson became a strong supporter of the hypothesis, but his chemistry was criticized by Berzelius, who opposed Prout's ideas. In 1833, Edward Turner, at the invitation of the British Association for the Advancement of Science, undertook a series of analyses and concluded that the hypothesis was not exact (although his results did not justify the assertion). The same conclusion was reached by Frederick Penny in Glasgow in 1839. A determination of the atomic weight of carbon by Dumas and Stas in 1839 to 1840 and the synthesis of water by Dumas in 1843 both supported the hypothesis. A series of researches by de Marignac of Geneva at the same time led to the conclusion that the principle might be exact if the unit were half as large. A further investigation by Stas in 1860 led him to "consider Prout's law as a pure illusion."

Although the general result of these studies showed that there was not always close agreement with the hypothesis, some experiments demonstrated an approximation so close that it seemed necessary to postulate the existence of an underlying regularity that further research must explain. Even as late as 1886, Sir William Crookes gave strong support to the theory in his presidential address to the chemistry section of the British Association. In it he proposed a primary substance *protyle* out of which the atoms were formed; hydrogen he considered the first and simplest. He recognized the integral value of the atomic mass of a great number of the elements but questioned the "absolute uniformity in the mass of every ultimate atom of the same chemical element. Probably our atomic weights merely represent a mean value around which the actual [integral] atomic weights of the atom vary within certain narrow limits. " In this way he accounted for those atoms of nonintegral atomic weights such as chlorine. This suggestion proved indeed prophetic, as will be seen when we consider the discovery of isotopes by Frederick Soddy. In the meantime let us look briefly at the life of the man whose suggestion so strongly stimulated the progress of atomic investigation.

William Prout was born at Horton, Gloucestershire, England, on January 15, 1785. He showed a marked interest in chemistry in his early schooling but eventually chose the practice of medicine as a profession. He obtained his M.D. degree at Edinburgh University at the age of 26 in June 1811. The following year he began his practice in London where he resided for the rest of his life. His interest in chemistry was shown not only by the papers that formed the basis of his hypothesis but also by his contributions in physiological chemistry. Of these, the most important was his demonstration that free hydrochloric acid is present in the stomach where it is an important factor in digestion.

Prout was elected a fellow of the Royal Society in 1819, and ten years later became a fellow of the Royal College of Physicians. After receiving that honor, he turned his attention almost completely to medical questions and to his practice. Toward the end of his life he became deaf and withdrew from active social life. He died in his home on Sackville Street, Piccadilly, on April 9, 1850, at the age of 65.

In reading Prout's paper it is interesting to observe that he incorrectly believed atmospheric air to be a chemical compound, in fact, N_2O. This seems strange since the oxides of nitrogen were then well known and the chemical properties of the compound N_2O are quite distinct from those of air. The relative weights of oxygen and nitrogen, taken as 10 and 17.5, are also unusually erroneous. However, accepting these values together with N_2O as the formula for air, we find that the relative weight of an air molecule would then be $2N + O = 2 \times (17.5) + 10 = 45$ and the parts by weight of oxygen in the air would be $10/45 = 22.22\%$ and of nitrogen $35/45 = 77.77\%$.

To find the specific gravities (weights per unit volumes) of oxygen and nitrogen relative to air we may ignore Prout's indicated calculation and verbalize it as follows: One fifth of a unit volume of air is composed of oxygen with relative weight 10, and four fifths of nitrogen with relative weight 35. The total weight is 45. One fifth of the air volume thus has weight 9, and accordingly the specific gravity of oxygen relative to air is $10/9 = 1.1111$. The nitrogen has relative weight 35, but an equal volume of air has weight $4/5 \times 45 = 36$. The specific gravity of nitrogen relative to air is thus $35/36 = 0.9722$.

To determine the specific gravity of hydrogen Prout employed the known specific gravity of one of its compounds, ammonia (NH_3), given relative to air by Davy as approximately 0.5902. Since three volumes of hydrogen combine with one of nitrogen to produce two volumes of ammonia, it follows at once that three times the specific gravity of hydrogen plus the specific gravity of nitrogen (found above) equals twice the specific gravity of ammonia. Inserting the numbers, the specific gravity of oxygen relative to hydrogen is thus $1.1111/0.0694 = 16$ and that of nitrogen $0.9722/0.0694 = 14$.

Note that Avogadro's hypothesis is not mentioned, although it had been published before Prout's paper was written. It is probable that Prout was unaware of it. In assuming that the relative specific gravities of the gaseous elements, hydrogen being unity, give the relative weights of the atoms, he tacitly assumed that equal volumes of gases have an equal number of particles. But he did not consider, as Avogadro had, that the particles might be polyatomic. Nevertheless, that one volume of hydrogen combined with a half volume of oxygen to produce one volume of water vapor suggested that oxygen, on the basis that the atomic weight of hydrogen equals unity, has an atomic weight of eight rather than 16, the weight ratio of equal volumes of hydrogen and oxygen.

References

J. R. Partington, *Annals of Science*, 4, 1939.

H. E. Roscoe and A. Harden, *The New View of Dalton's Atomic Theory*. New York: Macmillan, 1896.

3 *The Foundations of the Kinetic Theory of Matter*

Truth lies within a little and certain compass, but error is immense.

—Henry St. John, Viscount Bolingbroke, *Reflections Upon Exile*

Atoms in Motion: John Herapath (1790–1868)

While chemistry was yielding results leading to belief in the atomic composition of matter, another branch of physics supporting this same point of view was being developed in the kinetic theory of gases. Only Daniel Bernoulli had tried to fix such ideas quantitatively, but his brilliant efforts had attracted no serious attention because the concept of atoms was then highly speculative. By 1820, however, the outlook of science had been changed by John Dalton's atomism and, although some first-rank scientists, such as Sir Humphry Davy, still questioned the actual existence of atoms, nevertheless many were disposed to concede their reality. It is not surprising, therefore, that a new advocate of the kinetic atomic hypothesis concerning gases should have appeared: one John Herapath, a contentious, argumentative, quick-tempered schoolmaster and amateur scientist. We consider him now because he had a direct influence on others, specifically James Joule, who in time extended Herapath's ideas and calculated molecular velocities.

Herapath was born in Bristol, England, May 30, 1790. As the son of a malt maker he was put to work in his father's business. But mathematics and physics had captured his imagination and at the age of 21 he was already amusing himself by calculating lunar orbits with the help of Newton's *Principia*. The differences between his observations of the moon's position and the predictions of his calculations, even after using the methods of Laplace in the *Méchanique Céleste*, led him to examine the nature of gravitation. Newton in his *Opticks* had suggested that gravitation might arise from the variation in density of an elastic medium filling all space. Herapath supposed that the density and, hence, the pressure of this medium on the sun side of a planet was less than on the dark side. Therefore, the planet was attracted to the sun. This idea led him to consider the relationship between temperature, pressure, and density of the supposed ethereal medium. He rejected the idea that heat was an elastic fluid and that static, gaseous particles repelled each other. Instead, he adopted the kinetic view stated in his papers and began developing a kinetic theory of gases. The background for this development accounts for the strange title of his paper, "A Mathematical Inquiry into the Causes, Laws, and Principal Phenomena of Heat, Gases, Gravitation, etc." Whether it was

51

the implausible title or the contents of the paper that repelled its referee, we do not know. But, at any rate, it was denied publication in the *Philosophical Transactions* of the Royal Society. Never one to shrink from a controversy, Herapath aired the rejection in the *Annals of Philosophy*, where the paper appeared in 1821.

During the development of his theory in the period 1814 to 1820, he abandoned his father's business, married in 1815, and set up a mathematical academy at Knole Hill, Bristol. Apparently it did not prosper, for he moved to a London suburb in the latter part of 1820, settling there as a tutor of mathematics. This period of his life, extending from age 30 through 42, was spent principally as a teacher; he had sought an appointment as professor of mathematics in the University of London and had attempted to write a treatise on differential and integral calculus, but neither of these attempts succeeded.

In 1832, Herapath gave up teaching and entered the employment of the Eastern Counties Railway Company. Four years later, at the age of 46, he became part proprietor and manager of the railway magazine which he continued, as editor, under the title *Railway Magazine and the Annals of Science*. In 1839, he turned over the management of the magazine to his son, Edwin, and returned again to mathematics and physics, publishing two volumes of *Mathematical Principles of Natural Philosophy* in 1847. He was then 57 years old. He began a third volume but little progress was made on it. He died at the age of 77 at Lewisham, a London suburb.

Dr. Eric Mendoza of Manchester University, who has made a study of Herapath's life and work, comments on his theory as follows.

> It is not difficult to see that, from the point of view of physicists of the time [1820] . . . an insuperable difficulty [was] in the way of such a theory. It was that the collision between atoms must be perfectly elastic. . . . No such substances were known in nature, even steel balls colliding with one another would certainly run down after a short time. An even more serious difficulty was that for a particle to be perfectly elastic it must be able to deform and this meant that it must have a structure . . . but this was impossible if the particle were an atom, for by definition it could not be further divided. . . .

> The most important content of his work was a verbose but correct proof relating the pressure and volume of a gas to the velocities and masses of the rapidly moving atoms. [The formula states that *pv*, the product of pressure and volume, is proportional to the product of gas density and the mean squared molecular velocity.] John Herapath's achievement in deriving it was considerable. (Mendoza, p. 5)

Herapath's procedures in "proving" these relationships are set forth in the seventh and eighth propositions of his paper. The earlier propositions deal mostly with the mechanics of impact, discussions that can be found today in any good general physics textbook. We call attention to the last few lines of the paper in which the temperature resulting from a mixture of gases, initially at temperatures F and F_1, is calculated. Here absolute zero on the Fahrenheit scale enters into the calculation. There is no discussion of this calculation or explanation of it anywhere in the article. Evidently, the volume contraction for gases had been determined experimentally as 1/480 of the volume at 32°F.

The gravest theoretical error in the paper was the attempt to identify temperature with the momentum of the gas particles rather than with their kinetic energy. Although this error prevented any further development of his theory, the later elaboration of his ideas in his two-volume *Mathematical Principles of Natural Philosophy*, published in 1847, drew serious considerations from Joule, who acknowledged the value of Herapath's work to his investigations. The eventual discovery of Bernoulli's prior treatment of gas kinetics made Herapath's pioneer work seem less important and, gradually, reference to his name disappeared from the scientific literature. Lord Kelvin, however, in reviewing the history of kinetic theory wrote that "it was developed by Herapath and made a reality by Joule."

"Active Molecules"—Brownian Motion: Robert Brown (1773–1858)

It is a truism that among the many investigators working in a given discipline few ever make an outstanding discovery. It is therefore even more rare that a scientist should make a contribution of fundamental importance in a field quite apart from his own. But it is this distinction that must be accorded to Robert Brown for his discovery of what is now known as Brownian motion, a phenomenon arising from the imbalance of molecular impacts on a free microscopic particle. Brown, a botanist, made his discovery of "active molecules" in 1827 while attempting to determine the behavior of pollen grains suspended in water on a microscopic slide. He found the grains to be in active, chaotic motion and at first associated the motion with the vitality of the pollen. But in pursuing the investigation he soon found that all small particles under the same conditions behaved the same way. The reason for the motion was not then suspected since the kinetic theory of gases was only a nebulous suggestion; indeed, Dalton's atomic theory was still in its infancy. Herapath had just attempted to develop a kinetic theory of gases; but, although it had not been ignored, it had drawn only opposition and ridicule. A further attempt, which suffered much the same fate, was made about 20 years later by John Waterston. The real impetus for its acceptance came from a calculation of molecular speeds in 1851 by James Joule and from the work of other recognized scientists—August Kronig and Rudolf Clausius, and especially James Clerk Maxwell in 1859. Although these developments put kinetic theory on a firm basis, Brownian motion was not linked to these ideas. The first suggested connection between the two is probably due to Wiener in Germany, in 1863, to Fathers Delsaulx and Carbonelle, 1877 to 1880, and to Guoy, 1888.

Robert Brown was born on December 21, 1773, the son of a minister of the Scottish Episcopal Church, in Montrose, Scotland. His mother, Helen Taylor, was the daughter of an Episcopal clergyman. His early education was in the grammar school at Montrose; in 1787, when not quite 14, he entered Marischal College, Aberdeen, where he remained for two years, withdrawing when the family moved to Edinburgh. Although he spent several years at Edinburgh University studying medicine, he did not attain a degree. Both in Aberdeen and Edinburgh, he devoted much of his lesiure time to the study of plants in which he showed unusual aptitude. His first botanical work appeared when he was 18 and consisted of a list of all the Scottish plants not previously described in Lightfoot's *Flora Scotica*. Many of the plants were his own discovery.

In 1795, when 21, he obtained a commission in the Fifeshire Regiment, which was detailed to duty in northern Ireland where he spent most of the next five years. During this time he employed much of his leisure in botanical explorations and further established his standing in the field of botany. There was, however, one eventful break in his service that changed the course of his life. In the latter months of 1798 and the early ones of 1799, Brown was transferred to London on recruiting duty. While there, he was allowed to use the library and collections of Sir Joseph Banks, longtime president of the Royal Society, upon whom he made a very favorable impression.

In 1800, when Sir Joseph sought a naturalist for an expedition that was to sail the following year to explore the coasts of Australia, he immediately chose Brown. It is said that two days after Brown received Bank's letter, he resigned from the military service, left his regiment, and was on the way to England. The expedition departed from Portsmouth in the summer of 1801, and for four years Brown studied and classified the plants of the coastal regions of Australia and the island of Tasmania. When he returned to England in October of 1805 at the age of 31, he brought with him dried plants of almost 4,000 species, most of which were new to science. He was immediately elected librarian of the Linnaean Society, and this enabled him to continue his work on these new species. His first volume of results, *Prodromus Florae Novae Hollandiae et Insulae Van Diemen*, appeared in 1810 and brought further wide acclaim. At the close of that year, he became librarian to Sir Joseph Banks who, on his death, bequeathed his protégé the use of his library and collections for life. Seven years later, they were transferred with Brown's permission to the British Museum where he was made keeper of the botanical department. He continued in that position for 32 years until his death in his 85th year on June 10, 1858. His botanical writings were characterized by the minutest accuracy of detail together with the most comprehensive generalizations, a combination that gave his investigations an unusual stamp of completeness. These masterly publications brought him a reputation in his day as the greatest botanist England had produced.

Among his many honors were election to the Royal Society in 1811, several elections to its council, an honorary Doctor of Common Law degree from Oxford in 1832, along with John Dalton, Michael Faraday, and Sir David Brewster, election in 1833 as one of the eight foreign associates of the Academy of Sciences of the Institut de France, award of the Copley Medal of the Royal Society in 1839, president of the Linnaean Society 1849–1853, and membership in almost all the scientific societies of Europe. In private life, he was considered by friends to be warm, outgoing, and kindhearted, but to those outside his circle it is said that he was cold, distant, and reserved.

The Tragedy of a Genius: John James Waterston (1811–1883)

In the history of the men and women who have devoted themselves to discovery in atomic science it would be hard to find any whose rejection presents so great a tragedy as that of John James Waterston. His paper on kinetic theory, submitted to the Royal Society in 1845, gave a new and greatly extended mathematical treatment of that theory. It not only corrected the fundamental error of Herapath in assuming that temperature is proportional to the average velocity of the molecules but went much further, giving, for example, the first statement of the law of equipartition of energy among the different kinds of gas molecules in a mixture at thermal equilibrium. Waterston derived, in fact, practically all the consequences that follow from the now well-known equation that relates the pressure p exerted by a gas to the product of the number of molecules per cubic centimeter z, their mass m, and their mean squared velocity v^2, that is, $p = zmv^2$.

The judgments of the two referees to whom the paper was sent for an opinion on publication are illuminating and give some insight into the "professional's" view of the atomic theory of matter in 1845. One referee remarked that "the paper is nonsense, unfit even for reading before the Society." The other, more open-minded, noted that the paper "exhibits much skill and many remarkable accordances with the general facts . . . but the original principle [is] . . . by no means a satisfactory basis for a mathematical theory." The paper was not rejected outright but was relegated unpublished to the society's archives; only an abstract appeared in the *Proceedings of the Royal Society* for 1846. Had it been rejected, it could have been published elsewhere but, having been retained, it was consigned to practical oblivion.

In 1851 Waterston again attempted to draw attention to it in an abstract in the *Report of the British Association*. Waterston next referred to his gas theory in a paper "On a general Law of density in Saturated Vapors," published in the *Philosophical Transactions* of 1852, and then discussed his theory and that of Herapath at length in an article on the theory of sound published in the *Philosophical Magazine* in 1858. The reference in the paper on the theory of sound led Lord Rayleigh to read the paper in the society's archives in 1891 and to discover the tragic error the reviewers had made. By then, the accomplishments of others had far outdistanced the developments that Waterston had achieved. As J. S. Haldane, his biographer, remarks, "It is probable that in the long and honorable history of the Royal Society no mistake, more disastrous in its actual consequences for the progress of science and the reputation of British science. . .was ever made." In an effort to atone in some measure for the wrong that had been done, Rayleigh reprinted Waterston's paper in the *Philosophical Transactions* of the Royal Society in 1892.

We return to Haldane to glimpse something of the life of this unfortunate man. John James Waterston was born in Edinburgh, Scotland in 1811, the son of George Waterston, a sealing wax and stationary manufacturer, and the grandson of William Waterston. The latter married Catherine Sandeman, niece of Robert Sandeman, the founder of the religous sect known as the Sandemanians. Michael Faraday's father and Faraday himself were members of this church in London. Catherine's brother, George Sandeman, founded the London firm of wine merchants of that name, which is not only still in business but is familiar to many in the English-speaking world.

Young Waterston grew up in comfortable circumstances and in an atmosphere of culture. He was educated at the Edinburgh High School and Edinburgh University where he studied mathematics and physics under Sir John Leslie and was medalist for his year in Leslie's class. While studying at the university, he was also employed in a firm of civil engineers. At 21, he went to London to work at drafting and surveying for the expanding railway system in England. But he found this work too demanding for his scientific interests and for this reason he accepted for a time an appointment in the hydrographer's office in the Admiralty. Captain Beaufort, head of the office, recognized Waterston's ability and, sympathizing with his desire for a scientific career, suggested a teaching position as a naval instructor to the East India Company's cadets at Bombay. Waterston obtained the appointment and went to India in 1839, being then 28 years old. Here he found the leisure he had been seeking together with access to a scientific library at Grant College, Bombay. His first book, *Thoughts on the Mental Functions*, was, strangely enough, a work on the physiology of the central nervous system with application of molecular theory to biology. It was published anonymously in Edinburgh in 1843 and contained the elements of his later paper on kinetic theory. With a misleading title and speculative contents, it is doubtful if anyone read the book. Two years later, in December of 1845, Captain Beaufort forwarded to the Royal Society Waterston's developed kinetic theory. None of Waterston's efforts to get this paper published ever succeeded.

Waterston remained in Bombay until 1857 when he resigned and returned to Edinburgh. Although he was then only 46, his health had been permanently impaired as a consequence of a severe heat stroke suffered in India. This left him subject to attacks of dizziness in crowded rooms or in confined quarters while traveling. This infirmity best explains his singular reluctance to attend any scientific gatherings or to meet any of the scientists such as Maxwell, a fellow Scot vitally interested in the same ideas, who might have assisted him in gaining the recognition that his past work deserved. During the 11 years following his return from India, Waterston published 20 papers in the *Philosophical Magazine*, his last being dated 1868, the year that Sir David Brewster, one of the editors of the magazine, died. This coincidence led Haldane to suggest that Brewster's influence was responsible for the acceptance of Waterston's contributions. This seems a very harsh judgment because the papers were not ill-founded or of uncertain value. In 1878 Waterston sent two papers to the Royal Astronomical Society. Both were rejected. In the only retaliation open to him, Waterston then resigned from the Astronomical Society after having been a member for 26 years. If it did not seem incredible, it would almost appear that a conspiracy existed to shut him from all avenues of scientific expression. It is therefore to be expected that Waterston grew increasingly bitter as the years passed by. His nephew, George Waterston, wrote Lord Rayleigh after his uncle's death that he "seemed to me strangely contemptuous of scientific men with but few exceptions. He had not a word of complaint, nor did he speak of being neglected or ill-used. . . . He would not attend the meetings of the Royal Society of Edinburgh . . . and rather avoided the society of scientific men. He was of a most social, kind disposition, enjoying the society of young people." He never married and Haldane suggests that apart from the deep disappointments of his scientific life, which he took pains to conceal, his life in Scotland was happy. He lived the life of a popular bachelor, enjoying the best music, billiards, cigars, and chess. In his last days he lived in rooms on Gayfield Square off Leith Walk in Edinburgh. From here, on June 18, 1883, he set out for a walk in the morning air. By nightfall he had not returned; he was never again seen by anybody. It was as though he had vanished in thin air. His passing was but another episode in the same tragedy that had haunted his life.

Perhaps the best explanation of his disappearance is given by his nephew, in the letter to Lord Rayleigh referred to above: "My uncle . . . was fond of walking out by a new breakwater recently built at Leith, very well exposed to a fine sea breeze, but from its construction very dangerous to foot passengers. At this place the tide runs out very fast, and if he had fallen in he would have been carried out to sea. We know of no place near Edinburgh where he could so easily have disappeared, and no one who knew him thought of suicide as likely in his case."

For the modern reader more can be said about Waterston's paper than either he or Lord Rayleigh allowed in their introductions. Most elementary physics courses now give a brief treatment under equilibrium conditions of the kinetic theory of gases. In this treatment, one of the principal results obtained is a relation between the pressure p exerted by the gas, the number density z of the gas molecules (number per unit volume), the mass of an individual molecule, m, and v^2, the square of the molecular speed v averaged over all the molecules, or as it is usually denoted, the mean squared speed. Specifically this relation is found to be $p = zmv^2/3$. All of the relations that Waterston obtained can be described in terms of this formula. Had Waterston's views been generally known, chemistry, and atomic theory with it, would not have had to wait another 15 years before reaching the firm footing provided by Stanislao Cannizzaro and the brilliant theorist James Clerk Maxwell.

The Conservation of Energy, the Mechanical Equivalent of Heat: James Prescott Joule (1818–1889)

The principle of the conservation of energy is one of the greatest and most far-reaching in the whole structure of physics; nowhere has it been of more value than in the development of atomic and nuclear science. Like other great generalizations in science, it required a long period of gestation for its birth. If any one person can be credited with providing its scientific basis, that person is James Prescott Joule, for to him we owe the experimental determination of the mechanical equivalent of heat.

For the 18th-century scientist, heat was a subtle substance called "caloric" that permeated all bodies. When increased, it made the body hotter. It was one of the many fluids that afflicted the infancy of physics, and its existence was almost universally accepted. A few lone thinkers had ventured the thought that heat, instead of being a substance, was the effect of motion. But not until the middle of the 18th century was a clear distinction made between heat and temperature. The difference between these fundamental ideas was first pointed out by Joseph Black (1728–1799) in his lectures to students at Glasgow and Edinburgh universities.

To understand the development of the conservation of energy, one must know the meaning of thermal terms. One of the classical erroneous statements, often repeated, is that "heat is the sum total of all the energy of the molecules of a body."

James Prescott Joule (1818–1889)

This is precisely what heat is not. The sum total of all the energy of the molecules is patently nothing but the "internal energy" of the body. The temperature of any body is related to the energy of the *translational* motion of its molecules. To increase this motion, it is necessary to add energy to the body. If the transfer of energy occurs solely as a result of a temperature difference between the body and its surroundings then, by definition, *heat* has been transferred to it. On this basis, heat is properly defined as energy in transit due solely to a temperature difference.

In light of these clear distinctions, it is interesting to see how some of the celebrated minds who did not accept the caloric doctrine regarded the idea of heat. Isaac Newton, in the queries in Book III of his *Opticks*, asked in part in Query 5: "Does not . . . light act . . . upon bodies . . . heating them, and putting their parts into a vibrating motion wherein heat consists?" John Locke, who was undoubtedly influenced by Newton, wrote in 1706: "Heat is a very brisk agitation of the insensible parts of the object, which produces in us the sensation from which we denote the object hot; so what in our sensation is heat, in the object is nothing but motion." Daniel Bernoulli, as we have already seen in his deduction of Boyle's law, stated that "it is admitted that heat may be considered as an increasing internal motion of the particles." Lavoisier and Laplace in 1780 were even more explicit in their *Mémoire sur la Chaleur*: "Heat is the *vis viva* resulting from the insensible movements of the molecules of a body. It is the sum of the products of the mass of each molecule by the square of its velocity."

All of these statements are correct in identifying heat with energy, and Lavoisier and Laplace are amazingly specific and modern in their ideas. However, none of these assertions could be backed up by any experimental evidence, and what is more unfortunate, these views seem to have made no deep impression on the development of physical science.

The first experiments that pointed strongly to the inadequacy of the caloric theory and equally strongly to the energetic theory were performed by Count Rumford at Munich during his service as Minister of War to the Elector of Bavaria. Another was performed by Humphry Davy (1799), whom Rumford had hired to assist him while founding the Royal Institution in London. Davy's experiment consisted in rubbing together two pieces of ice to produce melting at reduced pressure in a container surrounded by ice. The ingress of caloric was thus minimized. Nevertheless the ice melted. Rumford's experiment was more convincing, although it too was not conclusive. However, Rumford's account of his experiment provides a valuable insight into the observations that led him to a more sophisticated view about the nature of heat.

Being engaged lately, in superintending the boring of cannon, in the workshops of the military arsenal at Munich, I was struck with the very considerable degree of heat which a brass gun acquires, in a short time, in being bored; and with the still more intense heat (much greater than that of boiling water, as I found by experiment) of the metallic chips separated from it by the borer.

The more I meditated on these phenomena, the more they appeared to me to be curious and interesting. A thorough investigation of them seemed even to bid fair to give a farther insight into the hidden nature of heat; and to enable us to form some reasonable conjectures respecting the existence, or nonexistence, of an igneous fluid: a subject on which the opinions of philosophers have, in all ages, been much divided.

In order that the Society may have clear and distinct ideas of the speculations and reasonings to which these appearances give rise in my mind, and also of the specific objects of philosophical investigation they suggested to me, I must beg leave to state them at some length, and in such manner as I shall think best suited to answer this purpose.

From whence comes the heat actually produced in the mechanical operation above mentioned?

Is it furnished by the metallic chips which are separated by the borer from the solid mass of metal?

If this were the case, then, according to the modern doctrines of latent heat, and of caloric, the capacity for heat of the parts of the metal, so reduced to chips, ought not only to be changed, but the change undergone by them should be sufficiently great to account for all the heat produced.

But no such change had taken place; for I found, upon taking equal quantities, by weight, of these chips, and of thin slips of the same block of metal separated by means of a fine saw, and putting them, at the same temperature (that of boiling water) into equal quantities of cold water (that is to say, at the temperature of $59^{1}/_{2}°F$) the portion of water into which the chips were put was not, to all appearance, heated either less or more than the other portion, in which the slips of metal were put.

This experiment being repeated several times, the results were always so nearly the same, that I could not determine whether any, or what change, had been produced in the metal, in regard to its capacity for heat, by being reduced to chips by the borer.

From hence it is evident, that the heat produced could not possibly have been furnished at the expense of the latent heat in the metallic chips. (Rumford, pp. 471ff)

The experiments of Rumford and Davy were concerned with the transformation of work into heat; the much more difficult problem of the transformation of heat into work was studied by Sadi Carnot in 1824. Carnot's researches display the work of a thoroughly original mind. His great contribution consisted in pointing out that the efficiency of thermal engines is not dependent on the substance that powers the engine but only on the difference in temperature through which this substance is taken in cyclical operation. The cyclical operation to achieve the maximum efficiency was also defined and it is known to every scientist today as the Carnot cycle. The great flaw in Carnot's work was the assumption that heat was a substance, caloric, that was conserved when mechanical work was produced. This assumption was generally accepted by the leading scientists of the time, including Michael Faraday, and accounts for the indifference that met the ideas of Julius Robert Mayer and Joule in the 1840s. From posthumous papers published in 1872, it appears that Carnot changed his ideas very radically before his death in 1832. But these ideas were unknown and did not influence the development of the theories of heat and energy.

The conservation of energy in a limited form was already understood through the formulation of mechanics by Isaac Newton and Christian Huygens. Thus, a pendulum moving without friction maintains a constant energy but not always of the same kind. At the end of its swing its energy is all potential, while at the middle it is all kinetic. The idea that mechanical energy could be transformed into internal energy and thus produce the same effect as "heating" a body and that a fixed ratio exists between mechanical work and thermal units (calories or Btu's) originated with an obscure German physician, Julius Robert Mayer, in 1842. Mayer was not trained in the physicial sciences and he carried on no experimental investigations. Using data then available he deduced a value (corrected) of 725 ft-lb per Btu. The details were published in a paper that appeared in *Liebig's Annalen* in 1842 and were accompanied by the first clear statement of the conservation of energy: "force [the term then also connoting energy] once in existence cannot be annihilated; it can only change its form."

From this statement it appears that the credit for the conception of both the mechanical equivalent of heat and of the conservation of energy should go to Mayer, whereas these great advances are usually credited to Joule and Helmholtz. The reason is that Mayer's work was inadequate and in some respects erroneous, as was pointed out in the early 1860s by Lord Kelvin and P. G. Tait, and by Sir G. G. Stokes. However, a subsequent paper of Mayer's in 1845 expanded his ideas with great power and insight. He discussed, among other phenomena, the idea that heat and energy output in man must be equated to his intake and that the process of the combustion of plants and coal was a process of recapturing the heat previously obtained from the sun. In the latter connection he proposed the contraction theory for the origin of the sun's heat; this theory was later elaborated by Kelvin. As a belated recognition for his services to physics, Mayer was awarded the Copley medal of the Royal Society in 1871. The main credit, however, for firmly establishing the existence of a mechanical equivalent of heat and, indirectly, the conservation of energy justly belongs to Joule for his long series of experiments on the mechanical equivalent.

James Prescott Joule was born on Christmas Eve, 1818, at Salford near Manchester, England. His father and grandfather were both brewers, and the family lived in comfortable financial circumstances. As a boy, Joule was not robust, and he was educated by tutors at home until he was 16. In 1835, he was sent with his brother Benjamin to study under John Dalton, who by then had been plied with honors, including the presidency of the Manchester Literary and Philosophical Society. Dalton taught them algebra and geometry and had started them on chemistry when he suffered his first paralytic stroke in the spring of 1837. Although Dalton's illness terminated the arrangement, his influence encouraged Joule to begin scientific investigations on his own. Using a room in his home as a laboratory, he began electrical and magnetic experiments, and at the age of 20 he published his first paper "On an Electromagnetic Engine" in Sturgeon's *Annals of Electricity*. Two years later in a paper published in 1840, he described the first known attempt to determine a unit of electric current. For this measurement he used a voltmeter, a device subsequently employed in the definition of the international ampere.

Joule presented his first paper before the Manchester Literary and Philosophical Society in November 1841, with Dalton as chairman of the meeting. It is reported that, on the conclusion of the paper, Dalton moved a vote of thanks to the author, the first such courtesy he extended in all his years as president of the society. Joule was elected a member two months later and was successively librarian in 1844, honorary secretary in 1846, a vice president in 1851, and finally president in 1860.

In 1841 Joule sent to the Royal Society a paper in which he announced, as a definite law, that the heat evolved by an electric current in a given time is proportional to the resistance of the conductor multiplied by the square of the current. To this day, this phenomenon is still referred to as "Joule heat." From his experiments with batteries and currents, Joule saw clearly that chemical energy in the battery is converted to electrical energy in the circuit and that this in turn is converted into heat. To pursue these ideas he set up an experiment to measure the heat produced by currents induced in coils rotated under water, together with the mechanical effort needed to turn the coils. By hindsight it is evident that this arrangement was overly complex. But with it Joule deduced his first value for the mechanical equivalent—838 ft-lb of work per Btu of heat. He reported this finding together with another result obtained from an independent method, the viscous flow of water through fine tubes (770 ft-lb per Btu), at the British Association meeting held in Cork, Ireland in August 1843. To quote his own words, "the subject did not excite much general attention."

The lack of precision in his findings led Joule to modify his apparatus; as a consequence he adopted the well-known arrangement of a paddle wheel churning water in a closed, insulated container. The first result, 890 ft-lb per Btu, was communicated to the British Association meeting in Cambridge in 1845. Joule must have felt that his experiments should be made known to a wider audience, for he sent to the editor of the *Philosophical Magazine* a letter describing the results of his experiments.

Two years later Joule, then 28, had new results obtained with the paddle wheel apparatus, and he forwarded a paper for the British Association meeting at Oxford, in August 1847. But the chairman suggested that, owing to the press of business, Joule should not read the paper but instead give a verbal description of his experiments. "This I endeavored to do," Joule wrote later, "and, discussion not being invited, the communication would have passed without comment if a young man had not arisen in the Section, and by his intelligent observations created a lively interest in the new theory." The young man was William Thomson, later Lord Kelvin, then only 23 years old. Thomson's comments made Joule's paper the sensation of the meeting, but few of those present, including Faraday, were persuaded of the correctness of the new views, which contradicted Carnot and the long entrenched caloric theory.

Another anecdote about Joule and Thomson is worth including at this point. Three days after the Oxford meeting, Joule married Amelia Grimes, the daughter of the collector of customs at the port of Liverpool. In his discussion with Thomson, there had been no reason to mention his impending marriage; nor had Thomson mentioned that he would depart shortly for a holiday in Switzerland. This occasioned an amusing incident, for a fortnight later Thomson, while walking in the valley of the Chamonix, saw a young man coming toward him carrying what looked like a stick. On closer approach the man turned out to be Joule with a long thermometer! His wife had preceded him in the *charabanc* (to which he would not entrust the thermometer), and he was walking to the top of a neighboring waterfall. If his ideas were right, there must be a difference in temperature of the water between the bottom and the top, owing to the dissipation of kinetic energy at the bottom of the fall. Evidently a wedding trip did not separate Joule completely from his science. This chance meeting encouraged the friendship that had begun at Oxford and led to a collaboration that lasted throughout Joule's life.

If, despite Joule's presentation at the Oxford meeting, scientists were reluctant to reexamine their views on caloric, or on heat and work, Hermann von Helmholtz's classic paper "Über die Erhaltung der Kraft" soon thundered an ultimatum that could not be ignored. As for Joule, he continued improving the precision of his measurements, obtaining the new value of 772 ft-lb per Btu. On June 21, 1849, he presented this result to the Royal Society and his paper was published in the *Philosophical Transactions* in 1850.

As a reward for his epoch-making studies, Joule was elected a fellow of the Royal Society in 1850 and awarded the Royal medal in 1852. During this period he performed his other famous experiment on the free expansion of a gas and followed it by systematic studies with Thomson on the change in temperature when a gas expands through a porous plug (the Joule-Thomson effect). These studies clarified ideas regarding real gases and led to the notion that the internal energy of an ideal gas is a function of the temperature only. In 1860 he was honored with the Copley medal for the same work with this citation: "The award of two medals for the same researches is an exceedingly rare proceeding in our Society— and rightly so."

Joule's establishment of the mechanical equivalent of heat was not his only contribution to the development of physical science. Among others, but one of the least known, was his calculation of the speed of gas molecules. This calculation in itself was not of great significance. What it did accomplish in the end was to draw the attention of scientists to a neglected area of physics that opened up new avenues to the world of atoms and molecules. Bernoulli had introduced kinetic theory in his remarkable paper of 1738, but at that time molecules and atoms had only the vaguest speculative pretensions to reality and his paper could hardly have been taken as more than an interesting exercise. Herapath independently "rediscovered" gas kinetics in the period 1816 to 1821, a time sufficiently after Dalton's chemical atomic theory for such ideas to be given more serious consideration. But his thinking was too new and his standing in science too uncertain to command the ear of established contemporary investigators. Waterston in 1846, also unknown but of sounder ideas, fared worse, for his paper was consigned to oblivion. To be respectable, to be accepted, and to fulfill the possibility of yielding a much broader basis for belief in atomistics, kinetic energy needed the backing of a scientist of secure standing. This was the service that Joule rendered. It is interesting to see that Joule, who must have been convinced of the reality of the kinetic hypothesis from his own work, was deeply influenced by Herapath and by the kinetic ideas of Sir Humphry Davy. By joining Herapath's hypothesis with those of so prominent a scientist as Davy, and by using the mixture as a basis for his own contribution, Joule conferred on this kind of speculation the stamp of respectability that kinetic theory so badly needed.

Joule's calculations of the velocity of hydrogen molecules was first published in the *Memoirs of the Manchester Literary and Philosophical Society* in 1851. Owing to the limited circulation of this journal, the paper remained generally unknown. Nevertheless, it found its mark in Germany, capturing the eye and the serious interest of Rudolph Clausius. In his hands kinetic theory was developed with sufficient accuracy and completeness by 1857 to command the attention of the scientific world and to encourage immediately another master mind, James Clerk Maxwell, who deepened its foundations and extended its scope. It would be inaccurate to leave the impression that Joule's paper was solely or even mainly responsible for the general growth of interest in gas kinetics. Much more was engendered by August Karl Kronig, in 1856, by a paper that had an influence entirely out of proportion to its substance. Nevertheless, it must be clear that the backing of Joule and Kronig was instrumental in drawing serious interest to gas theory. Joule's paper can be considered as hardly more than a suggestive and interesting calculation, but, as such, it served a valuable and useful function.

Osborne Reynolds, in a memoir on Joule prepared for the Manchester Literary and Philosophical Society in 1892, left behind these personal impressions.

> At fifty-one, [Joule] was rather under medium height, somewhat stout and rounded in figure, and his dress, though neat, was commonplace in the extreme. His attitude and movements were possessed of no natural grace, and his manner was somewhat nervous. He possessed no great facility of speech and conveyed an impression of the simplicity and utter absence of all affectation which characterized his life. It was not merely veneration arising from his fame that inspired members of the Society, but the inherent loveability of his character—kindly, noble, and chivalrous in the extreme, though modest and devoid of mere personal ambition. He was jealous for the interests of his friends and the Society in general, and in particular, jealous in the interest of everything truly scientific. Anything that looked like ostentation or quackery excited Joule's indignation, particularly when exhibited by those possessing the popular ear. On the other hand, he always noticed with encouragement the efforts of those who were yet unknown, and resented any attempt at the disparagement of their work— as though his own early experience had left him with a fellow-feeling for those who were struggling to get their views taken up.

All of the ideas for which Joule is remembered were explored and published before he was 34 years old. During the remaining years of his life he devoted himself to improving the accuracy of his work, establishing its place in the structure of physics, and in furthering the progress of science. His health began to fail in his middle 50s, and he was compelled to live quietly until his death at the age of 71 in 1889.

"I believe," he told his brother in 1887, "I have done two or three little things but nothing to make a fuss about." The judgment of his peers was by no means so modest. A tablet in Westminster Abbey honoring his "establishing the Law of the Conservation of Energy and determining the Mechanical Equivalent of Heat" is placed amid the memorials to the greatest names in British science.

The Range of Molecular Speeds in a Gas: James Clerk Maxwell (1831–1879)

In previous commentaries on Bernoulli, Herapath, Waterston, and Joule, the contributions that each of these men made to the development of atomic physics through the kinetic theory of gases have been reviewed. For reasons discussed previously, none of these contributions except Joule's proved effective in directing the serious attention of capable scientific investigators into this fruitful area of research. Joule's calculation of the mean molecular velocity in hydrogen at normal temperatures was originally published in the *Memoirs of the Manchester Literary and Philosophical Society*, a journal that was not widely read; the effect of Joule's contribution, therefore, was correspondingly small, although important. The reception given to kinetic theory changed markedly in 1856 as the result of a paper by Kronig; in the following year the first of several papers by Clausius placed the subject on an entirely different footing. Clausius's work caught the eye of James Clerk Maxwell in whose hands the treatment took on significant new dimensions.

In 1860 Maxwell published in the *Philosophical Magazine* a long mathematical paper entitled "Illustrations of the Dynamical Theory of Gases." Among other

James Clerk Maxwell (1831–1879)

new results there appeared a calculation for the distribution of molecular speeds—a signal advance in the knowledge of molecular motion—and a formula for the coefficient of viscosity of a gas that showed this quantity to be independent of pressure, a most unexpected and surprising result. This paper did much to establish Maxwell's contemporary scientific reputation. But his greatest contribution, and one that places him among the great physicists of all time, was his electromagnetic theory, including his deduction that light is propagated as an electromagnetic vibration. This work was first presented in 1864. Unfortunately, experimental electricity had not caught up with theory, and many years were to elapse before the full significance of this paper came to be appreciated. The demonstration of electromagnetic waves, for instance, was not made until 1888, when Heinrich Hertz showed their reality. This was a lapse of almost a quarter of a century. The importance of Maxwell's theory to the development of atomic physics was very real, for it showed how to treat the emission and absorption of radiation. Hence, later in this volume, some aspects of his electromagnetic theory are also presented. Before these achievements are examined, a brief account of Maxwell's life should serve to give us some appreciation of how he appeared to his contemporaries and the way in which he prepared and carried out his life's work.

When we read the story of genius, it is always interesting to recognize the influences that channel and mold its character. In Maxwell's case, it is clear that his outlook and bent were deeply conditioned by his father, with whom he had an unusually close and happy relationship. No man is an island unto himself, and to appreciate Maxwell both as a person and a scientist, it is necessary to know something of his family, particularly his father, John Clerk.

The Clerks of Penicuik had for years been a distinguished Scots family. A forebear, Sir John Clerk, was Baron of the Exchequer in Scotland from 1707 to 1755. According to R. T. Glazebrook, George Clerk, one of Sir John's sons, married a first cousin, Dorothea Maxwell, the heiress of Middlebie in Dumfriesshire, and took the name Maxwell. Succeeding Clerks who accepted any of the inheritance of

Middlebie also were obliged to assume the Maxwell name; among these was James Clerk Maxwell's father, John Clerk.

In 1826, John Clerk had inherited a new family home in Edinburgh at 14 India Street, had been admitted to the practice of law, and had married Miss Frances Cay of North Charlton, Northumberland. Being of an easy-going disposition, his practice was not particularly arduous, and he had time to follow a bent for practical mechanics, invention, and a curiosity about manufacturing processes. For several years the couple maintained their residence at 14 India Street, but a desire to establish a country estate led them to improve what remained of the Middlebie property and to add to it by purchasing an adjacent estate called Glenlair. No house existed on these properties, and during the construction of a suitable dwelling the couple lived intermittently in Edinburgh. There James Clerk Maxwell was born on June 13, 1831. Jamsie, as he came to be called, was soon taken to Glenlair, where he spent an active, inquiring childhood full of country fun and freedom. From the first, his inquisitiveness was very marked and his precocious, "Show me how it goes," or "What is the go of that?" must, at times, have been trying, though amusing. When the explanation was not completely satisfying, he would press the question with, "But what is the *particular* go of that?"

This happy, active, country childhood was rudely shaken at the age of eight by his mother's death from cancer. For two more years, father and son continued at Glenlair, but the growing necessity for appropriate schooling made it clear that young James should be educated in Edinburgh. The strong family ties of the Clerks provided a happy solution. His father's sister, Isabella, now the widowed Mrs. Wedderburn with a large house at 31 Heriot Row, opened her doors to this unusual boy and his perplexed parent. For the next nine years, James spent his winters there or with a maternal aunt, Miss Cay, deeply and happily absorbed in exploring the world of books, ideas, and physical science. When he had free time, his father was always at hand to encourage his interests, show him the "useful arts" or immerse him in church activities. In the summer, there were long, carefree months at Glenlair.

James's introduction to school life at the Edinburgh Academy was not without its trials, some of them amusing. Youth is seldom kindly disposed to differences in appearance and manners; being different in both, young James was soon a target for some of his rougher schoolmates. He arrived the first day dressed in a country jacket with a lace collar and square-toed shoes made to his father's design by a country bootmaker. On being asked who made his shoes, we are told that he made the following answer: "Div ye ken 'twas a man and he lived in a house in whilk was a mouse." Doubtless this reply baffled his schoolboy inquisitors, but it shows one of the sides of the puzzling personality which characterized Maxwell throughout his life. As Niven remarks, his speech was often "obscure in substance and the manner of expressing himself." Jamsie not only looked different from the other boys, but he was different, a fact that prompted his schoolmates to give him the nickname "Dafty."

Initially his school record was undistinguished, but gradually James began to excel in his studies. Around the time of his 13th birthday, before he had begun geometry, he became interested in solid figures and wrote to his father that he had made a "tetrahedron, a dodecahedron and two other hedrons whose names I don't know." At the end of the following school year, his talent in geometry and mathematics became clearly evident and he was awarded the annual mathematical medal. When he was 14, his father began taking him to meetings of the Edinburgh Royal Society, rather rigorous mental fare for one so young. Yet the meetings must have acted as an additional spur to his interest in mathematics for he began investigating the properties of ovals. His father took the drawings to Professor James D. Forbes at Edinburgh University, who realized that their construction presented novel features not in the mathematical literature. As a result, Forbes submitted the work to the Edinburgh Royal Society under the title "On the Description of Oval Curves and Those Having a Plurality of Foci." It was read at the March meeting in 1846.

During the next school year, Maxwell became interested in optics and began experimenting with Newton's rings and polarized light. As a result, he was taken to see William Nicol, the inventor of the polarizing prism. Stimulated by this visit, he made an examination of the colors of unannealed glass in polarized light, using a glass plate as polarizer. His drawings, in color, were sent to Nicol, who rewarded the young man with a pair of his prisms.

In the spring of 1847, young Maxwell completed his studies at the academy, being first in mathematics, first in English, and "nearly first" in Latin. The following autumn, at the age of 16, he began his studies at Edinburgh University. But his studies were by no means limited to his university instruction and, during vacations at Glenlair, he improvised a laboratory to carry out his own experiments. He remained at the university for three years, increasing his intellectual debt to Forbes, and developing his skills in mathematics and physics. His expanding powers were further demonstrated by the contribution of two more papers to the Edinburgh Royal Society. The first, on the theory of "Rolling Curves," was presented in the spring of 1849, and a second, "The Equilibrium of Elastic Solids," a year later.

In October of 1850, Maxwell, then 19, was entered by his father at Peterhouse College, Cambridge. Peterhouse was traditionally the college chosen by outstanding Scottish scholars, but after a few months Maxwell felt that his chances for an eventual fellowship were better at Trinity and transferred there at the end of his first term. Despite his years at the Edinburgh Academy and the university, there was still much of the eccentric in him. The mother of a school friend noted that "his manners are very peculiar . . . but I doubt not of his becoming a distinguished man." At Trinity, he was no recluse but was completely taken up with college life. His brilliance was quickly recognized, and no less a man than the great Cambridge tutor Hopkins characterized him as "the most extraordinary man I have ever met. It appears impossible for him to think incorrectly on physical subjects." At first, Maxwell lodged away from the college, but at the end of his second year he was elected a scholar and obtained rooms in Trinity.

Always an individualist, Maxwell experimented even with hours of sleep and studying. For a while, he slept from 5:00 to 9:30 P.M., studied from 10:00 to 2:00, exercised from 2:00 to 2:30 A.M., and slept until 7:00. The exercise consisted in running along the upper corridor, downstairs, along the lower corridor, then up to repeat the procedure all over again. This went on until the inhabitants of the rooms along his path took shots at him with hairbrushes, boots, or any other loose objects as he passed.

For the students the period from December 1852 to June 1853 was one of intense concentration for the coming tripos examinations. It was characteristic of Maxwell that he should have chosen to spend a few days of the Easter recess in Birmingham with a friend, using most of the time to inspect manufacturing processes in the local industries. In June he suffered a nervous breakdown. The effects lasted for some months, and he had not completely recovered when he took the tripos examinations in January 1854. These examinations, the culmination of his undergraduate study at Cambridge, were held in the Senate House where the temperature was so low that, at his father's suggestion, he wrote his examination with his feet and legs wrapped in a rug to ward off the chilling cold. When the results of the examination were announced, Edward Routh, an accomplished mathematician and later a noted Cambridge tutor, received the designation senior wrangler. Maxwell had finished second. In the more advanced competition for the Smith prize, the two were given equal standing.

After receiving his degree, Maxwell continued at Cambridge and, the following October, at the age of 24, was elected a fellow of Trinity. He assumed the duties of university lecturer and began serious work on the mathematical theory of electricity and magnetism. A few months later, in February, he was advised by his old mentor, Professor Forbes, that the chair of natural philosophy at Marischal College, Aberdeen was vacant and that he should apply for it. His father was now in poor health and the prospect of their being closer was certainly one of the factors influencing Maxwell's application. However, in April, while father and son were together at Glenlair, his father unexpectedly died. Although his situation had thus completely altered, Maxwell nevertheless accepted Marischal's subsequent offer and began teaching when the autumn session commenced in November. His success as a teacher, both initially and in later years, appears to have been rather indifferent. This result was not for want of concern. Some of his difficulty must be attributed to shyness and to his habit, in the momentary excitement of problems and explanations, of breaking off into ironical humor. It has been said that in scientific discussions his colleagues often were not sure whether he was serious or slyly joking.

The Adams prize, awarded by St. John's College, Cambridge, was offered in 1857 for the best paper on "The Motions of Saturn's Rings." The problem was "to demonstrate what type of structure adequately explained the motion and permanence of the rings." Maxwell entered the competition and submitted a paper that demonstrated that the only stable structure would be one composed of disconnected particles. It not only won the prize but earned the praise of Airy, the Astronomer Royal, who characterized it as "one of the most remarkable applications of mathematics" he had ever seen. This paper established Maxwell in the forefront of mathematical physicists. He was then 26 years old. But, in a sense, it did much more because it awakened his interest in the motions of groups of particles—the fundamental problem of the kinetic theory of gases. This interest soon led to his brilliant deduction of the distribution of molecular speeds in a gas at equilibrium at any temperature. This great step forward in the understanding of the behavior of the elementary particles of gases represents one of the major advances in the progress of the atomic theory of matter. The paper, "Illustrations of the Dynamical Theory of Gases," was published in the *Philosophical Magazine* in 1860.

About this time, Maxwell became engaged to the daughter of the principal of Marischal College, Katherine Mary Dewar. After a short engagement, they were married in June of 1858.

Two years later, Marischal College was absorbed into the University of Aberdeen, with the result that Maxwell's post was eliminated. Then 29 years old, he was appointed immediately to the professorship in natural philosophy at King's College, London. He held this post for five years, a period in which he was at the height of his powers. These years were spent in developing the electromagnetic theory, ideas on the dynamic theory of gases, and on the electromagnetic theory of light, which probably began to germinate in his thoughts around 1861. His greatest electrical paper, "A Dynamical Theory of the Electromagnetic Field," was read to the Royal Society of London on December 8, 1864. His experiments on the viscosity of gases, which produced the unexpected result that within wide limits the viscosity was independent of pressure, formed the basis for his Bakerian lecture to the Royal Society on February 8, 1866. The experiments on which this paper was based were carried out, as usual, in his own home, this time in a large garret running the length of his residence at 8 Palace Gardens, Kensington. The description of part of his procedure is interesting: "To maintain the proper temperature a large fire was for some days kept up in the room in the midst of very hot weather. Kettles were kept on the fire and large quantities of steam allowed to flow into the room. Mrs. Maxwell acted as stoker. . . . After this the room was kept cool for subsequent experiments by the employment of a considerable amount of ice."

Apart from his theoretical work, the years in London were notable for Maxwell's part in the experimental determination of the standard ohm and in the measurement of the ratio of the electromagnetic to the electrostatic unit of electricity. He gave public lectures and found time to develop an acquaintance with Faraday. But perhaps such concentrated activity was too much for his liking, for Maxwell resigned his post in the spring of 1865. Much of the remaining year was spent in London. Then, except for a trip to Italy, he devoted most of the period from 1866 to 1870 to working at Glenlair on his "Electricity and Magnetism" and his book on heat.

Maxwell's absence from a university post did not sever his academic connections. During this period he acted as examiner in the mathematical tripos at Cambridge and introduced new life into the examinations by posing challenging questions on heat and electricity and magnetism. His innovations emphasized that there was then no instruction in these subjects at Cambridge. As a consequence, a university committee that included Stokes and Routh was appointed to recommend revisions of the curriculum. The committee recommended that instruction be given in heat, electricity, and magnetism, and that a professor be appointed and a laboratory provided for experimental physics. In order to obtain the funds, the committee canvassed the separate colleges of the university regarding their willingness to be taxed for this purpose. It is not surprising that the usual narrow collegiate interests prevailed and that they answered with the greatest disinterest to an undertaking for which they felt no responsibility. But by good fortune the Chancellor of the University, the seventh Duke of Devonshire, was a man not only of great wealth but of unusual brilliance, having been second wrangler in the mathematical tripos of 1829 and first Smith prizeman. Moved by his understanding of the needs of the university and his connection with science, he offered the necessary funds in October of 1870 for the construction of a laboratory and the purchase of appropriate apparatus. This event marked the beginning of the Cavendish Laboratory. In later years, under J. J. Thomson and Ernest Rutherford, it was to become the scene of those great experiments that have furnished much of the foundation of modern atomic physics.

But, even with the money available, a man was needed to fill the new professorship and guide the building of the laboratory. The position was offered first to Lord Kelvin, who promptly declined. It was then offered to Maxwell, who accepted. He began his teaching in the autumn of 1871 at the age of 40.

In connection with his return to Cambridge, an interesting anecdote is attributed to Sir Horace Lamb and recounted by Sir J. J. Thomson in his *Recollections and Reflections*. As Thomson notes, it was the custom then for newly appointed professors to deliver an inaugural lecture and for the senior professors to attend as a compliment to the appointee. For reasons not clear, the announcement of Maxwell's inaugural was made in such a way that it did not come to the attention of the senior members of the university. In addition, his address was given in an obscure lecture room and not in the Senate House, the usual place for such events. The result was that only about 20 persons were present, principally young mathematicians who had taken or were about to take the mathematical tripos examination. A few days later, however, an announcement was made in the usual way, stating that Maxwell would commence his lectures on heat. Unaware that the inaugural lecture had been given, the great mathematicians and philosophers appeared at the appointed time and place with Adams, Cayléy, and Stokes seated in the front row. Maxwell, with a twinkle in his eye, spent the time explaining to his distinguished audience the difference between the Fahrenheit and Centigrade scales of temperature!

A rumor circulated afterward that Maxwell might not be entirely innocent in this confusion and that a combination of personal modesty and his mischievous sense of humor may have prompted him to avoid the usual formal introduction to his Cambridge colleagues. If we can believe Maxwell's involvement in this episode, it offers another interesting sidelight on his character. His standing at that time was principally based on his paper on the distribution of molecular speeds. None of the scientists of that day could know how great his contribution to electricity and magnetism was to become. His present-day reputation, which eventually came as a result of the confirmation of his electromagnetic theory, was then still many years in the future.

Aside from the reasons previously cited for the Duke of Devonshire's gift, there may have been another consideration in the Duke's mind, which was not evident until after the completion of the laboratory in June 1874. This was nothing less than the editing and publication of Henry Cavendish's studies in electricity. Cavendish, a relative of the Duke's forebears, had published during his lifetime only two electrical papers but had left in his effects some 20 packets of manuscript on mathematical and experimental electricity. In retrospect, it seems unfortunate that Maxwell acceded to the Duke's wish; the preparation of the manuscript proved a long and arduous task that extended over five years and was not finished until 1879. His absorption in this task was so complete that it practically precluded all professional work except his university duties. Although there is no doubt that the result did much to place Cavendish in his rightful place in the history of experimental physics, there is also no doubt that it occupied most of Maxwell's available time and energy in the short time he had yet to live. It was also during this period that his wife entered a long illness, and Maxwell, with his characteristic sense of devotion to those who were close to him, insisted on assuming her care. On one occasion he did not sleep in a bed for three weeks but, nevertheless, conducted his lectures and experiments as usual.

About Easter of 1879, Maxwell consulted his doctor about a stomach ailment that had troubled him during the preceding two years. The problem was not immediately diagnosed and the Maxwells, both ill, left in June for their customary stay at Glenlair. During the summer, he suffered great pain and became progressively weaker. In October, he was told that he had no more than a month to live. Rather than stay in isolation at Glenlair, he and Mrs. Maxwell returned to Cambridge where medical aid was more readily available. His condition rapidly grew worse, and he died at the age of 48 on November 5, 1879, a victim of cancer, the disease that had carried off his mother at the same age.

Maxwell's deduction of the distribution of molecular speeds in a gas is only a part of a long paper in which he presented a systematic development of the fundamentals of kinetic theory. His deduction of the distribution of the velocities of gas molecules was based on the accepted microscopic picture of independent elastic spheres of negligible size, each of the same mass and endowed with completely random rectilinear motion. Collisions of the molecules with each other and with the walls of the container were assumed to be perfectly elastic (that is, with no net loss of energy) and to be governed by the usual laws of mechanics. The mean squared velocity under these conditions is proportional to the absolute temperature.

In the earlier treatments of gas kinetics no one had attempted to examine the distribution of molecular speeds—probably because no promising avenue had appeared along which the problem might be attacked. It is here, of course, that Maxwell's genius lay, for he saw that by the application of the laws of probability a solution could be obtained.

Maxwell first considered the three mutually perpendicular components of the velocity of a molecule. He assigned to each of these components its own distribution function and then considered the distribution of all three components together as the product of these three separate distributions. This is another way of stating that the probability that a molecule has a velocity in a certain range is proportional to the probability that the x component of the velocity has the desired value, times the probability that the y component of the velocity has the proper value, times the probability that the z component has its proper value. If all the velocity components lie in the correct range, the velocity itself will lie in the desired range. As Maxwell pointed out, it is permissible to treat the three components of the velocity independently because the three components are mutually perpendicular and, therefore, do not affect each other.

Having expressed the probability for the molecular velocity to lie in a certain range as a product of the probabilities that the components of the velocity lie in the proper range, Maxwell then introduced the crucial step in his reasoning. He pointed out that no direction in space for the velocity is to be preferred over any other direction; all directions are equally probable. This means that the probability function cannot depend on the direction of the velocity and, hence, depends only on the magnitude of the velocity. Now, the magnitude of the velocity depends on (indeed, it is equal to) the sum of the squares of the components of the velocity. In other words, if v_x, v_y, v_z are the components of the velocity v in the three mutually perpendicular directions x, y, z, then the probability for these components to have a certain range of values depends only on $v_x^2 + v_y^2 + v_z^2$. But the probability is also given by a product of three terms, the first of which depends only on v_x, the second only on v_y, and the third only on v_z. If we equate these two expressions for the probability, as Maxwell does, we see that the probability must be an expression that can be written either as a product of three terms, each involving one, and only one, component of the velocity, or as a single expression that contains the sum of the squares of the three components of the velocity. It is easy to show that the exponential function is the only mathematical quantity that has this property.

The exponential function is defined as follows: Consider the following infinite sum: $1 + 1 + (1/2) + 1/(2 \times 3) + 1/(2 \times 3 \times 4) + \ldots$ where each new term in the sum is obtained from the preceding term by multiplying its denominator by the next integer. This expression is called the exponential infinite series and is designated by the letter e. The numerical value of this sum is very nearly equal to 2.718. Let us now raise e to some power x. We write this expression as e^x, and this is called the exponential function; it is also written as $\exp(x)$.

Maxwell's discovery is equivalent to saying that the probability that a molecule in the gas will be moving with a velocity whose x component lies in a small range around v_x is proportional to $\exp(-v_x^2/\beta)$, where β is some constant that depends on the mass of the molecule and the temperature of the gas. A similar expression holds for the y component of the velocity and for the z component. Maxwell evaluated the constant β in this expression by first using the expression for the probability to obtain the average value of the velocity of a molecule. This expression must, of course, contain β. But, according to the work of Clausius, it is also related to the gas law, which in turn is related to the temperature of the gas. One then finds that β is equal to $2kT/m$, where m is the mass of the molecule, T is the absolute temperature of the gas, and k is the Boltzmann constant.

If we take all of these things into account, we see that the probability that a molecule of mass m in a gas of absolute temperature T has a velocity that lies in a small neighborhood around the value v is $\exp(-mv^2/2kT)$. This is one of the most important formulas in physics and is the basis of statistical mechanics. To see the significance of this formula in its application to all physical systems and not only to gases, we note that $mv^2/2$ is just the average kinetic energy of a molecule in the gas. We may therefore say that the probability that, on the average, a molecule in the gas will be moving in such a way that its kinetic energy will have a value lying in a small neighborhood around E is just $\exp(-E/kT)$. This is a basic formula that was also derived by Boltzmann and can be applied to systems in equilibrium at the absolute temperature T. The probability for finding any one of these systems with a given amount of energy (or in some particular state of energy) decreases as the energy increases. In other words, states of high energy are much less probable than states of low energy, but the probability for states of high energy increases as the temperature of the system increases.

The great theoretical physicist Ludwig Boltzmann obtained the same expression for the probability of finding a particle in a given energy state, if it is a component of the system that is in statistical equilibrium at the absolute temperature T, by defining the observed macroscopic state of the system in terms of groups of microscopic states. Specifically, if there are N molecules (where N is very large) in a system and these molecules are free to move without affecting each other, and if the entire system is in equilibrium at the temperature T (that is, the temperature of the system remains constant for a sufficiently long time), then this total number of molecules may be divided into groups in each of which the molecules will be moving in very nearly the same way (of course, molecules belonging to different groups will be moving differently). Each such sum of the molecular groups defines a macroscopic state of the system, but the groups themselves are called microscopic states.

Now it is clear that the same macroscopic state can be obtained by different microscopic groupings. If two molecules belonging to two different molecular groups are interchanged, the number of molecules in each group remains the same so that the macroscopic state is the same as before. Consequently, each macroscopic state of a system corresponds to many different microscopic states, and we see that the most probable macroscopic state is the one that can be realized by the largest number of microscopic states. If one then counts the number of microscopic states that correspond to a given macroscopic state and then chooses the macroscopic state with the largest number of microscopic states as the most probable state of the system, one obtains Maxwell's formula. This is the way Boltzmann solved the problem and, in doing so, founded classical statistical mechanics.

References

Mendoza, Eric, *Memoirs and Proceedings of the Manchester Literary and Philosophical Society*, Session 1962–3.

Count Rumford, *Collected Works*, vol. 2, essay 9. Read before the Royal Society on January 25, 1798. Later published in *Philosophical Transactions*, 88, 1798, p. 80, and reprinted in *The Complete Works of Count Rumford*. Boston: American Academy of Arts and Sciences, 1870, vol. I.

4 New Confirmation of Chemical Atomic Theory

O world invisible, we view thee,
O world intangible, we touch thee,
O world unknowable, we know thee.

—Francis Thompson, *The Kingdom of God*

Polyatomic Molecules: Stanislao Cannizzaro (1826–1910)

Stanislao Cannizzaro was born in Palermo, Sicily, on July 13, 1826. His father was a government official and was for some years a magistrate and minister of police in that city. His mother, Anna di Benedetto, was a member of a noble Sicilian house. Young Stanislao showed an early aptitude for study and was a prize pupil with special abilities in mathematics. At 15 he entered the University of Palermo, intending to prepare for the practice of medicine. But after attending for four years he left without taking a degree. Following a short period in the laboratory of the physicist Melloni in Naples, he secured a position as assistant to Rafaelle Piria, professor of chemistry at Pisa. Piria was an able scientist and his influence on Cannizzaro was so profound that he won over his young assistant wholeheartedly to the study of chemistry.

It must be remembered that Europe at that time was in a state of political unrest that culminated in the 1848 revolutions. Conditions in Sicily, ruled from Naples by the Bourbon King Ferdinand II, were so oppressive that organized rebellion broke out in January 1848. This was the signal for uprisings all over Italy and Europe. Cannizzaro, then only 21 and full of patriotic fever, had already thrown over his work at Pisa and returned to Sicily to join the insurgent cause as an artillery officer. By March of the following year, the government was able to put down the rebellion and Cannizzaro, one of the last to capitulate, escaped by sea to Marseilles. Having made his way to Paris, he was able to devote himself to research in the laboratory of Chevreul. His exile was relatively brief for, two years later at the age of 25, he returned to northern Italy to accept an appointment as professor of physical chemistry at the National School at Alessandria. There he was so occupied with teaching that he had little time to devote to laboratory investigations. This may have been fortunate because his teaching required him to examine more closely the chaotic state in which chemistry was struggling and to ponder the basis of these problems.

In 1855, after four years at Alessandria, Cannizzaro was offered the chair of chemistry at Genoa, which he accepted despite the fact that there was no chemistry laboratory in the university. It was almost another year before he obtained space

Stanislao Cannizzaro (1826–1910)

to carry on his researches. Shortly after this move, Cannizzaro courted and married, in Florence, Henrietta Withers, the daughter of an English clergyman. The couple had two children, a daughter and a son who later became an architect in Rome.

During these years at Genoa, Cannizzaro was also digging deeply into the competing ideas regarding chemical combination and the foundations of atomic theory. At this distance in time it is hard to realize that the greatest impediment to a rational system of chemistry was the almost universal idea that the ultimate particles of the elements had to be monatomic. Cannizzaro's great contribution was to adopt a molecular, that is, polyatomic, view of the elements, and to show that the *atomic* weights of elements, prepared in volatile compounds, could be deduced by the application of Avogadro's hypothesis together with accurate combining weight data and vapor densities. This is the burden of his famous "Sketch of a Course of Chemical Philosophy," which appeared in 1858. He was then 32 years old.

The effect of the paper was not really felt throughout the chemical world until the Congress of Karlsruhe, which was held in September of 1860. Here the paper was distributed and Cannizzaro was able to expound his views to the assembled chemists. If many were not convinced, at least one distinguished German, Professor Lothar Meyer, wrote later, "It was as though scales fell from my eyes, doubt vanished, and was replaced by a feeling of peaceful certainty." The soundness of this new view was gradually appreciated, aided doubtless by Meyer's well-known treatise on *Modern Theories of Chemistry*, which appeared a few years later.

At this time political events in Sicily brought about a second short-lived rebellion, but the subsequent resistance under Garibaldi and his famous Thousand began the movement that resulted in the unification of Italy. Always the patriot, Cannizzaro returned to Sicily to join the new regime, although he did not serve in the army. In 1861 he was made professor of chemistry at Palermo. In the following ten years he served not only chemistry but education, becoming active in the establishment of municipal schools and, for a time, in the administration as rector of the university. In 1871, at 45, he left Palermo to accept the chair of chemistry at the University of Rome. Here again there was no laboratory and space and equipment had to be found. At the same time he was made a Senator of the Kingdom in recognition of his patriotic services. His position as a senator and as a member of the government led to his active participation in many official committees. One of the most important was the Council of Public Instruction of which he ultimately served as president. He was active in the organization of the Customs Laboratory, provision for public education in agriculture, and in the general advancement of science in Italy. Although these activities precluded much of his former investigation in chemistry, his interest in students and his zeal for teaching were always paramount. It has been said that "for him to teach was to live." This statement was literally true, for when his strength began to fail in his 83rd year and it no longer became possible for him to lecture, all of his ills increased and he passed away on May 10, 1910.

In order to understand what Cannizzaro accomplished, it is necessary to form some conception of the state of chemistry in 1858. It should be made clear at the outset that the valence of many of the elements was uncertain, and that this ambiguity led to much confusion regarding the constitution of compounds. Thus, J. J. Berzelius assigned the general formula RO to the chief metallic oxides, a designation accepted by C. G. Gmelin. Gerhardt and Laurent, on the other hand, assigned a formula R_2O. Regnault agreed with neither of these schools, assigning the formula RO to some metallic oxides and R_2O to others. He did this labeling on the basis of specific heat measurements of the oxides and the law of Dulong and Petit (1819). This law stated that the atomic heat (heat required to raise a weight of the element in grams equal to its atomic weight through a temperature change of one degree Centigrade) is approximately the same for all the elements, being about 6.2 calories. Exceptions to the law soon became known; by 1850, on the basis of Berzelius's system, they included silver, sodium, potassium, bismuth, carbon, and bromine. In Gmelin's system there were additional exceptions—gold, iodine, phosphorus, arsenic, and antimony. These exceptions in nearly all cases had an atomic heat of 12.4, although for most of the elements it was 6.2. In Gerhardt and Laurent's system the exceptions were the same as in Gmelin's system, except for the addition of sulfur. Regnault's proposal to make the exceptions fewer by regarding the oxides of silver, potassium, and sodium as R_2O did not meet with general acceptance.

These different systems resulted, of course, in a diversity of formulas for the chemical substances. Thus, water was variously HO, H_2O, H_2O_2, and so on. Worse still, the same formula might designate two different substances. Depending on the school to which a chemist belonged, H_2O_2 might mean water or hydrogen peroxide; C_2H_4, marsh gas or ethylene; and CuCl, cuprous or cupric chloride.

Cannizzaro resolved these difficulties, which pervaded the whole field of chemistry, by taking the point of view that the molecule of an element must, in general, be regarded as polyatomic and that the atom must be defined in terms of the molecule. According to Cannizzaro, "The different quantities of the same element contained in different molecules are all whole multiples of one and the same quantity, which, always being entire, has the right to be called an atom." This point of view, combined with Avogadro's hypothesis, accurate combining weight data (largely that of Berzelius), and precise vapor density data obtained by the method of J. B. Dumas, constituted the basic method by which a rational system was obtained.

Suppose we consider three gaseous elements: hydrogen, oxygen, chlorine, and five of their compounds. Since, according to Avogadro's hypothesis, the same number of molecules is present in each of the moles of the five compounds, say, of hydrogen, we shall assume that the compound having the least hydrogen is the one in which the hydrogen occurs as a single atom in the molecule. This assumption, of course, could be incorrect, but there are cross-checks on the assumption, as we shall see. The product of the density in grams/liter of each of the gases and the weight percentage of the element of interest in the compound gives the weight of the element in the compound. If the smallest of these values is divided into the previously obtained products, the result is a series of integers. On this basis, elemental hydrogen, which has twice the least weight, would be composed of molecules having two atoms. In the ammonia molecule there would be three hydrogen atoms, and so on.

If we examine next the compounds of oxygen, we see that water and carbon monoxide each contains the smallest amounts of this element and, hence, the conclusion is that these compounds contain only one atom of oxygen. Since elemental oxygen has twice the least weight present in the oxygen compounds, we conclude that it, like elemental hydrogen, is diatomic. Making use of the available hydrogen and oxygen data, we see that water consists of two atoms of hydrogen and one of oxygen and, therefore, its formula is H_2O. Dividing the weight of elemental oxygen by that of hydrogen, we find a ratio of 16. The ratio of the weights of the oxygen and hydrogen molecules is 16, and since each molecule contains two atoms, the relative atomic weights are O:H = 16:1. The chemical formula for various other compounds may be deduced by the same method as above. The same is true for atomic weights.

As Cannizzaro pointed out, atomic weight determination cannot always be made by the vapor density method. In the nonvolatile compounds an alternative method is to make use of specific heat data. If one considers cupric chloride, $CuCl$, and cuprous chloride, $CuCl_2$, a combining weight percentage of the former is found to be 63.96/36.04 and of the latter 47.02/52.98. Assuming in the former case 35.5, the value already found as the atomic weight of chlorine, simple ratios give that of copper as 63.0. In the latter instance, comparing the ratios $47.02/52.98 = x/2 \times 35.5$, we find that the atomic weight of copper is 63. If now the specific heat of copper 0.09515 (cal/g-deg) is multiplied by the value 63, we obtain 6.0 as the atomic heat. This result, in comparison with similar data where it is known that the element is monatomic, indicates that the monatomic assumption for copper is not contradicted.

The consistent results of these methods inspired the confidence of chemists. The procedures were shown to be equally valid in organic as well as inorganic chemistry. Although these branches had been considered almost separate sciences, reunification was now possible. The periodic system of the elements as originally proposed by Newlands was based on Cannizzaro's atomic weights. Indeed, until these weights had been settled, the overall relationship of the elements to each other must have remained obscure.

The Periodic Table of Elements:
Dmitri Ivanovich Mendeleev (1834–1907)

The introduction of the atomic theory of matter early in the 19th century made it relatively easy to understand the laws of formation of chemical compounds, even though there was no precise theory of the way atoms combine to form these compounds. The greatest advance in 19th-century chemistry occurred with the discovery by Mendeleev of the periodicity in the chemical properties of the elements as one passed from one element to another. Mendeleev presented his discovery in the form of the Periodic Table of Elements. To understand this table, we must introduce two fundamental concepts: (1) the atomic number, and (2) the atomic weight.

If we arrange all the elements in order, starting with the lightest and going to the very heaviest, the atomic number of an element is its position in this array. We assign the atomic number 1 to hydrogen, the atomic number 2 to helium, 3 to lithium, and so on. We shall see later that the atomic number is related to the number of electrically charged particles inside a neutral atom.

The atomic weight is a measure of the mass of a single atom on some arbitrary but convenient scale. The scale that has been in use until the present time is one in which the oxygen atom is arbitrarily given the atomic weight 16. On this scale hydrogen has the atomic weight 1.008 and helium the atomic weight 4.003. Mendeleev's great contribution consisted in the discovery that all elements in a given vertical column (in his periodic table) possess similar chemical properties. For example, lithium, sodium, potassium, and so forth, have similar properties and constitute the group called the alkali metals. Florine, chlorine, bromine, and so forth, constitute another group called the halogens, and helium, neon, argon, and so forth, belong to the family of noble gases. What is important about this arrangement of elements is that there is a repetition of chemical properties in a cycle of eight elements. This periodicity is not true for the entire chart of elements; as we go to heavier elements there is a change from a cycle of eight to one of 18.

The history of physics contains important discoveries that pointed to very deep consequences for the future of physics but which the discoverers themselves either failed to recognize or rejected outright. Among these discoveries are Kepler's laws of planetary motion, which foretold the law of gravity and the conservation principles, Balmer's formula for the wavelengths of four of the visible lines of the

Dmitri Ivanovich Mendeleev (1834–1907)

hydrogen spectrum, which pointed to integral quantum numbers, and, above all, Mendeleev's discovery of the periodicity of the chemical properties of the chemical elements, which clearly indicates a periodicity in the arrangement of electrons in the atoms according to their atomic weight. To the end of his life, Mendeleev rejected entirely the concept of the electron, without which his periodic table of the chemical elements cannot be understood. Mendeleev's attachment to Newtonian physics and to the great advances in early 19th century science stemmed from his traditional education and from the influence of his parents and his early high school and university teachers. It is remarkable that, in spite of these influences, he was a revolutionary democrat politically and courageous in the pursuit of his very offbeat scientific ideas.

The last of 14 children, whose father, Ivan Pavlovich Mendeleev, was a Russian literature teacher and whose mother, Maria Dimitrievna, owned a glass factory, Dmitri was a devoted son whose interests and curiosity encompassed not only all aspects of the natural world but also industrial matters and technological progress. After graduating from the Tobolsk Gymnasium at the age of 15 in 1849, where he acquired a love of history, mathematics, and physics, he matriculated in the physics and mathematics faculty of the Main Pedagogic Institute of St. Petersburg (now Leningrad). There he immersed himself in chemistry, zoology, and mineralogy. His detailed descriptions of the lectures of his chemistry professor are still preserved in the Mendeleev museum in Leningrad. The beauty of and the excitement stimulated by his professor's chemical demonstrations led Mendeleev to chemistry and he began his research career by doing chemical analyses of orthosilicate compounds and certain organic compounds. He graduated from the Institute at the age of 21 in 1855 with a brilliant record, but his support of democratic ideals was held against him by the Minister of Education when he applied for a teaching assignment. Consequently, he was given a post in the Simferopol Gymnasium, which had been closed owing to the Crimean War. Unable to teach in Crimea, he accepted a teaching post at the gymnasium in Odessa where he continued his chemistry research with special emphasis on the relationships between the crystal forms and chemical compositions of substances. This was the first step along the road that led him to the discovery of the periodic table of the chemical elements.

Mendeleev was active in many areas of technology that stemmed from his chemical research. In addition, he contributed to the science education of the public by writing popular articles on the natural sciences in such such journals as *The Ministry of Public Education*, *Natural Science News*, and *Industrial Notes*. His interest in the application of pure science to industry remained with him during his entire life.

By the time Mendeleev had completed his master's thesis at the University of St. Petersburg in 1856, his ideas about the formation of chemical compounds were set. These ideas excluded the concepts of the electron, of ions, and electrolytic solutions in general. But these erroneous beliefs did not interfere in any way with his search for the correlation between the chemical properties of atoms and their atomic weights.

In 1865 Mendeleev successfully defended his thesis for his chemistry doctorate. In it he argued that solutions are chemical compounds and defended the "principle of chemical atomism." Stimulated by the abolition of serfdom in Russia in 1862, he tackled some of the practical problems that confronted the national economy, concerning himself particularly with the petrology of the Baku oil fields. In 1865, having married Feozva Nikitichna Leschevaya some three years earlier, he settled down at the estate of Boblovo, near Klim; there the couple had a son and a daughter.

Mendeleev's appointment to the chair of chemistry at the University of St. Petersburg in 1867 triggered the breakthrough that he needed to complete his theory of the periodicity of the chemical elements. Required to give lectures in organic chemistry, but without a textbook that he could assign to his students, he began to write one himself. In undertaking this task Mendeleev found that he obtained a remarkable simplification in the description of the chemical properties of different elements if he arranged them in several groups according to their atomic weights; the different elements in each group, if properly chosen, then had similar chemical properties. He reported his discovery in a paper, "The Relation of the Properties to the Atomic Weights of the Elements," that was read to the Russian Chemical Society in March 1869 by a colleague. In the years that followed, Mendeleev expanded this work greatly and predicted the existence of chemical elements that were required to fill the gaps that were present in his original table. As each of these elements, in turn, was discovered, Mendeleev's reputation grew and the number of chemists and physicists who began to accept his theory increased rapidly. Mendeleev first referred to his discovery as the "periodic law of the chemical elements" in 1871; it has been described that way ever since. In the decade from 1870 to 1880, Mendeleev's law was thoroughly confirmed and began to appear in elementary and advanced chemistry texts.

During his entire academic career Mendeleev supported liberal and democratic causes in every area of society. As he stated, "I tried to fight against superstition . . . it took professors to act against the authority of professors." In 1894 he was awarded honorary doctorates by both Oxford and Cambridge and his 70th birthday was honored worldwide in 1904, as was the 50th anniversary of his great discovery, in 1905. He received the Copley medal of the Royal Society of London in 1905 and, at his death from heart failure on February 2, 1907, he was a member of scientific societies throughout the world.

5 *Beyond the Atom*

All things are artificial; for nature is the art of God.

—Sir Thomas Browne, *Religio Medici*

Atoms and Electricity: Michael Faraday (1791–1867)

The development of physics has often been greatly accelerated by the simultaneous, or at least coeval, work of an outstanding experimentalist and a brilliant theoretician. This was particularly true of Michael Faraday, probably the greatest experimental physicist of all time, and James Clerk Maxwell, who incorporated Faraday's theory of electromagnetic induction into the magnificent electromagnetic theory of light. Without Faraday's discoveries Maxwell could not have constructed his own theory, which in part expresses in mathematical language Faraday's concepts of tubes of force and his laws of electromagnetic induction.

That Faraday became the great physicist he was is all the more remarkable since he was almost entirely self-taught and approached his subject with scarcely any mathematical training. Nevertheless, his physical intuition and insight into nature more than compensated for this deficiency. That he even had a chance for a scientific career was in itself quite amazing since he came from a worker's family at a time when most children of the working class were lucky if they learned to read.

Born in London in 1791 to James Faraday, a blacksmith, Michael was apprenticed to a bookseller at the age of 13, since this was then a period of acute economic distress, and the Faraday family was at its lowest ebb. By that time he had learned reading, writing, and some arithmetic and was already probing into all kinds of areas. As he himself said, "I was a lively and imaginative person" and "facts were important to me. . . . I would trust a fact and always cross-examined an assertion." Although he read whatever he could lay his hands on, he preferred scientific books and was greatly influenced by Mrs. Marcet's *Conversations in Chemistry*. Even at that early age Faraday was not content to accept a statement but had to check it himself, for only in that way could he claim the fact as his own: "So when I questioned Mrs. Marcet's book by such little experiments as I could find means to perform and found it true to the facts as I could understand them, I felt I had got hold of an anchor in chemical knowledge and clung fast to it."

Michael might have remained a bookbinder all his life except for a chance occurrence and for his intense dislike of business and his equally intense love

Michael Faraday (1791–1867)

for science and philosophy. By the year 1810, Michael was doing so well in his apprenticeship that he had two boys working for him; but in spite of this he had money for little more than the bare necessities. Although he lived with his master, Mr. Ribeau, he was still very closely attached to his family. When a series of science lectures at the Royal Institution was announced, it was Michael's eldest brother Robert's savings that paid for Michael's attendance.

These lectures certainly had a great influence on him, for from then on he devoted all his spare time to science and philosophy and attended as many lectures as he could. Undoubtedly, this interest contributed greatly to his development as a devoted science lecturer later in his life. Fortunately, Mr. Ribeau and his wife were very sympathetic to Faraday's scientific interests and encouraged him as much as they could. He never neglected his craft during this period, although he must certainly have had many doubts about continuing his career as a bookbinder.

When Faraday completed his apprenticeship in 1812 and went to work as a journeyman bookbinder for a Mr. De la Roche, he soon made up his mind to leave the world of commerce and to become a scientist. De la Roche was very fond of Faraday as a person and worker, but he had no sympathy for Michael's interest in science and opposed it violently. This opposition was the final factor that contributed to Faraday's decion to leave his profession. In 1813 he described his change of career in the following charming passage.

> I was formerly a bookseller and binder, but am now turned philosopher, which happened thus: whilst an apprentice, I, for amusement, learnt a little chemistry and other parts of philosophy, and felt an eager desire to proceed in that way further. After being a journeyman for six months, under a disagreeable master, I gave up my business, and by the interest of Sir H. Davy, filled the situation of Chemical Assistant to the Royal Institution of Great Britain, in which office I now remain, and where I am constantly engaged in observing the works of Nature and tracing the manner in which she directs the arrangement and order of the world.

Faraday's association with Sir Humphry Davy was itself a matter of chance. A customer of Ribeau's had given him tickets for a few' of Davy's lectures at the Royal Institution, and Michael took full advantage of this opportunity, as indicated in his own account of the affair (in a letter written in 1829).

> You asked me to give you an account of my first introduction to Sir H. Davy, which I am very happy to do, as I think the circumstances will bear testimony to the goodness of his heart. When I was a bookseller's apprentice I was very fond of experiment and very adverse to trade. It happened that a gentleman, a member of the Royal Institution, took me to hear some of Sir H. Davy's last lectures in Albemarle Street. I took notes, and afterwards wrote them out more fairly in a quarto volume. My desire to escape from the trade, which I thought vicious and selfish, and to enter into the service of Science, which I imagined made its pursuers amiable and liberal, induced me at last to take the bold and simple step of writing to Sir H. Davy, expressing my wishes, and a hope that if an opportunity came in his way he would favour my views; at the same time I sent the notes I had taken of his lectures. . . .

> You will observe that this took place at the end of the year 1812, and early in 1813 he requested to see me and told me of the situation of assistant in the laboratory of the Royal Institution, then just vacant. At the same time he thus gratified my desires as to scientific employment, he still advised me not to give up the prospects I had before me, telling me that Science was a harsh mistress, and in a pecuniary point of view but poorly rewarding those who devoted themselves to her service. He smiled at my notion of the superior moral feelings of philosophic men, and said he would leave me to the experience of a few years to set me right in that matter.

> Finally, through his good efforts, I went to the Royal Institution, early in March of 1813, as assistant in the laboratory; and in October of the same year went with him abroad as his assistant in experiments and in writing. I returned with him in April 1815, resumed my station in the Royal Institution, and have, as you know, ever since remained there.

Davy's account of Faraday's request for work at the Royal Institution is quite amusing and clearly indicates the gentle quality of Davy's own character as shown by this brief exchange with an acquaintance of his, a Mr. Pepys: "Pepys, what am I to do? Here is a letter from a young man named Faraday; he has been attending my lectures and wants me to give him employment at the Royal Institution—what can I do?" "Do?" replied Pepys, "put him to wash bottles; if he is good for anything he will do it directly; if not, he will refuse." "No, no," replied Davy, "we must try him with something better than that."

As suggested in this passage, Davy recognized Faraday's merit even before he had observed him in the laboratory, where he made him an assistant and not a "bottle washer." This position certainly was due to the skill and care with which Faraday had written up the lectures he had attended. He carried over this painstaking accuracy and attention to detail to everything he did, and Davy soon recognized that Michael Faraday was to become an outstanding scientist.

Faraday performed his duties as assistant to various professors at the Royal Institution in a most excellent manner, allowing no detail in the lecture room to escape his attention and laying the foundation for his own career as a lecturer. But, during this assistantship period, the dominant idea in Faraday's mind was to perform his own experiments. When, in 1816, the year after he had been reappointed to the institution, he published his first paper in the *Quarterly Journal of Science*, his dream was realized. After that paper appeared, he rose very quickly and in 1823 became a fellow of the Royal Society.

This election marked the beginning of Faraday's greatest experimental period, which reached its peak in the discovery of electromagnetic induction, one of the most important scientific achievements of all time. Though he was extremely serious and devoted to his scientific work, Faraday's life, when he was not engaged in research, was filled with many gay and tender moments. His interests included poetry, the beauties of nature, all kinds of games, acrobatics, Punch and Judy shows, puzzles, and children, for whom he prepared very special and extremely instructive lectures. The quality of his attitude toward all people and his humility, even toward his work, is indicated by an excerpt from a letter to Sara Bernard, shortly before he married her.

> You know me as well or better than I do myself. You know my former prejudices and my present thoughts—you know my weaknesses, my vanity, my whole mind. You have converted me from one erroneous way, let me hope that you will attempt to correct what others are wrong. . . . Again and again I attempt to say what I feel, but I cannot. Let me, however, claim not to be the selfish being that wishes to bind your affection for his own sake only. In whatever way I can best minister to your happiness either by assiduity or by absence it shall be done. Do not injure me by withdrawing your friendship, or punish me for aiming to be more than a friend by making me less; and if you cannot grant me more, leave me what I possess, but hear me.

This tenderness toward, and understanding of, others, regardless of Faraday's own interests, certainly had its roots, to some extent, in his religous convictions and his sense of human dignity. He belonged, in his own words, to a "very small and despised sect of Christians, known, if known at all, as Sandemanians," who believed that organized religion was bound to degenerate into the opposite of true Christianity and that the words of Christ were sufficient guides for living. They practiced brotherhood, humility, and charity.

Woven into the colorful fabric of his life and scientific work was a thread of tragedy, for quite early in his life Faraday began to suffer from attacks of giddiness, headache, and loss of memory, which finally were fatal. From 1831 to 1840 these symptoms became quite severe. He himself, believing no doctor understood his ailment, was convinced that he was suffering from a decay of his physical and mental faculties and that there was no way to stop this deterioration. Nevertheless, in spite of this progressive disease, which at times left him momentarily paralyzed, his creative faculties remained as great as ever, and his productivity continued until he died on the afternoon of August 25, 1867, just one month short of 76 years after his birth.

Although Faraday's researches ranged over all phases of electrical and magnetic phenomena, covering almost every aspect of this branch of physics, he is best known for his work on electromagnetic induction. After Oersted discovered in 1819 that an electric current deflects a magnetic needle, the whole scientific world began an intense study of the electric current and its effects. But only a few, among them Faraday, looked for the inverse of the Oersted effect, namely, the effect of magnetism on an electric current. These investigations ultimately led Faraday to try to obtain an electric current from magnetism. In 1831 the following entry appeared as the first heading in his laboratory notebook: "Experiments on the Production of Electricity from Magnetism."

In his approach to this work, Faraday was greatly influenced by three things, the first of which was Newton's abhorrence of action at a distance. In a letter to Bentley, Newton wrote a remarkable passage to which Faraday constantly referred: "That gravity should be innate, inherent, and essential to matter so that one body may act upon another at a distance through a vacuum and without the mediation of anything else . . . is to me so great an absurdity that I believe that no man who has in philosophical matters a competent faculty of thinking can ever fall into it." This view led Faraday to introduce the concept of the field and the idea that the magnetic field between the magnet and the current affects the current in some way. He thus sought to explain interaction of magnet and current via a field, the concept that was to prove so useful to Maxwell.

A second influence was his own deficiency in mathematics. This gap in his education forced him to represent physical phenomena in terms of physical models rather than by means of abstract mathematical formulas. His idea of tubes of force stemmed directly from the need for a model. The model of the field with its tubes of force in turn led him to the explanation of induction as a cutting of the tubes of force. Whenever tubes of force are cut by a conductor, a current flows in the conductor; the faster the tubes are cut, the greater the electromotive force that is induced in the conductor.

A third influence was his conviction that all forms of physical action are basically one, that is, the concept of the unified field. In 1854, when he discovered a relationship between magnetism and light, he stated "I have long held an opinion, almost amounting to conviction . . . that the various forms under which the forms of matter are made manifest have one common origin: in other words, are so directly related and naturally dependent that they are convertible as it were into one another and possess equivalents of power in their action."

Following his discovery of electromagnetic induction—that the electromotive force induced in a loop of wire is proportional to the rate at which the magnetic flux through the loop is changing—Faraday began a series of experiments in electrochemistry that finally led him to the laws of electrolysis. These experiments are of great importance in the development of atomic physics. Before this work was done, the concept of the atom and of the molecule was widely accepted and used, but nobody had any idea as to the nature of the forces that bound two or more atoms in a single molecule.

Faraday suspected that electrical forces were at work in a molecule. He reasoned that if one could obtain an electric current from a voltaic cell, which operates by the chemical reaction of the electrodes with a chemical solution, then the atoms in solution must have electrical charges in them. To test this hypothesis, he sent various electric currents through different solutions and discovered that the same current flowing for the same length of time always decomposed the same quantity of material, or, to put it differently, the same number of ions of a given chemical compound. He then reasoned that each ion of the same kind has exactly the same charge. Later he discovered that all charges on ions are integral multiples of a single fundamental unit of charge and that one never finds fractions of this charge. He showed quite clearly that this invariance accounts for the reversibility of electrochemical phenomena and "that if the electrical power which holds the elements of a grain of water in combination, or which makes a grain of oxygen and hydrogen in the right proportions unite into water when they are made to combine, could be thrown into the condition of a current, it would exactly equal the current required for the separation of that grain of water into its elements again."

Throughout these descriptions of his experiments we again detect Faraday's humility, his sense of wonder, and his trust in experimental data. Thus he spoke of how "wonderful" it was "to observe how small a quantity of a compound is decomposed by a certain portion of electricity," or how large a quantity of electricity one can obtain from the combination of very small quantities of atoms. From this observation he concluded, correctly, that the electrical forces between ions are very large.

Finally, we cannot escape Faraday's sense of the beauty of nature and its laws, for he constantly spoke of "beautiful experiments," clearly indicating his belief that to him truth and beauty were one.

Electromagnetic Theory: James Clerk Maxwell (1831–1879)

As we noted earlier, Maxwell's pioneering contribution to atomic theory is to be found not only in his work on molecular velocities but also in his electromagnetic theory of light. Although Maxwell did not present his discovery as a contribution to the theory of atomic structure (the nature of the atom was hardly known at the time), we now know that without Maxwell's theory we could hardly understand how the atom absorbs and emits radiation.

During the 200 years or more from the time of Newton to the beginning of the 20th century, the science of physics developed very unevenly, with most of the emphasis in the 17th, 18th, and 19th centuries on mechanics. Comparatively little was done in optics, electricity, and magnetism—avenues that were to lead eventually to the inner world of the atom. This was quite natural, since Newton had already laid down the fundamental principles of mechanics. It remained for the great mathematicians and theoretical physicists of those centuries to put these principles into the most elegant forms possible. Moreover, the rapid development of celestial mechanics, following Newton's discovery of the law of gravity, attracted the outstanding mathematicians of that period—Pierre Laplace, Léonard Euler, Joseph-Louis Lagrange, Carl Friedrich Gauss, and others. Classical mechanics was particularly appealing to mathematicians because it offered a closed, completely self-contained rational system that could be developed axiomatically. The historical culmination of this remarkable development is contained in the equations of William Hamilton and, finally, in the famous Hamilton-Jacobi differential equation. Optics, electricity, and magnetism were laggards in this period because of the lack of a single fundamental principle around which to develop a complete system.

Maxwell was to change this state of affairs. In his day, as now, the laws of mechanics could be checked by reference to the motions of ordinary bodies. But the behavior of light was difficult to analyze without very special equipment, and electricity and magnetism appeared to be mysterious phenomena that were unrelated to any other observable events.

Various optical theories had already been introduced and, in time, as more and more optical data were collected, the wave theory as first propounded by Christian Huygens became generally adopted. This acceptance was primarily due to the work of Thomas Young and Augustin Fresnel, who discovered the interference and diffraction of light and explained these phenomena by using the wave theory. The acceptance of the theory that light travels as a wave was followed by a period of mathematical development; this, in turn, produced the equations that described the propagation of a beam of light in practically the same form in which they appear today. Before Maxwell's discovery, these equations were used even though the nature of light itself was unknown; one simply assumed that some disturbance was propagated from point to point in a medium in the form of a wave; then one could apply all the mathematical techniques of wave motion. In this way most of the problems of refraction, diffraction, and interference were analyzed and solved. Although this analysis was satisfactory from the point of view of the description of optical phenomena, physicists were not entirely satisfied with it because it had nothing to say about the nature of light itself. But this ultimate goal could not be reached until more was known about electricity and magnetism.

Although electricity and magnetism as separate phenomena were known to the ancient Greeks, Hans Christian Oersted first discovered in 1820 that there is a relationship between them. As almost everyone now knows, he found that a wire carrying current is surrounded by a magnetic field and therefore deflects a compass needle placed near it. This discovery opened the door to the rapid experimental development that culminated in the discovery by Michael Faraday in England and by Joseph Henry in the United States that electricity and magnetism are reciprocal phenomena in the sense that, under appropriate conditions, each of them can induce the other. These two investigators found that if a conductor (a copper wire, for example) is moved through a magnetic field, an electric current is induced in the wire. As these induction phenomena were studied and their laws more clearly understood, a fairly complete mathematical formulation of fundamental electromagnetic principles was developed. But, just as in the case of optics, the mathematics served only to describe the gross behavior of the electric and magnetic fields. The nature of electricity and magnetism was not yet understood.

Concurrent with studies of the relationship between magnetism and electricity, a great deal of research was done in electrostatics. With Charles Coulomb's discovery of the law of interaction between electric charges (mathematically similar to Newton's law), it was possible to develop the mathematical theory of static electricity. It became possible also to analyze the motion of conductors carrying currents and the mechanical interactions between such conductors. However, the relationship between electric charges and the behavior of an electromagnetic field was not clearly understood. In fact, all electric and magnetic phenomena were explained by the action, at a distance, of charges on each other. Since the concept of action at a distance dominated pre-Maxwellian physics, we shall consider its significance for a moment. Two electric charges (or two masses, or two magnets) do not have to be in direct contact to affect each other. Each charge induces an acceleration in the other even though the two charges may be quite far apart. This phenomenon, which seems mysterious if it is viewed merely as one charge exerting a force on the other without anything between them, is called "action at a distance."

At this point in the development of the theories of electricity, magnetism, and optics, James Clerk Maxwell, now regarded as the dominant figure of 19th-century physics, came on the scene. During the course of his life, as described earlier, he provided the basis for modern electromagnetic theory. As Maxwell pointed out in his famous paper on the electromagnetic field, all the theories on electromagnetic phenomena that had been developed up to the time of his own work assumed "the existence of something either at rest or in motion in each body, constituting its electric or magnetic state, and capable of acting at a distance according to mathematical laws." On the basis of this assumption, theories about the electromagnetic interactions of these bodies were developed. Maxwell concluded his description of contemporary 19th-century thinking with the following observation: "In these theories the force acting between the two bodies is treated with reference only to the condition of the bodies and their relative positions, and without any express consideration of the surrounding medium." What is crucial in this sentence is the reference to the surrounding medium.

Maxwell did not reject the hypothesis that charged particles inside bodies are the carriers of electricity and magnetism. He denied only that a theory could be successful if it disregarded the interaction of the charged bodies with the surrounding medium. The most successful of such theories that had been developed up to that time, that of Weber and Neumann, held that the force acting at a distance between two bodies must depend not only on their positions but also on their relative velocity. Maxwell, therefore, rejected this theory because it led to mechanical difficulties that made it undesirable as an ultimate theory.

In his paper, Maxwell outlined his plan of attack, saying that his attention would be on the surrounding medium: "I have therefore preferred to seek an explanation of facts in another direction, by supposing them produced by actions which go in on the surrounding medium as well as in the excited bodies, and endeavoring to explain the action between distant bodies without assuming the existence of forces capable of acting directly at sensible distances." Maxwell went on to say that "the theory I propose may therefore be called a theory of the *Electromagnetic Field* because it has to do with the space in the neighborhood of the electric or magnetic bodies, and it may be called a *Dynamic Theory* because it assumes that in space there is matter in motion, by which the observed electromagnetic phenomena are produced." Finally, "the electromagnetic field is that part of space which contains and surrounds bodies in electric or magnetic conditions."

The great departure taken by Maxwell from previous developments lies in the manner in which he introduces the electromagnetic field. Previously, physicists had considered space to be empty and spoke of lines or tubes of force between charged particles as mere mathematical fictions. Maxwell, however, imparted a reality to the electromagnetic field that went far beyond a mere mathematical convenience. This is one of those giant steps in science that herald a revolution in scientific thinking and mark the emergence of a great genius. Atomic theory could not have matured without Maxwell's contribution. The concept of the electromagnetic field has changed our entire way of looking at physical phenomena and has been of decisive importance in the development of the general theory of relativity, quantum electrodynamics, and the physics of elementary particles.

Interestingly enough, Maxwell thought it necessary to introduce a real ether or an ethereal substance as the carrier of his electromagnetic field. He believed that the "aethereal medium filling space and permeating bodies" was capable of motion and could transmit this motion to material bodies. He even believed that this ether had a small but finite density and that the energy of the electromagnetic field that was communicated to material bodies (in the form of the heat and motion of these bodies) was stored in the ether as potential energy or was present there in the form of the motion of the ethereal particles.

Although we know now that it is not necessary to introduce an ether in order to understand the propagation of light through empty space, Maxwell and the other physicists of his period thought that some kind of medium was necessary for the propagation of waves. According to Maxwell, the ether behaved like an elastic medium that was capable of "receiving and storing up two kinds of energy, namely, the 'actual' energy, depending on the motion of its parts, and 'potential' energy, consisting of the work which the medium will do in recovering from its displacement in virtue of its elasticity." According to this picture the propagation of waves occurred because there was a continuous alternate transformation from one of these forms of energy to the other, just as in a swinging pendulum or in a vibrating spring.

Starting from this idea of the ether as an elastic medium through which waves could be propagated according to the usual theories of wave motion, Maxwell set out to find the wave equation that governs the motion of electromagnetic waves. Reasoning that there was a continual oscillation of energy in the ether, he was convinced that the electromagnetic field moved about as a kind of undulatory phenomenon. Maxwell then proceeded to derive the equations that describe this wave phenomenon in terms of the electric field and magnetic field strength at a particular point. To do this, he analyzed the available experimental data related to the magnetic interaction of current-carrying conductors and to the induction of an electromotive force in a conductor when the conductor cuts across the lines of force of a magnetic field.

The laws governing the interaction of the magnetic fields surrounding two conductors were well known at the time of Maxwell. They had been formulated mathematically by André Ampère. Moreover, the laws of electromagnetic induction, as discovered by Faraday, were well known. These laws had also been expressed in mathematical form, although no one had yet been able to write down a single set of equations that could describe the behavior of the electromagnetic field as an entity consisting of an electric part and a magnetic part. There was no mathematical treatment that described the relationship in which each could be induced by the other and that could properly account for the oscillation of the energy in the electromagnetic field. In surveying what had been done previously, Maxwell was quick to see why previous attempts to obtain a unified theory of the electromagnetic field had been futile. The difficulty lay in the manner in which the electric current was treated.

Up to that time the current induced by an electromagnetic field had always been associated with a conductor. Maxwell's great contribution, and the one innovation that made possible his electromagnetic equations, was his idea of the displacement current, which he considered the natural extension of the current in a conductor. He arrived at his idea by considering what happens to a dielectric (or nonconductor) in the electromagnetic field.

In his famous paper on the electromagnetic field, Maxwell noted that under the action of an electromotive force the dielectric undergoes a change of state because energy is stored in it. This change is ordinarily called the polarization of the dielectric. It may be pictured as a kind of elastic strain resulting from the stress imposed upon it by the electromagnetic field. This strain, or polarization, is due to the displacement of the charged particles in the molecules of the dielectric. But, as Maxwell wrote, this displacement in itself is not a current, but, as he stated, "the commencement of a current." The next step in his analysis was to show that this idea of a displacement current must be introduced even when no material dielectric is present, as in the region between two condenser plates carrying a time-varying charge.

To comprehend this idea more clearly, consider the empty space between two condenser plates that carry a charge. As long as the charge is constant, there is a constant electric field between the two plates with the lines of force of the field at right angles to the plates. Now suppose that the two plates are connected by a conductor, so that the negative charge on one of the plates can flow through the conductor to the other plate. While this is happening, a current flows through the conductor. The conductor is surrounded by a magnetic field in accordance with Oersted's discovery. What is happening in the space between the two plates while the current is flowing in the conductor? The answer is that the strength of the electric field is changing. In fact, it is getting smaller. Making an analogy to what would happen to a dielectric if it were in the space between the two plates, Maxwell introduced the idea that there is a kind of current flowing between the two condenser plates while the current is flowing in the conductor. In other words, the space between the plates is to be treated as a kind of dielectric that is undergoing a change in its state of polarization. Hence, it is to be regarded as a natural extension of the conductor. Maxwell called this changing electric field between the condenser plates a displacement current and stated that it is measured by the rate at which the electric field is changing.

It is true that this displacement current is not a current in the same sense as is a flow of charged particles inside a current-carrying conductor. But it exhibits all the mathematical properties of a current and behaves like the natural extension of the current in the conductor. With his innovation, Maxwell was able to write down a set of equations for the electromagnetic field in which the electric field and the magnetic field enter in a symmetrical way.

From these equations he derived a single wave equation that describes the way any electromagnetic field is propagated. From this equation he then derived two more very important results that led to the identification of his theory with the electromagnetic theory of light. First, his equations demonstrate that the speed of propagation of the electromagnetic field is exactly equal to the speed of light. He showed also that the propagated electromagnetic disturbance is at right angles (transverse) to the direction of propagation. In Maxwell's words: "This velocity is so nearly that of light, that it seems we have strong reason to conclude that light itself [including radiant heat, and other radiations, if any] is an electromagnetic disturbance in the form of waves propagated through the electromagnetic field according to electromagnetic laws." Second, Maxwell's equations demonstrate that the square of the refractive index of a medium is "equal to the product of the specific dielectric capacity and the specific magnetic capacity." These two results together show clearly the relationship of light to the electromagnetic field. Later, Heinrich Hertz demonstrated that Maxwell's electromagnetic theory of light can be verified experimentally.

The idea of the displacement current leads quite naturally to the propagation of the electromagnetic field. If we consider the varying field between the condenser plates to be a current, then this current must be surrounded by a magnetic field. But since this displacement current changes with time, the magnetic field surrounding it is also changing. It, in turn, is therefore linked to an electric field, and so on. In other words, as the condenser is discharged, a chain of electric and magnetic fields is set up extending out into space. But the current flowing from one condenser plate to the other does not stop even after the entire charge has flowed from one plate to the other because of the inertia of the charge. Instead, the current oscillates back and forth and thus gives rise to an oscillating electromagnetic field that breaks away from the condenser and moves off into space.

Although we have described Maxwell's discovery in terms of the field between the plates of a condenser, his results are completely general and apply to electromagnetic fields everywhere: a varying electric field must be accompanied by a magnetic field and such a combination of an oscillating electric and an oscillating magnetic field (called an electromagnetic field) must move through space at the speed of light. In a vacuum, the direction of oscillation of the magnetic field and the direction of propagation of the two fields are mutually perpendicular to each other.

Cathode Rays—A "Fourth State of Matter": William Crookes (1832–1919)

The discovery of electromagnetic induction by Faraday in 1831 quickly led to successful attempts to use the phenomenon in technical applications. One outgrowth was the development of the spark coil, which consisted of a primary coil energized by a battery through an interrupter together with a separate secondary coil wound around the primary but insulated from it. The high voltage generated between the terminals of the secondary coil was often used for the production of a continuous spark discharge. At first these discharges were examined in air. The first experiment on the discharge produced at low pressure in a glass envelope seems to have been made by a rather obscure French investigator named Masson in 1853. The spark discharge, which in air is an unsteady and noisy phenomenon, at reduced pressure becomes a quiet, soft glow that completely fills the discharge tube.

The study of the low-pressure discharge became popularized through the intricate tubes made by Heinrich Geissler, an accomplished glass blower and technical assistant of Professor Julius Plücker at the University of Tübingen. Geissler's name soon became applied to all small discharge tubes used for spectroscopic or demonstration purposes. Geissler tubes were long used as an impressive scientific demonstration for beginners or in popular lectures on electrical science. Plücker, however, became interested in the phenomena that occur in the gas discharge as the pressure in the tube is progressively lowered. At sufficiently low pressures, he found evidence for the appearance of a beam of radiation emanating from the cathode or negative pole of the tube, striking the glass walls, and producing a greenish phosphorescence. Plücker is thus generally credited as the discoverer of what has come to be known as cathode rays. The study of these rays was destined to produce results of the greatest importance to the development of physics, the discovery of X rays by Wilhelm Roentgen and, from the investigations of J. J. Thomson, the discovery of the electron. These future developments opened the doors to the wholly unsuspected world of particles smaller than atoms, of highly penetrating radiations, and to the quantum nature of radiation.

Such prospects were by no means apparent in the early days of the discovery. Cathode rays were difficult to observe because it was difficult to produce the low pressures necessary to study them. The nature of the radiation was obscure since no information on the fundamental (particle) nature of electricity existed. After

Plücker's original paper, other researches were carried out by Hittorf and Goldstein. Perhaps the most important results achieved before 1878 were the discoveries that the unknown rays were emitted approximately at right angles to the cathode surface, that they traveled in straight lines away from the cathode after emission, and that they did not behave in the same manner as light, that is, the rays emitted normal to the cathode surface did not cross. Thus, a small object placed near an extended plane cathode produced a shadow. Had the cathode surface been a light source, the rays from each point on the surface would have been emitted in all directions and the outer portions of the light-emitting surface would have produced rays crossing beyond the small object, thus eliminating any shadow on a more distant screen.

This was the situation when Sir William Crookes began his cathode-ray researches, which were reported in his famous paper, "On the Illumination of Lines of Electrical Pressure and the Trajectory of Molecules," published in the *Philosophical Transactions* in 1879. This paper shows Crookes's genius for getting to the heart of a physical puzzle by very direct methods and his flair for clear-headed analysis of what was then a shadowy, amorphous, and ill-defined problem. His results were definitive, and they formed a solid base on which Thomson was able to proceed to one of the great discoveries of physics, the isolation of the electron.

But Crookes's reputation was not established by his paper on cathode rays. He had already established his position in science through his discovery by spectroscopic methods of the element thallium in 1861 at the age of 29. At that time he was already a free-lance scientist, inventor, and editor of a weekly journal, *Chemical News*, which he founded in 1859 and carried on until his death in 1919. He had the rare ability of organizing his work so that he could be effective in invention, editing, business affairs, and pure research at the same time. Personally, he was confident, optimistic, and even-tempered, devoted in his family life and considerate in his relations with others. In company he did not push himself, and he tended to be retiring, so that, as Sir Oliver Lodge remarked, his personality was not

William Crookes (1832–1919)

especially impressive. Photographs taken in his maturity suggest an unusually active and inquiring mind; the eye is particularly drawn to his long waxed mustache— a manifestation of an inner flamboyance uncommon among his fellow English scientists.

Like most scientists, Crookes did not come from a scientific background. His father, Joseph, was a London tailor whose Regent Street shop prospered enough to support, in turn, two wives and no fewer than 21 children. Five of these were by his first marriage and 16 by his second to a Northamptonshire woman, Mary Scott. William, the first child of this second marriage, was born June 17, 1832. While the home life of this large family appears to have been happy, young Crookes's intellectual stimulation seems to have come from his uncle, who conducted a thriving bookshop next store. Of these formative years, Crookes later wrote: "From my earliest recollections I was always trying experiments and reading any book of science I could find. A little older, I fitted up a cupboard as a sort of laboratory and caused much annoyance and trouble in the house by generating smells and destroying furniture. I don't suppose any of my family even knew the meaning of the word 'science.'"

In spite of unsettled economic conditions in England in the late 1830s and early 1840s, there was a growing feeling that technical education must be fostered to improve the industrial growth of the country. This conviction led to the founding in 1845 of the Royal College of Chemistry in London. Although Crookes's early schooling had been irregular, he entered the college in 1848 at the age of 16 and did well enough to earn a scholarship for the following year. In 1850 he was appointed junior assistant to the youthful and inspiring August Wilhelm Hofmann, who had been brought from Germany as professor of chemistry. At the end of the year, Crookes published his first paper, "On the Selenocyanides," an investigation marked by an unusual maturity in method, content, and style. He continued at the college until 1854, interesting himself in photography and the infant science of spectroscopy. During this period he also took time to attend Faraday's lectures at the Royal Institution and to acquire from them an enthusiasm for physics.

It is interesting to see how his original research in selenium compounds, when he was only 19, started a long chain of connected investigations that occupied most of his efforts in pure research. While examining the optical spectrum of impure selenium in 1861, he was struck by the presence of a bright green line which, on investigation, proved to stem from an impurity in the selenium. He very quickly showed that it originated in a previously unknown chemical element; wishing to indicate its color, he dubbed it thallium, from the Greek *thallus*, a budding twig.

This research secured his position in the scientific world and his election to the Royal Society in 1863. About 10 years later, he returned to the thallium research in an attempt to carry out an accurate determination of the atomic weight of the element. In doing so, his attention was directed to some puzzling discrepancies that appeared in the weighings. With customary thoroughness, Crookes found that these discrepancies originated in conditions affecting the scale—the air pressure in the enclosed balance and the heat radiation in the case. The disturbance was greatest at air pressures intermediate between the highest and lowest at which the weighings were conducted. A separate investigation of the relation between air pressure and heat radiation on an experimental enclosed vane led to the development of the radiometer, an instrument whose appearance caused a sensation in England and on the Continent. The radiometer, often seen in opticians' and jewelers' display windows, consists of a set of thin vanes mounted on spokes so as to form a mill rotating about a vertical axis. The vanes of the mill are blackened on one side and silvered on the other. When this assembly is mounted inside a partially evacuated glass bulb and is exposed to sunlight, the vanes will rotate in a direction away from the black surface. For this work Crookes was awarded the Royal Society's Royal Medal in 1875, although it was not until a year later that the action of the radiometer was explained. Sir Arthur Schuster at Manchester showed it to be produced by the free path of the molecules of the residual gas in the enclosure.

As a next step, Crookes conducted experiments with the radiometer vanes electrically charged. These experiments then led him to the use of cathode rays and to the research that is the subject of this sketch. The work was reported to the Royal Society on November 30, 1878. He was then 46 years old. These experiments represent the zenith of his scientific work and form the basis for his Bakerian lecture early in the following year. This lecture called the attention of the scientific world to the electrified-particle nature of the cathode rays.

Crookes's work did much to set what came to be called the British point of view on the rays. In discussing his experiments Crookes spoke very often of "projected molecules." He pointed out, on the basis of the newly developed kinetic theory, that in these experiments the trajectories are those of single particles. In concluding the lecture he ventured a sweeping prophecy from vistas only glimpsed in the investigations—that "the phenomena in these exhausted tubes reveal to physical science a new world." He could not possibly have guessed how right his judgment would turn out to be.

Seventeen years later, Roentgen discovered X rays emanating from the simplest type of Crookes tube, as the highly evacuated cathode ray discharge tube came to be known. How was it, then, that Crookes, with his keen eye and passion for thorough investigation missed one of the great discoveries that presaged 20th-century physics? Had he no clues? If we may believe a long-standing rumor, the answer seems to have been in the affirmative. It is said that Crookes had complained to the Ilford Company from time to time during his experimentation with cathode rays that batches of photographic plates sold to him were defective because, on development, they proved to be fogged. This is just the effect X rays from the discharge tubes would have had on plates kept in drawers nearby! Whether the rumor is true or only a calumny we do not know, for Crookes never alluded publicly to any such phenomenon.

Following his cathode ray investigations, Crookes returned to the study of the physical properties of gases. He showed by a precise experiment that Maxwell's kinetic theory prediction—that gas viscosity is independent of the pressure—was completely valid.

No one who has engaged in physical research, and understands all its frustrations and disappointments, could end the catalogue of these elegant investigations without a tribute to Crookes's assistant, Charles H. Gimingham. It was Gimingham's superb skill and devotion that contributed much to the success of Crookes's work during those exciting years. For his researches Crookes was awarded almost all the scientific honors that his native country could bestow: the Davy medal of the Royal Society, 1888; presidency of the Institution of Electrical Engineers, 1890; a knighthood in 1897; presidency of the British Association for the Advancement of

Science, 1898; the Copley medal of the Royal Society, 1904; an honorary degree from Cambridge University in 1908; the Order of Merit from King Edward VII in 1910; and, finally, the presidency of the Royal Society, 1913–1915.

In this brief biography nothing has been said of Crookes's interest in psychic phenomena and spiritualism. It happened in 1867, early in Crookes's career, that a younger brother Philip, to whom he was especially devoted, died as a result of yellow fever contracted in Havana. A fellow member of the Royal Society, who later became a well-known communications electrical engineer, Cromwell F. Varley, persuaded Crookes to try to communicate with his dead brother by spiritualist methods. In this way Crookes was introduced to beliefs that he held throughout his life and to phenomena that he tried to make understandable and to rationalize scientifically. He also made efforts to interest others in these phenomena as well as effects produced by various mediums, including the celebrated Daniel Home. The result of this activity was to arouse a storm of criticism and ridicule and eventually the active enmity of some influential members of the scientific community. By 1874 Crookes came to realize the hopelessness of bringing open-minded scientific opinion to spiritualist investigations and, as well, the threat they posed to his scientific career. He therefore dropped these activities, although it is clear that he never abandoned his beliefs. Almost a quarter of a century later, in 1896, he accepted the presidency of the Society for Psychical Research, and in the following year, on becoming president of the British Association for the Advancement of Science, he reiterated his belief in the reality of psychic phenomena. It is unfortunate for his reputation that in the popular mind he has been associated more with spiritualism than with science.

It must not be assumed that Crookes's later research was unfruitful. The discovery of radioactivity by Henri Becquerel in 1897 and of radium shortly after by the Curies renewed his interest. Although he was then in his mid-60s, he began a vigorous chemical study of uranium salts to determine their radioactivity. He found that if a uranium salt was precipitated from a solution by the addition of ammonium carbonate and then treated with excess reagent, a small amount of highly radioactive residue was obtained. The active substance in the residue, the first decay product of uranium, was called uranium X (now identified as a thorium isotope and protactinium). This was followed in 1903 by the discovery that single alpha particles, when projected on a zinc sulfide screen, produce starlike scintillations that are clearly visible with the aid of a suitable eyepiece. This device came to be known as the spinthariscope. The discovery that individual alpha particles may be counted by this scintillation technique had consequences of the first importance; in the hands of Ernest Rutherford and his school it led in the period 1909–1913 to the discovery that the atom has most of its mass concentrated in a very small central nucleus, and, in 1919, to the transmutation of elements.

During World War I, although in his 80s, Crookes served as a member of the scientific advisory board to the admiralty. He also continued his own research, forwarding his last paper on scandium and its spectrum to the Royal Society in December 1918. He died four months later at the age of 87 on April 4, 1919, at his home in London. As J. J. Thomson noted in his Royal Society anniversary address of that year: "[Crookes] had a singularly independent, original and courageous mind, he looked at things in his own way and was not afraid of expressing views very different from those previously considered orthodox. In not a few cases . . . time has shown that his ideas were much more in accordance with those which are now accepted than those commonly held by his contemporaries."

When Crookes began his experiments on cathode rays, they were regarded as very mysterious. The particle nature of the electric current was unknown and therefore could not shed light on the nature of the rays proceeding from the cathode of the discharge tube. To this was added the mystery of the origin of the phosphorescence of the glass brought about by the action of the rays.

Crookes's paper has the advantage of attacking the problem on the basis of a known phenomenon—that the rays are composed of electrified particles projected at high speed from the cathode of the tube. He backed up this point of view first by showing that in discharge tubes at high exhaustions the cathode rays follow straight lines and do not bend around corners. Furthermore, flat cathodes produce sharper shadows than point cathodes, a result to be expected from projected, mutually repelling particles, but not from light rays. If a mill is placed in the discharge tube it is observed to rotate as would be required by the impact of material particles, and a magnet in the vicinity of the tube causes the rays to deflect as charged material particles would. Focusing of the rays produces strong heating effects, agreeing with the material particle hypothesis. The green phosphorescence of the glass occurs because particles from the cathode strike the glass walls.

Crookes concluded the paper by making use of the kinetic theory of gases whereby each of the particles in the discharge might be expected to move without collision across the tube. Under such circumstances Crookes speculated that the particles might be in an "ultra-gaseous" state with quite different properties than they exhibit at higher pressures: "The phenomena in these exhausted tubes reveal to physical science a new world—a world where matter may exist in a fourth state [i.e., not solid, liquid, or gas], where the corpuscular theory of light may be true, and where light does not always move in straight lines, but where we can never enter, and with which we must be content to observe and experiment from the outside."

Crookes spoke with greater prescience than he realized, for "the phenomena in these exhausted tubes" did indeed reveal to science a new world. But the revelation was not immediately forthcoming. It took the researches of J. J. Thomson in 1897 to show that the new world which the tubes revealed was the subatomic world never before suspected. The electron was the first subatomic particle discovered, but it was soon followed by a host of others and an entirely new view of the constitution of the material world.

A Remarkable Regularity in the Hydrogen Spectrum: Johann Jacob Balmer (1825–1898)

In the year 1885 a remarkable paper by an obscure Swiss teacher of physics, Johann Jacob Balmer, appeared in the *Annalen der Physik und Chemie*; it was to have a profound influence on the development of atomic physics. Balmer's achievement in obtaining a mathematical formula that correctly gives the wavelengths of the spectral lines of the hydrogen atom was all the more remarkable since he based his calculation on only four lines in the visible spectrum.

Using the very accurate measurements of Ångström for these four hydrogen lines, Balmer obtained a simple arithmetic formula from which the wavelengths of

Johann Jacob Balmer (1825–1898)

the lines can be calculated. The essence of Balmer's discovery is that the wavelength of any line in the spectrum of hydrogen can be obtained by multiplying a certain numerical factor, which Balmer called "the fundamental number of hydrogen," by a series of fractions. His formula is given by $\lambda = b[m^2/(m^2 - n^2)]$ where b is the fundamental number and λ is the wavelength of any hydrogen line in Ångström units (10^{-8} cm). A given series of lines is determined by the choice of the integer n, and the individual lines in the series by the values of m. Thus, if $n = 2$ and $m = 3, 4, 5$, etc., the formula gives the series of lines in the visible hydrogen spectrum. This series is now known as the Balmer series. Other series were found later: for example, for $n = 1$ and $m = 2, 3, 4$, etc. (the Lyman series); for $n = 3$ and $m = 4, 5, 6$, etc. (the Paschen series); $n = 4$, $m = 5, 6$, etc. (the Brackett series), and $n = 5$, $m = 6, 7$, etc. (the Pfund series). The physical reason for the existence of these series was first clarified by Niels Bohr's model of the hydrogen atom.

Balmer's genius and the importance of his contribution lay not so much in deducing an empirical formula that gives the correct values of wavelengths of the lines that were then known but in generalizing the formula and showing that it can be expressed algebraically in terms of the squares of integers. Balmer pointed out that there must be other series of lines in the spectrum of hydrogen whose wavelengths can also be derived from his algebraic formula, although he did not attempt to find them.

Balmer's discovery gave a great impetus to spectral theory. All subsequent investigations into the origin of atomic spectra began with the fundamental assumption that the wavelengths of the spectral lines of all atoms can be represented by simple numerical relationships involving the squares of integers. In particular, Ritz, in the year 1908, introduced what is now known as the Ritz combination principle, which is a generalization of Balmer's discovery. The essential feature of this principle, which proved of great heuristic value, is the following: For any atom there exists a characteristic sequence of numerical terms (the numerical values of these terms vary from one type of atom to the other) such that the frequency (the speed of light divided by the wavelength) of any line in the spectrum of this atom is equal to the difference of two of the terms of the sequence. This means that the frequency of any line in the spectrum of the atom can be expressed in terms (differences) of the frequencies of other lines in the spectrum.

Ritz then went on to show that the fundamental sequence of numerical terms for any atom is obtained by dividing a certain numerical constant (which varies from one atom type to the next) by the squares of the integers taken in turn. This remarkable sequence, which stemmed from the basic work of Balmer, was of great importance in finally leading to the Bohr model of the atom. In the case of the hydrogen spectrum, the Ritz combination principle is particularly simple since the frequency of every line radiated by hydrogen is equal to the difference of two Ritz terms.

Niels Bohr, who was born in 1885, the very year that Balmer published his fundamental paper, was undoubtedly influenced by Balmer's numerical scheme, for it was clear to him that if the wavelengths (or the frequencies) could be represented in terms of integers, some type of quantum phenomenon is involved in the emission of radiation by an atom. With this as a starting point it was not difficult for a man of Bohr's genius to see how the concept of the photon could be used to introduce integers (the quantum numbers of the Bohr orbits) into the model of the atom.

Johann Jacob Balmer was a devoted Pythagorean from his boyhood and was convinced that the explanation of the mysteries of the universe lay in the correlation of observed phenomena with the appropriate combination of integers. Thus, any series of distinct events, such as the sharp emission lines in the spectrum of hydrogen, would have triggered his imagination and started him on a search for a Pythagorean relationship.

But Balmer was more than a mere numerologist, for he had an excellent educational background. He was born on May 1, 1825, in the small Swiss town of Lausen, Baselland, where he spent his early years and went to school. He later studied mathematics at the universities of Karlsruhe, Berlin, and Basel, and obtained his doctorate in mathematics at the latter university in 1869.

Balmer was interested not only in the formal aspects of mathematics but also in its philosophical implications and its relationships to the laws of nature. In a paper published in 1868, he dealt with the relationship of physical research to systems of world philosophy. Such problems undoubtedly stimulated his interest greatly in spectroscopy. In addition, he was a reasonably good mathematician and published papers on projective geometry.

Unlike most physicists at that time, Balmer did not carry out his research at a university because he never was appointed to a professorial chair. Although he began his academic career as a privatdozent (university lecturer)—the first rung on the academic ladder—at the University of Basel, he left the university to become a teacher at a girls' secondary school and did his famous work on hydrogen spectral lines there. This work was described in a series of papers that appeared in a Basel scientific journal and in the *Annalen der Physik und Chemie* in 1884, 1885, and 1897. Balmer died in 1897, just three years before Planck discovered the quantum of action, which was ultimately to justify Balmer's faith in the explanation of spectroscopy in terms of integers.

The Luminiferous Ether Receives a Mortal Blow:
Albert Michelson (1852–1931) and Edward Morley (1838–1923)

The last quarter of the 19th century was marked by what at that time seemed to be the final triumph of the undulatory theory of light over the corpuscular theory. With the beautiful experiments of Thomas Young and Augustin Fresnel on the interference and diffraction of light, and their brilliant theoretical analysis based on the wave theory, it appeared almost a certainty that a beam of light consisted of a periodic disturbance that was propagated through an all-pervading medium in the same way that waves are propagated across the surface of water. The nature of this light-propagating medium was not at all understood. But that some kind of medium was required to propagate light waves seemed self-evident. This idea was reinforced by Maxwell's discovery of the electromagnetic theory of light and his derivation of the wave equation for electromagnetic fields. It seemed obvious that the all-pervading medium invented by René Descartes, which Maxwell called the ether, was just what was required to produce and propagate electric and magnetic displacements in space. However, it soon became evident that one would have to assign very strange and, indeed, contradictory properties to the ether to account for astronomical observations.

To begin with, it was clear that the ether had to be enormously rigid in order to propagate a disturbance with the speed of light. On the other hand, it was obvious that the ether would have to permeate all material, since scientists knew that light moves through transparent solids, although at a speed slower than its speed in a vacuum. A question that naturally arose concerning the ether was what happened to it inside a body if the body was set in motion. This question was partly answered by Fresnel, who argued that in transparent bodies, at least, the ether is partly dragged along with the body when the body moves. This idea was verified by the experiments of Fizeau, who measured the speed of light in moving water.

However, this idea of the ether drag came into conflict with the aberration of starlight. If a star is observed along a line at right angles to the direction of the

Albert Michelson (1852–1931)

orbital motion from the earth, the light from the star comes to us from a slightly different direction, as though the star were somewhat displaced in the direction of the earth's orbital motion. The only way this can be explained, assuming that an ether exists, is to suppose that the earth does not drag the ether outside it. In other words, although we must suppose that the ether permeating a transparent body is partly dragged along when this body moves, we must not allow any of this drag to be communicated to the ether outside the body if we are to understand aberration. We must suppose, if an ether exists, that no matter how a body moves through this ether, the ether itself always remains at rest in the absolute space that Newton had introduced. In absolute space, it was assumed, the ether itself must be an absolute frame of reference relative to which the motions of bodies such as the earth could be measured. It was to test this particular point that Michelson and Morley , in 1887, devised their famous experiment to see whether a beam of light, moving at right angles to the direction of the motion of the earth, travels at a speed different from that of a beam moving parallel to the earth's motion. This experiment was therefore designed to measure the absolute motion of the earth through space, that is, relative to an absolute ether at rest in absolute space.

Albert Abraham Michelson was particularly qualified by his natural scientific ability and his training in optics to design and perform this piece of research, which was undoubtedly the most precise scientific experiment conducted up to that time. Born in Strelno, Germany, on December 19, 1852, he was brought by immigrant parents to the United States at the age of two. His father, a storekeeper, eventually settled the family in the mining town of Virginia City, Nevada, but young Albert was sent to San Francisco for his high school education. Lacking means for college, he attempted in June 1869 to obtain an appointment to the U.S. Naval Academy, but he was unsuccessful. It was at this juncture that he showed the quality of determination that was so characteristic of his later life. Deciding that he would somehow obtain a presidential appointment, he set out for Washington by any means available, for the first transcontinental railroad had just completed laying its track a month before. Once in Washington, since he had no way to obtain an interview with President

Edward Morley (1838–1923)

Grant, he waited on the White House steps until Grant appeared for his daily stroll. The result was that young Michelson was sent to Annapolis and admitted, apparently illegally since the ten presidential appointments allowed by law were all filled. At the academy Michelson excelled in science and mathematics. Graduating in 1873 at the age of 20, he served two years in the Caribbean before returning to the Naval Academy as an instructor in physics and chemistry.

At Annapolis he became interested in optics and began the research that brought him to scientific notice and which he continued at intervals throughout his life—his measurement of the velocity of light. His first apparatus represented an improvement of Foucault's rotating mirror method and his finding of 186,508 miles per second, in 1879, attracted the attention of Simon Newcomb, then president of the American Association for the Advancement of Science and an officer in the Navy Corps of Professors. Transferred to Washington, Michelson worked with Newcomb during the next year on an improved apparatus for velocity of light measurements.

Recognizing the need for further training in physics, Michelson obtained a leave of absence from the Navy and in 1880 went to Europe with his wife and children. He was abroad for two years studying at Berlin and Heidelberg in Germany and at the Collège de France and the École Polytechnique in Paris. It was while he was in Berlin that he made the first model of his famous interferometer and used it in an attempt to measure the earth's motion through the supposed all-pervading luminiferous ether. The null result aroused controversy throughout the scientific world. During this period in Europe he resigned from the Navy and accepted a position at the Case School of Applied Science in Cleveland but postponed his return to complete his work abroad. When he returned in 1882 he was ready for the experiment that established his position in the annals of physics.

Shortly after Michelson's arrival at Case, he met Edward W. Morley, professor of chemistry at Western Reserve University, which adjoined the Case campus. They soon became devoted friends and collaborators. The pair presented an interesting contrast. Morley, deeply religious, had been trained for the ministry but, finding no pulpit, had accepted a professorship of chemistry for which he was largely self-trained. He was energetic and talkative, and his long hair and expansive red mustache completed what can only be called a casual attitude toward dress. Michelson, on the other hand, perhaps because of his naval training, was always immaculately turned out. He had a reserved, dignified, rather handsome appearance and kept himself in trim by appropriate exercise. He was a strong individualist with an intense spirit, a strong drive, and a high opinion of his work and his worth. When crossed he could be blunt and imperious. Perhaps in keeping with his reserve he had an artistic side which found expression in music and painting. In experimental physics he had a passion for precision so that his work commanded high confidence in the scientific world. This passion was shared by Morley and made their mutual endeavors highly successful.

What is now called the Michelson-Morley experiment was begun about 1886, although previous experiments had been made by the team on the velocity of light in moving water. The idea of the experiment was essentially simple. Using Michelson's interferometer, a beam of light would be split, with one part sent out perpendicular to the earth's motion in space, to a fixed mirror and back, and the other parallel to the earth's motion, to a fixed mirror at the same distance and back. If the speed of light depended on the motion of the earth through the ether, then the two beams of light should not return at the same time. If they did not return at the same time, then the wave forms of the returning beams would not fit exactly, that is, be in the same phase, and this lack of phase would show in a displacement of the interference or fringe pattern in the field of view of the interferometer.

The difference in the time of arrival of the two light beams can be illustrated by the case of two equally fast swimmers who are sent the same distance and back again in a stream of water, with one swimmer going up and down the stream, and the other going across the stream and back. It can be shown algebraically that the swimmer going across stream will come back to the starting point before the swimmer moving parallel to the stream. This is what Michelson and Morley expected to find for their two beams.

To test this assumption, the interferometer apparatus was so arranged that it could be rotated in all directions, since the direction of the motion of the earth through space was not known. The results of the experiment gave no evidence at all that the earth was moving relative to the ether, since the two beams returned to the image plane of the instrument (that is, to the point at which they were focused by the observing telescope) at exactly the same time (within the limits of experimental error). As the two experimenters indicated on the last page of their paper, the actual measured difference in the times of return of the two beams "was certainly less than the twentieth part" of the displacement to be expected from the theoretical analysis of the problem.

In the last paragraph of their paper, Michelson and Morley described the crisis faced by classical physics because of the negative result of their experiment. Even if one retained the concept of the ether (which seemed to be required by the Maxwell theory), the Michelson and Morley experiment proved conclusively that there could be no measurable motion of the earth relative to it. On the other hand, the explanation of the aberration of light, as given by Fresnel on the basis of the wave theory of light, required the ether to be at absolute rest, relative to which the motion of the earth should be measurable. These results were in direct contradiction to one another.

The legacy of the Michelson-Morley experiment to atomic theory was tremendous, if indirect. It was the negative result of the experiment that, in part, led Einstein to one of the fundamental ideas upon which the theory of relativity rests, namely, that the speed of light is the same for all observers regardless of how they may be moving. With this assumption, it is easy to see why the two beams return to their starting point at the same time; moreover, it is no longer necessary to posit the existence of a troublesome ether. However, before the theory of relativity was introduced, with its revolutionary concept of a constant speed of light, another attempt was made to explain the Michelson and Morley experiment. This ad hoc explanation was that the distance between the two mirrors on the line parallel to the earth's motion had decreased (because of the earth's motion) by an amount that was just big enough to enable the parallel moving beam to return at exactly the same time as the beam moving transverse to the earth's motion. This is the famous Lorentz-Fitzgerald contraction hypothesis, which Lorentz derived by picturing matter as consisting of small charged spheres (electrons) that contract in the direction of motion because of the electrical forces acting on them. This theory was not satisfactory because it tried to explain observable phenomena in terms of invisible forces and ad hoc hypotheses which could not be tested experimentally. Einstein's theory removed this difficulty.

The Michelson-Morley experiment was concluded when Michelson was 34 years old. The work placed him in the forefront of American physics and he expected that the Case trustees would recognize this fact by substantially increased support. When the support did not materialize, Michelson decided in 1889 to leave Case and accept the offer made by the new Clark University in Worcester, Massachusetts. The dissolution of the Michelson-Morley partnership was a great disappointment to Morley, who despite other offers decided to remain at Western Reserve University. Michelson's stay at Clark University was of short duration, for the University of Chicago, then being created, offered him in 1892 not only a chair but the direction of the new Ryerson Physical Laboratory. Here he was able to carry on his work on a basis which, at last, he found acceptable. Michelson remained at Chicago until his retirement at the age of 77. During that period he received various honors including the presidency of the American Physical Society (1900), election as a foreign member of the Royal Society (1902), and the receipt of the Society's highest award, the Copley medal (1907). In the same year he was awarded the Nobel Prize "for his optical precision instruments and for the spectroscopic and meteorological investigations made with them." He was the first American scientist so honored. Michelson's outstanding researches included a new definition of the primary standard of length in terms of the red cadmium wavelength (6,348 Å), the production of precision-ruled diffraction gratings and the echelon grating, a study of the rigidity of the earth, the determination of the diameter of Betelgeuse in the constellation of Orion by interferometric methods and, finally, a redetermination of the velocity of light carried out with the stations at Mt. Wilson and Mt. San Antonio in California. During the course of this work, having retired from the University of Chicago to live in Pasadena, he died on May 9, 1931.

6 *The Beginnings of Modern Atomic Physics*

If the world were good for nothing else, it is a fine subject for speculation.

—William Hazlitt, *Characteristics*

The Discovery of X Rays: Wilhelm Conrad Roentgen (1845–1923)

The discovery of X rays was made by Wilhelm Conrad Roentgen at the University of Würzburg, Germany, on November 2, 1895. From the beginning it was clear that a momentous and revolutionary find had been made but no public announcement was given until about seven weeks later when Roentgen submitted a preliminary paper entitled "On a New Kind of Rays" to the Physico-Medical Society of Würzburg on December 28. It was printed immediately in the *Sitzungsberichte* (proceedings) of the society for 1895. As might be supposed, these proceedings were not widely circulated or read, so Roentgen, on receiving reprints, mailed them to a group of physicists and medical men in Europe, England, and America. It might have been expected that he would have sent his announcement to the *Annalen der Physik*, the leading German physical journal which had a worldwide circulation, but he did not. A reasonable explanation may be that Roentgen wanted not only to be very sure of his discovery but to establish the fundamental properties of the rays before others could do so. As time went on, however, the momentous character of the discovery became ever more apparent, and the necessity for communicating it the more pressing. Realizing that immediate publication could be had through a local scientific journal, it seems probable that he took advantage of this opportunity to establish his claim. Delay in the announcement and caution in publication were characteristic of Roentgen. Zehnder, one of his pupils as well as one of his biographers, remarks that once Roentgen "set himself a problem he always worked at it quietly and secretly without allowing anybody a glimpse into his method of working and thinking."

Once the news was made public, it produced a sensation throughout the civilized world. Popular interest was greatly heightened by reports that X-ray pictures could be taken through walls and opaque screens. This caused a widespread apprehension that all privacy might be destroyed. While this penetrating property of the rays was correctly described, privacy was certainly not endangered owing to the cumbersome equipment necessary and the fact that the pictures are not taken by reflected radiation. The medical uses of the rays for photographing through the flesh and in detailing the bony structure of the body were immediately recognized.

Wilhelm Conrad Roentgen (1845–1923)

Other uses both possible and fanciful appeared in the newspapers almost daily. Roentgen's second paper on X rays, dated March 9, 1896, also appeared in the *Würzburg Proceedings*.

Roentgen's discovery of X rays must, in most respects, be considered a lucky accident, but it should be remembered that chance favors the prepared mind. He had begun only a few weeks before to repeat some cathode ray experiments that had been reported by Philipp Lenard. In an interview in his laboratory Roentgen described the circumstances as follows.

> I was working with a Crookes tube covered by a shield of black cardboard. A piece of barium platino-cyanide paper lay on the bench there. I had been passing a current through the tube and I noticed a peculiar black line across the paper. . . . The effect was one which could only be produced, in ordinary parlance, by the passage of light. No light could come from the tube, because the shield which covered it was impervious to any light known. . . . I assumed that the effect must have come from the tube, since its character indicated that it could come from nowhere else. I tested it. In a few minutes there was no doubt about it. (Jauncey, 360)

These events occurred when Roentgen was 51 years old. The history of physics shows that most of the important discoveries have been made when their discoverers were relatively young. Whether age had any bearing on Roentgen's subsequent work is difficult to say, but it is interesting to note that he published only three papers dealing with the rays he discovered and these were written shortly after the event. Although he had published over 50 papers before the three on X rays, he published only three more between the latter part of 1897 and his death in 1923, and none of these was concerned with his rays.

Roentgen's first two communications about the rays, which had appeared in the *Würzburg Sitzungsberichte*, described the circumstances of the discovery and then several salient properties of the rays. First, he noted that all substances are more or less transparent to the rays but the denser substances, especially the heavy metals such as platinum and lead, are less so, although density is not the only factor to be considered. Next, detectors of X rays were considered; besides fluorescence, the sensitivity of both photographic plates and films was noted. (It may be of interest to remind the reader at this juncture of Sir William Crookes's complaint to the Ilford Company regarding their photographic plates during the period when he was experimenting with cathode rays!) Roentgen went on to say that X rays are not ultraviolet rays because they cannot be refracted by prisms or lenses made of a variety of materials. (Later experiments were to show that a slight refraction can be observed, and that X rays exhibit the properties of light waves of very short wavelengths.) Also X rays cannot be cathode rays because they cannot be deflected by a magnet, but they originate where cathode rays strike. Shadow pictures and pinhole photographs confirm the ray nature of X rays but interference phenomena can not be produced. The first communication closed with the suggestion that X rays are propagated as longitudinal vibrations in the ether, a guess that proved to be erroneous.

The second communication began with the observation that X rays make air conducting and discharge electrified bodies. Electrified bodies in dry hydrogen were also the subject of experiments with similar results. In highly exhausted spaces the discharge of a charged body proceeded much more slowly. All this, of course, is in complete accord with subsequent research that showed that the passage of X rays through gases ionizes them, producing electrons from the gas atoms and leaving the residual atom, the ion, with a positive charge. As both charged particles are highly mobile, an electrified body attracts charges of the opposite sign and soon becomes discharged. The paper closed with some observations regarding the use of apparatus to generate X rays and a comparison of aluminum and tungsten as sources of X rays when they are struck with cathode rays.

Roentgen was born in Lennep, near Düsseldorf in the Rhineland on March 27, 1845, the only child of a German father and a Dutch mother. Most of his childhood was spent in Holland. He received his early education at the gymnasium at Utrecht where, toward the end of his course of studies, he ran into trouble. Being caught in a youthful prank, he rather too emphatically refused to implicate others concerned. As a result he was expelled from the gymnasium. This misfortune prevented his entrance to the Hochschule at Utrecht. Then he learned that it was possible to enter the Polytechnic at Zürich, Switzerland, without a certificate. In 1863 he was accepted as a student in machine construction at Zürich, but he was hardly an eager student. The lake with its surrounding mountains often drew him from the laboratory on lengthy walks.

During his three years of study, he attended lectures on mathematics and on the mathematical theory of heat, the latter subject being taught by Rudolph Clausius. In 1866 he obtained the diploma of mechanical engineer. But his interest was in experimental physics and he continued his studies at Zürich under Kundt, an experimental physicist. He obtained his doctorate in 1869. Along with Kundt he went to Würzburg, Bavaria, in 1870, where academic customs seemed to annoy him. In 1872 Kundt was called to Strasbourg and Roentgen accompanied him. The university had just been organized, and Roentgen was free from bothersome traditions. In 1874 he was made a privatdozent at Strasbourg, and all obstacles to his career were then removed. In 1875 he became professor of mathematics and physics at the Agricultural Academy of Hohenheim. In 1876 he returned to Strasbourg as professor of theoretical physics. In 1879 he was appointed professor of physics and director of the Physical Institute at Giessen. In 1888 he was called to the chair of physics at Würzburg, as successor to Kohlrausch, and some seven years later made his discovery of X rays.

In an obituary article in the September 1, 1923, issue of the *Physikalische Zeitschrift*, Walter Friedrich of Freiburg described Roentgen as primarily an experimental physicist. He was the same type of physicist as Albert Michelson, for his interest was in experiments requiring extreme precision. As noted earlier, he was reluctant to publish his results. Because he was so very careful, he never had to revise the experimental results reported in any of his papers. His attitude toward theoretical physics was similar to Michael Faraday's in that he liked to think in terms of mechanical models. At the time of his great discovery, he believed in the elastic-solid theory of the ether. The idea of longitudinal waves in the ether were not abhorrent to him.

Roentgen's work on X rays was completed in a period of about 18 months in the 51st and 52nd years of his life, after which his active interest returned to those fields in which he had previously been interested. He thereafter avoided any further exploration into the new realm. We remember the names of J. J. Thomson, W. H. and W. L. Bragg, Max von Laue, and others. Roentgen was in his 50s and 60s while these men were doing such brilliant work in X rays. The discovery of X rays may have come too late in Roentgen's life. Thomson was not quite 38 years old, and the others were even younger at the time of the discovery. The discovery might have been too iconoclastic for Roentgen's mind. There is quite a lot to be said for the view that physics is a science for the young because an open mind is a prime requirement for advance.

In 1900, Roentgen left Würzburg and became professor of experimental physics and director of the Physical Institute at Munich. A year later, he received the first Nobel Prize in physics. Friedrich says that, although Roentgen was aware of the practical value of his discovery, he held himself remote from the fast-developing technology of X rays. Nor was it consistent with his modest scientific nature to exploit his discovery in a monetary way. According to his biographer, as Roentgen became older and his reputation grew, he turned all the more from this publicity

Walter Friedrich (1883–)

back to the companionship of the world of nature. At Munich he spent much time in the mountain village of Weilheim, where he owned a cottage. He hunted game in the mountains surrounding his cottage. With his colleagues he was reserved almost to the point of unfriendliness, yet he invited his students to the cottage to chat and play ninepins.

His wife died after a long illness in 1919, and he retired from his chair at Munich in the spring of 1920. On July 22, 1920, his last article—on photoconductivity—was sent to the *Annalen*. This article appears in the January 1, 1921, issue. It was written when he was 75 and is one of his outstanding contributions to physics. After a short illness, he died on February 10, 1923.

The Discovery of Radioactivity: Antoine Henri Becquerel (1852–1908)

In the biography of Daniel Bernoulli, the rarity of outstanding scientific ability recurring in several generations in one family was noted. The Bernoullis have already been cited as an exception, and to them we must add the Becquerels. The earliest of this family to distinguish himself was Antoine César. His intellectual activity extended from 1819 to 1879 and he was honored with the Royal Society's Copley medal in 1837 for his pioneer work in electrochemistry. His son Alexander Edmund published works extending from 1839 to 1883. Alexander Edmund's son Antoine Henri, the subject of this article and the most illustrious of the name, produced his first scientific paper in 1875 and his last in 1908. To this unequaled roster we must add Henri's son Jean Becquerel, who distinguished the family name in physics until 1953.

The first three Becquerels all contributed to the study of phosphorescence, and fluorescence, and Henri was acutely aware of the continuity of this research in his family. Working in the same laboratory and with the same instruments as his father and grandfather, he keenly felt the necessity of honorably discharging the responsibility of his lineage. Thus, in speaking of his own work on radioactivity, of which he was the discoverer, he minimized his contribution by saying, "These discoveries are only the lineal descendants of those of my father and grandfather on phosphorescence, and without them my own discoveries would have been impossible."

Henri Becquerel was born in Paris, December 15, 1852. His early education was received at the Lycée Louis le Grand and he entered the École Polytechnique at the age of 19, in 1872. Two years later he became a pupil at the École des Ponts et Chaussées. He was made engineer in 1877 and subsequently was for ten years chief engineer. In 1876, at 23, he became demonstrator at the École Polytechnique and two years later aide-naturaliste at the Musée d'Histoire Naturelle. In both of these positions he was subordinate to his father Edmund. He succeeded his father as professor in each of these institutions: at the Musée in 1892, where not only his father but his grandfather, as well as Gay-Lussac, had occupied the chair, and at the École Polytechnique in 1895. It is noteworthy that Henri's son Jean also succeeded to the professorship at the Musée, being the fourth in line to hold the post.

In addition to his researches on phosphorescence, Henri Becquerel's early researches were concerned with the Faraday effect, that is, the rotation of the plane of polarization of linearly polarized light under the action of a magnetic field. Subsequently, he devoted his attention to spectroscopy and the absorption

Antoine Henri Becquerel (1852–1908)

spectrum of didymium and neodymium. Following Roentgen's announcement of the discovery of X rays, that were associated at the time with the appearance of phosphorescence on the glass walls of Crookes tubes, Becquerel, following a suggestion of Henri Poincaré, attacked with renewed vigor the researches that his father had made on phosphorescence with the use of uranium salts. A more fortunate choice in the search for penetrating radiations could hardly have been made. Becquerel began his work sharing the popular hunch that the penetrating X rays were associated with phosphorescence. Using the double sulfate of uranium and potassium he exposed this salt to the sun for several hours. When the photographic plate on which it was placed was developed, the plate showed the outline of the salt, indicating the presence of a penetrating radiation. These researches led to the discovery of radioactivity. For this discovery he received the Rumford medal of the Royal Society in 1900 and the Helmholtz medal in Berlin in 1901. He was made a foreign member of the Royal Society and of the academies in Berlin and Rome, as well as the National Academy of Sciences in Washington. Other honors included honorary degrees from the universities of Oxford and Cambridge. In 1903 he shared the Nobel Prize in physics with Pierre and Marie Curie. In 1908 he was elected president of the Académie des Sciences, an honor which he enjoyed only briefly. Accustomed to such strenuous exercise as mountain climbing and ocean swimming, he had no intimation of the heart attack from which he died on August 25, 1908, at the age of 55, while on holiday on the west coast of France.

The Discovery of the Electron: J. J. Thomson (1856–1940)

Between the philosophical speculations of Democritus and the chemical experiments of John Dalton stretch 2,000 years of atomic theory which envisioned the smallest units of the elements as the ultimate particles of the universe. Indeed, until the classic experiments of William Crookes, who glimpsed a "fourth state of matter" in his cathode ray tubes, atoms were held to be homogeneous and unshatterable. In the hands of Roentgen the magic of the Crookes tube became even more apparent

Joseph John Thomson (1856–1940)

with his discovery of the X rays generated in its walls. But it was not in the spirit of magic that Joseph John Thomson, Cavendish professor of experimental physics at Cambridge University, set out to use the tubes in his study of electrical conduction in gases. Less than two years later his discovery of the electron swept away the whole fabric of atomic thought, threw a flood of light into the science of electricity, and shortly led to the invention of the electron tube, which as a circuit element has produced marvels then hardly dreamed of in the wildest fantasies of fiction.

It is interesting, but rather difficult, to understand Thomson's thoughts on his discovery; he tells us in his *Recollections and Reflections*, written a few years before his death: "At first there were very few who believed in the existence of these bodies smaller than atoms. . . . I had myself come to this explanation of my experiments with great reluctance." Why was he reluctant to be the first to peer into the world beyond the atom? We do not know, for he never seems to have explained the reasons for this remark. Thomson was in many ways a very different man than many of the great scientists. There was nothing of the remote academician, secluded seer, or eccentric about him. He came from a home in moderate circumstances and his education was subsidized almost from the beginning. He always rubbed elbows with the world and maintained, in the best sense, a worldly, commonsense, and warmly human point of view. Looking back on a long, influential, and eminently successful life, he had the wisdom and humility to observe: "I have had good parents, good teachers, good colleagues, good pupils, good friends, great opportunities, good luck and good health . . . had it not been for the sacrifices made by my mother, I could not have completed the course [at Owens College] or come to Cambridge. . . . The events that determined my career . . . coming to Trinity and all that it has meant to me—were sheer accidents." To this very human acknowledgment might be added the opinion of Lord Rayleigh, one of his biographers, who said of him that much of his success as director of the Cavendish Laboratory was due to his concern for his students and interest in their achievements. Thomson's biography, based on his own *Recollections and Reflections* with some additions from Lord Rayleigh's writings, follows.

Joseph John Thomson was born in a suburb of Manchester on December 18, 1856, the son of Joseph James and Emma Swindells Thomson. His father, a man of Scottish descent, carried on a family business as a bookseller and publisher. When ready for regular instruction, young Thomson was sent to a small, local private school according to the prevailing custom. His father had not intended him for university training but had wanted to apprentice him to a firm of locomotive builders with the idea that he should become what we now call a mechanical engineer. Engineering in those days was learned less by formal schooling than by the apprentice method, and firms accepting apprentices generally required a substantial fee for the privilege. When young Thomson was ready for apprenticeship at 14, it was found that the firm had a long waiting list. In attempting to decide what should be done, his father discussed the situation with a friend, who suggested that the youngster, while waiting, might very profitably attend the local college. As Thomson points out, this decision was one of the great turning points in his life, for his father accepted the advice.

Owens College, where Thomson enrolled, was a unique institution in England, having been founded by a Manchester merchant for instruction in subjects taught in the English universities but requiring no religious tests. The staff consisted of an unusual number of very able men, among whom were Thomas Barker, professor of mathematics, former senior wrangler and fellow of Trinity College, Cambridge; Balfour Stewart, whose teachings and writings inspired many students in physics; Osborne Reynolds, whose fame in engineering is perpetuated by the Reynolds number criterions in hydrodynamics; H. E. Roscoe, one of the outstanding English chemists; Stanley Jevons, author of a widely used book on logic; Adolphus Ware, professor of history and English and later vice-chancellor of Cambridge University; and, finally, James Bryce, professor of law, who later became Lord Bryce, ambassador to Washington, great friend of the United States and the author of the celebrated work, *The American Commonwealth*.

It was natural that a young student directed toward engineering should be most influenced by mathematics and physics. Young Thomson excelled in both of these and achieved an excellent record. At the end of his second year, the early death of his father at the age of 39 completely altered the family outlook, and the large fee required to commence his intended apprenticeship was now out of the question. It was therefore decided that he should finish his last year at the college and obtain a certificate in engineering. But even this ambition required the assistance of a scholarship, which fortunately the college awarded. By the time the year had elapsed, Professor Barker's esteem for his pupil had increased so much that he persuaded young Thomson to stay on after receiving his certificate and continue with mathematics and physics so that he might compete for an entrance scholarship at Trinity College, Cambridge. This was an idea that Thomson had never considered but one that he eagerly adopted.

His first try the following year was unsuccessful, but the next year, 1876, when 19, he was admitted to Trinity. As an undergraduate he was a student of the great mathematical coach E. J. Routh, and of such distinguished professors as Cayley, Adams, and Stokes. He took the tripos examinations in January of 1880 and, like James Clerk Maxwell before him, came out second wrangler. Joseph Larmor, who subsequently attained eminence in physics and the Lucasian professorship at Cambridge, was first. Later in the year, Thomson took the fellowship examinations for Trinity and was elected on his first try, an unusual and unexpected accomplishment.

It was at this time that he commenced work in the Cavendish Laboratory, embarking on mathematical researches on the passage of electricity through gases and on electric charges in motion—subjects that were inspired by Crookes's experiments and that were to occupy him throughout his life. At this time, Lord Rayleigh had just assumed the direction of the laboratory as Cavendish professor, Maxwell having died the previous October. Under Rayleigh's direction, Thomson completed several papers, the most notable being the demonstration that a moving charged particle should behave as if it possessed additional mass, the added amount being in the surrounding field. Despite his absorption in these studies, Thomson joined the competition for the Adams prize in 1882. His paper, "Treatise on the Motion of Vortex Rings," won the award. Such a subject has a strange sound today, but it must be remembered that at that time Lord Kelvin's vortex theory of matter was the only one available to the mathematical physicist seeking to account for matter in the universe. These successes led to Thomson's election as university lecturer in 1883 and in the following year to membership in the Royal Society. In that year, Rayleigh resigned the Cavendish professorship in accordance with his original stipulation that he would serve in the post for a period of five years. Thomson, then only 28 years old, sent in his name as an applicant, without, he tells us, "serious consideration of the work and the responsibility involved." To his surprise, and somewhat to his dismay, he found himself the successful candidate.

As in everything else that Thomson undertook, it was soon clear that as Cavendish professor he would be outstandingly successful. The number of science students in the university increased rapidly. Thomson's own stature was greatly enhanced in 1893 by the publication of his book, *Recent Researches in Electricity and Magnetism*, a work that spread his reputation throughout the scientific world. But the event that provided him with the opportunity to develop one of the great schools of physics and to bring world renown to the Cavendish Laboratory was the institution of a new policy at Cambridge in 1895. It provided that graduates of any other university in England or abroad were eligible for admission as research students, and after two years' residence and the submission of an approved thesis, were eligible for the B. A. degree. The introduction of this rule produced a profound effect on the excellence of the laboratory for it brought many able students from overseas who made substantial contributions to research. In fact, there appeared in the first group to be admitted a talented and industrious student who likewise was destined to create for himself an outstanding place in the history of physics. This young man, Ernest Rutherford from Christchurch, New Zealand, would follow Thomson as Cavendish professor.

It must be remembered that the year 1895 was one of the epoch-making years in physics. In November of that year, Roentgen announced his discovery of X rays—a discovery that began the era of modern physics. Thomson immediately began experimenting with gases exposed to X rays and discovered the conductivity that the radiation engendered. He studied the formation of ions, the saturation current, ion mobilities, and recombination. But his most important research was the investigation of the nature of the cathode rays that appeared when the gas discharge occurred at low pressure.

In this study he was only one of the many who were immensely stimulated by the mystery surrounding both cathode rays and X rays. One of the first investigators to report on the nature of cathode rays was Jean Perrin in Paris, who showed that the cathode-ray beam carried negative charges and that a magnetic field deflected the beam, as would be expected for such charges. By actually catching the negative charges in a Faraday cage, he had gone one step further than Crookes, whose magnetic deflection of the rays agreed with the presence of negatively charged particles.

What Perrin did not show was that the cathode rays and the negative charges were one and the same thing. It remained for Thomson not only to clinch this fact but to take the immensely more significant step of demonstrating that the negative particles were the same in all the gases used in generating cathode rays and that these particles, or "corpuscles" as he called them, were subatomic particles. These demonstrations constitute the discovery of the electron.

The discovery of the electron in 1897 required a complete revision of the physical view of matter—the nature of the positive charge, the constancy or the range of variation of atomic charges, and the disposition of positive and negative charges in the atom. These were only some of the larger questions raised by the discovery. To clarify some of them, Thomson and his students set about the determination of the charge carried by ions, using C. T. R. Wilson's cloud method. The cloud method was later refined by Robert Millikan to become the famous oil drop experiment. X-ray scattering experiments by Charles Barkla eventually showed that the number of electrons in the light elements was about half the atomic number; finally, Thomson himself developed a model for the atom that is known by his name. In it, the electrons were embedded at equilibrium positions in a sphere of uniformly distributed positive charge. These were the principal researches that occupied him until 1906, when he was awarded the Nobel Prize.

Following these investigations, Thomson turned to the study of the positive charges occurring in the gas discharge. In order to study the positive rays, he found it convenient to change the crossed electric and magnetic field arrangement used in the isolation of the electron to one in which these fields were parallel. Under such conditions, the positive ions produce parabolic traces on a sensitive plate placed at right angles to the original path of the rays. The first gas analyzed was neon; among the traces produced were those corresponding to elements of mass 20 and 22, but no trace was found corresponding to 20.2, the chemical atomic weight of neon. Careful investigation led to the conclusion that the particles producing traces at mass 20 and 22 indeed came from the neon gas in the tube. Frederick Soddy had already showed as a result of his radioactivity researches that atoms of stable lead existed as chemically similar particles but with different atomic weights. These particles he designated as isotopes (equal place). Thomson's experiments now showed that isotopes were not limited to the radioactive series, but the advent of World War I made it impossible to continue the researches.

During the period 1914–1918, Thomson served in advisory and committee work with the Board of Inventions and Research under Lord Fisher. In 1915 he was elected president of the Royal Society, succeeding Crookes, the man who had so greatly influenced his whole life's work. Early in 1918, Dr. H. Montague Butler, the master of Trinity College, died, leaving the position vacant. Trinity is the only one of the Cambridge colleges in which the master is not elected by the fellows but is appointed by the Crown. The Prime Minister, then Lloyd George, offered the post to Thomson, who accepted. The following year, at the age of 63, Thomson relinquished his Cavendish professorship after 35 years of service during which he had made the Cavendish Laboratory one of the most outstanding centers of physics in the world. As Lord Rayleigh suggests, his success in this endeavor was the result of "his personality, fertility in suggestion, kindly interest in pupils and genuine enthusiasm for their success."

His retirement as Cavendish professor and the assumption of the mastership of Trinity did not sever his connection with physics, for he continued his university lectures and his contributions to research throughout his life. Most of his later papers deal with the properties of the electron. It is remarkable that he found time in the midst of his administrative responsibilities to keep abreast of the rapidly expanding developments in physics, which, in the mid-1920s, through the researches of Louis de Broglie, Werner Heisenberg, and Erwin Schrödinger, brought a completely altered concept of matter—the wave nature of particles. Thomson readily championed this point of view; it is interesting that its experimental confirmation in 1927 was made by his son G. P. Thomson and by Davisson and Germer in the United States. His little monograph, *Beyond the Electron*, which dealt with the wave nature of matter, was published in 1929. His last paper, written in 1939, one year before his death, was entitled "Electron Waves." During his lifetime he published more than 200 research papers. He wrote 16 books, 11 under his own name alone, one with his son, and four textbooks with J. H. Poynting.

Through generations of research students at Cavendish his influence in physics spread around the world. In the years that followed no fewer than 81 of his students held professorships in universities throughout the British Isles, Europe, Canada, the United States, Russia, South Africa and India. What it meant to work under his direction is perhaps best described by Professor F. W. Aston, who wrote

> Working under him never lacked thrills. When results were coming out well his boundless, indeed childlike, enthusiasm was contagious and occasionally embarrassing. Negatives just developed had actually to be hidden away for fear he would handle them while they were still wet. Yet when hitches occurrred, and the exasperating vagaries of an apparatus had reduced the man who had designed, built and worked with it to baffled despair, along would shuffle this remarkable being, who, after cogitating in a characteristic attitude over his funny old desk in the corner, and jotting down a few figures and formulae in his tiny tidy handwriting, on the back of somebody's Fellowship thesis, or on an old envelope, or even the laboratory cheque book, would produce a luminous suggestion, like a rabbit out of a hat, not only revealing the cause of the trouble, but also the means of cure. This intutive ability to comprehend the inner working of intricate apparatus without the trouble of handling it appeared to me then, and still appears to me now, as something verging on the miraculous, the hall-mark of a great genius. (Aston, 9)

His personal characteristics, especially in his later years, are perhaps best described by Lord Rayleigh, who said of him

> He had no undue fear of giving himself away and was quite ready to express opinions offhand which had not been carefully matured. His readiness of speech and stimulating personality made him very much in demand as a speaker. . . . He had no command of foreign languages and indeed refused to make any attempt to converse even in French. In this matter he relied on the help of Lady Thomson. He had of course complete facility in reading French and German, but never wrote or spoke in these languages. . . . As master of Trinity, his relations with the junior members of the college were very easy. He took the keenest delight in their athletic performances and nothing delighted him more than the opportunity of going to a good football match or watching the performance of

Trinity men on the river. He was sincerely pleased to be asked as a guest to an informal luncheon club by undergraduates, and he almost seemed to think more of the honour of being asked to this than of the many and impressive dignities which had been conferred upon him by universities and learned bodies all over the world. (Rayleigh, 587–97)

Before Thomson had begun his study of the cathode rays, electrical discharges through low-pressure gases had been studied for a number of years; with the improvement of vacuum techniques it became possible to study discharges between electrodes that could be considered as being surrounded by a vacuum. Under these conditions, the discharge from the cathode presented itself as a well-defined beam. One of the most important questions that faced physicists in the mid-1890s concerned the nature of this beam.

Thomson himself clearly stated the problem in the opening paragraph of his 1897 paper, "Cathode Rays." That very competent physicists were aligned on each side of the corpuscular-wave controversy over cathode rays is evidence that there must have been serious difficulties in arriving at the final decision. Thomson himself was convinced that these rays were negatively charged corpuscles. He refers in his paper first to the work of Jean Perrin, who showed that when cathode rays pass into an electroscope, the electroscope becomes charged. Thomson repeated these experiments of Perrin's in such a way that there could be no doubt that the cathode rays consisted of negatively charged particles. In this paper his most important contribution is the determination of the ratio of the electric charge on the cathode ray particles to the mass of these particles. This he accomplished by first deflecting the particles in an electric field directed at a right angle to the beam. He then brought the beam back to its original direction by subjecting it to a magnetic field at right angles to the electric field. Since the effect of the electric field is just balanced by that of the magnetic field, a mathematical relationship can be found for the ratio of the charge to the mass in terms of the geometry of the tube and the known electric and magnetic fields.

Thomson investigated a number of different gases and found that the ratio of the charge to the mass of these particles was independent of the cathode that was used. The immediate consequence of Thomson's discovery was the recognition that the ratio of the mass to the charge of these cathode particles is smaller by a factor of about 2,000 than this same ratio for the lightest known ion, the hydrogen ion in electrolysis. To Thomson this fact signified only one thing, that these particles in the cathode rays were a new form of matter that did not fit anywhere in the table of elements. He correctly interpreted the cathode rays as "matter in a new state, a state in which the subdivision of matter is carried much further than in the ordinary gaseous state: a state in which all matter . . . is of one and the same kind; this matter being the substance from which all chemical elements are built up."

The Discovery of Polonium and Radium: Pierre Curie (1859–1906) and Marie Sklodovska Curie (1867–1934)

No one can read the life story of Pierre and Marie Curie without being deeply impressed by the almost superhuman dedication they brought to their partnership in science. Before their marriage in 1895, Pierre had already achieved a name in physics by his discovery, some 14 years earlier with his brother Jacques, of the phenomenon of piezoelectricity. This effect is the appearance of positively and

Pierre Curie (1859–1906)

negatively charged surfaces on certain properly prepared crystals, such as quartz, when the crystal is placed under compression or tension. Conversely, if such a quartz crystal is placed between metal plates with opposite electrical charges, a mechanical stress is induced in the crystal. If an alternating voltage is applied to such plates, and the free period of vibration of the crystal is the same as the applied voltage, then resonance is produced. This stable resonating arrangement has found widespread application in controlling the frequency of vacuum tube oscillators, and thus is of the greatest importance in the broadcasting and communications industries. In addition to this discovery, Pierre published one of the most useful researches in the history of magnetism, dealing with the temperature variation of the magnetic susceptibility of a wide range of substances. Although many scattered susceptibility measurements had been made on a variety of substances, no general conclusions could be drawn from the results, and the temperature variation of the susceptibility of magnetic materials was little known. As a result of Pierre Curie's investigations it became clear that dia- and paramagnetism were distinctly different phenomena and ferromagnetism was somehow related to paramagnetism since the former passed into the latter at a fixed temperature which is now called the Curie temperature.

Pierre was born in Paris on May 15, 1859, the son of a physician. He received his early education at home and then at the Sorbonne. Between the ages of 19 and 23 he was an assistant in the laboratory of the Faculté des Sciences. Following that, he was appointed chief of the laboratory at the School of Physics and Chemistry of the city of Paris. It was here that he carried out his research on magnetism, without assistants and with the most meager facilities. Pierre was an idealist and a dreamer and completely immersed in his work. His love of physics was for its own sake, with no eye either to professorships or awards. When proposed by the director of the School of Physics for a decoration he wrote, "If you obtain this distinction for me, you will put me under the obligation of refusing it, for I have quite decided never to accept any decorations of any sort" (Curie 1937, 126). Such was the

Marie Sklodovska Curie (1867–1934)

situation when, early in 1894, Pierre met Marie Sklodovska, a student of physics at the Sorbonne.

Marie, the daughter of Polish schoolteachers, was born in Warsaw on November 7, 1867, the youngest in a family of four girls and one boy. From the beginning she showed outstanding intelligence. As a result of family financial reverses it was necessary for her at the age of 15 to seek a position upon completion of her secondary school training. For six years, from September 1885 to September 1891, she served as a governess in order to save money for a university education.

At that time, Polish laws forbade the higher education of women. Marie had a married sister, Bronya Dluski, practicing medicine in Paris, and arranged to live with her and her husband to continue her study of physics and mathematics at the Sorbonne. But the hoped-for arrangement proved impossible—the interruptions to her studies were frequent and the distance she traveled to reach the Sorbonne was too great. To solve the problem she moved into the cheapest quarter near the university on a budget of 100 francs a month, a sum that had to cover everything, including university fees. The only solution was to spend practically nothing for food, or for her coal stove, even in the bitterest months of winter. More than once she fainted from hunger. Yet nothing mattered to her except learning. Her resolution carried her through to July 1893 when, at the end of her strength, she took first place in the master's examination in physics. She then went home to Poland to recuperate.

The following fall, Marie was able to return to the Sorbonne with a small scholarship to complete a master's degree in mathematics. It was during this time that she met Pierre Curie. A year later they were married, and marriage meant the opportunity for both to work at Pierre's laboratory at the School of Physics and Chemistry—an arrangement that was only interrupted for a short time in 1897 by the birth of their daughter Iréne.

Becquerel's discovery and his initial researches in the field of radioactivity appeared in publications covering the brief span of three months—from late February to late May 1886. Following this, he seems to have returned to investigations in magneto-optics. Inexplicably, this new field lay quiescent for about a year and a half. But the mystery of the rays intrigued Marie and early in 1898 she began studying the invisible radiation from uranium. It must be remembered that Becquerel had arrived at the correct conclusion that the rays originated in uranium regardless of the state of chemical combination of the metal. Although he had tested other phosphorescent sulfides, including those of calcium, strontium, and zinc, and found no activity, it had not occurred to him or to anyone else to make a systematic search of the elements to determine whether or not uranium was unique in emitting penetrating and ionizing rays. But the idea of a systematic search did occur to Marie and she soon discovered that thorium and its compounds behaved in the same manner as uranium. Her tests were made by placing a uniform layer of the material in question on the lower of two parallel plates spaced 3 cm apart. The potential difference between the plates was kept at 100 volts and the current between them, resulting from the activity, was measured with an electrometer. A few days after her paper appeared announcing the radioactive nature of thorium, the same discovery was reported by G. C. Schmidt in Germany.

Turning next to an examination of the ores of uranium and thorium, she found that two uranium ores, pitchblende and chalcolite (a hydrous uranium copper phosphate), showed an activity greater than that attributable to the uranium that each contained. Evidently an unknown active element was present in the ore. At this point husband and wife joined forces to find the source of the unknown activity.

The research of the Curies was guided by the activity of the chemically separated substances as determined in their plate apparatus. Pitchblende, more active than uranium, was submitted to chemical treatment and the high activity was found to accompany bismuth sulfide. The most effective method found for separating the active element from the inert bismuth was to heat the mixture in vacuo at 700° C. The active part—400 times more active than uranium—tended to deposit on the cooler part of the tube. The Curies proposed the name polonium for it in honor of Marie's native country.

In the course of the chemical separations, another activity was found to be even stronger than that associated with bismuth—an activity accompanying an element showing chemical properties resembling barium. This element was eventually named radium. Their research was completed with the help of an assistant, G. Bemont, at the end of 1898.

But in the initial researches the presence of polonium and radium had only been inferred from the intensity of the radioactivity in a chemical precipitate. In order to clinch the discovery it was necessary to isolate each of the elements and determine their atomic weight. This could only be accomplished by concentrating the very small percentages of the active substances contained in the pitchblende ore. To obtain finite amounts of the new elements it was necessary to begin with tons of material, but tons of pitchblende delivered in Paris would cost much more than they could afford. This difficulty was overcome by an initial gift from the Joachimsthal mine in Bohemia of a ton of pitchblende residue after the uranium had been extracted.

Once the ore was obtained its processing required a large laboratory to carry out all the necessary chemical treatments. A search showed that no laboratory space was available except an abandoned wooden shed with leaking skylights situated across the courtyard from the tiny laboratory which Marie had been using at the School of Physics and Chemistry. It was in these quarters, stifling hot in summer and freezing cold in winter, that the fanatically dedicated couple endured 45 months of almost unremitting labor to prepare a sample of pure radium chloride and to determine its atomic weight. This success was achieved in 1902, but its price was almost complete exhaustion for Pierre and Marie, especially since both, while continuing the research, had taken on new teaching duties in order to obtain sufficient income for their own and their child's support.

It should be of special interest to students of physics and chemistry that Marie Curie had begun her work on radioactivity in order to present a thesis at the Sorbonne for her doctor's degree. In her four years of epoch-making work on radioactivity and radium, there had been no time to go through the formalities of such an application. At last, in June 1903, the examination was taken and the degree conferred. Six months later Marie and Pierre shared the distinction of winning the Nobel Prize in physics with Henri Becquerel, the discoverer of radioactivity. Marie was too ill to go to Stockholm to receive the prize, so neither was present at the ceremonial session. Characteristically, they refused all invitations for lectures or for ceremonial honors; they avoided receptions and social functions. Only official scientific occasions could bring them out.

Following the award of the Nobel Prize, the Curies continued their research on the properties of radium. In December 1904, their second child Eve was born. Pierre, long unsuccessful in seeking a chair in physics at the Sorbonne, now had one created for him in 1905. His election to the Academy of Sciences, denied earlier, was voted, but still the election, strangely enough, was a close contest with a relatively unknown scientist. In the midst of success and the brightest outlook for the future, Pierre was killed on April 19, 1906, in a street accident. Crossing the street in the rain he was struck by a heavy horse-drawn van. He was then 46 years old.

Marie succeeded to Pierre's chair at the Sorbonne and continued the work they had begun together. In 1911 she was awarded an unprecedented second Nobel Prize, this time in chemistry. Until recent years she was the only person to obtain this double honor. Such recognition evidently required from French academic circles a tangible act to provide adequately for her scientific work. Consequently, the Institut du Radium came into being in July 1914. World War I precluded official work in the physics section of the institute and it was not until 1919 that research was begun. At that time Marie was joined by her elder daughter Iréne, who later became the wife of Frédéric Joliot. With Joliot, Iréne shared the 1935 Nobel Prize in chemistry for induced radioactivity.

Marie Curie continued her selfless devotion to science almost to the end. Overtaken by pernicious anemia induced by years of overwork and radiation exposure, this greatest woman of science died peacefully on July 4, 1934. Albert Einstein, in a moving tribute, said of Marie Curie: "Her strength, her purity of will, her austerity toward herself, her objectivity, her incorruptible judgment—all these were of a kind seldom found joined in a single individual. . . . The greatest scientific deed of her life—proving the existence of radioactive elements and isolating them—owes its accomplishment not merely to bold intuition but to a devotion and tenacity in execution under the most extreme hardships imaginable, such as the history of experimental science has not often witnessed" (Einstein, 227–28).

The Discovery of the Alpha and Beta Rays from Uranium: Ernest Rutherford (1871–1937)

Henri Becquerel's discovery that uranium emits a penetrating radiation or radiations was quickly followed by the recognition of similar properties in thorium by Marie Curie and by Schmidt and shortly after by the isolation of polonium and radium by the Curies. But the nature of the radioactive radiations remained obscure. Ernest Rutherford, a young graduate student from New Zealand working with J. J. Thomson at the Cavendish Laboratory, Cambridge, was intrigued by the mystery and began researches to solve the problem. Becquerel had shown that the radiations make the air that they traverse conducting, in the same manner as X rays. The conduction of the air resulting from the passage of X rays was just the problem that Rutherford had been studying under Thomson's direction. He was therefore well fitted to apply his knowledge to the solution of the uranium problem.

In his paper entitled "Uranium Radiation and the Electrical Conduction Produced by It," which was published in 1899, Rutherford first discussed the comparative merits of photographic and electrical methods for making his investigations. He attempted next to see if the rays from uranium are refracted in passing through a prism of glass, aluminum, or paraffin wax in a manner analogous to light rays through a glass prism. A photographic plate was used to determine if any bending of the rays could be detected. No effect was observed. He next described the process of ionization, showing what kind of results should be obtained if uranium radiation makes the air conducting by this process. The results confirmed the ionization theory.

To test the complexity of the radiation Rutherford made use of the apparatus set up to prove the ionization theory. He found that by placing successively an increasing number of sheets of very thin aluminum foil over the uranium source, the ionization current produced in the contiguous air continually decreased by a constant ratio, a result that follows when there is absorption of a homogeneous

Sir Ernest Rutherford (1871–1937)

radiation. Using thicker sheets of foil, the same result was secured until a thickness of 0.002 cm was reached. At this point the addition of another 0.004 cm was shown to have only a small effect. The conclusion was therefore drawn that the radiation was not homogeneous, at least two different kinds being present. The first radiation, having a high absorbability, or a low penetrating power, he called the alpha radiation, and the second, more penetrating, he called the beta radiation. Each of the separate radiations was found to be homogeneous. The nature of the alpha and the beta radiations, however, was not discovered. Rutherford concluded this section of his paper with the observation that "the cause and origin of the radiation continuously emitted by uranium and its salts still remain a mystery."

Because Rutherford was one of the greatest experimental physicists of all time, the reader desiring a deeper understanding of this great man should consult the magnificent biography entitled simply *Rutherford*, written by his friend, Prof. A. S. Eve, who later, like Rutherford, was also McDonald professor at McGill University. The biographical material that follows is based on Eve's book and will be devoted to Rutherford's early life and training—up to the time of his discovery of the alpha and beta rays from uranium.

Ernest Rutherford was born on August 30, 1871, at Brightwater about a dozen miles south of Nelson, one of the principal towns on the north coast of South Island, New Zealand. Both his parents, James Rutherford and Martha Thompson, had been brought to New Zealand from Scotland as children. The couple, as was common then, had a large family, seven sons and five daughters. Ernest was the fourth child and second son. Four years after his birth, the family moved to the town of Foxhill, about ten miles from Brightwater, where his schooling began. This period of residence lasted seven years, until 1882, when the family moved again, this time to Havelock on the northeast coast. Young Rutherford was then 11 years old. At Havelock he won a scholarship for his outstanding marks, and the award made it possible for him to go to Nelson College. This scholarship was the deciding event in his life. Nelson College had excellent teachers, especially one W. S. Littlejohn, a Scot who taught Rutherford science and mathematics. It was not only in these studies that Rutherford excelled; he was soon first in his class in all subjects. But about this time Rutherford's family, having suffered tragedy and reverses at Havelock, moved to Pungarehu on the North Island, where the elder Rutherford began a profitable flax-growing and exporting business. In 1889, shortly after this move, the 18-year-old Ernest won a scholarship to Canterbury College, a part of the University of New Zealand at Christchurch. Here he spent five years, earning a B.A. and then an M.A. degree, the latter with a first both in mathematics and physical science. In his fifth year, deeply impressed with Hertz's experimental discovery of electromagnetic waves, he set up a Hertz oscillator and a receiver equipped with a magnetic detector of his own design in the college basement. With this apparatus he was able to detect electromagnetic waves over a distance of about 60 feet with building walls intervening.

It was just at this time that Cambridge University opened its doors to research students. By attending for two years and writing an acceptable thesis, a student could obtain the Cambridge B.A. degree. One of the 1851 Exhibition Scholarships at Cambridge was open to students at the University of New Zealand and Rutherford had the good fortune to win it. Eve's biography notes that "when his mother came to tell him of his good fortune, he was digging potatoes. He flung away his spade with a laugh, exclaiming: 'That's the last potato I'll dig.'"

In September 1895, when he had just turned 24, Rutherford arrived in Cambridge as the first of the new research students at the Cavendish Laboratory and the first such student to be accepted by J. J. Thomson. Rutherford continued with his electromagnetic-wave researches begun at Christchurch and improved his apparatus sufficiently to transmit signals a distance of more than a half mile. This feat attracted a great deal of favorable attention and earned him a reputation as an outstanding research student in a very few months.

In the spring of 1896 Rutherford began studying with Thomson the ionization of the air produced by the passage of X rays. Later this investigation was extended to ions produced by the action of ultraviolet light on zinc. The work continued during 1897 and into 1898. Toward the latter part of this period he became interested, as noted previously, in the manner in which uranium produces conductivity in the air. Before he completed the paper presenting the results of his research, he was offered in the summer of 1898 the position as McDonald professor of physics at McGill University in Montreal, Canada. He decided to accept and left Cambridge early in September to take up his new responsibilities. He submitted his paper from McGill University.

The Discovery of Gamma Rays: Paul Villard (1860–1934)

Paul Villard was born in Lyon, France, on September 28, 1860. He was the only son of parents of English origin; his father was an artist and musician. As a youth, young Villard was allowed considerable freedom, so that he found his early education at the lycée in Lyon confining and irksome. This attitude carried over into his later education at the École Normale Supérieure in Paris, which he entered at the age of 21. After receiving his certificate from the École Normale he taught at several lycées before accepting a post in the faculty of sciences at the University of Montpellier. His work at Montpellier greatly encouraged his interest in science and, having a small income sufficient to meet his needs, he soon resigned his appointment to devote his full energies to advanced study in chemistry at the École Normale Supérieure in Paris.

After obtaining his doctorate he continued at the École, more and more completely immersed in research. He never married and his life was centered around his laboratory work. He was active in the study of cathode rays and X rays. It was in the course of investigating the radiations from radium to see whether or not a penetrating radiation like X rays might be emitted that he discovered gamma rays. Villard was active also in invention but sought no financial reward from his labor. The honor that he most prized was election to membership in the French Academy of Sciences in 1908. During his later years he suffered from ill health, apparently related to his exposure to radiation against which he seems not to have taken sufficient precaution. He died in Bayonne on January 13, 1934.

Villard was engaged in investigating the properties of reflected and refracted cathode rays and of the similar rays obtained from a radium source and summarized his results in a paper entitled "On the Reflection and Refraction of Cathode Rays and Deflectable Rays of Radium," which was published in 1900. The first part of the paper dealt with the behavior of a cathode ray beam incident on a thin insulated metallic plate, as well as the nature of the reflected radiation. Villard had then used a very thin foil of aluminum (0.02 mm) to obtain a refracted beam, that is, radiation emerging from the face opposite to the incident radiation. This emerging radiation he characterized as an apparent refraction, but he considered that the phenomenon was actually a new emission, or secondary radiation. Next he examined the refraction using the rays from radium as a source but with a much thicker (0.3 mm) metal plate as the refracting material. It was in this part of the experiment that he discovered that radium emits a very penetrating radiation that is not deflected by a magnetic field. Hence the rays carried no electric charge. These rays soon came to be known as gamma rays and were later found to be like very penetrating X rays.

The Transformation of the Elements: Ernest Rutherford (1871–1937) and Frederick Soddy (1877–1956)

Early experiments gave no indication that the emission of alpha and beta particles from radioactive elements produced any chemical change in these elements or that the activity decreased with time. Then, late in 1899, the Curies discovered "excited radioactivity," produced by radium on nearby bodies. Early in 1900 Ernest Rutherford discovered thorium emanation, an active gas, and noted that the same excited radioactivity, or active deposit, as it came to be called, could be traced to this thorium gas. But a new and surprising fact was uncovered by Rutherford; the activity decreased exponentially with time!

Following these investigations Rutherford left Canada in April for New Zealand to visit his parents and to marry his fiancée Mary Newton. An event of considerable significance in the study of radioactivity occurred during Rutherford's visit to New Zealand. In May 1900, Sir William Crookes announced the discovery of uranium-X. He prepared it by first treating a uranium solution with ammonium carbonate, then dissolving the precipitated uranium carbonate in an excess of the ammonium carbonate. A final light precipitate resulted. This substance was filtered and constituted uranium-X, a material that appeared at the time to have all the activity of the original uranium. The chemistry of radioactive substances thus began to be of fundamental importance.

On his return to Montreal in the autumn, Rutherford realized that he would need an able collaborator to assist him with chemical analyses. His extraordinary luck, which has often been commented upon, came to his aid. A very able young chemist, Frederick Soddy, had been added to the chemistry staff during the summer. Soddy, a vigorous and adventurous young man of 23, trained at Oxford, had applied for a professorship which had become vacant at the University of Toronto. Not content with waiting to hear the outcome of his application, he had impulsively decided to appear in Toronto in person to bolster his candidacy. On arriving in New York from his ocean crossing he started at once by train for Canada. On the way he was chagrined to read in the newspaper that the vacancy had been filled. Later, by the merest chance, he decided to stop briefly in Montreal to have a look at the new McDonald Laboratories at McGill University. Here the senior professor of chemistry showed him the chemistry facilities and offered him a demonstratorship on the spot; Soddy accepted. It was in this way that Soddy became associated with McGill and, very shortly thereafter, with Rutherford.

Early in 1901 Soddy gave up his appointment to spend full time on radioactivity research. He and Rutherford collaborated on a paper entitled "The Cause and Nature of Radioactivity" which was published in 1902 in *Philosophical Magazine*. It contained the epochal announcement of the instability of atoms, which chemistry had always considered immutable, and the transformation theory of the radioactive elements, that is, the change of uranium and thorium atoms by radioactivity into other chemical elements. These daughter substances are also radioactive, decay in activity exponentially, and have a definite half-life.

The essence of their paper began with the demonstration that thorium nitrate when treated with ammonia yields a precipitate, thorium hydroxide, and a filtrate, ammonium nitrate. If the precipitate is treated with nitric acid, thorium nitrate is produced in water solution. This thorium nitrate is again reacted with ammonia producing a purer thorium hydroxide and ammonium nitrate. Chemically the processes may be written as follows: $Th(NO_3)_4 + 4NH_4OH = Th(OH)_4 + 4NH_4NO_3$; $Th(OH)_4 + 4HNO_3 = Th(NO_3)_4 + 4H_2O$. The ammonium salts were driven off by ignition and the residue was found, on a weight basis, to be several thousand times as active as the thorium from which it was derived. This active substance was designated Th-X. Its activity was found to decay exponentially after the first two days and the thorium hydroxide was found to recover its activity with time. Subsequently the thorium hydroxide that had recovered its activity was converted to the nitrate and precipitated with ammonia. Examination of the filtrate showed that Th-X of the original level of activity was obtained; therefore, the investigators inferred that this substance was continuously generated by the disintegration of thorium. Their paper gave supporting evidence for the conclusion, including the measurement of the rate of production of Th-X. The latter, in turn, was considered to decay to thorium emanation and this to the excited radioactivity. The activity of thorium itself was also investigated.

From their experiments, Rutherford and Soddy reached the general conclusion that radioactivity is a manifestation of subatomic change—a phrase that seems prophetic in view of the nuclear atom, which Rutherford was to propose nine years later. Thus, if radioactivity occurs because of a change in the atom, that is, by the expulsion of an alpha or beta particle, the atom itself changes and a new substance is born, a fact that was proved chemically. Consequently, all radioactive elements were considered to be undergoing spontaneous transformation into new elements; the atom could no longer be viewed as the immutable entity that chemistry had previously perceived it. Thus, the most sweeping changes in the contemporary outlook on matter were introduced.

The Quantum Theory of Radiation: Max Planck (1858–1947)

Near the end of the 19th century, physicists were almost unanimous in the conviction that all the fundamental laws had been discovered and that there remained only mopping-up actions in the form of more precise experiments. It was genuinely felt that no matter how refined experimental apparatus became, there would be no serious departures from the theories known at the time. Thus, it was taken for granted that ultimately all astronomical observations would take their proper place in the Newtonian gravitational system, even though at that moment the motion of the planet Mercury could not be accurately accounted for by Newton's theory. Likewise, it was assumed that all the problems concerning radiation would one day yield to the laws of Maxwell. Although no one had any notion about the way radiation and matter are related or by what process matter emits and absorbs radiation, no one had any doubts that these questions would be answered in time by the proper application of Newton's laws of motion and Maxwell's theory of the electromagnetic field.

At this same time, physicists were encountering special difficulties in trying to understand the nature of the energy emitted in hollow enclosures by the walls surrounding such enclosures. This problem, which at first sight may have seemed uninteresting, became the rock upon which the ship of classical physics foundered.

In solving this problem, Max Planck was led to the discovery of the quantum of action and the concept of the photon. To understand the difficulty of the problem, we note first that when a body is heated it radiates energy; as its temperature rises, not only does the amount of energy emitted per second increase rapidly, but the quality of the emitted radiation changes visibly. For example, when the temperature is low, the emitted energy is concentrated mostly in the long wavelengths and the body glows with a cherry-red color; as the temperature of the body increases, the cherry-red color gives way to a yellow and, finally, a blue-white color.

The late 19th- and early 20th-century physicists, such as Gustav Kirchhoff, Lord Rayleigh, James Jeans, and Willy Wien, had explained most of the observed data relating to this type of radiation in terms of the classical electromagnetic and thermodynamic principles. But one observation stubbornly refused to yield to classical physics—the spectral distribution of the total amount of emitted energy. It was found that if one separated the emitted radiation into its various component colors by means of a spectroscope, the results obtained (the amount of radiant energy in each color) did not agree with the predictions of the theory. It was in dealing with this flaw in classical physics that Planck presented his revolutionary idea of the quantum of energy.

To study the properties of the radiant energy in an enclosure, Planck started out by picturing the material in the walls as being composed of simple harmonic oscillators, because Heinrich Hertz had already shown how such bodies emit and absorb energy. Planck's idea was justified since Kirchhoff had proved that the nature of the radiation in an enclosure does not in any way depend on the material of the walls but only on their temperature. Using this simple model, Planck analyzed the way in which such oscillators would emit into and absorb energy from the enclosure. This energy—referred to as black-body radiation because it is the same

Max Planck (1858–1947)

as the energy emitted by a perfectly black body—is distributed in a definite way among all possible wavelengths. A certain fraction of the total radiant energy is concentrated in each color. The exact amount depends only on the temperature of the enclosure and the wavelength being considered.

The problem that Planck faced was to determine the mathematical form of the relationship between the concentration of energy in a particular color, the temperature of the walls, and the wavelength of the particular color being considered. As Planck noted in his Nobel lecture, he expected to discover this law for the distribution of black-body radiation within the framework of classical electrodynamics. This was his anticipation, in spite of his knowledge that previous similar attempts to achieve the same results by other physicists had failed. Planck proceeded by developing the most general laws of the emission and absorption of a linear harmonic oscillator. He saw at once, however, that there was nothing about the classical electrical properties of an oscillator that would cause it to absorb and emit radiation in such a way as to give a result that agreed with experiment.

We may understand the nature of the difficulty if we consider a harmonic oscillator vibrating with a definite frequency. It emits and absorbs electromagnetic radiation of this same frequency, according to classical electromagnetic theory. As a result, Planck could not account for the distribution of the total energy among the various frequencies, since he could see no way in which two oscillators vibrating at different frequencies could influence one another and establish a condition of equilibrium that depended only on the temperature. However much he tried to invent some method by which this result might be achieved, he found himself up against the fact that each oscillator must interact in a reversible manner with the radiation field. A distribution of the radiation over the entire spectrum could not occur since this distribution would mean that the oscillator would have to re-emit radiation of all frequencies even though it could absorb only one frequency.

In other words, Planck first tried to find some asymmetrical relationship between the rate at which an oscillator emits energy and the rate at which it absorbs energy. He felt that if he then expressed this relationship in terms of the temperature, he would obtain the condition for equilibrium between the oscillators and the radiation—and that then the energy-distribution formula would drop into his lap. This, however, proved to be a vain hope since, as Ludwig Boltzmann pointed out, all the effects considered by Planck could, according to the laws of classical mechanics, work in exactly the reverse direction, so that emission and absorption would be completely symmetrical. This symmetry led Planck finally to discard the electromagnetic approach to the problem (the approach through the laws of radiation) and to consider the laws of thermodynamics, with which, as he said "I felt more at home." Since Planck had done a good deal of research into the second law of thermodynamics, he decided to attack the problem from that direction with the aid of the concept of entropy.

The second law of thermodynamics is essentially a qualification of the first law of thermodynamics, which itself is an extension of the principle of the conservation of energy to include heat as well as mechanical energy. The second law does not deny that there must always be an energy balance under all conditions, but it severely restricts the conditions under which heat can be turned into mechanical energy (work). It is, of course, always possible to change work completely into heat; the reverse, however, is not true, and heat can never be changed completely into work without leaving some kind of compensating change elsewhere in the universe. There are irreversible processes in nature so that a system left to itself evolves only along certain directions and not along others; certain processes are completely excluded.

The second law defines the directions in which a system may move or the states that it will reach if left to itself; these states are calculated by introducing a function of the state of the system that is called entropy. This important quantity, as Boltzmann first pointed out, is a measure of the probability that a system will evolve in a certain way; the higher the entropy of a particular state, the greater is the likelihood of ultimately finding the system in that state. To say of a system composed of two bodies in contact that the entropy increases when heat flows spontaneously from the hotter to the cooler body is the same as saying that heat must always flow spontaneously to the cooler body in such a system.

The entropy of a system is similar to the energy of a system in that one cannot give it an absolute value (at least this was the case when entropy was first introduced). One can only determine the difference between the entropy of a system in some standard or reference state and the entropy in another state. In fact, when the concept of entropy was introduced, it was defined mathematically as the change in the energy that takes place in a system, divided by the absolute temperature at which the change occurs. Since the entropy is thus related to the energy change within a system, Planck felt that he could deal with the energy distribution of the radiation emitted by harmonic oscillators by starting with the entropy of a collection of such oscillators. His belief was reinforced by the work of W. Wien, who at just about the same time had discovered a law for the spectral distribution of the radiation emitted by a black body, which agrees very well with the observations for the high frequencies, or short-wavelength end of the spectrum. Planck was convinced that he could find a universal and simple relationship that would express the most general distribution law, if he could relate the entropy of the radiation to the energy by means of Wien's law.

Planck found, by combining Wien's law with the mathematical expression for entropy, that a certain simple quantity (the R in the text of his Nobel address) varies directly with the energy—it is a linear function of the energy. Planck at first thought that this quantity was the universal law that he was seeking. But it turned out that Wien's law disagreed with the data for black-body radiation for the long wavelengths. Planck thus realized that he would have to look further. He was aided in his search by the experimental work of Rubens and Kurlbaum, which showed that the quantity R for long wavelengths varies as the square of the energy, rather than directly as the energy. A classical formula giving this result for long wavelengths had already been derived by Lord Rayleigh. Planck therefore decided to set up an algebraic formula for R, consisting of a sum of two terms; one term was to depend on the first power of the energy, and the other on the second power with two coefficients that were to be determined. He determined these coefficients by choosing them so that the new formula went smoothly into (became) the Wien formula for the short wavelengths, and the Rayleigh formula for long wavelengths. If the wavelength in Planck's formula is allowed to become very small, the formula approaches the Wien formula. As the wavelength becomes large the Planck formula approaches the Rayleigh formula.

Although the formula Planck thus obtained agreed with the distribution of the energy in black-body radiation over the entire spectrum, he was not entirely happy with it since, as he said, "even if this radiation formula should prove to be absolutely accurate, it would after all be only an interpolation formula found by happy guesswork and would thus leave one rather unsatisfied." He therefore was concerned from the day of its discovery with the problem of giving it physical meaning.

To deal with this problem, Planck started again from the concept of entropy. But this time he related it to the probability that a certain state would occur in a system. In the case of radiation, this was the probability for the distribution of the energy emitted by a black body among the various frequencies of the spectrum. This relationship between probability and entropy had already been treated by Boltzmann, who had formulated the principle that entropy is a measure of the physical probability of finding a system in a given state.

To use this relationship between entropy and probability to obtain the general law of black-body radiation, it was necessary to start with a formula for the absolute entropy. But up to that time, as already noted, only differences of entropy were considered to have any meaning. This absolute definition of entropy was now introduced by Planck in such a way that the constant in the usual formula for the entropy of a system goes to zero as the absolute temperature of the system approaches zero. The application of this absolute formula to the determination of the spectral distribution of black-body radiation led to the formula that Planck had previously obtained by his empirical ad hoc methods. But there were two constants in the final radiation formula that had to be properly interpreted before the entire procedure could be given physical meaning.

One constant was fairly easy to interpret, since it turned out to be just twice the average kinetic energy of a harmonic oscillator divided by the absolute temperature of the ensemble of oscillators that were in equilibrium with the black-body radiation. This constant, k, which is called the Boltzmann constant, plays a very important role in the kinetic theory of gases and is equal to two-thirds the average kinetic energy of molecules in a gas divided by the absolute temperature of the gas. This constant also plays a very important role in the relationship between probability and entropy; it is just the ratio of the entropy of a system in a given state to the logarithm of the probability of the system being in that state.

The second universal constant that appears in Planck's formula required a good deal more thought on Planck's part before he properly interpreted it. The initial difficulty arose because he attempted to fit this constant into the framework of classical physics and the wave properties of radiation. He soon realized, however, that he could not complete this task without completely destroying the results of his theory and, therefore, the agreement that he had found with the observational data. He finally, and quite reluctantly, concluded that this new constant represented a drastic departure from classical physics and would have to be explained in terms of an atomism in radiation that was quite revolutionary.

This universal constant h, known as Planck's constant, has the very small numerical value 6.6×10^{-27}; it has the dimensional properties of an energy multiplied by a time so that it must be related to what is called action in classical physics. The importance of this concept for a given system in classical physics lies in the assumption that all systems tend to move along paths for which the total action, as with the planets, is a minimum. Moreover, in classical physics the action is always assumed to vary continuously, representing a continuous change in the system as it moves along its orbit. But the existence of the constant h indicated to Planck that in the future one would have to revise one's entire thinking concerning the way in which events occur in nature. Instead of a classical construction based upon the assumption that there exists a "continuity of all causal chains of events," one would have to introduce a quantum description of nature based upon the concept that action itself is atomistic and therefore quantized or discontinuous. When Planck obtained his quantum formula of radiation, few physicists were prepared to accept its full implications, that the wave picture of electromagnetic radiation as it had been developed by Maxwell was not complete and that it would have to be replaced by a wave-corpuscular picture.

The full significance of the quantum of action h is only now apparent when we see that it appears in all atomic, nuclear, and high energy processes. The presence of h in Planck's formula distinguishes it from the classical radiation formula and we find in general that all quantum formulae are characterized by the presence of this constant. Just as Planck's formula changes into Rayleigh's classical law as h goes to zero, so all atomic formulas become classical formulas as h goes to zero. In a universe in which h decreased steadily, ultimately becoming zero, quantum phenomena (the discontinuous structure of radiation, of action, and of energy in general) would disappear and classical science would be valid. From all that we know today, it is clear that the variegated structure of the atom and such stable structures as the electron and the photon are possible because h is finite. If h were zero, atoms as we know them could not exist and such things as organic chemistry and life itself would disappear. The importance of the quantum of action h for life processes is indicated by the continuity and remarkable stability of the gene. If action were not quantized, even small environmental changes would alter the genetic structure; but the existence of a quantum of action means that genes retain their structures unless enough energy (in the form of a high-energy photon, for example), is absorbed to disrupt this structure. Thus, genes can only change their structure discontinuously and not gradually.

That h is a very small number indicates that the discontinuities in nature are very minute. To the unaided eye, therefore, action (defined by the physicist as the momentum of a body multiplied by the distance it moves with the same momentum), energy, radiation, all appear continuous. Only to the physicist who peers deeply into matter and observes the behavior of the tiny individual components of matter (electrons, atoms, protons, neutrons) are the discontinuities apparent and the need for the finite h and a quantum theory obvious.

The atomicity of action means that the emission and absorption of radiation by matter is discontinuous. Thus radiation of frequency ν can be absorbed or emitted only in bundles (or quanta) $h\nu$. The larger the frequency (the bluer the light), that is, the shorter the wavelength, the more energy there is concentrated in a bundle. In the emission and absorption of black-body radiation by oscillators at a given temperature, the very high-frequency oscillators play a very small role because a great deal of energy per quantum is required to excite such an oscillator. Precisely because the roles of the high-energy oscillators are practically eliminated in black-body radiation by the quantum of action, Planck's theory gives the correct spectral distribution of black-body radiation.

Planck was hesitant to extend his quantum concept beyond what was required to give his radiation formula. He was sure that energy is emitted and absorbed in little packets but was not convinced that the packets are permanent features of radiation. When Einstein developed his theory of the photon, however, he showed that energy is not only emitted and absorbed in bundles but that these photons, as Einstein called them, exist as unchanging entities after they have been emitted.

Like most of his outstanding German contemporaries in science, philosophy, art, literature and music, Max Planck was the product of a great educational and cultural tradition. The many-sided influences of his home and school led him to music as well as to science; if science had not won his complete devotion, he could have become an excellent pianist and organist. His predilection, however, was for mathematics and the mathematical sciences, particularly physics. In general, he was strongly attracted by logical harmonies and structures so that he was also inclined toward classical philology because of its grammatical harmony.

Born in Kiel, Germany on April 23, 1858, Planck spent a very happy childhood with three siblings and with devoted parents, who carefully nutured his intellectual and musical talents. When Planck chose a scientific career rather than a musical one, they strongly encouraged him in his choice. In October 1874, Planck matriculated at the University of Munich to study mathematics, but he did not continue along that path because he was strongly attracted to theoretical physics. Planck was prompted in this transformation from mathematician to physicist by his deep interest in cosmology. But he continued studying higher mathematics which became an indispensable tool in his researches in theoretical physics to which he committed himself entirely when he went to the University of Berlin and began studying Clausius's treatise on the mechanical theory of heat. During this period thermodynamics became his great love, and he chose the second law of thermodynamics as the subject of his doctoral thesis. He was strongly attracted to thermodynamics because he considered the first and second laws of thermodynamics to be the most general principles in physics and, therefore, laws upon which one could build a trustworthy theoretical structure.

Planck was launched on his academic career on May 2, 1885, when he was appointed professor extraordinarius at the University of Kiel; this position was below a full professorship because theoretical physics was not considered to be on the same high level as experimental physics in German universities at that time. Planck was deeply interested in all fields of physics and so he gave lecture courses in a range of topics even though he had few students. With an assured academic post and an adequate salary he married his fianceé Marie Merck and began raising a family.

His scientific career began in earnest when he was appointed to the University of Berlin to succeed the renowned physicist Kirchhoff on November 25, 1888, and to become the director of the Institute of Theoretical Physics which had been established for him. Planck quickly advanced from assistant professor to full professor (ordinarius) in 1892 and remained in that post until October 2, 1926.

The importance of Planck's theoretical work was quickly recognized by his contemporary theoreticians. He established a close relationship and an extensive correspondence with Von Helmholtz, Wien, Boltzmann, Hertz, Sommerfeld, and Albert Schweitzer. At the same time he was in close contact with theologians, historians, philosophers, and musicians; outstanding violinists and pianists of that time were often invited to give concerts in his home.

Planck's major work, which culminated in the discovery of the quantum of action and ultimately led to the quantum theory and quantum mechanics, began with his application of thermodynamics to the study of black-body (thermal) radiation. This study led him in 1900 to his famous radiation formula which is basic to an understanding of atomic physics, electromagnetic field theory, and stellar radiation. Shortly after this period, as editor of the *Annalen der Physik*, the most prestigious journal of physics in Germany at that time, Planck was instrumental in bringing Albert Einstein's work to the attention of the scientific world by publishing Einstein's revolutionary papers in the *Annalen* even though, at that time, Einstein was an unknown young man. A decade later, Planck's intensive lobbying efforts on behalf of the University of Berlin would persuade Einstein, by that time recognized as an outstanding physicist, to leave Switzerland for Germany.

Planck received many honors during his lifetime, the most distinctive of which was the Nobel Prize in physics in 1918, but his proud life was one of the most tragic in the history of scientists. His first wife, Marie, died in 1909, and his son Karl was killed in World War I. His two daughters Margarete and Emma died in childbirth, two years apart, and his older son, Erwin, from his first marriage, was executed by the Nazis in 1944 on suspicion that he was involved in the plot to assassinate Hitler. In 1911 Planck married his first wife's niece, Marga Von Hoesslin, who bore him a son Hermann. During the last two years of his life Planck lived in Göttingen where he was honored with the founding of the Max Planck Society for the Furthering of Science. He died on October 4, 1947, at the age of 89.

Mass Changes with Velocity: Walter Kaufmann (1871–1947)

With the discovery that cathode rays consist of negatively charged particles (electrons), physicists began an intensive study of the properties of electrons; one of the most interesting and important questions dealt with their mass. With the equipment that was available immediately after the electron had been discovered, only the ratio of its charge to its mass could be measured directly. Only after Robert Millikan had measured the charge on the electron, first with charged water drops in 1909 and then with his famous oil-drop experiment in 1910 and 1911, was it possible to obtain a precise value for the mass of the electron.

Although the electronic mass could not be measured directly, some important conclusions could be drawn from observations, particularly since the various applicable theories pointed to some unusual properties of the mass. The problem that arose in connection with the mass of the electron is essentially the following one: Since an electron has an electrostatic field surrounding it because of its own charge, we must picture this field as moving along with the electron. Moreover, if the electrostatic field is set moving, it should, in principle, be accompanied by a magnetic field according to Maxwell's electromagnetic theory. Indeed, Rowland in 1878 had demonstrated experimentally that a moving charge is accompanied by a magnetic field whose lines of force form concentric circles about the line of motion of the charge. Setting an electron in motion requires a greater push than setting an uncharged particle in motion, if we consider the situation in terms of Newton's laws of motion and Maxwell's electromagnetic theory.

Let us consider an electron and an uncharged particle of the same mass at rest, and let us accelerate these particles by applying the same force to both of them. According to Newton's second law of motion, the force applied to either of these particles, divided by the acceleration imparted to this particle by the force, is the mass of the particle. In the case of the uncharged particle, this ratio (the way a particle responds to force) was referred to as the true mass of the particle.

The situation for a charged particle is much more complicated because of the electrostatic and the magnetic field. The same force that imparts a given acceleration to the uncharged particle cannot impart the same acceleration to the electron because, to begin with, the entire electrostatic field of the electron must also be set moving. Moreover, the moving electron immediately finds itself surrounded by a magnetic field that (according to the laws of induction) is always so directed as to oppose the force acting to accelerate the electron. In other words, the electron behaves as though it were more massive when it is set moving than when it is at rest. When William Kaufmann undertook his experiments on the variation of the mass of an electron with velocity, physicists differentiated between what they called the true mass and the apparent mass of the electron. The true mass referred to the mass of the electron when it was not in motion and the apparent mass to its mass in virtue of its motion.

A theoretical investigation by J. J. Thomson had indicated that the mass of a moving charge should depend on its velocity and that the apparent mass should increase with velocity. In 1897, Searle, using a special model of the electron according to which the charge is pictured as distributed over an infinitely thin surface shell, obtained a simple expression for the apparent mass as a function of the velocity. Another question which arose in connection with this problem was whether or not the so-called true mass was not itself entirely of an electromagnetic nature. As Kaufmann pointed out in 1901 in his paper on the variation of mass with speed, the theoretical conclusions about the relative contribution of true and apparent mass to the total mass depend on the model of the electron that one uses. Today questions concerning the true and apparent mass do not arise, since only the total mass of the electron has physical significance.

In 1901, when Kaufmann performed his experiments on the change of mass of the electron with velocity, the theory of relativity had not yet been developed. But the electron theory of H. A. Lorentz had yielded, as early as 1895, the same relation later obtained by Einstein. This was entirely a theoretical result, however, and it was not certain that the speed of light was the maximum attainable speed of any object. Nevertheless, physicists such as Kaufmann knew that the speed of light played some kind of limiting role in the behavior of charged particles, for he stated that "electrons cannot move with speeds in excess of that of light, at least over a path that is large compared to their dimensions because during such motion the electrons would radiate energy until their speeds were reduced to that of light." After Kaufmann demonstrated that the apparent mass of the electron increases with speed, he remarked that his results showed that the "increase in apparent mass is such that it approaches infinity as the speed approaches the speed of light," a completely correct conclusion.

Since Kaufmann had no high-energy accelerators to obtain electron speeds sufficiently high to show an appreciable increase of the mass, he used Becquerel rays, the electrons emitted by radioactive atoms, now called beta rays. These rays were much more energetic than the cathode rays that were available to him. In the first few paragraphs of his paper, Kaufmann gave arguments in support of the belief that Becquerel rays are the same as cathode rays despite their much higher speeds. Then he outlined his experiment and described his apparatus.

William Kaufmann was born on June 5, 1871 in Elberfeld, Germany, where he spent his early schooldays. When he was ready for advanced instruction he entered the University of Munich. He continued there until 1894, earning his Ph.D. degree at the age of 23. Leaving Munich, he went to the University of Berlin, then to Göttingen, and from there to Bonn. It was during his period at Göttingen that he did his most important experimental work, in particular the experiment which yielded the dependence of the mass of the electron on its speed, a classic investigation that is still cited in the textbooks of modern physics. His research was marked by great proficiency in experimentation, especially in the techniques for obtaining the high vacuums necessary for cathode ray discharge tubes. His most notable contribution to this art was the construction of the first rotary high-vacuum pump; it was very artfully made of loops of glass tubing through which separate columns of mercury forced trapped volumes of gas out of the vacuum space. Although the pump was extremely fragile, unwieldy, and temperamental, Kaufmann used it with great success in his celebrated electron mass research.

The Electron Theory of Matter: Henrik Anton Lorentz (1853–1928)

The physics of the late 19th and early 20th centuries was dominated by the investigations and writings of a few giants, among them Lord Kelvin, Lord Rayleigh, J. J. Thomson, Max Planck, and Albert Einstein. Near the very top of the list was the great Dutch physicist H. A. Lorentz, who laid the basis for the modern electron theory and wrote *The Theory of Electrons*; one of the great classics of scientific literature, this book served many generations of students as the bridge between the macroscopic world of the electromagnetic field and the microscopic world of the electron.

Henrik Anton Lorentz was born on July 18, 1853, in Arnhem, Holland, the son of Gerrit Frederik Lorentz and Geertruida van Ginkel, whose first husband had died shortly after the birth of her first son. Although Henrik's mother died when he was four years old, she left a deep impression on him; he visited her grave whenever he was in Arnhem. Fortunately, his stepmother was kind so that his boyhood was a happy one.

Although the Lorentz family was fairly prosperous and cultured, there was only one children's book in the household and it became Henrik's most precious possession; on each page he carefully placed his initials H. A. L. When he began school at the age of six, his teachers quickly became aware of his unusual abilities, for he ranked first in all of his classes. He progressed so rapidly (partly because he taught himself) that at the age of nine he mastered logarithms by studying a table that he had bought in the marketplace with his own money.

He carried over his excellence in scholarship to his work in high school where he was an outstanding student, doing everything he undertook with such ease and assurance that he never considered this period of his life as anything but a very happy one. The natural, calm self-assurance that he developed never left him.

Lorentz at this time developed a great interest in history and in the humanities. He spent many hours in the pleasant hills of Arnhem discussing the Reformation with his friends and teachers, and he read extensively, mostly English authors; he was particularly fond of Macaulay, Carlyle, Scott, Thackeray, and Dickens. Since he had an excellent memory, he knew Dickens practically word for word. His remarkable memory and analytical powers helped him master the many languages that he soon learned (Dutch, English, French, German, Greek, and Latin) although

Henrik Anton Lorentz (1853–1928)

his English teacher felt that his English sounded a bit too much like Dickens. He had the amazing ability to discover the grammatical rules and idiomatic usages of a language just by careful reading, without any formal instruction. In German he read Goethe and Schiller, and in French he spent many hours with Voltaire. His interest in learning French led him to attend a French church regularly.

It is evident from the course of his studies in these early years that Lorentz developed his great gifts in mathematics and physics without the help of anyone, for he had no one to turn to until he entered Leyden University in 1870, where many had already heard of his skill and were waiting to see and meet this "highly gifted, dark little person."

Lorentz passed his examination for the degree of Candidate in Mathematics and Physics (equivalent to our B.A. degree) summa cum laude at the age of 18; but Professor Van Geer, who administered the preliminary oral examination in mathematics, although satisfied with Lorentz's performance, expressed some disappointment. On being questioned about this sentiment, it turned out that Van Geer had mistakenly examined Lorentz for the doctoral degree.

Lorentz left Leyden and returned to Arnhem in the following year to become a high school teacher. At the same time, he prepared himself for the doctorate examination, studying all the necessary subjects on his own. He passed this examination summa cum laude in June 1873. During this study period he immersed himself in the work of Maxwell; it influenced the rest of his scientific life. He devoured Maxwell's original papers as soon as he could lay his hands on them. That he mastered them is evident from his superb doctoral thesis, *On the Reflection and Refraction of Light*, which he defended publicly at Leyden on December 11, 1875, at the age of 22. This dissertation showed that Lorentz had not only mastered Maxwell's theory but had learned how to use and apply it as a powerful tool. There is no doubt that this work convinced Lorentz that Maxwell's theory could give a description of nature only if it were complemented by an electrical theory of matter that would show how the electromagnetic field hypothesized by

Maxwell interacts with matter. This idea was the germ of his electron theory which we discuss below.

Because of this excellent dissertation a new chair of theoretical physics was created for Lorentz at Leyden. After giving his inaugural address, "Concerning Molecular Theories of Physics," on January 25, 1878, he assumed his teaching duties and at the age of 25 became the first professor of theoretical physics in the Netherlands.

Lorentz was not only happy in his academic work but also in his family. After marrying Aletta Kaiser in 1881 he settled down to a very happy and productive life. Two daughters and two sons, one of whom died in infancy, were born to the couple and contributed greatly to Lorentz's happiness.

Lorentz was an excellent and sympathetic teacher, completely devoted to his students. To work under him meant being assured of a doctoral degree in a relatively short time; he worked right along with each student, directing the research and even helping to write the final thesis as well. Understandably, he was revered by his students.

Lorentz was as kind as he was brilliant and helped people whenever he found them in need. It did not matter whether it was the famous Professor Kamerlingh Omnes, whose lectures Lorentz took over when the former became ill, or a young bricklayer apprentice whom Lorentz helped to become a master mason—all shared his kindness.

Once Lorentz settled down to his career, great scientific achievements followed each other in rapid succession and hardly any branch of physics was left untouched by his remarkable contributions. His greatness as a physicist and human being was quickly recognized and every honor in the scientific world, including the Nobel Prize, was awarded him. His collected works, issued in nine volumes between 1934 and 1939, epitomize one of the great transitional periods of physics and are still an exciting intellectual adventure to the student of physics.

On February 4, 1928, at the age of 76, this genius died in Haarlem, and all the world mourned. In the words of the physicist Paul Ehrenfest, who spoke the funeral oration, we see his greatness: "Henrik Anton Lorentz dead! Since death was powerful enough to close the eyes of Lorentz, the ancient question arises in our hearts: 'What after all may be the meaning of our human lives?'"

After Maxwell had offered his equations of the electromagnetic field and Heinrich Hertz had verified them experimentally, physicists began to investigate the relationship of the electromagnetic field to the matter from which it emanates. Until the profound investigations of Lorentz, the approach to this problem had been a macroscopic one: the electromagnetic field was pictured as being coupled to matter through certain macroscopic parameters, such as the dielectric constant of a substance, the conductivity, the magnetic permeability, and so on. In this way, physicists were able to treat many problems, such as the behavior of light in a refractive medium, electromagnetic induction, and electric currents in matter. But, as Lorentz remarked, this procedure could "no longer be considered as satisfactory when we wish to obtain a deeper insight into the nature of the phenomena."

Lorentz obtained a deeper insight into this relationship by postulating the existence in the interior of matter of large numbers of extremely small charged particles, electrons. According to this picture, which, from our modern point of view, is quite naïve, the electron is understood to be a small hard sphere with an electric charge distributed uniformly over its entire surface. Aside from their being charged spheres, the electrons of Lorentz were to be treated like any other particles, and the same mechanical laws were to apply to them as apply to ordinary bits of matter. In particular, they were supposed to move, when acted upon by a force, according to Newton's laws of motions: if a force were applied to an electron, it was supposed to experience an acceleration proportional to the force and inversely proportional to its mass.

Lorentz went beyond picturing the electron as a particle that would behave in an electromagnetic field according to the accepted laws of motion. He ascribed definite properties to the interior of this particle and introduced the assumption that the state within such an electron is similar to the conditions found outside it. Indeed, to derive some of the results he was seeking, he postulated that the interior of the electron is permeated by the ether and that electromagnetic fields exist in its interior just as they do in the space surrounding it. Lorentz was very careful to point out that the application of electron theory to the analysis of the behavior of matter in its interaction with radiation was a daring approach to the solution of these problems but that it was also in the tradition and, indeed, an extension of the molecular and atomic theory that had proved so fruitful in chemistry and molecular physics.

We must remember that when Lorentz advanced his theory, the existence of electrons had not yet been universally accepted by physicists, in spite of J. J. Thomson's investigations into the conduction of electricity through gases. These developments had come at a time when some outstanding physicists were still skeptical about the existence of atoms and molecules, so that the introduction of a theory of electrons was a very bold step.

Lorentz's electron was drastically different in concept from the amorphous, imprecise wave-particle that we now have in mind when we speak of an electron. According to Lorentz, the electron was to be considered as an electric charge distribution with its density having a definite value in the interior of the electron and sinking gradually to zero across a thin outer surrounding shell. The noteworthy property of the Lorentz electron is its description in terms of ordinary geometrical concepts that make it possible to speak of its interior.

Since Maxwell's electromagnetic theory says nothing about the existence or the properties of electrons, Lorentz found it necessary to extend Maxwell's theory to take the electron into account. This he did by simply adding some terms to Maxwell's equations. Wherever Maxwell's equations contain the time rate of change of the electric field, Lorentz added to this quantity the product of the charge density of the electron and its velocity. He altered the Maxwell equation, which states that the lines of force in an electric field must all be closed loops, by imposing the condition that such lines of force must begin and end on an electron. To study the motion of an electron in an electromagnetic field, Lorentz had to introduce an additional formula—his famous expression (the Lorentz force) for the force exerted by the electromagnetic field on a moving electron. This expression consists of two terms, one giving the action of the electric field on the electron and the other giving the action of the magnetic field. With these two simple changes in Maxwell's equations and the expression for the force, Lorentz forged an extremely powerful tool for the analysis of the interaction of radiation and matter.

Lorentz's procedure was so effective that with it he explained such diverse effects as the dispersion of light, the absorption and emission of radiation, and the various Zeeman effects; he also used it to derive the famous Einstein-Lorentz transformation equations. One other remarkable result that he obtained with the aid of this simple theory was the expression for the force of an electron on itself, that is, "the force to which an electron is subjected on account of its own field." Lorentz obtained this expression by picturing each point of the electron as exerting a force on a point P "of the ether occupied by the particle (electron) at the time t for which we wish to calculate the force." It is remarkable that this expression still endures even though its derivation is based on a model of the electron that physicists now recognize as incompatible with modern quantum theory.

In his treatise on the electron, Lorentz first presented the hypothesis that had also been proposed by Fitzgerald to explain the negative result of the Michelson-Morley experiment. As already noted, the Michelson-Morley experiment indicated that the speed of light, as measured by an observer on earth, is independent of the motion of the earth. This conclusion contradicted the long-accepted Newtonian concepts of space and time. To explain this result, Fitzgerald had suggested that the arm of the Michelson interferometer lying parallel to the direction of motion of the earth contracted by just the right amount to give the null effect that had been observed.

Lorentz showed that his theory of the electron would quite naturally lead to the contraction hypothesis, since he had demonstrated that an electron in motion changes its shape from a sphere to an ellipsoid, with the long axis of the ellipsoid perpendicular to the direction of motion of the electron. Having obtained this result for the electron, Lorentz generalized it further to include the spacings between molecules in a solid structure.

> We can understand the possibility of the assumed change of dimensions, if we keep in mind that the form of a solid body depends on the forces between molecules, and that, in all probability, these forces are propagated by the intervening ether in a way more or less resembling that in which electromagnetic actions are transmitted through this medium. From this point of view it is natural to suppose that just like the electromagnetic forces, the molecular attractions and repulsions are somewhat modified by a translation imparted to the body, and this may very well result in a change in dimensions.

Lorentz considered the contraction of an electron in motion (which he assumed led to the contraction of all bodies in motion) as a very real physical shrinkage and not merely the result of change in the way space and time appear to moving observers. According to Lorentz, then, the contraction in the size of an electron occurs as the result of the electron's motion and not as a result of the relative motion between electron and observer. With the development of the theory of relativity, this one-sided way of looking at the effect of motion on spatial dimensions had to be discarded and replaced by a symmetrical picture in which only relative motion is important.

Yet, through the idea of a contracting electron, Lorentz established a very important criterion for the introduction of physical theories and hypotheses. This supported the legitimacy of introducing a hypothesis, however unlikely or strange it might appear, to further the development of science—just so long as the hypothesis is not known to be wrong. He remarked, after considering certain objections that had been raised against a contracting electron: "Notwithstanding all of this, it would, in my opinion, be quite legitimate to maintain the hypothesis of the contracting electron, if by its means we could really make some progress in the understanding of phenomena."

He then went on to assert a belief, which has not found acceptance in modern physics, that the electron may possess some complex internal structure: "In speculating on the structure of these minute particles we must not forget that there may be many possibilities not dreamed of at present: It may very well be that other internal forces serve to insure the stability of the system, and perhaps, after all, we are wholly on the wrong track when we apply to parts of an electron our ordinary notion of force." Although no internal structure of the electron has yet been found, it is becoming more and more evident that the next great development in our knowledge of the nature of matter will deal with precisely this question.

Lorentz used the theory of electrons to obtain another very important result, now referred to as the Lorentz transformations. Many of the problems that arose during the years when Lorentz was pursuing his investigations dealt with the effect of the earth's motion on the behavior of light. Lorentz proposed to treat these problems by transforming the electromagnetic equations of the electron to a coordinate system not attached to the earth but to one moving with the electron. In carrying out these transformations, Lorentz found it necessary, in discussing the behavior of an electron in the moving system, to picture a change as taking place not only in the dimensions of the electron but also in time as measured by a clock moving with the electron. Lorentz called this time the local or proper time of the electron. By following this procedure he obtained a set of transformation equations that Einstein later developed independently from a more general point of view; these are the basic equations of the special theory of relativity. The transformation equations were introduced by Lorentz in order to keep Maxwell's electromagnetic equations the same in a fixed and a moving coordinate system.

The remarkable thing about Lorentz's theory of the electron is that by means of simple classical procedures he obtained results still correct today, to a high order of approximation. However, it is precisely because of some higher-order disagreements between observation and the simple Lorentz theory of electrons that today's more sophisticated theories had to be introduced. Aside from the fact that Lorentz explained many experimental phenomena by means of electron theory, his work is important as a significant departure from all previous treatments of the interaction between charged particles. Before him, charged particles had been assumed to be capable of acting on each other at a distance, with the forces depending on the distance between these particles as well as on their charges and their state of motion. The Lorentz treatment, however, uses the idea of the medium or field. The electrons are pictured as interacting with each other, not directly but locally with the medium or field in which they are embedded. Local interaction with the field is then pictured as being propagated in all directions with a definite speed. According to Lorentz, the medium, or the field (in a sense it is equivalent to the Maxwellian ether) is unaffected by the motions of the electrons and is essentially ordinary space with certain additional electrodynamic properties. The way in which this electrodynamic space differs from the older idea of the ether is that each point in it is characterized by four quantities—three components of a vector (the vector potential) and a scalar (the scalar potential)—which completely define the electromagnetic field at that point.

Einstein's Legacy: Albert Einstein (1879–1955)

For the man in the street, the name of Einstein is a byword for the ultimate in intellectual capacity. At a slightly more sophisticated level, his name is synonymous with the theory of relativity; but what this theory asserts still remains a mystery to the majority of people. Perhaps it is easiest to say that in the educated imagination Einstein has become the modern Newton—the preeminent physicist. The comparison is very apt because both Newton and Einstein looked at the physical world in the most fundamental terms, and both produced a profound reorganization of the scientific outlook.

There are other parallels, for in their youths both were unlikely candidates for the laurels of genius. Newton was a farm boy whose intellectual stature was evident only after he had completed his undergraduate study at Cambridge. At one point in his life, Einstein was what we would today call a high-school "drop-out." His remarkable abilities went unrecognized in the university, and he remained an unknown entity until he began to publish the profound ideas on which his fame now rests. Newton as an undergraduate also seems to have been almost unknown, yet he made a deep impression on Isaac Barrow, Lucasian professor of mathematics at Cambridge. In a few years Barrow understood his pupil's genius so well that he resigned his chair, recommending Newton as his successor. Einstein, on the other hand, seems to have impressed no one. He was not offered an appointment even as an assistant—the lowest rung on the academic ladder. His epoch-making paper on relativity, published in 1905 when he was 26, was appreciated by the scientific world only after a lapse of years. If the story is true, it was an obscure professor in the old Polish university at Kracow who first recognized the significance of this work. Calling the paper to the attention of his students, he announced: "A new Copernicus has been born."

Albert Einstein's parents, Hermann and Pauline Koch Einstein, were freethinking German Jews of modest means living in the small city of Ulm. Here Albert was born on March 14, 1879. A year later the family moved to Munich, where Hermann operated an electrochemical business. It was in Munich that the boy grew up. As a child Albert was backward and took so long in learning to speak that his parents feared that he was not normal. Not only did his speaking come late, but he was a silent, dreamy child averse to physical play and characterized by a certain slowness that was irritating to many of his elders. His parents, indifferent to religion, began his schooling by sending him to a nearby Catholic elementary school where he was the only Jew in his class. But the discipline of the German classroom, standing at attention when spoken to, and the mechanical drill, all annoyed the boy and he found school unpleasant. At ten he entered the Luitpold Gymnasium but liked it no better. Here, however, he had the second intellectual experience of his youth (the first was the wonder instilled by a magnetic compass which his father had given him at age five). He discovered geometry in a book that he read entirely by himself at the age of 12 before the subject was presented at school. For Einstein, the orderliness and the logic of the theorems made an impression that was never lost.

Albert Einstein (1879–1955)

When Albert was 15, his father's business failed. In the belief that Milan offered better opportunities, it was decided that the family—father, mother and Albert's younger sister Maya—would move there. Because Albert had not completed Gymnasium and a diploma was necessary for university admission, he was left behind to live in the care of orthodox cousins. Unhappy at school, without close friends, and in a home atmosphere in which he could not be at ease, he soon found his situation unbearable. The direct action of leaving school and presenting himself to his family in Italy would have been a decisive move entirely foreign to his nature. Instead, he decided to ask a doctor to state that he had suffered a nervous breakdown and must return to his parents.

As it turned out, this ruse was unnecessary; one of his instructors advised him that he would never amount to anything and that he ought to leave school because his indifference was demoralizing both to his teachers and to the other students. And so it was with vast relief that Albert, released from his unhappy situation, set out for Italy and temporary freedom.

But Albert's father was hardly more successful in Milan than in Munich. Consequently, Albert was soon forced to think about supporting himself. His interest in science and mathematics suggested an eventual teaching career, and with financial assistance from more affluent relatives he applied for entrance to the famed Swiss Federal Institute of Technology in Zurich. Having no diploma, it was necessary to take the entrance examination, a step that produced another failure. Thus, he was entered in the high school at Aarau to make up his deficiencies. From there he was able to enter the Polytechnic Institute without further trouble.

Once admitted and with courses dealing almost completely with mathematics and physics, one might expect that Einstein would have outdistanced all his fellow students. But this was not the case. Finding in the roster of subjects such a multitude of mathematics courses that he was unable to decide which were fundamental, he solved the problem deciding that he had no pressing need for higher mathematics. Physics, too, was greatly subdivided, but here he was able to make his way. Once again he found the discipline of classes irksome, so he read in his room. A friend who did attend classes took notes; with their aid Einstein was able to pass the required examinations.

Graduating in 1900, he started looking for employment as a teacher. What he desired most was a position as assistant in the institute, but he found no professor sufficiently impressed with what he had done, nor were other regular teaching positions available. He lived by doing the odd jobs of the intellectual world—substitute teaching and tutoring backward students. After two years, he was rescued from this precarious existence by the help of his good friend Marcel Grossman, who had supplied him with the lecture notes at the institute. Through Grossman's family, a job was found for him as a junior patent examiner in the Swiss patent office at Berne. Here at last Einstein found a satisfactory haven, time to think and to work at his own pace. Since the pay was sufficient to support a family, he married a former fellow student with whom he had studied physics, a reserved and rather stolid Hungarian woman, Mileva Maritsch. He was then 23 years old.

In the next three years, Einstein prepared three remarkable papers, all of which appeared in the *Annalen der Physik* early in 1905. The first of these papers revived the corpuscular theory of light by introducing the revolutionary idea of the free photon—the atom of light—as an explanation for radiation phenomena (other than black-body radiation) and especially as an explanation for the photoelectric effect. His ideas in this latter field were verified by Robert Millikan's experiments between 1912 and 1915. The second paper was a mathematical theory of Brownian motion that provided a further proof of the reality of gas molecules on the basis that particles suspended in a fluid should behave as large gas molecules. This prediction was verified by the beautiful experiments of Jean Perrin in 1909. Finally, the third paper, his first publication on the theory of relativity, dealt with that branch of the subject now called the special theory, which has been most useful in atomic physics. The papers on radiation and special relativity presented new and revolutionary approaches to physics. It is remarkable that Einstein developed these ideas entirely on his own and without the benefit of discussion with academic colleagues. Long after these papers were published Einstein remarked that, until he was about 30, he had "never seen a real theoretical physicist." He did not seem to need to test his ideas in the academic arena, nor did he ever seem uncertain as to the validity of his revolutionary innovations.

In 1909, as his work began to be recognized, he was appointed professor extraordinary at his old school in Zurich, after a short apprenticeship as privatdozent at Berne. The following year he was appointed to the chair of theoretical physics at Prague. His wife, however, disliked Prague; when he was offered a promotion at the Polytechnic in Zurich, he returned there in 1912. This appointment, too, was short-lived, for the following year he accepted (with many misgivings because of his unhappy youth in Germany) the post as director of the Kaiser Wilhelm Physical Institute in Berlin. With this move came a disruption in his family life because the austere Mileva divorced Einstein and remained behind in Switzerland with their two sons.

The following year Germany became involved in World War I. Only Swiss citizenship saved Einstein from being regarded as a traitor for his refusal to participate in any effort connected with the war and for his determined pacifism. About this time he married his cousin Elsa, a union which for both was destined to be a lasting and happy relationship. Despite the war he continued with his work and in 1916 published his general theory of relativity. Owing to the hostilities it was impossible to carry out the measurements needed to provide a test of the theory. Thus, the prediction that light from distant stars would be deflected in the gravitational field of the sun had to wait until eclipse expeditions could be formed in 1919. The favorable results gained him worldwide renown. The Nobel Prize in physics was conferred on him in 1921, not for the theory of relativity but for "his contributions to mathematical physics and especially for his discovery of the law of the photoelectric effect." Although Einstein gained fame principally for his theory of relativity, this citation was necessary because of the clause in Alfred Nobel's will stipulating that the prizes were to be awarded for discoveries conferring great benefit on mankind.

In 1921 Einstein visited the United States to further the Zionist movement for a Jewish national homeland in Palestine and for the establishment of the Hebrew National University in Jerusalem. In 1922 he traveled to the Far East and in 1925 to South America. In the winters of 1930 to 1931 and 1932 to 1933 he was visiting professor at the California Institute of Technology at Pasadena. Following this last visit, the consolidation of the Hitler movement in Germany made it inadvisable for him to return to Berlin, and he and his wife settled temporarily at Ostend, Belgium. It was fortunate that he did not reenter Germany. In a short time all of his property was confiscated, and it seems probable that he would have been arrested. In the autumn of 1933 Einstein accepted a professorship at the Institute for Advanced Study at Princeton, where he remained until his death in 1955. One of his most significant acts during this period was his famous letter to President Roosevelt, written largely on the urging of Leo Szilard, advising the president of the feasibility of a superbomb made from uranium and the danger of a German lead in this awesome endeavor. Ironically, it was this man of peace who paved the way for the development of nuclear explosives, which increasingly threaten all humanity.

Any appraisal of Einstein must acknowledge his greatness as a man as well as his greatness as a physicist. He was the supreme humanist, standing on the side of justice for all regardless of the cost to himself. He believed deeply in the worth and dignity of every human being, and he fought against the evils that people everywhere suffer at the hands of those who misuse power. One of his biographers has remarked that Einstein was a man of the sort who appears but once in a century—it would be more accurate for a physicist to say once in two centuries, for this was the approximate time that separated Newton and Einstein.

No man, with the possible exception of Newton, has so influenced the science of an era as Einstein influenced the physics of the 20th century. His papers, whose full significance for the structure of the atom only became evident many years after they were published, are models of excellence and simplicity, with much of the development based more on physical reasoning than on mathematical formulation. Although, at first sight, the relationship between his work and the course of atomism is not always evident, a closer analysis shows that much of what we accept today about atoms would be untenable if any of Einstein's basic discoveries were discarded.

Without his concept of the photon as an unchanging entity, such phenomena as the photoelectric effect, the photoionization of atoms, fluorescence, and photoluminescence could not be understood. Without the special theory of relativity such things as nuclear energy, the electron spin, the fine structure of spectral lines, positrons, and antimatter, in general, could not be understood. Today physicists are beginning to feel more and more strongly that a complete picture of the structure of matter is possible only if Einstein's general theory of relativity, published in 1916, is properly incorporated into atomic theory.

Einstein began his work with a series of papers, between 1902 and 1904, in which he developed the essential features of statistical mechanics, without fore-knowledge of the similar work that J. Willard Gibbs had already completed in 1901. In 1905, he applied his statistical mechanics to the analysis of Brownian motion, using the kinetic theory of gases. In the same year and in the same journal, the *Annalen der Physik*, two more of Einstein's fundamental papers appeared. In one, he developed the theory of the photon and used it to explain the photoelectric effect, Stokes's law, and photoluminescence, among other phenomena. In the second, he began his series of famous papers on the theory of relativity. In the ensuing ten years, Einstein presented his theory of specific heats in which, for the first time, the quantum theory was made to account for the observed behavior of the specific heats of solids as the absolute temperature approaches zero. In this same creative period, he published his proof that energy and mass are equivalent, deriving the famous equation $E = mc^2$. He also published an analysis indicating that the equivalence of gravitational and inertial mass is not a mere accident of nature but the basis of a profound physical principle that leads to a new theory of gravity. He also extended the application of statistical mechanics to all physical systems and showed that the relationship of statistical mechanics to thermodynamics is valid under the most general conditions. Shortly thereafter, he published his general theory of relativity, his general statistical derivation of Planck's law of radiation, in which only atomic processes of absorption and emission are assumed, and his first paper on cosmology, which ushered in modern cosmology. Finally, he recognized the importance of the wave properties of particles and the need to consider these properties in the statistical mechanics of such particles.

The remarkable thing about all of Einstein's investigations is that they are all of a fundamental nature. In some cases, they opened up totally new realms in science. His papers demonstrated an uncanny ability to penetrate to the heart of the most obscure problem. There seems never to have been any doubt in his mind that he had the correct answers, even when confronting the most profound questions that disturbed his contemporaries. His principal concern, it appears, was to supply the answers in an understandable manner, with as little formalism as possible. All of these papers are marked by bold departures from the accepted paths and by confident applications of new and untried ideas. In his papers on the specific heats of solids, for example, he departed immediately from the idea, accepted until then, that the atomic vibrators composing these solids obey the ordinary classical laws of mechanics; instead, he assumed that they are governed by the quantum theory. This was, indeed, a revolutionary step of the first magnitude, since it had previously been thought that quantum effects were to be ascribed to radiation only. Because Einstein obtained more nearly correct results for the specific heats of solids with the quantum theory (impossible with classical physics), physicists realized that the quantum theory would have to be applied to all atomic processes. The next great advance in this direction was made by Niels Bohr in his monumental work on atomic spectra.

Had Einstein restricted himself to any one of the fields listed above, his contributions would still have marked him as one of the great physicists of our time. His paper about photons and the explanation of the photoelectric effect, for which he was given the Nobel Prize, is remarkable because, although it deals with one of the most puzzling problems physicists had to cope with, it is amazingly free of complex mathematical formulas. Indeed, except for one or two equations, the analysis is carried out with elementary algebra that could be followed by a good third-year high school student. This paper is also of special interest because it shows that statistical mechanics, which up to then had been applied only to systems of particles, can also be applied to radiation in a container. Such radiation acts in many respects like a perfect gas. By carrying out the analogy between the statistics of molecules in a gas and the statistics of radiation in a container, Einstein established the existence of photons as unchanging entities under all conditions.

Although Max Planck had introduced the quantum of action to account for the spectral distribution of black-body radiation, the concept of the photon was not a very popular one and was rejected by most physicists at that time. Planck himself felt very uncomfortable about the photon, and considered it more or less as a useful device to derive the correct radiation equations. He was inclined to picture the photon as having reality (if it had any at all) only during the processes of absorption and emission. At all other times, then, radiation had only a wave structure and character. Einstein departed completely from this tentative position that sought a compromise between classical physics and the new quantum hypothesis. He went over entirely to the quantum theory. He stated his position very clearly in the paper we are discussing and accepted the concept in the following words: "It appears to me, in fact, that the observations . . . can be understood better on the assumption that the energy in light is distributed discontinuously in space." Having stated this revolutionary position unequivocally, Einstein demonstrated the correctness of his assumption by applying to a container of black-body radiation, at the absolute temperature T, the statistical laws of Boltzmann and Maxwell, and those that he himself had developed in connection with the behavior of systems of particles. He carried through the analysis in two broad steps. He first considered freely moving particles, electrons and molecules, bound resonators, and harmonic oscillators intermingled with the radiation in the container and interacting with it, as well as interacting among themselves, all in dynamic equilibrium. He treated this ensemble of radiation, particles, and resonators according to the classical electromagnetic theory of Maxwell—the continuous distribution of energy in black-body radiation— and the classical theory of the electron. Therefore, since thermal equilibrium exists in this ensemble (the temperature is constant), it follows (from the classical laws of thermodynamics and statistical mechanics) that the mean kinetic energy of the resonators must equal that of the free particles, the molecules, and electrons. The kinetic energy of the free particles is given by the equation $E = (3/2)kT$ (k being the Boltzmann constant) and hence is proportional to the absolute temperature T. In this derivation there is implicit the concept that each oscillator can have all possible energies consistent with the condition that it be in equilibrium with the freely moving molecules at the given temperature. By equilibrium we mean that, on the average, each molecule and each resonator has the same energy.

Einstein next considered the interaction of the radiation and the resonators when equilibrium exists; he assumed that Maxwell's theory governs this interaction. Therefore, the resonators can absorb and emit radiant energy continuously (classically). Under these conditions the mean energy of a resonator of frequency ν must be directly proportional to the energy density of the radiation of that frequency and inversely proportional to the square of the frequency itself. On equating this result for the mean energy of a resonator with that found by the methods discussed above, Einstein found that the energy density of the radiation in the container is proportional to the absolute temperature and the square of the frequency of the radiation.

Einstein pointed out that this result is clearly incorrect, since the density of radiant energy in the higher frequencies would grow larger and larger if the frequency were continually increased, which is contrary to the observed facts. It is clear, then, that the idea of a continuous distribution of radiant energy in the container, in accordance with Maxwell's theory, is incorrect, although, as Einstein showed later in his paper, this is approximately true for very high temperatures and long wavelengths, that is, for large radiation densities.

To show that this result for the long wavelengths becomes more and more accurate with increasing absolute temperature, Einstein started out from the correct Planck formula for the radiation density and considered it for large values of the absolute temperature divided by the frequency of the radiation. Under these conditions he showed that Planck's formula is the same as the one developed from the classical theory of Maxwell (it is proportional to the square of the frequency and the absolute temperature). This led to two conclusions: first, the classical theory of radiation (the wave theory with the continuous distribution of energy) is valid if we are dealing with dense quantities of radiation of long wavelength; second, the constants appearing in Planck's formula must be related to the gas constant and Avogadro's number since these constants appear in the classical formula. In fact, by equating the classical formula for the density of the radiation to that obtained from the Planck formula for high temperatures and long wavelengths, Einstein derived the mass of the hydrogen atom (the reciprocal of Avogadro's number). This result, to Einstein, was a clear indication that Planck's formula is correct and that the classical formula is only a correct approximation under certain conditions.

In the next part of the paper Einstein considered the more important question of the nature of radiation when the density of the radiation is small and when the wavelength is very short. Under these conditions, one may replace Planck's formula by a somewhat simpler one, which was first introduced by Wien but which, in Einstein's words, "is completely satisfied experimentally for large values of v/T" (with v the frequency and T the temperature).

Einstein now compared the behavior of the radiation in this case with that of a perfect gas by introducing the entropy of the radiation which he easily calculated from the Wien formula. He then showed that the entropy of the radiation expressed in terms of the volume of the radiation is of exactly the same mathematical form as the formula for the entropy of a perfect gas. To obtain the entropy of the gas in the form that is suitable for comparison with the entropy of the radiation, Einstein applied to the gas an important principle that was first stated by Boltzmann: the entropy of a system is related to the probability for the state of the system. In doing so, Einstein gave a definite physical meaning to Boltzmann's equation relating entropy and probability; he defined the probability of a state as the length of time during which the system remains in this state, relative to some standard time.

In form, the equation Einstein derived for the entropy of an ideal gas is identical, in its dependence on volume, with the formula he derived for the entropy of monochromatic radiation in a container. Therefore, he concluded that the chance of finding, at any given moment, in some smaller volume of the container its total monochromatic radiation is expressed in a formula that is the same in form as the probability that the molecules in a perfect gas, distributed through a volume V_0, all will be found in a smaller volume V at any given moment.

Einstein's formula for the entropy of a dilute, short-wavelength, monochromatic radiation gas of frequency ν leads, via Boltzmann's relationship between probability and entropy, to the probability $(V/V_0)^{E/h\nu}$. This is the probability for finding all the radiation in the smaller volume V_0, where E is the energy of the radiation and h is Planck's constant of action. Since the probability for finding all the molecules, n, of a perfect gas in the volume V_0 at any time is $(V/V_0)^n$, Einstein concluded that the two exponents in these two formulas, $E/h\nu$ and n, must be the same. Since n is the number of molecules in the gas, it follows from this reasoning that $E/h\nu$ is the number of distinct particles (quanta or photons) of radiation in the radiation. Since E is the total energy of the radiation, it follows that each quantum or particle of radiation has an amount of energy $h\nu$.

This is exactly the content of Planck's quantum hypothesis, but Einstein's results go further than Planck's hypothesis in that they show that radiation always consists of quanta or photons. Einstein stated in his paper: "Monochromatic radiation of small energy density (within the validity range of the Wien radiation formula) behaves in thermodynamic theoretical relationships as though it consisted of distinct independent energy quanta of magnitude $h\nu$."

In the last part of the paper Einstein applied these conclusions to the explanation of two effects: Stokes's rule and the photoelectric effect. We shall consider here only the second of these two phenomena. It had been known for a number of years before Einstein's work, as the result of Heinrich Hertz's experiments and observations, that when light strikes a metal surface electrons are emitted. This is known as the photoelectric effect. The energy of the emitted electrons does not depend on the intensity of the light used but only on the color of the light. This cannot be understood on the basis of Maxwell's classical electromagnetic theory. According to this theory, the intensity of the beam incident on the metal surface should determine the energy of the ejected electrons, which is proportional to the square of the speed with which each electron comes off the surface.

Einstein explained the phenomenon very easily, however, by means of his photon hypothesis:

> According to the concept that the exciting radiation consists of energy quanta with energy content $h\nu$, the production of cathode rays by light can be understood as follows. Quanta of energy penetrate into the surface layer of the body and their energy, at least in part, is transformed into kinetic energy of electrons. The simplest explanation is that a quantum transfers all its energy to a single electron. . . . If each quantum of energy of the exciting light gives up its energy to an electron independently of all the other quanta, then the velocity distribution of the electrons, that is, the quality of the produced cathode ray, is independent of the intensity of the exciting radiation; on the other hand, the number of electrons leaving the body, all other conditions being the same, will depend on the intensity of the exciting radiation.

During the next three years Einstein came back to the problem of the nature of radiation and, in a series of papers as brilliant and simple as the one discussed above, established the existence of the photon beyond any doubt. At the same time, he showed that the statistical methods that worked so well in the analysis of Brownian motion can be applied with the same success to radiation. He used these statistical methods to analyze the statistical fluctuations of radiation in a volume. If one considers a small part of a large volume containing radiation, the energy of the radiation in this small volume fluctuates from moment to moment. Einstein calculated this fluctuation using Planck's formula for the energy density of the radiation.

He showed that if E is the mean energy of the radiation of frequency ν in a volume, then the square of the fluctuations of this energy is equal to $h\nu(E/V) + (8\pi\nu^2 d\nu/c^3)^{-1}(E/V)^2$ where $d\nu$ is the frequency range and V is the total volume of the radiation. This is a very remarkable result, as Einstein first pointed out, for it can be shown that although the second term in this expression can be derived from classical electromagnetic theory, the first term, $h\nu(E/V)$, cannot be so derived and can only be accounted for by assuming that radiation has particle properties.

Einstein had thus indicated mathematically when energy can be expected to behave as a group of particles and when as a wave. The two terms in this expression show that the wave properties are dominant when the frequency is small (long wavelengths) and the energy density E/V is large. On the other hand, the first term (hence, the particle properties of radiation) dominates when E/V is small (dilute radiation) and the frequency is large (short wavelengths).

Not content with this analysis alone, Einstein went on to consider a small mirror suspended freely in black-body radiation and analyzed its motion as though it were a Brownian particle in a gas. As the mirror is bombarded by photons from all sides, it fluctuates back and forth and carries out a Brownian motion. Again Einstein found that this motion consists of two parts, one of which can be explained by means of Maxwell's classical electrodynamic theory. But the other effect can only be understood if one accepts the quantum hypothesis.

Max Born, in an essay honoring Einstein on his 70th birthday, wrote that volume 17 of the *Annalen der Physik* of 1905 is "one of the most remarkable volumes in the whole scientific literature. It contains three papers by Einstein, each dealing with a different subject and each today acknowledged to be a masterpiece, the source of a new branch of physics. These three subjects, in order of pages, are [the] theory of photons, Brownian motion, and relativity." When these papers were published, Einstein was still a clerk in the Swiss patent office, and his only contact with the mainstream of physics and the great physicists of that period was through some of their original papers and the standard treatises by such men as Ernst Mach, Gustav Kirchhoff, Hermann Helmholtz, Heinrich Hertz, and others that were available at the time.

Considering the magnitude of Einstein's achievement in writing any one of the three papers mentioned by Born, we are amazed that this feat could have been done without direct contact with other physicists of that period. If one reads Einstein's *Autobiographical Notes*, it becomes clearer as to just why he departed from the classical approach to space and time and how he arrived at his theory of relativity. Remaining outside the influence of the dominant physicists of the late 19th and early 20th century, Einstein was free to speculate to his heart's desire and to wander along forbidden intellectual paths. Moreover, he was fortunate in beginning to probe at a time when serious disagreement existed between classical theory and experimental data and when the old mechanistic ideas (that all the phenomena of nature could be explained by means of Newtonian mechanics) were being challenged by the success of Maxwell's theory.

The work of Faraday, Maxwell, and Hertz had introduced into the concept of the universe an entity that could exist in space quite independently of palpable matter, namely, the radiation field. As Einstein remarked in 1949

> The factor which finally succeeded, after long hesitation, to bring the physicists slowly around to give up the faith in the possibility that all of physics could be founded upon Newton's mechanics, was the electrodynamics of Faraday and Maxwell. For this theory and its confirmation by Hertz's experiments showed that there are electromagnetic phenomena which by their very nature are detached from every ponderable matter—namely the waves in empty space which consist of electromagnetic fields. If mechanics are to be maintained as the foundation of physics, Maxwell's equations had to be interpreted mechanically. This was zealously but fruitlessly attempted, while the equations were proving themselves fruitful in mounting degree. One got used to operating with those fields as independent substances without finding it necessary to give one's self an account of their mechanical nature; thus mechanics as the basis of physics was being abandoned, almost unnoticeably, because its adaptability to the facts presented itself finally as hopeless. (Einstein, 25–27)

It was precisely in analyzing the incompatibility between electromagnetic phenomena, or light, and the classical laws of mechanics that Einstein realized that a fundamental change was necessary in our concepts of space and time. His primary interest in pursuing this analysis was to understand the manner in which light interacts with rapidly moving media. Put differently, how would an observer moving with a very great speed see radiative processes unfold themselves?

Most remarkable about Einstein's work was his willingness to give up the most cherished ideas of physicists, to set out boldly along new paths, and to reformulate problems in terms of the most elementary ideas. He stated in his *Autobiographical Notes* that he realized shortly after 1900, that is, after Planck's trailblazing work, that

> ... keither mechanics nor thermodynamics could (except in limiting cases) claim exact validity. By and by I despaired of the possibility of discovering the true laws by means of constructive efforts based on the known facts. The longer and the more despairingly I tried, the more I came to the conviction that only the discovery of a universal formal principle could lead us to assured results. The example I saw before me was thermodynamics. The general principle was there given in terms of the theorem: the laws of nature are such that it is impossible to construct a *perpetuum mobile* (of the first and second kind). How then could such a universal principle be found? After ten years of reflection such a principle resulted from a paradox upon which I had already hit at the age of sixteen. (Einstein, 53)

And here Einstein described the first of a series of gedanken (carried out in the mind of the scientist) experiments for which he is famous. These experiments, although extremely simple, were powerful tools in his hands and revealed to him the basic physical ideas involved in the phenomenon he was analyzing. The gedankenexperiment that occurred to Einstein at the age of 16 dealt with the way in which a beam of light would appear to an observer traveling with the speed of light.

> If I pursue a beam of light with the velocity [of light in a vacuum], I should observe such a beam of light as a spatially oscillatory electromagnetic field at rest. However, there seems to be no such thing, whether on the basis of experience or according to Maxwell's equations. From the very beginning it appeared to me intuitively clear that, judging from the standpoint of such an observer, everything would have to happen according to the same laws as for an observer who, relative to the earth, was at rest. (Einstein, 1949, 53)

This idea was the beginning of the principle of invariance (the laws of nature should appear the same to all observers moving with uniform speed with respect to each other) that Einstein was to make the basis of all his work in relativity. In this gedankenexperiment, Einstein had already formulated the idea that an observer could not travel with the speed of light, for if he did so, he would observe a stationary electromagnetic field.

A more profound analysis of the unique role of the speed of light in nature finally convinced him that this feature of light could be understood only if one revised one's concepts of space and time. He stated it very clearly in the following paragraph:

> One sees that in this paradox [the way a beam of light would appear to an observer traveling with the speed of light] the germ of the special theory of relativity is already contained. Today everyone knows, of course, that all attempts to clarify this paradox satisfactorily were condemned to failure as long as the axiom of the absolute character of time, viz. , of simultaneity, unrecognizedly was anchored in the unconscious. Clearly to recognize this axiom and its arbitrary character really implies already the solution of the problem. The type of critical reasoning which was required for the discovery of this central point was decisively furthered, in my case, especially by the reading of David Hume's and Ernst Mach's philosophical writings. (Einstein, 53)

Einstein eventually realized that further progress could be made in resolving the above paradox only by first carefully analyzing such apparently simple ideas as the distance and the time interval between two events.

In classical physics, that is, before the advent of the special theory of relativity, two observers moving with uniform speed with respect to each other could compare their separate descriptions of events in the universe by means of simple mathematical formulas that connected their two coordinate systems (transformation of coordinates). Any law of nature expressed mathematically by one observer in terms of his coordinate system could be translated into the mathematical language of the other observer by means of these transformations. Since all observers moving with uniform speed with respect to each other must be taken as equal in the eyes of nature, the transformation of a law from one coordinate system to another should leave the law unaltered (principle of relativity or the principle of invariance). Yet these transformation equations do not leave the constancy of the speed of light an experimentally verified law. That is, they fail to show that the speed of light (based upon the Michelson-Morley experiment and Einstein's gedankenexperiment) remains unaltered when one passes from one system to another. In other words, the constancy of the speed of light for all observers, regardless of the speed with which they are moving with respect to the earth, is in direct contradiction to the transformation equations of classical physics.

Einstein became aware of this phenomenon and realized that if one accepted the constancy of the speed of light as an experimentally and, in terms of Maxwell's equations, a theoretically established law of nature, then one would have to replace the classical transformation equations by "relations of a new type [the Lorentz transformation]." Einstein saw further that any set of relations (transformation equations) that enables one to pass from a description of the laws of nature and events in one coordinate system to a description in another coordinate system involves a very definite concept of space and time; it has meaning only in terms of the specific way in which the measurements of distance and time are introduced.

The Newtonian way of looking at space and time, as absolute and independent entities in our universe, was based upon a very definite way of interpreting measuring rods and clocks. As long as one insisted on the correctness of this classical picture, it was impossible to fit the constancy of the speed of light into the framework in which the laws of nature are taken as independent of the choice of the inertial frame of reference. The only way these two ideas can be made compatible is by introducing a new hypothesis concerning "the actual behavior of moving measuring rods and clocks, which can be experimentally validated or disproved." Einstein was led to a reevaluation of the role of measuring rods and clocks in establishing the laws of physics. He saw that their behavior in motion would be affected if we accept the constancy of the speed of light.

In classical physics it was taken as an a priori truth that if different observers in the universe were to order the events of history along a time axis and were to specify their positions and the distances between them, then this would always be the same arrangement regardless of how the observers were moving with respect to each other. According to this picture, space and time are pictured as entirely independent of each other. Einstein's great contribution was to demonstrate that the constancy of the speed of light brought with it a new picture of space and time in which the two are fused into a single continuum with the space and time parts having different aspects for different observers.

By combining the space and the time intervals between the two events into a single space-time interval, Einstein showed how the classical transformation equations would give way to the Lorentz equations. All these developments were presented in a small book by Einstein written in 1916 entitled *Relativity, the Special and the General Theory*. The transformation equations Einstein obtained from this general analysis of space and time (they are the basis of the special theory of relativity) are identical with those Lorentz obtained by analyzing the behavior of his theoretical electrons. But the Lorentz derivation did not lead to a revision of the concepts of space and time, since Lorentz always considered his equations applicable only to electrons; indeed, from the way Lorentz derived these equations, there was no justification for drawing more comprehensive conclusions from them. Lorentz, in fact, long persisted in the idea that his transformation equations were a peculiarity of electronic behavior and that the classical transformations were valid in the general case.

One of the most important consequences of the special theory of relativity is the equivalence of mass and energy, which Einstein expressed in his famous equation $E = mc^2$. This equation states that if an amount of mass m is transformed into energy by some process or other (for example, by thermonuclear reactions in stellar interiors or in the explosion of a hydrogen bomb) the amount of energy released is equal to the amount of mass that disappears multiplied by the square of the speed of light. This means we get vast amounts of energy by destroying only small amounts of mass. Of course, every time we obtain energy from any type of chemical reaction, that is, the burning of coal, some mass is destroyed, but the amount is much too small to be measured. To see how much energy is released when mass is transformed into energy, imagine that just one gram is destroyed. The energy released is 900 million trillion ergs or 20 trillion calories; enough heat to melt about one billion pounds of ice.

In stellar interiors, energy is released by a series of thermonuclear reactions in which the nuclei of the light elements are built up into heavier nuclei with a transformation of mass into energy. Thus, inside the sun four and a half million tons of mass are transformed into energy every second as a result of a thermonuclear process in which hydrogen is continuously being transformed into helium. Einstein's mass-energy equivalence relationship is important not only for an understanding of energetic processes but also for all branches of atomic, nuclear, and high-energy particle physics. It is interesting, therefore, to see just how this basic equation is derived from fundamental principles.

One can obtain this equation from a straightforward application of the principles of relativistic mechanics to a moving particle. This is how Einstein derived it originally, shortly after he published his first paper on relativity in 1905. We can see, without going into a detailed, rigorous analysis, why such a relationship between mass and energy must exist if we note that radiation, that is, energy, as has been observed experimentally, exerts a pressure when it strikes a surface. In other words, it exerts a push against the surface. We know from Newton's laws of motion that a push is equivalent to the transference of momentum to the surface at a certain rate. In other words, energy exerts a pressure against a surface because this energy carries momentum with it. But we know that energy and momentum are related to each other through the speed at which the energy moves. This simply means, in the case of light, that its energy divided by its speed (the speed of light c) is just equal to the momentum of the light, which we may write as the mass of the light multiplied by its speed. If we equate these two expressions, we then obtain the mass-energy equation. This is not a rigorous derivation, but it does indicate the direction of Einstein's thinking when he derived the formula.

In his paper on the equivalence of mass and energy, Einstein gave a very simple and elegant derivation of the mass-energy formula, based on a few elementary properties of light. Among these are the conservation of momentum for light and the aberration of light. This last effect is merely the apparent change in the direction of motion of a beam of light as seen by a moving observer. If an observer is moving at right angles to the direction of propagation of a beam of light, the observer sees the light approaching him from a direction that is slightly tilted (with respect to the true direction of the light) in the direction of the observer's motion. We observe the same phenomenon when running in the rain. If we stand still, and if the rain comes down vertically, we hold our umbrella upright, but if we are running, we must tilt our umbrella forward because the rain appears to come from a slightly forward direction.

To derive the mass-energy equation Einstein considered two pulses of radiation (equal in energy content) impinging from opposite directions on the opposite faces of an absorbing body. He then analyzed the absorption as viewed by an observer fixed with respect to the body and one moving perpendicular to the direction of motion of the radiation. For the fixed observer nothing happens because the actions of the two pulses of light are equal and opposite. But for the moving observer, the aberration of the light introduces a net transfer of momentum to the body at right angles to the motion of the light. However, since the speed of the body as seen by the moving observer must be the same before and after the radiation pulses are absorbed, the transfer of momentum must mean the mass of the absorbing body is increased to give the additional momentum after the absorption. (Momentum can be increased by either increasing speed or mass.) Thus, mass and energy are equivalent. The equation is obtained by applying the principle of the conservation of momentum.

In 1905, when Einstein published the first in his series of great papers on molecular physics, he had just completed the examinations for his doctorate and had already written two other papers on the foundation of thermodynamics and statistical mechanics. At that period of his life, while he was still working as an examiner in the Swiss patent office, he was interested in three branches of physics: molecular physics, radiation theory, and theories of space and time. Although he was out of touch with his contemporaries, and, indeed, with the mainstream of physics, he unerringly surmised that some of the greatest discoveries in physics were to be made in these three fields.

While working on his theory of the photon and on his papers on the special theory of relativity, Einstein saw quite clearly that a deep insight into the behavior of molecular systems could be gained most easily by analyzing such systems statistically. He realized that even without any particular model of a molecule, one can derive the known (observed) macroscopic behavior of molecular systems by applying the laws of statistics to such systems. This field of physics, now called statistical mechanics (of which the kinetic theory of gases is a special case) had already been applied by Maxwell and by Boltzmann to gaseous systems; Willard Gibbs had written down the basic equations for systems of particles in liquids as well as the general equations of statistical mechanics. Einstein, however, was not aware of the papers of Gibbs, which had been published in 1901 in a rather obscure journal. Thus, before doing his work on Brownian motions, he developed the principles of statistical mechanics ab initio in two fundamental papers.

In undertaking his analysis of Brownian motion, even before he knew that such notions had been observed long ago, Einstein was prompted by his desire to relate the existence of molecules to some directly observable phenomena that could be explained only in terms of their random motions. Although kinetic theory had been used by both Maxwell and Boltzmann to derive the laws of perfect gases, there were still many physicists who rejected the molecular concept since there was no direct experimental evidence for the existence of molecules. Einstein saw directly in the Brownian motion a way of proving the existence of molecules directly, although when he wrote his first paper on the subject he was not too certain about the exact nature of Brown's discovery.

After the botanist Robert Brown had published his observations of the erratic, irregular motions of fine pollen grains dispersed in water, various theories were advanced to explain this motion. But only M. Gouey in 1888 and F. M. Exner in 1900 attempted precise investigations of the Brownian phenomenon. Although both of these men ascribed the motion to molecular action, their analyses were inadequate and incorrect.

Einstein reasoned that if the observed Brownian motion was caused by the submicroscopic thermal motions of the molecules and that, if this Brownian motion could then be explained by applying statistical laws to molecular motions, two important goals could be achieved: on the one hand, one would have a means of determining the domain of applicability of classical thermodynamics (down to what size particles); on the other hand, one would also have a means of determining exactly molecular and atomic dimensions. As Einstein had noted in his first paper, "had the prediction of this movement proved to be incorrect a weighty argument would be provided against the molecular-kinetic concept of heat."

In the first paper, Einstein considered the state of equilibrium of suspended particles irregularly dispersed in a liquid. They are subject to a constant force in a given direction, and move about irregularly, owing to the uneven molecular bombardment. He then showed, by applying to the system the second law of thermodynamics (the entropy principle), that the suspended particles can be in equilibrium only if the applied constant force is balanced by the osmotic pressure. From this Einstein concluded that particles suspended in a liquid differ from the molecules of substances dissolved in the liquid only by their size and in no other observable way. Hence, suspended particles diffuse in a liquid just as the molecules of a solute do. By applying this type of reasoning, Einstein then derived an expression for the coefficient of diffusion of the Brownian particles in terms of the coefficient of viscosity of the liquid. He also derived the equation of diffusion and from this equation expressed the mean displacement of a suspended particle (under the bombardment of the molecules of the liquid) in terms of the coefficient of diffusion.

By introducing the coefficient of diffusion into this expression for the average displacement, he obtained an expression for the average displacement in terms of Avogadro's number (the number of molecules in one mole of substance), the viscosity of the liquid, and the size of a suspended particle. In this way Avogadro's number can be measured directly from the observed viscosity, the observed average displacement of the suspended particles, and the observed sizes of the particles. This was the first observational determination of Avogadro's number.

In later papers, Einstein applied his theory of Brownian motion to various problems in molecular physics such as the determination of molecular dimensions, problems of diffusion, the departure of the motion of suspended bodies from classical thermodynamics, and so on. All of these papers are marked by their extreme simplicity and deep physical insight. In all the cases that Einstein considered, his results are in complete agreement with observation. These papers were of great importance in the development of physics since they showed quite clearly the power of the statistical approach. Einstein used this same statistical analysis later in his theory of the photon and his derivation of the Planck radiation formula, setting a pattern and precedent for his contemporaries and those who followed him.

Although the gravitational force is the weakest of all the forces known in nature and therefore plays a negligible role in the dynamics of an atom, we have included a discussion of general relativity in this volume because it must be taken into account if a correct picture of the structure of fundamental particles, such as electrons and protons, is to be obtained. In spite of its being so small, the gravitational force is the most ubiquitous of all the forces and is a property not only of material particles such as electrons but also of photons and neutrinos. Moreover, as we penetrate into a particle, we may expect the gravitational force to become more and more important and, finally, to dominate all other forces, playing a very important role in the particle's interior.

In addition, there is a cosmological aspect of the structure of matter that can be understood only if the gravitational properties of the universe are understood. Many physicists have long felt that the existence of electrons and protons is but one aspect of the much larger problem of the structure of the universe. They think that these basic particles must, somehow or other, be derivable from a single unified theory that gives not only the gravitational field but all the other force fields associated with the known particles of nature.

Finally, the argument that the gravitational force is small, and therefore need not be considered in atomic theories, is valid only to the extent that one wishes to obtain numbers from the theory that are in agreement with present observations. But we know from the previous crises in physics that new conceptual developments must be sought at the very borderline of the observations. In time, as observational techniques improve, we may hope to detect the contribution of the gravitational force to the dynamics of the atom; any departure of observation from theory will be an important indicator of the new direction that theory will have to take. Even now, physicists are introducing a quantum-mechanical version of Einstein's gravitational field equations. In what follows we discuss Einstein's theory of gravitation (the general theory of relativity) from the classical point of view, as it was originally proposed by Einstein, and not from the quantum-mechanical viewpoint.

Of the many remarkable theories concerning the nature of matter and the structure of the universe that have been developed in the 20th century, the most astounding is probably the general theory of relativity as propounded by Einstein. As the creation of a single mind, it is undoubtedly the highest intellectual achievement of humanity. For the first time in the history of science the geometry of space-time and the laws of nature were demonstrated as interdependent. That the theory was immediately recognized as an amazing departure from the traditions of physics is indicated by Arnold Sommerfeld's reaction to a letter he received one day from Einstein. In the letter, Einstein communicated great excitement on discovering that he could deduce the observed perihelion motion of Mercury, which could never be explained by Newtonian theory, as well as Newton's theory of gravitation itself, as a first approximation from a non-Euclidean description of space-time in the neighborhood of matter. On reading this statement, Sommerfeld was amazed. He wrote back that he was incredulous and could hardly believe such a result possible. To this letter, Einstein sent the following reply on a postcard dated February 1916: "Of the general theory of relativity you will be convinced, once you have studied it. Therefore I am not going to defend it with a single word."

The great mathematician Hermann Weyl stated in the preface of his book *Space, Time and Matter* that "Einstein's theory of relativity has advanced our ideas of the structure of the cosmos a step further. It is as if a wall which separated us from Truth has collapsed." Weyl's statement was prophetic because the general theory has already unraveled many snarls in the realm of cosmology and has given us a consistent and powerful means of investigating the nature and structure of the universe. Moreover, it is becoming increasingly obvious that Einstein's theory is indispensable in studying the extreme states of matter that are found in very dense stars such as white dwarfs and black holes. It has become evident that no acceptable scheme of the structure of fundamental particles, such as electrons, can be depicted without immediately introducing the concept of general relativity.

Let us now examine the basis of the general theory of relativity and just what led Einstein to this geometrical formulation of one of the most important laws of nature—the law of gravity. First, we should reconsider the special theory of relativity and see why Einstein found it necessary to extend or generalize this earlier statement of the principle of relativity. In the special theory, Einstein pointed out that, if no gravitational fields are present and if the observers under consideration are not accelerated, all inertial frames of reference (that is, all frames of reference that are moving with constant velocities with respect to one another) are equivalent and equally permissible for expressing the laws of nature. In other words, there can be no absolute frame of reference, such as a stationary ether, in our universe in which the laws of nature assume an especially simple or correct form. This means, taken together with the constancy of the speed of light, that absolute time, absolute space, and absolute motion must be discarded. Only the space-time continuum described by a four-dimensional space-time geometry is absolute.

We begin to see why Einstein called this the special or restricted theory of relativity and why he found it necessary to develop a more general version. According to the special theory, all inertial frames of reference are equivalent in the eyes of nature. Yet these inertial frames, as a group, are singled out from all other frames of reference, such as those of accelerated systems, as the correct ones in which to formulate the laws of physics. Now since inertial frames alone are employed, one should be able to formulate the law of gravity so that it is the same in all such frames.

But Einstein quickly saw that this goal was impossible to reach. The special theory shows that the inertial mass of a body, that is, the mass that resists the action of forces and is also responsible for the apparent inertial forces that arise from acceleration, varies with speed so that it is different in different inertial frames. But we know from experiment that the inertial mass of a body is exactly equal to its gravitational mass (the mass that is responsible for the weight of a body). This means that the gravitational mass of a body varies from inertial frame to inertial frame, and, since the gravitational force depends on the gravitational masses of bodies, there is no way to formulate a law of gravity that has the same form in all inertial systems. Einstein, therefore, gave up the attempt to incorporate the gravitational field into special relativity and set out to extend his principle so that gravitational forces could be incorporated into it.

In this attempt to generalize, Einstein was guided by the very condition that made special relativity invalid for gravitation, namely, the equality of inertial and gravitational mass. This equality leads to the principle of equivalence, the starting point of Einstein's derivation of the general theory of relativity. We recall that as long as inertial frames of reference are given special preference in nature, gravity cannot be represented in the framework of relativity. This is because the gravitational field induces acceleration in bodies, so that there is no inertial frame in which a body in a gravitational field can be considered as moving with uniform motion. Einstein, therefore, saw the need to generalize the theory so that no particular frame of reference or set of frames would be given preference over other frames. We shall now see how Einstein used the principle of equivalence to do this task.

If noninertial frames of reference, such as accelerated frames (rotating coordinates, for example), are to be put on a par with inertial frames in describing nature, then there should be no physical way of distinguishing between inertial frames and accelerated frames. This notion at once seems contrary to our experience because everybody knows that when he is in an accelerated system, he can feel himself being pushed or pulled as the velocity of the frame of reference, let us say a train, changes as the result of the acceleration. The principle of equivalence allowed Einstein to solve this enigma. He reasoned that since inertial mass and gravitational mass are equal, all bodies in a gravitational field behave as though they were really not in a gravitational field but, instead, were reacting inertially to the acceleration of a noninertial frame of reference; whereas all bodies reacting inertially in accelerated systems behave as though they were not in accelerated systems but in gravitational fields.

In other words, the equality of inertial and gravitational mass implies that inertial and gravitational forces are indistinguishable; therefore, according to Einstein, these forces must be treated as equals so long as we are dealing with very small regions of space. In any small enclosure there is no experimental way of distinguishing between gravitational forces and forces arising from accelerations of the entire enclosure. This may be interpreted in another way by saying that in any small region of space it is always possible to eliminate a gravitational field through an appropriate acceleration of the system (for example, by means of a freely falling elevator or an orbiting space capsule). This is the famous principle of equivalence.

In small regions of a gravitational field, objects may behave as they do in a region free of gravitation if one introduces in place of an inertial coordinate system a coordinate system that is accelerated with respect to the inertial system. Conversely, by means of acceleration one can reproduce in a field-free space the effects that are found in an inertial system at rest in a gravitational field. Therefore, both frames of reference are equally valid for the description of nature. Indeed, this means not only that inertial and noninertial frames of reference are equally valid but that the concepts of absolute acceleration and gravitational force are no longer tenable, since what one observer may take as a gravitational force is just as validly perceived as an inertial force by another observer. Therefore, the basic demand of the special theory of relativity, the invariance of the laws under the Lorentz transformation, is too narrow. Invariance of the laws must be postulated relative to non-linear transformations of the coordinates in the four-dimensional continuum.

Einstein acknowledged that the condition imposed by his special theory of relativity is invalid. He had assumed that mathematical descriptions of nature were to be taken as laws only if their forms remained unchanged in going from one inertial frame of reference to another by way of a Lorentz transformation. But he recognized that this was not the case when one deals with gravitational fields or accelerated systems, such as one that is rotating. To take into account gravitational fields and accelerated systems, we must require that the mathematical expressions for the laws of nature remain unchanged if we pass from one frame of reference to any other frame by the most general type of coordinate transformation we can imagine.

To see how this reasoning led Einstein to his mathematical formulation of the general theory we shall consider first the physical significance of the principle of equivalence and the principle of general covariance, that is, the invariance of the laws of nature under general transformations. Since the principle of equivalence asserts that no distinction can be made between an observer in an inertial frame in a gravitational field and an observer in an acclerated frame in field-free space, the whole idea of an absolute gravitational force between two bodies is untenable. Therefore, gravity must be treated in some other way. If we discard the idea of gravitational forces, we run into difficulty with Newton's laws of motion: we know that a body moving in a gravitational field does not travel in a straight line, and hence, according to Newton's first law of motion, must be under the action of a force. But the principle of equivalence denies the existence of such a force so that Einstein appears to be up against a contradiction.

Einstein overcame this obstacle by altering Newton's first law of motion; he removed from it any reference to forces. We remember that Newton's first law states that a body moves in a straight line with constant speed unless a force acts on it. In other words, the presence of a force acting on a body is indicated by the departure from straight-line motion. But we can get away from the difficulty by stating that the departure from straight-line motion (straight in the Euclidean sense) is not due to a force but to a difference in the geometry. In other words, we replace the usual statement of Newton's first law by the statement that a body that is free to move, that is, not in contact with any other body or subjected to electromagnetic forces, will move in a "straight line" characteristic of the geometry of the space in which it is moving. If a gravitational field is present, the geometry of the space is not Euclidean and a straight path is curved as represented in Euclidean space. This means that what Newton called a gravitational force is to be considered as the manifestation of the non-Euclidean character of space. In this way Einstein was led to his geometrical interpretation of gravitation.

In carrying out his program of geometrizing gravitation, Einstein was guided by what he had already done in special relativity because he knew that the general theory would have to be an extension of the special theory. Moreover, he was aided by the analogy between what he planned and Gauss's theory of surfaces. The special theory of relativity grew out of the need to replace the hitherto separate three-dimensional space manifold and the one-dimensional time manifold by a single four-dimensional space-time manifold (a space-time interval) that is the same for all observers moving in inertial frames of reference. Time and space in special relativity theory are merged into a single four-dimensional space-time manifold by a natural extension of the theorem of Pythagoras from three to four dimensions. This result was achieved by considering the separation between two events in space and time and merging these separate space and time intervals into a single space-time interval in four-dimensional geometry.

If, then, one is treating conditions governed by the special theory (inertial frames of reference and no gravitational fields), the space-time interval that one obtains is essentially the theorem of Pythagoras in four-dimensional Euclidean space. Thus, we may say that the special theory of relativity is given by the geometry of four-dimensional Euclidean, or flat space. Recognizing this fact, Einstein at once suspected that one could obtain a general theory of relativity that would take into account gravitational fields and accelerated frames of reference by replacing the four-dimensional Euclidean geometry of space-time by a suitable four-dimensional non-Euclidean geometry. In this work he was guided by the following reasoning: If we consider a flat two-dimensional surface and allow a particle to move freely on it, the particle will move along a straight line. However, if we have a two-dimensional curved surface along which a particle is constrained to move, the particle naturally moves in a curved orbit, following the curvature of space itself. It follows, then, that we may compare a particle moving freely (relative to an inertial frame) in space free of gravitational fields to the particle on the flat surface. In addition, we may compare a particle moving freely in a gravitational field, or relative to a noninertial frame, to the particle moving on the curved surface. All that was required, then, for a generalization of the theory that would account for gravitational fields and noninertial frames was to replace the four-dimensional Euclidean space-time interval of the special relativity theory with a non-Euclidean four-dimensional space-time interval.

Einstein used the theory of surfaces that had been developed by Gauss and then extended to many dimensions by one of Gauss's most famous students, Riemann. Einstein showed that by using the appropriate mathematics (the tensor calculus) it is possible to obtain a set of geometrical quantities in four-dimensional space that describes the non-Euclidean properties of this space and, at the same time, describes the behavior of masses in gravitational fields or in accelerated frames of reference. The mathematical properties of the real gravitational field or of fields generated by accelerated frames of reference (for example, a rotating frame of reference) are contained in a set of tensor equations that are referred to as Einstein's field equations. Einstein showed that, to a first approximation, these field equations give Newton's law of gravity.

Although one must solve Einstein's field equations to see exactly how a mass moves in a gravitational field (this solution was first obtained in 1917 by Karl Schwarzschild for planetary motions around the sun), Einstein derived the various effects by applying the principle of equivalence. In his book *The Meaning of Relativity*, Einstein applied the principle of equivalence to a rotating disk to show that for an observer on such a disk the geometry is of the hyperbolic non-Euclidean kind. It is easy to see why this is so since rods and clocks placed at various points on the disk travel at different speeds and thus contract and slow down by different amounts.

In the case of a body in a gravitational field, the principle of equivalence can also be applied, since gravitational fields can be duplicated by appropriately accelerated frames of reference. One sees what happens in a gravitational field by simply applying the contraction of rods and the slowing down of clocks to observers in frames of reference that are allowed to fall freely in gravitational fields. In this way, such things as the bending of beams of light, the slowing down of clocks, the shrinking of rulers, and so on, in gravitational fields can be derived.

To see how the principle of equivalence is used, we apply it to a ray of light grazing the sun. According to this principle, the gravitational field of the sun may be reproduced by an elevator accelerating away from the sun at the same rate as a freely falling body on the sun. We may see how the gravitational field of the sun affects the ray of light if we imagine it passing through the accelerated elevator. An observer in this elevator would see the light entering one side of the elevator at a certain height above the floor. But, by the time the light moved across the elevator, the elevator would have accelerated upward, so that to the observer it would appear that the light had accelerated—that is, had fallen downward. In short, light must fall in a gravitational field. Hence, the sun bends a ray of light that is grazing it. This phenomenon has been verified in many solar eclipses.

What is remarkable about Einstein's treatment of the gravitational problem is not the brilliance of the mathematics or the clever and sophisticated use of physics but, rather, his deep insight into the very heart of the matter and his intuitive knowledge that the gravitational field must be related to non-Euclidean geometry. As in all of his papers, where a very sophisticated type of mathematics might be expected for a rigorous analysis of the problem, Einstein obtained his final results with a simple algebraic analysis of elementary physical processes. One is always amazed, especially in the case of the general theory of relativity, at the daring qualities of Einstein's generalizations and his unerring penetration to the heart of a problem.

References

Aston, F. W., *The Times*, London, September 4, 1940.

Curie, Eve, *Madame Curie*. New York: Doubleday, 1937.

Einstein, Albert, "Autobiographical Notes," in Paul Schilpp, *Albert Einstein, Philosopher-Scientist*. Evanston, Ill. : Library of Living Philosophers, 1949, vol. 7.

Einstein, Albert, *Out of My Later Years*. New York: Philosophical Library, 1950.

Jauncey, G. E. M., *American Journal of Physics*, 13 (1945), p. 360.

Lord Rayleigh, *Obituary Notices of Fellows of the Royal Society*, vol. 3, 1941, pp. 587–97.

7 *New Ideas and New Measurements*

> *The hidden harmony is better than the obvious.*
>
> —Heraclitus, *Fragments*

The "Thomson" Atom: J. J. Thomson (1856–1940)

Following the discovery of the electron it was clear that a complete revision of atomic theory was required. The atom could no longer be regarded as the ultimate unit of matter. Thomson's experiments had shown that, regardless of the gases used to produce the cathode ray discharge, the same subatomic particle, the electron, always appeared. Since this particle carries a negative electric charge and the atom, as a whole, is uncharged, questions immediately arose as to the number of electrons per atom, the nature of the positive charge, and the spatial relation of the latter to the electron or electrons present.

The very small electronic mass determined by Thomson's experiments, approximately 1/1,000 the mass of the hydrogen atom, at first suggested that the hydrogen atom might contain some 1,000 electrons. This contemporary thinking was clearly set forth in the closing pages of Ernest Rutherford's book, *Radioactive Transformations*, published in 1906. Rutherford pointed out that atom models had already been suggested, the first by Lord Kelvin.

Kelvin proposed, in 1902, an atom model consisting of a sphere of uniformly distributed positive electricity in which discrete electrons were embedded so that equilibrium was obtained when these charges were at rest. A year later J. J. Thomson published calculations on the stability of a model in which electrons, arranged uniformly around a circle within the positive sphere, rotated at high speed. A further paper by Thomson, appearing early in 1904, reexamined Kelvin's static atom model at considerable length. Much of this paper, with additions, appeared in Thomson's book, *The Corpuscular Theory of Matter*, published in 1907. In this work static electrons are placed one by one in a positive sphere and the stability is examined. Somehow Kelvin's proprietary claim to this atomic scheme was lost, so that in later years the arrangement became known as "the Thomson atom."

But while Thomson was examining and elaborating the original Kelvin scheme, Kelvin himself went on to other more complicated models. Finally, in December 1905, he proposed a Boscovichian atom that had alternating shells of "vitreous and resinous" electricity with "the total vitreous greater than the resinous." The electrons were embedded in the vitreous (positive) shells and could therefore, if unstable, be ejected with varying speeds as demanded of electrons issuing from radioactive atoms. Still another model, proposed in 1904, was that of Nagaoka who, harking back to James Clerk Maxwell's paper on Saturn's rings, suggested that the atom might consist of a number of electrons revolving with nearly the same velocity in a ring about a positively charged center. Rutherford noted this suggestion in his famous paper of 1911 in which he proposed the nuclear atom. All of these atom models had varying degrees of plausibility; they could account qualitatively for various atomic properties but not for all. Thomson, however, was perhaps the most persistent in his search for a model that would give both qualitative and some quantitative agreement with experiment.

Suppose one begins with the question: How many electrons are there per atom? Thomson obtained an answer to this question from several sources. The first came from experiments on the scattering of electrons made to pass through thin sheets of metal. Philipp Lenard, for instance, had shown some years previously that cathode rays can pass through thin metal windows and ionize the air outside the tube in which they were generated. By comparing a computed value of electron scattering with that observed experimentally, Thomson found that the number of electrons per atom needed to produce the observed scattering should be approximately the same as the atomic weight of the scattering material, assuming unit atomic weight for hydrogen. (Except for hydrogen, this result was approximately two times too large.) The second source of information was the dispersion of light by hydrogen. Here a calculation showed that the number of dispersion electrons per atom of hydrogen must be closely equal to unity. The third source was X-ray scattering experiments. When a beam of X rays passes through matter, the atoms both absorb and scatter the rays; hence, the amount transmitted decreases as the thickness of the material increases. From early X-ray scattering measurements the number of electrons per atom was found to be of the order of the atomic weight. Later, more accurate measurements by Barkla showed that for the light atoms, except hydrogen, it was more nearly half the atomic weight. As a consequence of all this evidence, it was apparent that hydrogen, the least massive of all the atoms, consisted probably of one electron, and an equal amount of positive charge. Heavier atoms were presumably obtained by adding one electron for every unit of positive charge.

Results from kinetic theory had shown that the diameter of an atom was of the order of 10^{-8} cm. From the scattering experiments it was known that an electron was not much deflected by passing through thin foils many atoms thick, so the conclusion was reached that the density of positive charge must be low. Accordingly, Thomson, in making a model for hydrogen, the simplest atom, had some basis other than Kelvin's proposal for assuming that positive charge, equal to that of the electron, occupied the whole atomic volume with uniform density.

Having made these tentative choices, the question of stability demanded examination. Where was the electron in such an atom? Elementary electrical the-ory shows that if the electron is assumed to be at the center of the positive sphere, any displacement of it will result in vibrations about the center. These would con-tinue indefinitely if the electron did not lose energy; but since a vibratory motion about the center is an accelerated motion, and classical electromagnetic theory required that accelerated electrons must radiate energy, the electron would natu-rally be brought to rest. Hence, the undisturbed atom would be a static atom, and if disturbed, would produce dynamically stable vibrations, dying away with time. If the disturbances were sufficiently violent, the electron would be ejected, resulting in a hydrogen ion. All this seemed in accord with experience. But a little further investigation showed that despite its good beginning, the model had at least one serious defect. The radiation emitted by the vibrating electron should, according to the theory, consist of light of a single wavelength, appropriate to the far-ultraviolet region of the spectrum. Experimentally, one observed quite unaccountably a spectrum in the visible region consisting of several discrete wavelengths. Other series of lines also existed in the infrared and ultraviolet.

Despite this defect, Thomson went ahead to examine the stability of the multielectron atom. He showed in his paper that, when an atom contains two electrons, stability is obtained by keeping the size of the sphere of positive electricity constant. As regards the two electrons placed inside the sphere, equilibrium is obtained when they are on a line through the center of the sphere and equidistant from it, the distance being half the radius of the sphere. As the number of electrons increases to four, they can no longer be in static equilibrium in a planar arrangement; instead they are located at the corners of a regular tetrahedron. Stable arrangements with greater numbers of electrons up to 100 were then discussed. Thomson was also able to show that the electron arrangements in his scheme of "atom-building" suggested an explanation of the periodic properties of the chemical elements. The stability of experimental configurations, using magnetized needles thrust through corks and floated on water, iron spheres floating on mercury, and elongated conductors floating vertically in water was then briefly noted as a result of the work of other investigators. These experiments supported the idea that a number of corpuscles, if confined to a plane, will arrange themselves in a series of rings as Thomson's calculations indicated.

One of the main props for the Thomson atom was its support of alpha-particle scattering experiments. It is ironic that this aspect of his model on closer investigation led to its downfall!

The Determination of Avogadro's Number: Jean Perrin (1870–1942)

Jean Baptiste Perrin was born at Lille, France, on September 30, 1870. After his early education in Lille, he attended the École Normale Supérieure in Paris and became a teacher in physics there from 1894 to 1897. At the age of 26 he was awarded a doctorate of science degree from the University of Paris and was also appointed a lecturer in physical chemistry. He became professor of physical chemistry at 40 in 1910 and held this post for 30 years, except for the period of World War I, during which he served as an officer with the engineers. He was elected a foreign member of the Royal Society in 1918. In 1923 he became a member of the French Academy and in 1926 was awarded the Nobel Prize in physics. He was then 56 years old. Because of his eminence he became, ten years later, under secretary of state for scientific research in the government of Premier Léon Blum. Two years afterward he was elected president of the Académie des Sciences and in the same year president of the department of scientific research in the French government. He retired in 1940 and, after the German invasion of France, lived in Lyon. In December 1941 he came to the United States to teach in the Free French University

Jean Perrin (1870–1942)

in New York. At that time he made his home with his son, Dr. Francis Perrin, who was then visiting professor of physics and mathematics at Columbia University. His work in the United States lasted only a few months, being terminated by his death in New York on April 17, 1942.

It was not until after the beginning of this century that Einstein's theory of the Brownian motion and Jean Perrin's application of the kinetic theory of isothermal atmospheres provided new ways to establish Avogadro's constant for determining the number of molecules in a gram molecular weight of an element or compound, for example, in 32 grams of oxygen. Perrin's method and the details of his experimental work appeared in his 1908 paper, "Brownian Motion and Molecular Reality." The paper is a model of scientific writing and deserves to be read not only for its importance but for its admirable clarity and direct style.

Once the importance of Avogadro's number had been established by Stanislao Cannizzaro, it became increasingly urgent to ascertain its exact numerical value. Perrin's method is simple in principle and may be readily explained. The kinetic theory of gases predicts that in a stagnant atmosphere maintained at a constant temperature T, the number of molecules, each of mass m, composing the atmosphere will decrease with height above the earth's surface according to the relation $n = n_0 e^{-mgh/kT}$. In this expression n_0 is the number of molecules per unit volume at the earth's surface; n is the corresponding number at any given height h; e is the base of natural logarithms; g is the acceleration of gravity; k is the Boltzmann constant; T is the absolute temperature.

The Boltzmann constant k is defined as the ratio of the ideal gas constant per mole, R, and Avogadro's number, N. It should be recognized that mgh represents potential energy of a molecule at a height h above the earth's surface. Since $3kT/2$ is the mean kinetic energy W of the molecules, kT is given by $2W/3$. By taking the logarithm of both sides of the equation, and rearranging and making the substitution for W, we get one of the relations noted in Perrin's paper: $(2/3)W \log n_0/n = mgh$.

Since experimentation at various heights in the atmosphere is difficult, and the realization of an isothermal atmosphere highly improbable, it is necessary to simulate the conditions for experiment. It occurred to Perrin that particles much larger than a molecule, but small enough to show Brownian motion when immersed in a fluid at constant temperature, might behave as molecules do; they might show the same height-density characteristics as required by the theory. Accordingly, he carried out experiments using emulsions containing microscopic particles of mastic or gamboge gums, obtained by centrifuging solutions of these materials. The particles were found to remain in suspension and to be distributed exponentially with height when counted using a microscope focused in the liquid.

It was thus possible with such emulsions to use the isothermal-atmosphere theory to determine N. Only one modification was necessary to the foregoing equation. Gum particles suspended in a liquid are buoyed up by it so that the net downward force is the difference between the gravitational force and the buoyant force. If Δ represents the volume of the uniform particles, δ their density, and ∂ the density of the solution, the net downward force is given by $\Delta(\delta - \partial)g$. Making this substitution, we obtain the equation given by Perrin: $(2/3)W \log n_0/n = \Delta(\delta - \partial)gh$. In order to compute Avogadro's number, the quantity W is expressed in terms of R/N instead of k, that is, $3RT/2N$. If we also change the indicated natural logarithm to the base 10, we finally get $2 \times 303(RT/N) \log_{10} n_0/n = 4/3 \pi a^3 g(\delta - \partial)h$, where a is now the radius of the gum particles. The final result obtained, using this relation, was 7.05×10^{23} (or 70.5×10^{22}, as Perrin gives it). This quantity is substantially larger than the presently accepted value of 6.025×10^{23} because of the error associated with the many different measurements that must be made with this method.

Referring to the formula above, used to calculate N, two factors appear to be especially susceptible to relatively large errors, namely, the value of a, the radius of a gum particle, and the ratio n_0/n of the number of particles at different heights. Perrin's paper lists three methods for finding the radius of the particles. Even the method that he considers best in his paper does not totally resolve the problem, however, as it contains uncertainties not easy to estimate. The same is true of the determination of n_0/n. Both of these quantities in the ratio involve small numbers of particles which must be identified at a glance (when the visual method is used) as clearly in the focal plane. Another way of expressing the same uncertainty is to say that the height h used for determining a given value of n_0 is not accurately known.

The value of Avogadro's number as now accepted is obtained from the relation that essentially defines the quantity of electricity called the Faraday. This quantity is the product of the electronic charge and Avogadro's number, that is, $Ne = F$. Now F can be determined with good precision by measuring the amount of charge necessary to deposit 1 gram mole of univalent ion in electrolysis. Dividing this number by the measured value of the electronic charge yields N. The electronic charge was determined by Robert Millikan around 1912, very soon after Perrin's determination. Using Millikan's value in the expression for the Faraday, N was found to be 6.062×10^{23} with an estimated precision of 0.1 percent. Despite all of Perrin's care in his experiment the difference between his value and the presently accepted one would indicate how difficult it is to use detailed indirect measurements and obtain high precision. Actually, Millikan's value for the electronic charge had small errors also, so that the present value of N is taken to be 6.025×10^{23}.

The Alpha Particle and Helium: Ernest Rutherford (1871–1937)

Ernest Rutherford's desire to be closer to the center of science than a position at McGill University afforded was realized in 1907 when Professor Arthur Schuster, head of physics at Manchester, retired in Rutherford's favor. Manchester was to be the site of Rutherford's greatest triumph, the discovery of the nuclear atom, and from the beginning he found everything there to his liking.

One of the finest researches that Rutherford completed at Manchester in collaboration with T. Royds, a research student, was the proof that alpha particles are indeed ionized helium atoms. This had been suspected for some time and Rutherford's correspondence as early as 1905 expresses this conviction. Various lines of evidence had indicated the correctness of this point of view, including charge-to-mass ratio (e/m) measurements. The direct proof was still lacking, however, until Rutherford conducted an elegant and simple experiment. It consisted of compressing a small quantity of the emanation of radium into a closed glass capillary surrounded by a glass enclosure tipped with a spark discharge tube. The walls of the capillary were sufficiently thin to allow the alpha particles emitted by the emanation to shoot through them, but they were nevertheless quite gas-tight to the passage of helium itself.

The alpha particles entering the evacuated glass enclosure surrounding the capillary were stopped by the thicker walls of the enclosure. Tests on the discharge tube located on top of the enclosure were made to ensure the absence of helium. After several days spark tests showed faintly visible yellow helium lines and after six days most of the visible helium spectrum was present in the discharge.

Atoms of Electricity: Robert Andrews Millikan (1868–1953)

The idea that electricity, like matter, might be composed of discrete units seems to have originated with Benjamin Franklin about 1750. He wrote "The electrical matter consists of particles extremely subtile, since it can permeate common matter, even the densest, with such freedom and ease as not to receive any appreciable resistance." Despite this categorical statement, no experimental evidence existed for this point of view until Michael Faraday's discovery of the laws of electrolysis more than 80 years later. Then it was found that the transference of a fixed quantity of electricity through a given electrolyte, regardless of its concentration, always caused the appearance of a definite amount of material at the cathode of the electrolytic cell. The complete significance of this discovery could not then be appreciated because the atomic theory of matter was not on a firm basis, despite the demonstrated value of John Dalton's theory in advancing the science of chemistry. Moreover, Faraday himself was very cautious about the atomic theory; if he entertained the theory at all, he thought of atoms as proposed by Roger Boscovich, that is, as point centers of force. The idea of a fundamental electric charge appears to have been foreign to Faraday's thinking as it certainly was to the great master of electrical theory who followed him, James Clerk Maxwell. In fact, in speaking of the discrete charge explanation of electrolysis, Maxwell states that "it is extremely improbable that . . . we shall retain in any form the theory of molecular charges . . . "

Nevertheless, the hypothesis of discrete charges was not completely neglected. G. Johnstone Stoney, in particular, emphasized it in 1881 and not only estimated

Robert Andrews Millikan (1868–1953)

the size of the fundamental unit as 3×10^{-11} esu (electrostatic units), more than ten times too small, but also designated this quantity of charge as the "electron." Indeed, he went further and proposed as basic units of measurement not length, mass, and time but the (presumably) more basic natural units, the velocity of light, the coefficient of universal gravitation, and the elementary electric charge. Inasmuch as Faraday's laws of electrolysis cannot do more than make statements about the average charge of the great number which must pass through an electrolytic cell in order to deposit appreciable amounts of matter, Stoney's hypothesis of an elementary electric charge could not have been taken too seriously. As an example of how wrong he could have been, the reader need only recall the constancy of atomic weight determinations together with Dalton's dictum that the atoms of a given substance were "all alike in weight, figure, mass, etc." and compare it with what was found when isotopes were discovered.

Even the discovery of the electron by Thomson did not require that there be a single fixed elemental electric charge. His experiments measured the ratio of charge to mass, e/m, of the cathode ray particles; the fact that particles appeared in all gases with the same charge-to-mass ratio did not exclude some compensating adjustment of charge and mass, however unlikely that might be. While it was assumed that the charge e (as well as the mass m) of all the particles was constant, direct experimental measurements were needed not only to show the constancy of the charge but also to find its value. Once the magnitude of e was determined, then the mass of the electron could be deduced from the known value of the e/m ratio. The knowledge of both these constants was of the utmost importance for the further understanding of atomic structure and the development of atomic theory.

The necessity of determining the value of e was, of course, immediately clear to J. J. Thomson following the e/m measurements. He, as well as others in the Cavendish Laboratory, made several attempts to determine its value. All were based on the application of C. T. R. Wilson's discovery that a sudden expansion of air in a chamber saturated with water vapor will produce a cloud, the droplets of which condense around ions present in the moist air. In an experiment by H. A. Wilson in 1903, the ions were produced by X-ray irradiation of the saturated air. The top of the cloud produced by the expansion in the cloud chamber was timed as it fell between two horizontal metal plates, first under the action of gravity alone and then under the combined action of gravity and an imposed electric field acting on the ions. From this a value of e was deduced but its precision was low, the separate determinations fluctuating between 2.0×10^{-10} and 4.4×10^{-10} electrostatic units (esu of charge).

Attracted by the possibility of greatly improving the accuracy of the measurement of e, and thus finding m, and of the importance of the whole problem, Robert Millikan took up the experiment at the University of Chicago in 1909. His original idea was to use Wilson's method but to improve its accuracy by including a measurement in which the top of the cloud was held steady by the application of a sufficiently high voltage. This required the building of a 10,000-volt small cell storage battery which at that time was an undertaking of some difficulty. When the experiment was repeated with the application of this high voltage the workers found, instead of holding the cloud steady, it forced the immediate disappearance of the cloud. The ions in the cloud were pulled to the charged plates by the strong electric field. However, instead of destroying the experiment, this event unexpectedly disclosed a much more accurate way of carrying it out, for it was found that, in the field of the observing telescope, a few drops remained for some time and appeared to be suspended almost motionless. These were the few for which the upward force of the electric field just balanced the downward pull of gravity. Thus, rather than experimenting on a cloud of drops, it was possible to make measurements on a single charged drop and to keep this drop under observation for some time. This change in the experimental procedure ultimately made it possible to show unambiguously that nature supplies electric charge only in one fixed size, that all charges, no matter where they may occur, are only multiples of this charge, and that electric charges of any other magnitude do not exist.

To see how this fundamental electric charge is determined, we shall review Millikan's experiment very briefly. Suppose that oil droplets (water droplets evaporate) are sprayed into the air space between two horizontal metal plates. Once in this space the droplet is a free body and will fall in the normal way under the influence of gravity. However, since the droplet is very small it will soon reach a terminal velocity v_1 because of air resistance, and its subsequent downward motion will be at constant speed. In fact, the motion of any object moving at low speed in a resisting medium assumes a speed proportional to the force driving it. In the case of the oil droplet, the driving force in free fall is its weight w_1, which by Newton's second law is given by $w = mg$, where m is the mass of the droplet and g is the acceleration of gravity. Suppose next that the plates are charged so that the upper of the two has positive polarity and the droplet negative. The electric field F between the plates then acts to drive the droplet upward, the net upward force being $Fe_n - mg$, where e_n is the number of electronic charges carried by the drop. This force determines an upward terminal velocity v_2.

Since in this range of speeds the "constant" of proportionality between force and terminal speed is, in fact, constant, the two proportional relations $mg \approx v_1$ and $Fe_n - mg \approx v_2$ may be divided, giving $v_1/v_2 = mg/(Fe_n - mg)$. If this equation is solved for e_n we find $e_n = (mg/Fv_1)(v_1 + v_2)$. If now the drop changes its charge by capturing a charged ion, then $e'_n = (mg/Fv_1)(v_1 + v'_2)$. Hence the value of the charge on the ion e'_i is the difference of the last two relations or $e_i = e'_n = (mg/Fv_2)(v'_2 - v_2)$. Since in a given experiment mg and Fv_1 are constants, the value of e_i is proportional to $(v'_2 - v_2)$. The latter quantity was found to have a minimum value of which all other values were multiples. It thus became clear that ionic electric charge was somehow regulated by the laws of nature to occur in the aforementioned single fixed size.

The main work on the determination of the electric charge was finished in 1912, although subsidiary investigations to increase its accuracy were continued until 1917 when Millikan left his academic work at Chicago to devote his energies to organizing scientific work for the United States government during World War I. During the period of 1912 to 1917 he also carried out another important ivestigation, the verification of Albert Einstein's photoelectric equation, together with a determination of Planck's constant. Einstein's photon theory was not taken very seriously at that time and this research did much to convince physicists of its importance.

During World War I Millikan's most important responsibility was the organization of antisubmarine research. He ultimately became Chief of Science and Research, Division of the Signal Corps, U.S. Army. He was also instrumental during the war period in founding the National Research Council. Following the war, he returned briefly to the University of Chicago, but a very strong bid for his services was made by the California Institute of Technology; in September 1921 he took the position there as director of the Norman Bridge Laboratory. He was then 53 years old and had been at Chicago for 25 years. At Caltech he had practically a free hand but, of necessity, much of his time was devoted to administration. Through his contact with physicists both in the United States and in Europe he was able to build an outstanding physics department, strengthened by the addition of Einstein, H. A. Lorentz, Arthur Sommerfeld, and others as visiting professors. In 1923 Millikan was awarded the Nobel Prize in recognition of his work on the electronic charge and the photoelectric effect. In these years he began research in the new field of cosmic rays. Perhaps the high point in his cosmic ray work came in 1932 when Carl D. Anderson, working under Millikan's direction, discovered the positive electron, or positron. This breakthrough was followed a few years later by Anderson's discovery, with S. H. Neddermeyer, of the first of the heavy electrons, the muon.

Millikan continued at Caltech until his retirement in 1945. He died in Pasadena at the age of 85 on December 19, 1953. His influence on American physics was profound and reflects his personal characteristics of independence, vigor, and industry, qualities that were instilled in him as a youth. He was a man of great self-assurance and conveyed the impression of an able and forceful business executive. These qualities were so evident that an amusing canard circulated about him among the graduate students in physics in several of the country's leading universities. It was said that the unit of pomposity was the "kan" but that this unit was so large as to be impractical for ordinary use. Therefore the practical unit was taken as 1/1,000 of this value or the "millikan."

Millikan was born in Morrison, Illinois, on March 22, 1868, the son of a struggling Protestant country minister. When he was five the family moved briefly to McGregor, Iowa, and again in 1875 to Maquoketa, Iowa, where he grew up with two brothers and three sisters in the typical fashion of small-town life of that period. He finished high school at 17 and went to Oberlin College in 1886. In his sophomore year he was entrusted with the teaching of physics, a subject that he had never studied. He continued teaching until 1893 when he took his M.A. degree two years after graduation. He was then 25 years old.

Having been offered a fellowship in physics, Millikan spent the next two years at Columbia University and obtained his Ph.D. there in 1895. After three terms of postdoctoral study in Germany at Jena, Berlin and Göttingen, he returned to take a position as assistant to Albert Michelson at the University of Chicago in September 1896. His first 12 years there, from 1896 to 1908, were spent largely in teaching, textbook writing, and getting several research problems under way, none of which yielded very promising results. In 1909 he began his outstanding work on determining the electronic charge.

8 Two Far-Reaching Discoveries

In this life we want nothing but facts, Sir; nothing but facts.
—Charles Dickens, *Hard Times*

The Discovery of Cosmic Rays: Viktor F. Hess (1883–1964)

When we look at the stars on a clear night, we might conclude that most of the energy in interstellar space consists of electromagnetic waves, because the stars constantly pour out such energy. But our conclusion would be wrong because in each cubic centimeter of space there is about as much energy in the form of very energetic charged particles as there is in the form of light. When these ions reach the earth, we call them cosmic rays.

There are many unsolved mysteries about these particles, among which are their origin, their age, and their relationship to the stars and galaxies. We know today that these rays appear to strike the earth with equal intensity from all directions of space and with energies ranging up to a billion trillion electron volts, far larger than any other known energies in nature.

The history of the study of cosmic rays goes back to the work of Viktor F. Hess, a Viennese physicist who was puzzled by certain discrepancies in the behavior of gamma rays emitted by radioactive sources in the earth's crust and in its atmosphere. He suggested, in a paper written in 1911, that most of the penetrating radiation observed at the earth's surface is gamma radiation emitted by terrestrial radioactive atoms; he pointed out, though, that not all of it could come from the earth because its intensity appeared to fall off only very gradually with increasing distance from the terrestrial source, as measured in balloon flights. Although Hess felt quite certain about his hypothesis, he decided, before pursuing it further, to check the absorption of gamma radiation by the earth's atmosphere, for he could not exclude entirely the possibility that gamma radiation was not absorbed as effectively by air as he had thought. A smaller absorption coefficient than had been assumed would give a greater radioactive gamma ray intensity with increasing distance from the earth's surface than that calculated previously.

His absorption measurements, in agreement with those made by previous investigators, showed that the penetrating gamma radiation from the earth's radioactive material falls off very quickly with increasing height (at 500 meters only a small percent of this terrestrial radiation survives). Hess decided that it was worthwhile to

Viktor F. Hess (1883–1964)

check his hypothesis by a series of balloon-flight measurements. In 1912 the results of these measurements were reported in a classic paper, "Penetrating Radiation in Seven Free Balloon Flights," in *Physikalische Zeitschrift*.

To measure the cosmic radiation, Hess used three ionization chambers. Two of these had walls with thicknesses of 2 millimeters, so that such things as beta rays and soft, that is, nonpenetrating gamma rays and X rays were excluded. The third chamber was made of very thin walls to check on any soft rays and beta rays (electrons) that might have been present.

The data from all the balloon flights were consistent in showing that, after one deducted from the measurements radiation produced by the earth, the intensity of the penetrating radiation increased with height. Hess further demonstrated that the intensity of this penetrating radiation was the same at night as during the day and did not decrease during an eclipse. He therefore concluded, correctly, that it did not originate in the sun but came from regions beyond the solar system.

Since Hess's discovery of cosmic rays, a great deal of work has been done in this field. Originally, the exact nature of these rays was not known, but we now know that they are mostly very energetic protons intermixed with small quantities of heavy positive ions and electrons and positrons. When these very energetic particles strike the earth's atmosphere, they give rise to vast showers of other atomic and subatomic particles. Thus, they create positron-electron pairs, mesons of all kinds, and many of the strange particles, so puzzling to physicists today, called hyperons. They also create proton-antiproton and neutron-antineutron pairs.

Since, for many years, the energies of cosmic ray particles far exceeded anything that could be produced in laboratories on the earth, people interested in high energy physics quite naturally worked with cosmic rays. Thus, Carl Anderson discovered the positron and the mu-meson by analyzing the various tracks on cloud chamber photographs, and the pi-meson was discovered in cosmic rays in 1947 by C. F. Powell and G. P. Occhialini. Other new particles have also been discovered in these rays, but today, with the use of high-energy accelerators, physicists are turning more and more to the laboratory for detailed studies of new particles.

The origin of cosmic rays is still essentially an unsolved problem, although various theories have been advanced to account for them. Enrico Fermi proposed the theory that positively charged particles in the Milky Way, such as protons, continuously collide with clouds of gas-carrying magnetic fields and are thus accelerated to very high energies. It has been suggested, also, that cosmic rays arise from supernovae explosions and other very energetic distant events such as quasars (sources of high frequency radio waves). Finally, some physicists and astrophysicists believe that cosmic rays were born with the universe itself.

Viktor Franz Hess, who discovered cosmic rays, was born on June 24, 1883, in Waldstein Castle near Peggan in Steiermark, Austria; his father Vinzens Hess was a forester in the household of Prince Öttingen-Wallerstein. After studying at the gymnasium at Graz, Hess entered the University of Graz and received his doctorate in 1910. He began research in the field of radiation at the Physical Institute in Vienna under Professor Von Schweidler.

The techniques and instruments needed for the study of radioactivity were also suitable for the detection of cosmic rays. It was quite natural, therefore, that Hess progressed from the study of radioactivity to cosmic rays. Since the entire earth is radioactive, any attempt to assess its total radioactivity leads inevitably to the detection of cosmic radiation. Shortly after beginning his work in radioactivity, Hess, in 1911, began to measure cosmic radiation and in 1912 conclusively demonstrated, through tests made during balloon flights, that these penetrating rays come from interstellar space.

From 1910 to 1920, Hess was assistant at the Institute of Radium Research of the Viennese Academy of Sciences. In 1919 he received the Lieben prize for his discovery of cosmic rays and a year later became assistant professor of experimental physics at the University of Graz. Hess remained at Graz for only a year and then took a leave of absence for two years to work in the United States where he became director of research for the U.S. Radium Corporation in New Jersey. At the same time, he acted as a consulting physicist for the Bureau of Mines of the U.S. Department of the Interior in Washington, D.C.

Upon completion of this work he returned to Graz where he was appointed full professor of experimental physics in 1925. Six years later, Hess accepted the chair at Innsbruck University and, at the same time, became director of the newly established Institute of Radiology. Under his guidance and initiative, the station at Hafelekar Mountain was founded for the further study of cosmic rays. Just before World War II Hess came to the United States and was appointed professor of physics at Fordham University. He became a United States citizen in 1944 and remained at Fordham until his death in 1964.

The Cloud Chamber: Charles Thomson Rees Wilson (1869–1959)

It would hardly seem credible that researches in atmospheric physics dealing with the formation of clouds and mist should lead to the invention of one of the most important tools of atomic physics. Nevertheless, it was in pursuit of studies on the formation of clouds that C. T. R. Wilson, starting in 1894, began the development of a most wonderful instrument that became known as the Wilson cloud chamber. With this device some 16 years later the marvel of seeing the path of charged atomic particles and of atomic collisions was realized through the trails of foglike droplets that the particles left in their wakes. Even more important than seeing such events, the apparatus can depict them photographically. Once photographed, tracks left by the particles in the chamber are recorded for future careful study.

In the hands of skilled experimenters the cloud chamber became a powerful tool for studying atomic collisions and analyzing the results of such collisions. It was the instrument by which much of the science of cosmic radiation was unraveled, and it was through such studies that Carl D. Anderson discovered the positive electron or "positron" in 1933. A few years later, with S. H. Neddermeyer, his cloud-track studies disclosed the first of the heavy electrons—the muon.

With the same device another prophecy of quantum physics, pair production— the generation of an electron-positron pair from the conversion of a high-energy photon—was made evident. These phenomena represent only a sampling of the highly useful results achieved by the use of the cloud chamber. It is therefore no wonder that Ernest Rutherford once characterized the instrument as "the most original and wonderful in scientific history." But, as is the case with most inventions, technological advance led to improved ways of performing old tasks; consequently, the cloud chamber has given way in modern high-energy physics to the bubble chamber and the spark chamber. But whatever path improvements may take, it is to the pioneer who showed the way that most honor should be given and it is in this spirit that all physicists honor the memory of C. T. R. Wilson.

"C. T. R.," as he came to be called, was the son of John Wilson, a sheepfarmer whose family had lived on the land for generations in the neighborhood of Edinburgh, Scotland. His mother, Annie Harper WIlson of Glasgow, was John Wilson's second wife; Charles, their son, was the youngest of eight children and was born on February 14, 1869. When he was four his father died and the family moved to Manchester. In 1884, at the age of 15, he entered Owens College, from which he graduated at the age of 18. After an additional year at Owens he entered Sidney Sussex College, Cambridge, on a scholarship and decided to study physics rather than medicine, which had been his previous interest. He took his Cambridge degree at 22 and remained at the University for another year as a demonstrator in chem-

Charles Thomson Rees Wilson (1869–1959)

istry and physics. Then, feeling that his prospects at Cambridge were uncertain, he accepted a position as assistant master at the Bradford Grammar School.

Before leaving for this post he spent a few weeks at the Meteorological Observatory at the summit of Ben Nevis, the highest of the Scottish hills. Here, in September 1894, he saw the wonderful optical phenomenon of the sun playing on the clouds around the hilltop and their changing forms in various kinds of weather. Wilson was so impressed that he resolved to study cloud formation in the laboratory, a decision that determined the course of his life's work in physics.

Although he enjoyed his relations with the students in his new secondary school position, he soon came to realize that such work was incompatible with his urge to do research. Even though he had no prospect of employment, he left the school and returned to Cambridge. Fortunately, he soon obtained a demonstratorship and a connection with the Cavendish Laboratory where he enjoyed the advice and guidance of its director, J. J. Thomson. Early in 1895 Wilson assembled an apparatus for expanding moist air under controlled conditions in which the expansion ratios could be measured. The formation of clouds and drops was then investigated as a function of the expansion ratios.

Early in 1896, following the news of the discovery of X rays, he found that the cloud chamber would produce a fog that took many minutes to fall when exposed to the rays. It became clear from this experiment that ions would act as condensation nuclei for moisture in the chamber. Other experiments in succeeding years with "uranium rays" and ultraviolet light produced similar results. In support of the ion-condensation theory it was also found that if the ions were removed with an electric field, no cloud was formed.

The quality of Wilson's work must have deeply impressed his seniors, for he was elected a fellow of the Royal Society in 1900, as well as fellow of Sidney Sussex College and University lecturer and demonstrator. He was then 31 years old.

Not until the spring of 1911 did the idea occur to him that the condensation of droplets on ions might be used to show the path of an ionizing particle. An alpha particle projected into the air dislodges electrons from atoms in or close to its path. In the cloud chamber these electrons, as well as the positive ions formed by their detachment, act as condensation centers for water vapor if the degree of supersaturation is sufficient. This supersaturation is achieved by suddenly expanding saturated air through a valve connected to the volume of the chamber cylinder or by using a retractable piston in the cylinder. The sudden expansion produces a temperature drop in the chamber air without immediate precipitation, thus supersaturating the volume. The apparatus is generally arranged so that a photograph is taken of the cylinder directly following the expansion; the mobilities of the ions are sufficiently small so that in this time the tracks do not become diffuse. An electric field maintained in the chamber volume sweeps out the ion droplets, clearing the chamber for the next expansion. Wilson, in his Nobel lecture, noted his immediate success: "The first test was made with x-rays" ... and ... "the cloud chamber filled with little wisps and threads of cloud," showing the path of electrons ejected from the atoms of air by the rays. This was the high point of Wilson's work in physics.

Wilson continued his teaching and research at Cambridge until his retirement at 65 in 1934. During these years he was honored with the award of the Hughes medal in 1911, the Royal medal in 1922, and the Copley medal in 1935. In 1925 he was appointed Jacksonian professor of natural philosophy and in 1927 shared the Nobel Prize in physics with Arthur H. Compton.

In 1908, at the age of 39, Wilson married Jessie Fraser Dick. The couple had three children, a boy and two girls. To those who knew C. T. R. at Cambridge perhaps his most notable characteristics were his patience and even temper. He was essentially a lone worker; unlike other leaders in Cambridge research, he had no students until well into the 1920s. His unassuming manner and angular features, particularly in his later years, suggested his descent from the generations who had lived on the Scottish land before him. Indifferent to acclaim and honors, he was a gentle and serene man, whose work proceeded from a genuine love of the natural world and from the revelation that he had experienced at Ben Nevis in his youth.

Two years after his retirement at Cambridge, C. T. R. returned to Scotland, settling in the village of Carlops in the region of his birthplace near Edinburgh. Here he enjoyed a vigorous and healthful old age. He died at 90, after a brief illness, on November 17, 1959.

9 *The Nuclear Atom*

They are ill discoverers that think there is no land, when they can see nothing but sea.

—Francis Bacon, *The Advancement of Learning*

Strange Results from Alpha-Particle Scattering: Hans Geiger (1882–1945)

Entirely new vistas into the structure of matter were opened in October 1897 by J. J. Thomson's discovery of the electron. This subatomic particle, less massive than hydrogen, the lightest known element, somehow had to be a part of all atoms. But if it were a part of all atoms, then these atoms must have a structure. It was the task of physics to puzzle out what this structure might be. The first suggestion that appeared in the scientific journals came from Lord Kelvin in 1902; he envisioned an arrangement in which diffuse positive electricity was spread homogeneously through the volume of the spherical atom. The exact nature of this positive electricity was not delineated, but it seems to have been pictured as a kind of viscous material resembling jelly. In this sphere of positive charge, the small discrete negative charges, the electrons, were supposed to be embedded; moreover, the positive charge was of such magnitude as to counterbalance exactly the negative charges of the electrons so that the whole atom was electrically neutral.

This arrangement was investigated more carefully the following year by J. J. Thomson and the stability calculated in detail for various numbers of electrons. As a consequence the model came to be called the "Thomson atom" and has generally been known by this name ever since. It must be clear, however, that no direct physical experiments on atoms led to this model; instead, it was a product of sophisticated imagination and was shaped to fulfill criteria of mechanical and electrical stability. One of the ways in which its credibility could be tested was to use the recently discovered alpha and beta particles from radioactive atoms. If a stream of these particles was projected at a piece of metal foil, physically thin but, in reality, many atoms thick, the particles incident on the foil would have to be deflected by the atoms or pass through them. An analysis of the scattering could then be made to provide information on the structure of the foil atoms.

Upon his arrival in Manchester, Ernest Rutherford found in the laboratory a young German, Hans Geiger, who had come to work for a few years in England. Geiger was a very able experimental physicist and, at Rutherford's suggestion, began work on an electrical device to count individual charged particles. Out of this

Hans Geiger (1882–1945)

work came the famous Geiger counter which contributed so much to experimental atomic physics, to particle physics, and to the study of the cosmic rays in the years that followed. Another problem that was attacked was a suitable alpha-ray source to provide a fine beam for scattering experiments from thin foils. A young undergraduate, Ernest Marsden, assisted in the scattering experiments; the pair devised a simple experimental arrangement by which it was possible to obtain results very quickly.

Radium emanation enclosed in a conical glass tube capped with a piece of thin mica supplied alpha particles that were deflected, at an angle, to a thin metal foil. Those alpha particles deflected by the foil so that they emerged on the same side as the incident particles, that is, at about 90° to the original direction, were intercepted by a zinc-sulfide screen. A thick lead plate prevented particles from the source from reaching the screen directly. Each alpha particle impact on the screen could be detected by the scintillation it produced. By observing the screen with a low-power microscope the scintillations could be counted and the dependence of the number thus deflected on the atomic weight of the atoms of the foil noted.

Several results were immediately apparent: First, large deflections of alpha particles did occur; second, the number so deflected increased as the atomic weight of the foil increased. The effect of increasing the thickness of the scattering foil was then examined and it was found, as was known for beta rays, that the scattering was not a surface but a volume effect. The fact that a thickness of 6×10^{-5} cm of gold was sufficient to produce a deflection of 90° or more was a matter of great surprise. Finally, with an altered arrangement, the number of particles scattered by a platinum foil through an angle of 90° or more was determined. The result showed that about 1 in 8,000 was so deflected.

Geiger next carried out an experiment aimed at determining the average deflection of alpha particles that traversed various thin metallic foils whose absorption was equivalent to several centimeters of air. As was expected, the average deflection measured was only a few degrees. Nevertheless, the large deflections found in the Geiger and Marsden experiment disturbed Rutherford; as Marsden noted in his remarks at the Rutherford Jubilee International Conference at Manchester in 1961, "he thought over these remarkable results for many weeks." Rutherford knew that such deflections were too big to be consistent with the Thomson model of the atom, and he began looking for an alternative to explain the data.

Hans Geiger was born in 1882 in Neustadt, Germany. He was the son of a professor of Indo-Germanic languages at the University of Munich and grew up in an academic atmosphere. His doctoral research, under Wiedemann at the University of Erlangen, was concerned with the conduction of electricity in gases and his doctorate was awarded in 1906. He then went to Manchester to work with Arthur Schuster; following Rutherford's arrival there in 1907, he entered into a most fruitful collaboration on the latter's researches. In the five years of their association, Geiger wrote papers on the statistical nature of radioactive decay, the determination of the number of alpha particles emitted per second from a gram of radium, the Geiger range law for charged particles passing through matter, the well-known Geiger-Nuttall relation between the decay constant of a radioactive nucleus and the range of the emitted alpha particles, as well as the paper on the scattering experiment.

Geiger was 30 years old when he left Manchester in 1912 to become director of the newly established laboratory of radioactivity at the Physikalische-Technischen Reichsanstalt in Berlin. Geiger remained at this position until 1925 when he accepted the chair in physics at the University of Kiel. It was here, in 1928, that he developed, with his student W. Müller, the famous instrument that has become a household word, the "Geiger counter." In 1929, he went from Kiel to Tübingen where he began research on cosmic rays. In 1936 he moved again, this time to the Institute of Technology in Berlin; he remained there until rheumatoid arthritis caused his death at 63 on September 24, 1945.

Geiger was a man of warm and humane feeling, with a genuine concern for the students who worked under his direction. The testimony of his students indicates that working under him was both an exciting and heart-warming experience. He was not only a leader in research but he lectured widely on many different scientific topics for popular audiences. In 1937 he became editor of the *Zeitschrift für Physik*, one of the outstanding physics journals of that time. He was a devoted man of science who left his mark on the history of physics.

The Nuclear Atom: Ernest Rutherford (1871–1937)

By the time Ernest Rutherford was ready to leave McGill University to come to Manchester, his research into the nature of the emanations from radioactive atoms and his theories of radioactivity had made him quite famous. Not only was he invited to give lectures at the outstanding American universities, but he was also in great demand as a popular lecturer and as an author of popular articles. Although, as he stated in various letters to his mother, he tried to discourage these requests because they were a drain on his time, he seldom denied them, and he took great delight in entertaining his audiences. He enjoyed counting the house and, in one letter to his mother, remarked that, "I had the largest audience they had ever raised at McGill. They were stored everywhere, including some who were looking through a ventilator in the top of the roof."

In 1904 he was invited to give the famous Bakerian lecture to the Royal Society of London; during this lecture he expounded his theory of the chain reaction of radioactive products and laid down the basic principles of the arrangement of elements in radioactive families. While in England during that year, he gave a lecture to a full audience of the Royal Institution. He proposed the radioactive method of estimating the age of the earth and demonstrated that Lord Kelvin, at that time the grand old man of British physics, was wrong in concluding that the earth was only a few million years old and could, at most, exist for another hundred million years. Lord Kelvin had arrived at his result on the assumption that all the energy from the sun had come from gravitational contraction, which Rutherford knew was false. On seeing Lord Kelvin in his audience, Rutherford, on the spur of the moment, softened his blow to Kelvin's analysis by stating that Kelvin had based his theory on the hypothesis that "no new source of energy would be discovered. That prophetic utterance refers to what we are considering tonight, radium." In telling the story of this lecture later, Rutherford remarked that Kelvin, who had been asleep up to that point in the lecture, "sat up and beamed upon me."

In that same year, Rutherford received the Rumford Medal of the Royal Society for his theory of radioactive decay. He was happy about this award, for not only was it a great distinction but it also carried a monetary award of £70, which was quite welcome at a time when professors' salaries were not very great.

Following upon these honors, he was invited to give the Silliman lectures at Yale, which meant that he would have to be away from his wife three more weeks than he had meant to be. In an amusing letter to her, he then asked whether she preferred an additional $2,500 or three more weeks of his company. Two years later, in 1906, he was awarded his first honorary degree by the University of Philadelphia. In writing to his mother about it, he remarked that he was rather youthful for such honors "as they are usually the special perquisites of septuagenarians. They [honorary degrees] don't worry me much I can assure you but one is supposed to value them very highly—I imagine the esteem is largely dependent on whether you feel you deserve them or not."

During this period of intense scientific and academic work, Rutherford always found time for play. He was not the sort of man who did nothing but physics in every waking moment of his life. He enjoyed many activities and was an omnivorous reader. He was fond of light reading and while at McGill kept four libraries busy supplying him with novels, detective stories, historical fiction, biographies, and books of general interest. He was also fond of bridge and golf, and he thrived on open-air activities. He felt that spending all one's time at a desk was stultifying, and he would often be heard advising his colleagues to leave their offices and "go home and think."

On May 20, 1905, two years before he left McGill, Rutherford received a letter from Sir William Ramsay of University College, London, supporting the application of a young German physicist, Dr. Otto Hahn, for a research position at McGill. Rutherford welcomed Hahn to McGill and introduced him to the research techniques in radioactivity that were later, in 1939, to lead Hahn to the discovery of uranium fission.

In 1907, when Rutherford decided to accept the Langworthy chair of Professor of Physics at the University of Manchester, he was recognized by the greatest physicists of the time as their equal. He had already published about 50 papers, each of which was of first-rate importance. But his greatest achievements at Manchester were yet to come. He was very happy to be in Manchester, which was then one of the great centers of scientific research, and his constant contact with the top physicists of the day was a source of great inspiration to him as well as to them. Arriving at Manchester, Rutherford immediately pitched into his research, teaching, and lecturing activities, most of which stemmed from his work on radioactivity. This was just the kind of life Rutherford wanted. In a letter to his mother dated October 29, 1907, he reported that everyone was kind to him and that he was enjoying his life thoroughly with a good many "outside lectures in hand."

Shortly after coming to Manchester, in fact six months after his arrival in 1908, his great scientific achievements were recognized by the Turin Academy of Sciences, which awarded him the Bressa Prize for his discovery of the "mutability of matter and of the evolution of the atom." This award was followed by an honorary degree from Trinity College, Dublin, and then, in 1908, the Nobel Prize for chemistry, which is certainly one of the great ironies in the history of physics and chemistry. In his acceptance speech at Stockholm, Rutherford declared that he had dealt with many kinds of radioactive transformations with different periods of time but that the quickest transformation he had met was his own—a transformation from a physicist to a chemist in a single moment.

After returning from Stockholm, Rutherford became interested in the properties of the thorium family of radioactive elements, which was the principal subject of Hahn's experimental investigations. Hahn was at Berlin then, and he had working with him at that time a young assistant, Fräulein Lise Meitner, who was to play a prominent role some 30 years later in the uranium-fission problem. Rutherford and Hahn exchanged a series of letters on the properties of the thorium family, and Rutherford, as usual, was greatly interested in the emanations from these radioactive atoms. He became greatly engrossed at that time in the passage of alpha particles through matter and was very much impressed with the ease with which these particles passed through thin foils of metal and of very thin glass. It was at this time, two years before he published his paper on the nuclear atom, that his ideas on the structure of matter began to crystallize. The passage of alpha particles through matter clearly indicated to him the possibility of using alpha particles as a probe and also suggested to him the possibility that the atom itself is mostly empty space. This idea came to him in 1909, a period of great happiness for Rutherford who went calmly along his way enjoying his research work, his teaching, his lecturing, his family life, and his regular correspondence with his mother, whom he kept informed about most of his activities. At this point in his life, Rutherford became interested in a new form of recreation—motoring. He bought a Wolseley-Siddeley automobile and spent many hours with his family driving around the countryside.

Once Rutherford became convinced of the possibility of probing the structure of the atom with alpha particles, he followed the work of Geiger on the large-angle scattering of alpha particles with great interest and began to formulate his theory of the nuclear atom. Although he ultimately accepted the picture of a positively charged nucleus, he was at first inclined to the idea of a negatively charged nucleus. As he stated in a letter to Bragg, "I am beginning to think that the central core is negatively charged . . . ," but soon after that he changed his mind. In 1911, Rutherford had worked out most of the details of his theory of scattering of alpha particles from a nuclear atom and communicated his ideas in a general form in a letter to Bragg. In this letter, he also discounted Crowther's analysis, which arrived at scattering results in agreement with the J. J. Thomson atom. Rutherford put this down to the "use of imagination and the failure to grasp where the theory was inapplicable."

The thing that impressed Rutherford most about the alpha-particle scattering was that some of the alpha particles, even though only a few, returned almost back toward their source, behaving the way a bullet might if it bounced directly back when fired at a sheet of paper. To him this outcome could mean only one thing: there was an enormous force in the atom. He announced this idea at a Sunday supper at his house in Manchester in 1911 when he was dining with a few scientists, including the theoretical physicist, C. G. Darwin, who later recounted that he was present "half an hour after the nucleus was born." Here Rutherford presented the idea that it would take 100 electron charges on the gold nucleus to give the observed result.

Darwin goes on to say, in his recollection of this event, that he was surprised at Rutherford's mathematics and the way he worked out his results, for Rutherford was neither a mathematician nor a theoretical physicist. In looking for some kind of picture that would show him the effect of a large force on a body, he probably thought of a gravitationally controlled comet swinging around the sun and coming directly back again—just like the alpha particles. This naturally led him to the idea of a hyperbolic orbit and, possibly, first suggested the idea of a negative nucleus attracting the positive alpha particle the way the sun attracts a comet. Of course, he soon saw that the only important thing in the dynamics of the problem was the inverse square law of force and that the orbit of the alpha particle must be a conic section whether the force was attractive or repulsive. There is evidence that at this stage of his analysis he went back to Isaac Newton's *Principia* and used Newtonian mathematics to obtain his final results.

Having conceived the idea of a hyperbolic orbit for the alpha particles, Rutherford recalled a theorem in geometrical conics which he had learned at school. The theorem relates the eccentricity (that is, the shape) of the hyperbola to the angle between the asymptotes. This is the content of an equation in Rutherford's paper on the scattering of alpha and beta particles. Using this mathematical relation and the principles of conservation of momentum and energy, he then obtained a complete solution of the alpha-particle scattering problem.

That very day Rutherford asked Darwin to check his conclusion and also to work out the results if the law of repulsion were an inverse cube law. Rutherford was particularly interested in the question, "How close to the nucleus can an alpha particle approach?" He got this answer from his theory and the measurements of the scattered alpha particles. The answer, 3×10^{-10} cm, showed him how small and compact a nucleus is, and thus the nuclear atom was born.

Rutherford was supremely confident of his conclusions and was sure that he had the right answer, which not only explained the scattering of alpha particles but also gave a correct picture of the structure of the atom. This is evident in Geiger's account of events. One day in 1911, Rutherford, very sprightly and happy, came into Geiger's laboratory and informed Geiger that he knew what the atom looked like and how to explain the large deflections of the alpha particles. Geiger began his crucial experiments to test Rutherford's analysis on that same day and thus verified one of the greatest contributions to physics. This must be considered the very peak of Rutherford's research and is certainly to be counted among the greatest of all scientific achievements.

An interesting insight into the general attitude of British physicists of that period toward new scientific theories is given by Rutherford's remarks about the theory of relativity at the Brussels conference in 1910. When, at a luncheon, Rutherford was teasing Willy Wien about relativity, Wien, after explaining that Newton's law of the addition of velocities was wrong, remarked that "no Anglo-Saxon can understand relativity." "No!" laughed Rutherford, "they have too much sense." Later, Rutherford's attitude changed, for he welcomed the great contributions to physics stemming from the theories of Max Planck, Albert Einstein, and Niels Bohr without which his own theory of the atom would collapse.

In the Thomson model of the atom the positive electricity was considered to be somewhat like a jelly in which the discrete electrons were embedded. The electrons were held stationary in their equilibrium positions by the attraction of the positive electrical fluid surrounding them and the repulsion of the other electrons. Radiation from such an atomic model occurred whenever the atom was disturbed because the electrons would be forced out of their equilibrium positions and set vibrating. As they vibrated they radiated energy until they came to rest again. This picture is in complete accord with the classical electromagnetic theory of the electron as developed by James Clerk Maxwell and H. A. Lorentz. For this reason the Thomson model was favorably received by Thomson's contemporaries.

Thomson's purpose in developing this model was to explain the "scattering of electrified particles in passing through small thicknesses of matter." In scattering experiments, the crucial criterion for the atom model is the angle through which a charged particle is deflected from its original direction of motion as it passes through a metal foil used as the scatterer. Thomson assumed that the angle of deviation suffered by the charged particle was always caused by a large number of collisions with many atoms. Any single collision played only a minimal role in the total deviation, which was a cumulative effect. It can be shown that on the basis of the Thomson model the total deviation is not the average deviation produced in a single collision multiplied by the total number of collisions; rather, the multiplicand is the square root of this sum of collisions. If each collision resulted on the average in a deviation of $1°$, 100 collisions would give rise to a net deviation of only $10°$.

Rutherford pointed out the importance of this fact by calling attention to the observations of Geiger and Marsden. They had found, in their experiments with alpha particles passing through a layer of gold foil about 6×10^{-5} cm thick, that they could be scattered through an angle of $90°$ or more. If only small deviations occurred in each encounter, the alpha particle would have to undergo about 10,000 of the lesser collisions to produce such a large total deviation. This was highly improbable, as Rutherford pointed out, because of the extreme thinness of the gold foil. Rutherford contended that such large deviations must have been caused, therefore, by single direct collisions. He then proceeded to analyze the theory of single collisions on the basis of a model of the atom that was radically different from the Thomson model.

In this Rutherford model the positive electricity is not distributed over a large volume but, instead, is concentrated in a very small nucleus at the center of the atom. As Rutherford pointed out in his 1911 paper on the scattering of alpha and beta particles, the actual analysis is the same whether one assumes that the positive charge is concentrated at the nucleus and the electrons are on the outside, or vice versa.

A model of this sort cannot be in static equilibrium, since the electrons would all be dragged into the nucleus if they were not moving in stable orbits around it. Yet this kind of dynamical equilibrium is in serious contradiction with classical electrodynamics. Rutherford was aware of this problem but chose to ignore the difficulty for the time being. He stated that the "question of the stability of the atom proposed need not be considered at this stage. . . . "

By very simple but elegant arguments and with the most elementary mathematics, Rutherford showed that his model of the atom gives rise to the kind of deviations during single collisions that Geiger and Marsden had observed. The paper itself is exemplary in its simplicity, yet so profound that none could doubt that Rutherford's ideas must serve as the basis for a new and correct picture of the structure of the atom.

We should note, however, that the picture of the atom that Rutherford drew was still very tentative and vague. He speculated that not all of the positive charge was in the nucleus; that "a small fraction of the positive charge may be carried by satellites extending some distance from the center." Although the values he obtained for the charge on the nuclei of different metals are all too large, for example, 100 for gold, he correctly concluded that the nuclear charge should be "approximately proportional" to the atomic weight of the atom. But Rutherford was not sure that this would hold for the light elements and indicated that for these his simple theory of atomic collisions was no longer applicable.

Although throughout most of his paper Rutherford did not specifically mention the planetary theory of the atom, it is clear from his refence to the work of Nagaoka that he had this planetary model in mind. It is here that we have the starting point of modern atomic theory. Contrary to popular belief, however, the experiments on the scattering of alpha particles used by Rutherford were not his own but those of Geiger and Marsden. Rutherford's great contribution lay in showing that the Thomson model of the atom cannot possibly explain the large number of large-angle scatterings, whereas the nuclear model can.

It was shown that, at least from the standpoint of scattering, atoms behaved as though most of their mass was concentrated in a very small nucleus, although the Rutherford model did not distinguish whether this nucleus was positively or negatively charged. It soon became evident from many lines of evidence that the nucleus did indeed have a positive charge, that the nuclear model was in accord with other experiments, and that the electrons were charges of approximately the same size as the nucleus but moving in orbits such that the diameter of the atom was of the order of 10^{-8} cm. Thus the nuclear atom was discovered.

Atomic Structure: Niels Bohr (1885–1962)

Niels Bohr, at various periods of his life, was referred to as the "Great Dane," the "spirit of modern physics," the "symbol of modern physics," the "main architect of our work," and so on, in recognition of his outstanding contributions, which dwarfed all but those of Einstein. We may, indeed, refer to these two figures as the two great masters whose work formed the basis of most phases of modern physics. Bohr was more than a mere discoverer of new theories; he was a great teacher, a source of new ideas, a penetrating critic, a bold innovator, and a stimulating companion who probably inspired more scientific papers for his students and visiting scientists than any other physicist.

Physics and Bohr were both fortunate in the date of his birth, October 7, 1885, for he reached maturity and began his research career in 1905, just when the new and exciting concepts of the quantum theory and the theory of relativity were

Neils Bohr (1885–1962)

emerging from the tentative pen of Planck and the bold imagination of Einstein. For physics, what a heroic and golden time the 57 years of Bohr's scientific career spanned. In this period atomic theory as we now know it was created, mostly through his efforts and inspiration, and the causal approach of classical mechanics was replaced by the Bohr complementarity scheme of quantum mechanics.

Niels Henrik David Bohr was born in Copenhagen, the son of an outstanding professor of physiology at the University of Copenhagen. He grew up in an environment of great learning and culture, surrounded by people devoted to science, humanism, and mankind. Such outstanding men as the physicist Christiansen, the philosopher Hoffding, and the philologist Thomsen were constant visitors at the Bohr home; Niels was a devoted listener of the four-way conversations that were held almost every Friday night. This early exposure to humanistic tenets probably did much to influence Bohr's later warm regard for people. In the words of Professor J. Rud Nielson, one of his closest associates, Bohr was one of the "wisest and most lovable of men."

When Bohr began his high school studies, Planck had just announced the quantum theory. By the time Bohr entered the university in 1903, he was already grappling with these challenging new ideas. But he was by no means the sedentary student who devotes himself entirely to books. He was much too restless and active for such a life and was very fond of bicycling, sailing, skiing, and soccer. Indeed, he and his younger brother Harald, who was to become an outstanding mathematician, were both talented soccer players. A description of Niels Bohr's ebullience and physical exuberance is given by J. Rud Nielson.

When I think of Bohr as he appeared nearly fifty years ago, the speed with which he moves comes to my mind. He would come into the yard, pushing his bicycle, faster than anybody else. He was an incessant worker and seemed always to be in a hurry. Serenity and pipe smoking came much later. He was always friendly and less remote than most Danish professors in those days. (Nielson, 23)

Bohr never quite finished with a problem regardless of how long he worked on it or how many papers he published. He pursued each subject year after year, with dogged tenacity, looking at the questions from every angle, and considering the same points over and over again, until he had polished away the rough edges and achieved what, at least momentarily, appeared to be a finished product, only to come back to it again a few months or a few years later. This approach to his research work was already evident in the first paper that he published in 1906 at the age of 21. This paper was of both a theoretical and experimental nature, and he could not pull himself away from the laboratory to write up his results. Each time he completed some phase of the work, he saw some new features of the problem that he felt he must understand. Finally, his father sent him away from Copenhagen to his grandparents; here Bohr wrote the paper. The paper dealt with the surface tension of water, and it won him the gold medal of the Danish Academy of Science.

The American physicist John A. Wheeler, with whom Bohr collaborated on the very important problems of nuclear structure and nuclear fission just before World War II, describes this persistence of Bohr as follows:

> This place [a special room in Bohr's beautiful home] satisfied his definition of a work room, "a place where nobody can keep you from working." One of the most delightful features was a set of drawers, about 25 in number. Each was perhaps an inch deep and contained a draft manuscript having to do with one or another issue of physics. Each topic ripened from draft to draft—sometimes over many years—until the point was reached where, in Bohr's judgment, publication was at last appropriate. Among the manuscripts for which the basic idea reached far into the past, one of the most celebrated dealt with angular momentum and its exchange in atomic and nuclear transformation processes. It never reached the point of publication. However, it, like other drafts in this collection, defined conclusions and stated issues, and it furnished the starting point for the development of new ideas. (Wheeler, 44)

After Bohr had received his doctorate in 1911 for a brilliant dissertation on the electron theory of metals he went to Cambridge to work with J. J. Thomson and a few months later to Manchester to work with Rutherford. He remained in Manchester until 1913, the year in which he published the first of his epochal papers on atomic and molecular structure. He then returned to Copenhagen to begin his family life, for he had married Margarethe Nølund the year before; he lectured at the university, having been appointed a docent.

The scientific climate in Denmark at that time was not conducive to the full flowering of Bohr's genius; little was heard of his work and few of the top physicists were interested in what he was doing. Professor Knudsen, who had verified the kinetic theory experimentally, on being asked to explain Bohr's theory, remarked that he could not because he had not read Bohr's papers. Bohr must have felt this lack of interest for he returned to Manchester in the fall of 1914 as a reader in mathematical physics.

In 1916 Bohr again returned to Copenhagen, this time to stay. A professorship in theoretical physics had been created for him. From that time on, the level and general tone of physics in Denmark improved steadily. Although he had to work in a small room and had no paid assistant (he did most of his work at home), Bohr developed the basic ideas of the famous correspondence principle in the first two years of this period. A detailed account was published as a *Memoir* of the Danish Academy.

Bohr's greatest influence on world physics began with the dedication, on March 3, 1921, of his famous Institute for Theoretical Physics, which he had proposed as early as 1917. To spend a year or two at this exciting center of physics became the goal of every young physicist. Many of today's top names in physics will be found on the rosters of that institute. Bohr's institute not only attracted the brilliant youngsters but also the well-established physicists of all countries. It was not unusual to find a few dozen famous men working there at the same time including Kramers from Holland, Klein from Sweden, Dirac from England, Ehrenfest from Holland, Heisenberg from Germany, Brillouin from France, Pauli from Austria, and Gamow and Landau from Russia.

After Bohr received the Nobel Prize in physics in 1922, one honor after another was heaped on him, and he was in constant demand as a lecturer. He would prepare his lectures, as he did his papers, with the greatest care, polishing every sentence over and over again, right up to the very moment that he began speaking.

Much of his lecturing took him to foreign countries. In 1933, he came to the United States for the first time where he lectured at the University of Chicago and the California Institute of Technology in Pasadena. This was the first of a series of trips to this country; he spent most of his time in the United States (except when he worked at Los Alamos during World War II) at the Institute for Advanced Study at Princeton. He thought very highly of many American physicists, particularly J. Robert Oppenheimer and John C. Slater at MIT, about whom he commented, "There is probably no one in Europe who knows as much about atomic problems as Slater." In his remarks about American physics, he clearly indicated his devotion to the idea of the universal quality of the human mind and to the idea that there is no monopoly of knowledge by any one ethnic group or country. He went on to say: "Who knows where science will flourish most highly a hundred years from now; perhaps in Japan; perhaps in China? We know nothing about such things."

Although Bohr was quick to see positive elements in his colleagues, he was also apt to see the flaws and the irrelevancies. In spite of considering A. H. Compton an excellent physicist, for example, Bohr felt that Compton's philosophy was too primitive and should have no place in physics. Bohr remarked: "Compton would like to say that for God there is no uncertainty principle. That is nonsense. In physics we do not talk about God but about what we can know. If we are to speak about God we must do so in an entirely different manner" (Nielson, 27).

In Denmark itself Bohr was greatly admired. The government and the public alike made every effort to demonstrate their esteem. The brewer Carl Jacobsen in 1932 donated a beautiful mansion as a lifetime residence for Denmark's most renowned scientist and scholar. This was a happy occasion for Bohr, for he could now entertain large groups of people and even had room enough to have a number of guests (usually scientists with whom he was collaborating) stay at his house for extended periods.

During this very fruitful period Bohr began to pay greater and greater attention to the underlying philosophical and epistemological aspects of science, and of physics in particular. Thus, he became more and more concerned with his principle of complementarity or the so-called "Copenhagen interpretation of the quantum theory," and by dint of his great persuasive ability he swung over the great majority of physicists to this point of view. He did this not by publishing any single paper but by constant discussion with the gifted group of physicists assembled at his institute. In the words of Victor Weisskopf, who was one of the youngest and most perceptive of that group,

> It was his [Bohr's] greatest strength to assemble around him the most active, the most gifted, the most perceiving physicists of the world. . . .
>
> It was at that time and with those people that the foundations of the quantum concept were created, that the uncertainty relation was first conceived and discussed, that the particle-wave antinomy was for the first time understood. In lively discussions, in groups of two or more, the deepest problems of the structure of matter were brought to light. You can imagine what atmosphere, what life, what intellectual activity reigned in Copenhagen at that time. Here was Bohr's influence at its best. Here it was that he created his style, the "Kopenhagener Geist," the style of a very special character that he imposed on physics. We see him, the greatest among his colleagues, acting, talking, living as an equal in a group of young, optimistic, jocular enthusiastic people, approaching the deepest riddles of nature with a spirit of attack, a spirit of freedom from conventional bonds and a spirit of joy that can hardly be described.
>
> In that great period of physics, Bohr and his disciples touched the nerve of the universe. The intellectual eye of man was opened on the inner workings of Nature that were a secret up to this point. The concept of quantum state was cleared up, its fundamental wholeness, its indivisibility which, however, has that peculiar way of escaping ordinary observation because the very act of such observation would obliterate the conditions of its existence. Bohr, whose penetrating analysis contributed so much to the clarification of these problems, called that remarkable situation "complementary." It defies pictorial description in our accustomed classical terms of physics. . . . (Weisskopf, 59).

Bohr's concern with presenting his theory of complementarity to the world in as clear and understandable a manner as possible led him to profound problems of philosophy, language, and pedagogy. He always sought better ways of expressing the same idea and he outlined plans for an all-encompassing text that would contain as little formalism as possible and would strongly emphasize the physics. At that time, he said: "I believe that I have come to a certain stage of completion in my work. I believe that my conclusions have wide application also outside physics [this theory of complementarity]. . . . I should like to write a book that could be used as a text. I would show that it is possible to reach all important results with very little mathematics. In fact, in this manner one would in some respects achieve greater clarity" (Nielson, 27).

Bohr never did write such a book for this was a period of great research activity. He became involved in the famous controversy with Einstein in a series of confrontations and dialogues in which Einstein propounded ingenious quantum-mechanical paradoxes and Bohr, just as ingeniously, resolved them, using Einstein's own relativity theory and principle of invariance to accomplish it. The result of these confrontations was a paper, published by Bohr in honor of Einstein's 70th birthday, which will remain one of the great classics in the history of physics and philosophy.

This was also the period in which quantum mechanics was being applied to the electromagnetic field. Here Bohr pursued the subject very vigorously in close collaboration with Pauli, Heisenberg, and Dirac, who had already made important independent contributions. In particular, in a famous paper published jointly with L. Rosenfeld, Bohr established the basic principles of field quantization and showed that the Heisenberg uncertainty relations applied to the measurements of field quantities just as they did to the measurement of dynamical quantities. It was characteristic of Bohr's concern with presenting the best possible product that one is capable of that, after having worked on this paper with Rosenfeld for more than a year, he was still fearful, on the very eve of its publication, that it might be all wrong and that he could have done a better job.

Then World War II intervened. Bohr's institute, already more than a center of physics, had become a refuge for all those scientists seeking to escape Nazi persecution in the period from 1933 to 1940. When Denmark fell, the institute had to be abandoned because Bohr refused to collaborate with the Nazis. He barely escaped to Sweden and from there, via England, to the United States where he met many of the scientists he had previously helped to slip out of occupied Europe and for whom he had obtained positions in the United States. Bohr's escape from Denmark is itself a very dramatic story. The Danish police had warned him that the Nazis were looking for him, and he managed to cross the Sound of Sweden just ahead of them. He was then flown under the greatest security to London and from there to Los Alamos, where he was known simply as Mr. Nicholas Baker. To mention his true name there was strictly forbidden, for the knowledge that he was there would have revealed what was going on in Los Alamos. Throughout the war, his gold Nobel medal remained in Copenhagen, unknown to the Nazis; it was dissolved in a bottle of nitric acid. After the war, the metal was recovered and recast.

Bohr had already contributed greatly to the exploitation of atomic energy. In January 1939, just before he embarked from Denmark for one of his visits to the United States, Frisch and Meitner had told him about the discovery of uranium fission by Hahn and Strassmann. Bohr, as usual, had been impatient to get to the heart of the problem and he could hardly restrain his excitement when John A. Wheeler met him as he walked off the ship. They immediately began a theoretical analysis of the fission process. In a series of brilliant deductions, using his liquid-drop model and his concept of the compound nucleus, Bohr showed conclusively that the isotope U^{235} and not U^{238} undergoes fission. This was the first step in an irreversible process that led to the atomic bomb and to the nuclear chain reaction.

Bohr acknowledged the need for the atomic bomb when he arrived in the United States during the war, but he immediately saw that it would have to be controlled internationally if it were not to lead ultimately to the destruction of the world. He pleaded passionately with both Presidents Roosevelt and Truman to set up plans and political machinery to forestall a nuclear arms race because he was well aware, early in the game, that the Russians would very quickly develop the bomb. In 1957, after seeing a Russian film on nuclear energy, he commented: "The Russians got the first nuclear power plant, and they have the largest cyclotron. . . . It was perfectly absurd to believe that the Russians cannot do what others can. . . . There was never any secret about nuclear energy. . . . It is also absurd to expect that the Russians will put up with everything and give up any position of power they may possess . . ." (Nielson, 29).

When questioned about the horribly destructive hydrogen bomb and Edward Teller's contributions to it, Bohr remarked that "Old physicists who have turned administrators might not think of this solution. However, if you had asked a good class of physics students, two or three of them would have suggested this solution. Anyway, the Russians did the very same thing" (Nielson, 30).

Great as Bohr was as a scientist, he was an equally great humanist. A concern for all people and a hatred for injustice were primary ideals even in his early years. He felt that nothing could be achieved in righting the wrongs of society without sacrifice. During the depression, when he first visited the United States, he was amazed at the great wealth in the midst of poverty and remarked that in Denmark "the burdens of the depression are fairly evenly distributed over all layers of the population.... Of course one cannot improve conditions for all layers of society without renunciation on the part of some.... It is sheer folly to believe that one can achieve anything in this world without renunciation ..." (Nielson, 26).

Bohr also, very early in his career, saw the need for international cooperation in all human endeavors and was repelled by the type of national jockeying that went on at international conferences. He rejected this kind of interplay in the following words.

> This is the way it goes with international meetings as long as we have independent countries. The aim of most delegates to such meetings seems to be to obtain as many advantages as possible for their own country and to cheat and deceive the others as much as possible.... We must be internationalists, and in science we succeed fairly well.... All peoples and races are essentially alike; the difference is in their traditions and backgrounds.... Every valuable human being must be a radical and a rebel, for what he must aim at is to make things better than they are.... (Nielson, 26)

Bohr found great happiness and fulfillment not only in his science but also in his family to which he was intensely devoted. His personal life was not free of tragedy. In September 1934, the oldest of his five sons (he had no daughters) was drowned in a sailing accident and those who were present did all they could do to restrain Bohr from throwing himself into the sea in a futile attempt to save the boy. His grief was intense and for weeks he was inconsolable. Finally, the pressure of new scientific discoveries and his sense of the continuity of life restored his equilibrium. He must have found solace in the writings of the great religious teachers because he remarked at one point that these teachers had achieved their influence through their power to console those who had suffered great sorrow.

When Niels Bohr died on November 16, 1962, in the midst of active scientific work, his task was almost done; with him a scientific era may have disappeared— the era of the great men who created modern physics. His name is already written under those of Newton and Einstein on the rosters of science.

Let us now look more closely at Bohr's contributions. The year 1913 was a crucial one in the history of atomic theory and for our understanding of the structure of the atom and of matter. In July of that year one of the greatest of all scientific papers appeared in the *Philosophical Magazine* of London, Edinburgh, and Dublin under the authorship of N. Bohr. The title of this paper, "On the Constitution of Atoms and Molecules," indicated that it might contain some interesting ideas but hardly heralded the shattering impact its contents were to have on future concepts of atomic structure.

Before this paper on the structure of the hydrogen atom appeared, two important discoveries had been made: one dealing with the properties of radiation and the other with the interaction of atoms with charged particles, that is, with ions. Although, at first sight, these two developments do not appear to be directly related, they were really the key to Bohr's revolutionary theory of the atom.

We recall that Planck introduced the concept of the quantum of energy and the quantum of action to derive the correct formula for black-body radiation. But, except for Einstein's extension of this idea, which led him to introduce the photon and to apply it to the explanation of the photoelectric effect and the specific heats of solids, no one else had applied Planck's quantum concept to anything but the behavior of radiation. The next great step was to be taken by Bohr who saw quite clearly that Planck's discovery of the quantum of action and Einstein's concept of the photon could be combined with Rutherford's discovery of how alpha particles are scattered by atomic nuclei to derive a self-consistent planetary (that is, nuclear) atomic model.

Bohr saw that, without introducing Planck's ideas into the theory of atomic structure, there was no way to obtain a stable planetary atom. On the other hand, he saw that introducing the Planck constant of action would do two things: lead to a description of stable electronic orbits and demonstrate the kind of discrete line spectrum that is observed.

Let us now consider why Rutherford's discovery required a drastic departure from the laws of classical electrodynamics. We saw that to account for the scattering of alpha particles Rutherford found it necessary to introduce a model of the atom in which the positive charge is concentrated in a nucleus at the center. But, as we noted, the ideas of Rutherford were still somewhat tentative since he only hinted at a planetary model; he dealt only with the nucleus of the atom. In any case, the arrangement of the electrons in precise orbits was not clearly postulated, and the whole question of the stability of an atom constructed according to such a model presented what appeared to be insurmountable difficulties. An electron can be in stable equilibrium with a central positive nucleus only if the electron revolves around the nucleus in a closed orbit (circle or ellipse), for only then is the electrical force of attraction balanced by the centrifugal force acting outwardly on the electron. But, as we have already noted, an electron moving in this way must, according to classical electrodynamical theory, lose energy by radiating, so that it must ultimately spiral into the nucleus.

We see that we can overcome this difficulty if we introduce the kind of discontinuity into atomic processes that Planck introduced in analyzing the emission and absorption of radiation by the walls (actually, the atoms in the walls) of the container housing the radiation. The spiraling of an electron into the nucleus of an atom is clearly a consequence of the continuous change in its motion, as permitted by the classical laws of dynamics. These laws allow an accelerated electron to lose energy continuously by radiation so that there is no way, in terms of classical electrodynamics, for the electron to remain in any one orbit. The slightest change in its state of motion can, according to such a picture, cause it to gain or lose a slight amount of energy, and vice versa.

But if the electron can change its state of motion only discontinuously, that is, in discrete steps, it must then stay in a particular orbit until it emits or absorbs enough energy in one single process to go from one orbit to another. This then leads to discrete orbits, and transitions from one such orbit to another give rise to a discrete spectrum.

Niels Bohr was aware of this phenomenon when he began his historic work. He was in his early 20s when Rutherford published his paper on the scattering of alpha particles by heavy nuclei. Bohr was then working in theoretical physics at the Manchester Laboratory under Rutherford and was greatly stimulated by the

Rutherford nuclear model and its implications for the quantum theory. Realizing its importance, he set out to see if he could solve the problem of stability by introducing the new quantum concepts. He recognized that the Rutherford model would "meet with difficulties of a serious nature arising from the apparent instability of the system of electrons: difficulties purposely avoided in atom models previously considered, for example, in the one proposed by Sir J. J. Thomson." And yet Bohr felt sure that the difficulty present in the Rutherford model could be eliminated by introducing a quantity involving the dimension of length.

His reasoning went somewhat as follows. With the Thomson model of an atom one automatically obtains a fundamental length about equal in magnitude to the size of the atom because this is the diameter of the positive sphere of electricity in which the electrons are supposed to be embedded. The appearance of such a length in the Thomson theory of the atom means that this kind of atom cannot collapse to a dimension less than this and, therefore, is stable. Bohr noted in one of his papers that "such a length does not appear among the quantities characterizing the second [that is, the Rutherford] atom, viz. the charges and masses of the electrons and the positive nucleus; nor can it be determined solely by help of the latter quantities." He means that no quantity having the dimensions of a length can be constructed from the charge and masses and be numerically of the right size for an atom. All such lengths would be much too small.

But here Bohr experienced one of those miraculous insights that so often occur to the creative mind. In looking around for some way to introduce a length of atomic dimensions into the Rutherford model of the atom as naturally as possible, he observed that a new quantity had been introduced into physics seven years previously by Max Planck—Planck's constant, or, as it often is called, "the elementary quantum of action." By the introduction of this quantity, the question of the stable configuration of the electrons in the atoms is essentially changed since this constant is of such dimensions and such a magnitude that it, together with the mass and charge of the particles, can determine a length of the order of magnitude required. If Planck's constant is squared and the result divided by the product of the mass of the electron and the square of its charge, the number thus obtained is 10^{-8} cm, which is the size of an atom.

This type of general reasoning used by Bohr to obtain a correct theory is characteristic of all of his work. Einstein, Enrico Fermi, H. A. Lorentz, and the other outstanding figures of atomic theory had the same ability to arrive at an important result without complex mathematical analysis and involved physical arguments. Bohr did not have to analyze all the aspects of the Rutherford model of the atom before discovering what was needed to make it work. His observation, that in the Rutherford model, as it stood, no quantity of the nature of a length and of the right order of magnitude was present, showed him what was wrong and where he would have to look to find the missing length. He was further strengthened in his beliefs by the ease with which the Planck theory of radiation had cleared up many difficulties that had bothered physicists for years; in Bohr's words, the elementary quantum of action had previously demonstrated "the inadequacy of classical electrodynamics in describing the behavior of systems of atomic size." Since the instability of the Rutherford atom was associated with the classical theory of the radiation of electrons moving in closed orbits, why could not the Planck theory straighten things out in this respect as well? Bohr was convinced that it could.

Using elementary algebra, the simple concepts of classical electrostatics, and the Newtonian laws of motion, Bohr derived a simple expression for the frequency of an electron moving in a circular orbit; he noted that, according to classical theory, this electron would have to sink into the nucleus because it would lose energy continuously. Then he introduced Planck's constant, imposing the condition that the electron must not radiate continuously but rather in the form of "distinctly separated emissions." The amount of energy radiated during such emissions must be some integral value of Planck's constant multiplied by the frequency of the atom in its orbit. Thus Bohr introduces Planck's constant into the Rutherford model of the atom and at the same time shows that the stability of, and the radiation by, the atom depend in some way upon the integers. The presence of these integers in conjunction with Planck's constant shows the departure of the theory from classical dynamics and electrodynamics.

We briefly discuss here Bohr's procedure, which is also outlined in an address on the spectrum of hydrogen delivered in 1913 before the Physical Society in Copenhagen. This address is particularly interesting because it clearly indicates the tentativeness of the quantum theory at the time and the incompleteness of its acceptance. Even Bohr, who was to put the Planck theory to its most important use, stated in this address: "In formal respects Planck's theory leaves much to be desired; in certain calculations the ordinary electrodynamics is used, while in others assumptions distinctly at variance with it are introduced without any attempt being made to show that it is possible to give a consistent explanation of the procedure used" (Bohr, 6). Nevertheless, Bohr clearly recognized that "energy quanta" had come to stay and summed up his feelings as follows: "It is therefore hardly too early to express the opinion that, whatever the final explanation will be, the discovery of 'energy quanta' must be considered as one of the most important results arrived at in physics, and must be taken into consideration in investigations of the properties of atoms and particularly in connection with any explanation of the spectral laws in which such phenomena as the emission and absorption of electromagnetic radiation are concerned" (Bohr, 7).

After a brief survey of the laws of radiation and Planck's contributions to these laws, Bohr proposed to combine these ideas with the results of Rutherford's alpha-particle scattering experiments and in this way derive a formula for the frequencies of the spectral lines of hydrogen. He chose hydrogen to work with because he assumed it to be (and correctly so) the simplest atomic structure, having just one electron revolving around a single proton. Moreover, the spectrum of hydrogen had been studied more thoroughly than that of any other atom and the numerical simplicity of Balmer's formula for the spectral lines suggested that a simple algebraic analysis might solve the problem.

Bohr began, then, by picturing an electron as revolving around a proton according to the classical laws of Newtonian mechanics, the centrifugal force on it, resulting from its orbital motion, being balanced by the electrostatic pull of the proton. This leads to the same kind of orbit for the electron around the proton as the planets have around the sun: elliptical orbits, first discovered by Kepler.

But here a difficulty is encountered because a charged particle behaves quite differently from a planet; it would, according to Maxwell's electrodynamic theory, lose energy continuously by radiation, causing the electron's orbit to become smaller and smaller, finally shrinking to the size of the proton. Bohr recognized this and introduced the Planck concept of the quantum to prevent this catastrophe.

Bohr now pictured the electron as being capable of moving in only a discrete set of orbits; while the electron is in any of these orbits, it does not radiate energy, even though classical electrodynamics demands that it must. This use of discrete, nonradiating orbits accomplishes two things. First, it leads to a set of distinct spectral lines since a discrete set of frequencies is associated with the discrete orbital motions of the electron; discrete, nonradiating orbits lead to a stable nuclear atom since one must then have the lowest orbit which is, then, the closest the electron can get to the nucleus. Moreover, discrete orbits can immediately be correlated to a quantum picture of nature since changes in the atom can only occur in discontinuous steps, the electron changing from one fixed orbit to another; this must, therefore, involve discrete changes in the energy of the atom or in the action of the electron in the emission or absorption of a quantum.

To account numerically for the actually observed spectral lines, Bohr proceeded as follows. He first noted that in the hydrogen atom the energy of an electron moving around the proton in a Keplerian orbit depends only on the size of the orbit (that is, upon the semimajor axis of the ellipse in which the electron is moving); this fact is in accordance with the Newtonian laws of motion. We may think of this energy as the negative of the work that would have to be done to tear the electron completely out of this orbit and bring it to a very great distance from the proton, as occurs in ionization. Thus, a discrete set of orbits of different sizes means a discrete set of energy states for the electron (or for the atom) since the energy of the electron depends on the size of the orbit.

Bohr now introduced a crucial step in his argument. He postulated that the radiation or the absorption of energy occurs only when the atomic system passes from one energy state to another and that each such change in the energy state means the emission or absorption of a quantum of energy $h\nu$, where h is Planck's constant of action and ν is the frequency of the emitted radiation. Since one can assign a definite amount of energy to the electron in any orbit, according to the laws of classical mechanics (just as this can be done for each planet in the solar system), one can calculate all possible energy changes and hence obtain the frequencies of all possible quanta. These should then correspond to the observed spectral lines.

Bohr described these ideas in a general way in his essay on the spectrum of hydrogen, where he introduced the concept of "stationary states" to describe the discrete orbits.

> During the emission of the radiation the system may be regarded as passing from one state to another; in order to introduce a name for these states, we shall call them "stationary" states, simply indicating thereby that they form some kind of waiting places between which occurs the emission of the energy corresponding to the various spectral lines. As previously mentioned the spectrum of an element consists of a series of lines whose wavelengths may be expressed by [a] formula. . . . By comparing this expression with the relation given above it is seen that—since $\nu = c/\lambda$ where c is the velocity of light—each of the spectral lines may be regarded as being emitted by the transition of a system between two stationary points in which the energy apart from an additive arbitrary constant is given by $chF_r(n_1)$ and $chF_s(n_2)$ respectively. Using this interpretation the combination principle asserts that a series of stationary states exists for the given system, and that it can pass from one to any other of these states with the emission of a monochromatic radiation. We see, therefore, that with a simple extension of our first assumption it is possible to give a formal explanation of the most general law of line spectra. (Bohr, 11)

From this description we note a very important point. Each spectral line involves two orbits and hence two different frequencies or two energies, not one as we would have in classical physics.

To complete his derivation of the Balmer formula for the hydrogen spectral lines, Bohr still had to introduce a scheme for picking out among all possible orbits the particular discrete set associated with the hydrogen atom. Why should the electron be limited to just these orbits and no others? What magical property do these orbits have that gives them their preferred character? To understand this and follow Bohr's reasoning we note first that Kepler's third law of planetary motion also applies to the motion of the electron: it can be derived by equating the Coulomb electrostatic pull of the proton on the electron to the centrifugal force on the electron. According to this law, the square of the period of the electron (the period is time in fractions of a second taken to go once around the orbit) is proportional to the cube of the radius of the orbit. This means that the square of the frequency of the electron (frequency being the number of orbital trips made in one second and hence the reciprocal of the period) depends on the cube of the energy since the energy is related to the size of the orbit. Thus frequency is related to energy.

But from classical physics one can also show that the period of an electron or its frequency and the size of its orbit can be related to its angular momentum. The angular momentum of a particle is a quantity that is obtained by multiplying the radius of the orbit of the particle and its momentum. Since Planck's constant h is in the nature of an angular momentum, Bohr simply assumed that the angular momentum of an electron in an orbit can only equal $h/2\pi$, or $2h/2\pi$, or $3h/2\pi$, and so forth, since $h/2\pi$ is an indivisible unit of angular momentum. In this way Bohr related the sizes of his discrete orbits to integers and to Planck's constant. From this he then obtained a formula, in terms of Planck's constant and the integer assigned to an orbit, for the energy of the electron.

As Bohr noted in his 1913 paper "On the Constitution of Atoms and Molecules," the theory as he developed it is pretty much of a hybrid, for the dynamic equilibrium of the electron in one of its orbits (Bohr's "stationary states") is treated by means of ordinary classical mechanics and electrostatics, whereas the quantum theory is used to describe what happens when the electron jumps from one stationary state to the other, which is accompanied either by the emission or absorption of a single quantum of homogeneous radiation (radiation of a definite frequency).

By using these simple though revolutionary ideas Bohr eliminated the difficulties that plagued all previous theories. In particular, he referred to the work of Nicholson, who had worked extensively on a model of the atom similar to the Bohr model. Nicholson had attempted to obtain the Rydberg-Ritz formula for the hydrogen spectrum but had failed because, as Bohr points out, "The frequency of lines in a line spectrum is identified with the frequency of vibration of a mechanical system in a distinctly indicated state of equilibrium." And, as Bohr went on to say, although Nicholson did relate the radiation from his atom to the Planck theory, it could not be a correct picture because such systems are incapable of emitting homogeneous radiation in discrete quantities.

By means of simple algebra and elementary physical concepts, Bohr obtained the correct expression for the frequencies of the spectral lines in the Balmer series in the spectrum of hydrogen. In addition, he showed that his formulas contain the frequencies of the other series in the hydrogen spectra, such as the Paschen and Lyman series. Bohr also showed in this paper that certain series of lines that were thought to be due to hydrogen must, indeed, be due to helium, and he obtained a simple formula for the helium spectrum that is algebraically similar to the formula for the hydrogen spectrum, differing from it only because the charge on the nucleus of helium is twice the elementary electron charge itself (the charge on the hydrogen nucleus, the proton).

After Bohr obtained the correct formula for the frequencies of the spectral lines, he gave a critical analysis of the assumptions that he made to obtain the formula. He was particularly concerned with the assumption that different stationary states correspond to the emission of different numbers of energy quanta or photons and that the frequency of the radiation emitted by an electron that falls from a state of rest at a great distance from the nucleus to a nearer orbit is equal to half the frequency of the motion of the electron in this final stationary state.

Bohr first pointed out that the assumption of the emission of different numbers of quanta in association with different stationary states is not really necessary for a correct application of the theory; the assumption that one quantum corresponds to each transition of an electron is all that one requires. Here he used what has since been called the "correspondence principle," for he considered how the electron would behave in passing into an orbit where it moves very slowly. Under these conditions he could compare the results of his theory with the results obtained from the classical electrodynamics since, for very small frequencies, the classical picture is correct, and the Planck theory of radiation passes over to the Maxwellian picture.

By using the correspondence principle, Bohr showed that one may correctly assume that only one quantum of energy is emitted, regardless of the orbit into which the electron moves. He also used this principle to validate the assumption that the frequency of the emitted radiation is half the frequency of the motion in the final orbit. One other important point comes out of these considerations. As Bohr pointed out, and as we have already noted, one can show that the result of the calculation in this paper may be "expressed by the simple condition that the angular momentum of the electron around the nucleus [on the assumption that the orbits are circular] in a stationary state of the system is equal to an entire multiple of a universal value, independent of the charge on the nucleus." This is a statement of a very important principle later generalized by Wilson and Sommerfeld and contains within it the germs of Heisenberg's principle of indeterminacy.

It is interesting to note in connection with Bohr's paper how close Nicholson had come to discovering these results. As Bohr pointed out: "The possible importance of angular momentum in the discussion of atomic systems in relation to Planck's theory is emphasized by Nicholson." There can be no doubt that Bohr was influenced by and owed a good deal to Nicholson.

Bohr was greatly influenced and encouraged by Rutherford, who thought very highly of him and his work. They spent many hours together in 1912, and, in a letter dated July 24, 1912, Bohr thanked Rutherford for "your suggestions and criticisms [which] have made so many questions so real for me, and I am looking forward so very much to try to work upon them in the following years." While Bohr was preparing his paper, Rutherford advised that Bohr be as brief as possible, consistent with clarity, for as he said: "Long papers have a way of frightening readers. It is the custom in England to put things very shortly and tersely in contrast to the Germanic method, where it supposed to be a virtue to be as long winded as possible." On March 6, 1913, Bohr sent what he called the first chapter of his paper to Rutherford with the request that the latter communicate it to the *Philosophical Magazine*. Bohr was quite concerned about his use of both classical mechanics and the quantum theory to obtain his results and hoped "that [Rutherford] will find that I have taken a reasonable point of view as to the delicate question of the old mechanics and of the new assumptions introduced by Planck's theory of radiation."

Rutherford recognized the great importance of Bohr's paper and immediately sent it on for publication, but he had many questions about the basic ideas. Thus, in a letter to Bohr on March 20, 1913, he stated, after praising Bohr's ingenious derivation of the hydrogen spectral lines, that "the mixture of Planck's ideas with the old mechanics makes it very difficult to form a physical idea of what is the basis of it all." He then went on to point out that he considered a "grave difficulty in [Bohr's] hypothesis. . . . How does an electron decide what frequency it is going to vibrate at when it passes from one stationary state to the other? It seems to me that you would have to assume that the electron knows beforehand where it is going to stop."

That it took great courage to publish such a revolutionary theory at that period is clear from a letter that G. Von Hevesy, an outstanding Hungarian chemist working on relativity, wrote to Rutherford in the fall of 1913. He described his meeting with Einstein at a science congress a few months earlier, shortly after the publication of Bohr's paper, and stated that Einstein "told me that he had once had similar ideas but he did not dare to publish them." Einstein considered Bohr's paper to be "of the greatest importance" and "one of the greatest discoveries."

In the last few paragraphs of his paper, Bohr reached an important conclusion, not verified until a number of years later, that indicated that the quantum would have to be extended to processes outside the atom. He noted that in a gaseous mixture of free electrons and electrons bound inside atoms, the bound electrons do not have the same energy that the free electrons have on the average, as determined by the temperature of the gas. Bohr correctly pointed out that this is so because the bound electrons cannot absorb energy continuously but only in discrete quanta. Hence, collisions between free and bound electrons lead to a new exchange of energy.

Like most of Bohr's important contributions his paper on the constitution of atoms and molecules is marked by very little formal mathematics but by very penetrating physical arguments and bold assumptions of a revolutionary nature. Its great value lies in its unification of atomic theory with the Planck theory of radiation.

The Quantum Theory Is Tested: James Franck (1882–1964) and Gustav Hertz (1887–1975)

James Franck, one of the outstanding German experimental physicists of the decade prior to World War I, was born in Hamburg on August 26, 1882, and studied at both the universities of Heidelberg and Berlin. Soon after receiving his doctoral degree he went to the Kaiser Wilhelm Institute of Physical Chemistry at Berlin-Dahlem, where he was one of the departmental heads, and began his investigations into atomic structure.

At the Kaiser Wilhelm Institute he collaborated with Gustav Hertz and completed basic experiments on the collisions of electrons with atoms, demonstrating

James Franck (1882–1964)

that an atom can take on energy from collisions only in discrete amounts, in agreement with Bohr's theory. In 1920 Franck was called to the University of Göttingen as full professor of experimental physics. He served there from 1920 until 1935 when he left Germany because of the Nazi racial laws.

While at Göttingen, Franck established one of the outstanding atomic laboratories in the world, which attracted leading postdoctoral students from all countries. The United States, in particular, owes a great debt of gratitude to Franck for training and inspiring many of the best American experimental physicists.

Although Franck received the Nobel Prize for physics in 1925 for his electron-collision experiments, much of his best work was done in the study of molecular structure and later in photochemistry, particularly after he came to Johns Hopkins as professor of physics in 1935. In 1938 he was appointed professor of physical chemistry at the University of Chicago and played an important role in the development of atomic energy. He was the leader of a group of scientists on the Manhattan Project who felt that the atomic bomb should not have been dropped before warning Japan that the United States had such a weapon. To this end, he prepared and circulated the famous "Franck Petition," urging President Truman to demonstrate the bomb before authorizing its use. After the war Franck devoted almost all of his research time to the study of photosynthesis. He died in 1964, while still active in his scientific work.

Gustav Hertz was born on July 22, 1887 in Hamburg and studied physics at the universities of Göttingen, Munich, and Berlin, obtaining his doctorate from the latter institution. He served as an assistant in physics at Berlin from 1913 until the beginning of World War I. After serving in the war and being severely wounded, he came back to Berlin as an unpaid lecturer, privatdozent, in 1917 and began the collaboration with Franck that led to the famous Franck-Hertz experiments and the Nobel Prize for physics.

Hertz served as professor of physics at the University of Halle from 1925 to 1928 and then in the same rank at the Technical Institute of Berlin-Charlottenberg. He resigned from this chair in 1934 for political reasons and became director of a

Gustav Hertz (1887–1975)

research laboratory for the firm of Siemens. Although Hertz was of Jewish descent, he chose to remain in Germany during the Nazi regime and managed to survive the catastrophe of the Final Solution. He was taken prisoner by the Soviet Army in 1945 and was sent to the Soviet Union to contribute his scientific talents to the Stalinist regime. Later he returned to what became East Germany and in 1955 was appointed professor of physics at the University of Leipzig. He remained on the faculty there for six years until his retirement in 1961.

After Niels Bohr had introduced his quantum model of the planetary atom and had used it to derive the correct formula for the Balmer lines of hydrogen, experimental physicists began to devise various ways of probing the atom to see if they could obtain some insight, other than that offered by the spectral lines, into the nature of the discrete orbits and stationary states. Now, there are not many ways by which one can try to get a "look at the inside" of an atom; among the accessible methods, only two were available to physicists when Bohr announced his theory. Both involved bombarding the atom with particles: photons, on the one hand, and material particles, such as electrons or atoms, on the other. Since bombarding an atom with photons is essentially the same thing as studying its spectrum, only collisions between atoms and material particles, such as electrons, seemed to offer a possible new source of information. Consequently, James Franck and Gustav Hertz, who were pioneers in this field, turned to electron collisions to study the interior of an atom.

Franck and Hertz devised a very simple instrument consisting of a long wire surrounded by a wire-mesh cylinder whose axis coincided with that of the wire. Surrounding the wire-mesh cylinder, and very close to it, was an external solid foil cylinder. The apparatus was operated as follows. The atoms, to be studied in the form of a gas or vapor under low pressure, were placed in the cylinder surrounding the wire. A current was then sent through the wire until, glowing, it became hot enough to emit electrons. A positive voltage was established between the wire and the mesh so that the electrons were attracted to the mesh. These electrons moved through the vapor to the mesh and passed through it. After passing through the mesh, the electrons reached the surrounding foil where they were collected. The external foil cylinder was connected to the ground through a galvanometer so that the number of electrons striking the foil could be measured. Finally, a constant retarding voltage (to decelerate the electrons passing through the mesh) was placed between the mesh cylinder and the outer foil cylinder. This retarding potential could be altered at will so that the number of electrons striking the foil could be controlled.

Suppose now that a certain voltage is placed between the mesh and the glowing, conducting wire. What do we find at the electron-collecting foil cylinder? That depends on a number of things: the accelerating potential between the wire and the mesh; the kind of gas in the cylinder; the retarding potential between the mesh and the external foil. If no gas is present in the cylinder and if the accelerating potential is smaller than the retarding potential, no electrons reach the outer foil and the current in the galvanometer is zero. This means merely that the electrons coming from the wire and passing through the mesh are not moving fast enough to overcome the retarding potential and thus reach the foil. If the accelerating potential is slowly increased until it is exactly equal to the retarding potential, or slightly larger, a current will suddenly be observed in the galvanometer.

We now consider a gas, let us say mercury vapor, present in the cylinder. A gradual increase in the accelerating potential is applied. What effect do the gas atoms have on the electrons? As long as the accelerating potential lies below a certain critical value, which is different for different gases, the situation is exactly the same as though no gas were present. We must keep in mind that the atoms exist only in certain discrete energy states and they can pass from a lower state to a higher (that is, from the ground state to an excited state) only by absorbing a discrete amount of energy, which must be furnished by the colliding electron. Furthermore, only a discrete amount of energy can tear an electron out of any one of the atoms and thus ionize the atom. If now an electron coming from the hot wire collides with a mercury atom, it can give up energy to this atom either by making the atom move faster or by exciting the atom internally. But if the potential that accelerates the electrons as they come from the wire is less than the smallest excitation energy of the mercury atom, the electron, according to the quantum theory, with its discrete energy levels, cannot excite the atom; at most, it can only increase its velocity during a collision. Since a mercury atom is very massive compared to an electron, the colliding electron has little effect on the motion of the atom; the electron bounces off with no loss of energy like a ball bearing bouncing off a massive wall. Such collisions are called elastic collisions and have no effect on the stream of electrons coming from the hot wire. In other words, as long as the accelerating potential is below a certain critical value necessary for excitation, the electrons collide with the mercury atoms without loss of energy and the current in the galvanometer of the apparatus is just as though no mercury vapor were present.

Now suppose that the accelerating potential is increased slowly until the electrons acquire just enough energy to excite or to ionize the mercury atoms. Then, according to the quantum theory, these electrons should lose all their energy; their collisions with atoms are inelastic, that is, they do not bounce off the atoms. Franck and Hertz found that this is precisely the case and so demonstrated the existence of discrete energy levels. They discovered that for an electron to excite the mercury atom, it must have no less than 4.9 volts of energy. As soon as the electrons were accelerated to 4.9 volts of energy, the current in the detecting galvanometer fell to zero because the electrons lost all their kinetic energy by inelastic collisions and therefore could not reach the external foil cylinder against the retarding potential. Of course, electrons that were torn out of the mercury atoms could not reach the external cyclinder either because they could not acquire enough kinetic energy from the accelerating potential.

Consider now what happens when the accelerating potential is greater than 4.9 volts. Each electron still suffers an inelastic collision, but it does not lose all of its kinetic energy, only that part represented by falling through 4.9 volts potential difference. In other words, it still has some energy left and can reach the detecting cylinder. As the accelerating potential is increased steadily beyond 4.9 volts, the current in the galvanometer begins to increase again (after having fallen to zero at 4.9 volts). Now it reaches a greater intensity because added to the initial stream of the electrons from the wire are the electrons torn from the mercury atoms during the collisions. If the accelerating potential is steadily increased from 4.9 volts to twice this value, the current in the galvanometer again suddenly drops to zero because now each electron has just enough energy to excite two atoms in two separate collisions (it loses just 4.9 volts of energy in each collision). When it does so, it loses all its energy. At this higher accelerating potential the electrons acquire enough energy to ionize the atoms closer to the wire than previously. After these initial ionizing collisions near the wire, the electrons (now with practically no energy) are speeded up again before reaching the mesh and suffer ionizing collisions a second time. There is thus a second maximum in the galvanometer current when the accelerating potential is increased beyond twice the ionization potential.

Franck and Hertz then went on to show that as the accelerating potential is increased, a new maximum appears in the galvanometer current for each integral multiple of the ionization potential. They applied this technique not only to mercury but to other atoms as well. Moreover, they demonstrated that electrons moving with the ionization energy could also excite the mercury resonance line $\lambda = 2536$ angstroms. In addition, they showed that if one multiplies the frequency of this line by Planck's constant h, one obtains exactly the ionization potential of 4.9 volts.

The work of Franck and Hertz was important at this stage of the development of atomic theory because it was not clear from Bohr's theory of atomic spectra alone whether the quantum theory could be applied to ordinary mechanical energy of motion or whether it was limited to the emission and absorption of radiant energy. These experiments demonstrated that a particle like an electron would transfer its energy in a collision only in multiples of a fundamental quantum. From this point on it was clear that the quantum theory would have to be taken into account in all processes. This is precisely what Bohr had predicted in the last few paragraphs of his fundamental paper, which we discussed previously, and is also in line with what Einstein had insisted on at the first Solvay Congress in 1911.

The Discovery of Isotopes: Frederick Soddy (1877–1956)

One hundred years of chemical atomic theory steadfastly corroborated John Dalton's hypothesis that the atoms of the chemical elements were immutable and the atoms of any given element were identical in mass, shape, and size. Actually, atomic theory in the first half of the 19th century left much to be desired with respect to rigor. Many leading physicists and chemists, among them Humphry Davy and Michael Faraday, had grave reservations concerning it. However, the reorganization of chemistry accomplished by Stanislao Cannizzaro about 1860 went far to promote a wide acceptance of the reality of atoms. Coincidentally, further confirmation was provided by the advances made in kinetic theory by Rudolf Clausius and James Clerk Maxwell, and later by Ludwig Boltzmann. Although no serious scientist questioned the immutability of atoms, here and there a voice such as Sir William Crookes's was heard speculating that the atoms of a given element might not be identical and that the atomic weight of an element might be the average of the weight of several unequally massive atoms having the same chemical properties. The basis for such a belief could easily have been the lingering echo of William Prout's hypothesis, together with the recognition that some atomic weights, such as chlorine, were by no stretch of the imagination integral.

The first breach produced by experimental evidence in the century-old atomic credo came in 1902 from the researches of Rutherford and Soddy. They showed that the radioactive elements were characterized by an ephemeral life and that their atoms were transmuted by radioactivity into others of different chemical properties. This discovery led in about a decade to the classification of the decay products into well-ordered radioactive decay series. Thus, thorium atoms of atomic weight 232 (Th^{232}) were found to decay through a long series of transformations to lead of atomic weight 208, while uranium atoms (U^{238}) decayed to lead of atomic weight 206.

As experience with radioactivity increased, there began to emerge more clearly in several minds, and apparently most clearly in Frederick Soddy's, the realization that the radioactive transformations produced atoms of the same chemical species but of different weights. The first awareness of this situation seems to have come to Soddy in 1910 as a result of the chemical nonseparability of radium and mesothorium (Ra^{228}). In his Nobel lecture (1921), Soddy stated: "From this date [1910] I was convinced that this nonseparability of the radioelements was a totally new phenomenon, quite distinct from that of the most closely related pairs, or groups of elements, hitherto observed in chemistry, and that the relationship was not one of close similarity but of complete chemical identity" (Howorth, 184). The accuracy of this statement has been confirmed recently by Cranston, one of his early associates, in an article published in the *Proceedings of the Chemical Society* for April 1964. We now recognize such chemically inseparable atoms as isotopes. But they did not receive this generic name until 1913 when the term was introduced by Soddy in an article, "Intra-atomic Charge," which appeared in the December 4, 1913, issue of *Nature*. This article was actually more concerned with the establishment of the concept of atomic number. The Royal Society obituary article on Soddy notes that the name isotope was suggested to him in 1913 by Dr. Margaret Todd, an Edinburgh M.D., during a conversation that took place in the home of Soddy's father-in-law, Sir George Beilby, in Edinburgh, Scotland.

Soddy himself was an interesting personality. Born in Eastbourne, England, on September 7, 1877, the son of a successful grain merchant, he was brought up in a home of deep Calvinistic tradition. In this strict atmosphere truthfulness was the essential thing and deference to personal feelings was of very secondary consideration. This attitude, impressed on him as a child, persisted throughout his life. His education was begun in the Eastbourne schools and at Eastbourne College where he first became interested in chemistry. Like the majority of scientists, he was the first in his family to show such a bent. From Eastbourne he went on to the University of Aberystwyth, Wales, principally to prepare for Oxford where he was enrolled at 19 as a student in Merton College. As an undergraduate he was active in the university Junior Chemical Society. He finished his undergraduate studies after two years, in 1898, with first honors in natural science and then stayed on at Oxford until the spring of 1900 when, at the age of 23, he applied for a professorship in chemistry at the University of Toronto. The details of this application and its results have already been recounted above. The subsequent collaboration with Rutherford was an unlikely partnership between two strong personalities, but it proved to be a happy one because of the complementary abilities of the two men and the contributions each was able to make to the joint effort. Although their work together was finished in the short span of three years, their papers revolutionized the outlook on radioactivity. Alexander Fleck evaluated their accomplishment in the following tribute.

> It is difficult at this distance of time to visualize the immediate effect of these papers on the scientific world of that day. Up to then the ideas of possible explanations of radioactivity rested on a form of continuity of emission of energy analogous in some way to the wave theory of light. Continuity seemed to be the basic idea of the Curies.... The effect of the Rutherford and Soddy papers ... was that of a wave which swept over the scientific world and carried away most if not all of the vague alternatives. (Fleck, 205)

Soddy left McGill University in 1903 and worked for a year in London with Sir William Ramsay. In keeping with his love of adventure, this time was followed by a lecture tour in western Australia before he took up his duties as lecturer in chemistry at Glasgow University in the fall of 1904. His 10 years at Glasgow cover most of the period of his productivity in chemistry, including his study of isotopes. Although Soddy had foreseen the existence of the isotope in 1910, he continued to develop his concepts in the following three years. In 1914, he accepted a professorship of chemistry at Aberdeen. During his stay there he became concerned with investigations relating to England's part in World War I. In 1918, Soddy was appointed Dr. Lee's professor at Oxford and three years later, at the age of 44, received the Nobel Prize in chemistry for his work on isotopes.

The war years, with their material destruction and death of valued friends, shifted his interests from chemistry—he became more and more preoccupied with the ills of the world. Economic and financial ideas and the place of money in the social structure of civilization, a theme in which he had shown a pronounced interest as far back as 1906, now assumed a much larger place in his life. Much of his intellectual endeavor after going to Oxford was devoted to an attempt to base a monetary system on energy quantities. The movement gained few disciples and commanded little popular following. In August 1936 his wife, to whom he was deeply devoted, died suddenly. The loss was so crushing that, coupled with the social difficulties connected with his unorthodox economic views and other personal frictions at Oxford, Soddy decided to retire. And so at 59 years of age his resignation from Dr. Lee's chair was accepted and he severed his formal connection with the academic world. The following year he embarked on a long tour of India and the Far East, partly with a view to developing a process for the treatment of monazite sand. On his return he lived briefly at Enstone near Oxford, then moved to Brighton. He continued to write on economic and social questions and on mathematics. He died at Brighton on September 22, 1956 at the age of 79.

It is evident from this short biography that Soddy was a man of strong opinion and unorthodox views. He had a great sense of adventure, which found outlets in mountain climbing and in travel. His greatest accomplishment was undoubtedly the discovery, in collaboration with Rutherford, of the transmutation of the elements through radioactivity. But his group displacement law is also a first-rate scientific achievement. This law describes the chemical change occurring when an alpha particle or a beta particle is emitted and is inextricably bound up with the concept of atomic number and the discovery of isotopes. As an example of his vision the reader may be even more deeply impressed by the following remarks, which he voiced in an address in Glasgow in 1906 in connection with his researches in radioactivity.

> We cannot regard the existence of this large internal energy as a peculiar property of the element radium . . . yet uranium since it produces radium with evolution of energy must possess all the internal energy of radium and more. . . . But there is no saying . . . how or when a discovery may not be made which will unlock this great store of energy bound up in the structure of the element. We are starting the twentieth century with the prize in full view.

In a meeting of the Royal Society on February 27, 1913, Soddy spoke on the existence of atoms of different atomic weight having the same chemical properties, that is, identical spectra and nonseparability by any known process. Sir Arthur Schuster, who was responsible for bringing Rutherford to Manchester, correctly pointed out in a 1913 letter to *Nature* that Soddy must have meant "any known chemical process," since atoms of similar chemical behavior but differing mass can be separated by diffusion, centrifuging, or by electric or magnetic methods. Schuster then asked whether an "ionium-thorium" mixture would in fact show only the thorium spectrum; perhaps, he suggested, the spectra of ionium and thorium are identical. Actually we know now that ionium is a decay product of uranium; having an atomic number of 90, the same as thorium, it would behave in a fashion chemically identical to this element although its mass is 230 as compared with 232 for thorium. In modern terminology ionium is an isotope of thorium, Th^{230}. Schuster correctly made the point that different elements cannot have the same spectrum. The confusion here is that ionium and thorium are the same element.

In his reply to Schuster, Soddy admitted that in speaking of nonseparability he was thinking only of chemical means. He pointed out, however, that the periodic law requires that just one element may appear in each place in the periodic table but that radioactive series show that different elements, not necessarily of the identical atomic mass, do occupy the same place. He then argued that these different elements are chemically identical. The basis for designating substances as different elements rested on their different radioactive disintegration rates. Thus thorium (Th^{232}) has a half-life of 1.39×10^{10} years, while ionium (Th^{230}) has a half-life of 8×10^4 years. The basis for putting such substances in the same place in the periodic table was their chemical inseparability and identical spectra.

The latter point, however, Soddy admitted, was based on only one case, ionium and thorium. He went on to point out that Th-X and radium would be a crucial test of identical spectra. In this connection, we know now that Th-X is Ra^{224}.

Nothing has been mentioned in these comments regarding any search for isotopic species in the stable elements. Although the idea of isotopes had first occurred in connection with the radioactive elements, J. J. Thomson's researches with positive rays disclosed that neon consisted of two isotopic species. Thus the actual recognition of isotopes for both radioactive and stable atoms appears to have taken place almost simultaneously.

The Positive Rays: J. J. Thomson (1856–1940)

The discovery of the electron, in great measure owed to the brilliant analysis of J. J. Thomson, was quickly followed by the discovery of the proton. This discovery stemmed directly from the work by Goldstein on what he called canal rays. Thomson, in his Bakerian lecture of 1913, "Rays of Positive Electricity," described Goldstein's observations and deductions, saying that "Goldstein [in 1886] observed that when the cathode in a vacuum tube was pierced with holes, the electrical charge did not stop at the cathode; behind the cathode beams of light could be seen streaming through the holes. . . . He ascribed these pencils of light to rays passing through the holes into the gas behind the cathode; and from their association with the channels through the cathode he called them *Kanalstrahlen*."

That the particles in these rays are positively charged is shown by their motion toward the cathode; this was proved experimentally by W. Wien, who deflected them in strong magnetic fields. Wien also demonstrated from the ratio of their charge to their mass, that these positively charged particles are more than a thousand times more massive than electrons.

The step from the recognition that canal rays consist of positively charged massive particles to the discovery of the proton was not so easy as it might appear, for, as Thomson notes, "The composition of these positive rays is much more complex than that of cathode rays [that is, the electrons], for, whereas the particles in the cathode rays are all of the same kind, there are in the positive rays many different kinds of particles."

The problem that arose, then, was to develop an experimental method to separate out the various particle components (that is, to obtain a mass spectrum) in the canal rays. In his Bakerian lecture, J. J. Thomson described a simple device, involving electric and magnetic fields, that is the forerunner of the modern mass spectrometer. Thomson pointed out in his address that a stream of charged particles moving in a straight line can be deviated by applying at right angles to this stream parallel electric and magnetic fields. The simultaneous action of the electric and the magnetic fields causes the beam to be deflected from its original direction. The electric field deflects the beam in one direction, say left or right, the magnetic field deflects it up or down.

A simple algebraic analysis shows that both the magnetic deflection and the electric deflection of a particle in the beam depend on the ratio of the electric charge to the mass of the particle and on the speed of the particle. If, then, the electric and magnetic fields are kept at constant strength, no two particles will strike the receiving surface at the same point unless they have the same initial speed and the same ratio of charge to mass. In this way, a spectrum of the canal rays is obtained on the surface and each point in the spectrum represents a particular value of the speed and of the ratio of charge to mass.

In his analysis of this procedure, Thomson pointed out that all the particles that have a given value of the ratio of charge to mass strike the surface along a parabola so that each kind of particle has its own parabola. Thus, if the canal rays are allowed to strike a photographic plate, one can see at a glance, by counting the number of parabolas recorded on the plate, how many kinds of particles are in the beam.

The parabola corresponding to hydrogen ions (protons) is the one that is deflected most. Since the ratio of charge to mass is known, this ratio can easily be read off for all other kinds of particles using the positions of their parabolas relative to the position of the hydrogen parabola.

Using this type of analysis, Thomson discovered that the particles in the beam of canal rays could be divided into ionized (he called them electrified) atoms and ionized molecules. Since his experimental arrangement permitted him to determine what multiple of the unit charge was on any ion, he could determine the mass of this ion as compared to the mass of a proton from the position of the ion's parabola relative to the parabola of the proton.

From the positions of the parabolas on the photographic plate, Thomson was thus able to catalogue the atomic and molecular weights of the various elements and compounds in the tube.

Transmutation of an Element: Ernest Rutherford (1871–1937)

Although Ernest Rutherford's paper on the nuclear atom may be considered as the high point of his scientific career, he was still to do a great deal of research work of the very highest order. The period immediately after Bohr's fundamental paper appeared was one of great scientific activity, and new discoveries were being announced in all the great European laboratories. Rutherford was busy giving lectures, attending conferences, and corresponding with scientists all over the world. Wherever he went, he advocated seeking all possible ways of obtaining very high voltages. "I think it is a matter of pressing importance, at the present time," he said, "to devise electrical machines of the highest possible voltages." With the nuclear atom an established fact, Rutherford reasoned that one could probe the positively charged nucleus (the neutron had not yet been discovered) only with very high energy ions, and very high voltages would be needed to produce them. Although very energetic alpha particles were available from radioactive nuclei, only very weak beams (that is, very few alpha particles per second) could be obtained for any one experiment because of the scarcity of radioactive elements.

In 1914, the British Crown recognized Rutherford's great scientific achievements by knighting him, which pleased but also somewhat embarrassed him for, as he wrote to Hevesy, "it is, of course, very satisfactory to have one's work recognized by the powers that be, but the form of recognition is also a little embarrassing for a relatively youthful and impecunious Professor like myself. However, I will trust it will not interfere with my future activities." This honor in no way affected Rutherford's attitude toward his work, his colleagues or his students. He went about his laboratory, as always, concerning himself with the details of everyone's project and announcing himself by his loud laugh long before he arrived to discuss the details of an experiment. He took great pride in the successes achieved by his coworkers and was very happy that "the laboratory had [not even] one piece of bad work to its discredit."

During this period, Rutherford was in constant correspondence with scientists over the entire world. He was particularly interested in the work of Hahn and Lise Meitner who were studying the nature of gamma rays. Hahn and Meitner argued that these were electromagnetic waves emitted by the radioactive nucleus when it readjusts itself to a new equilibrium position after emitting an alpha particle or a beta particle. This point of view proved to be true and was supported by Rutherford who was busily trying to measure the wavelengths of gamma rays. After Rutherford's visit to the United States, where he had given the first William E. Hale lecture to the National Academy of Sciences on "The Structure of the Atom and the Evolution of the Elements," he returned to Manchester to continue with his gamma-ray experiments.

By this time Henry Moseley, whom Rutherford had nurtured at Manchester and who had subsequently gone to Oxford, was obtaining his phenomenal results on the X-ray spectra of the elements. He kept Rutherford constantly informed on the progress of his work, even though Rutherford no longer had any direct contact with it. This point emphasizes one of the special qualities possessed by Rutherford—the knack of imparting to everyone who worked with him the feeling that he was a kind of father-confessor for all scientific problems—always available for consultation and advice. Everyone working in the Manchester laboratory in those years recalls that it was a very happy family with Rutherford as father. Geiger summed it up in these words: "Nothing was so refreshing or so inspiring as to spend an hour . . . alone with Rutherford. . . . I would be loath to part with the memory . . . spent in fellowship with a mastermind."

With the advent of World War I, the character of the Manchester laboratory began to change and the family of research students was quickly broken up, with many of the younger men going into the war industries. Interestingly enough, during those early war years Rutherford was in fairly constant communication by mail with the scientists of the Central Powers. Soon Rutherford himself was involved in war research work and diverted part of his laboratory to finding a way of detecting submarines. He also was active as a member of the Board of Inventions and Research. During the last year of the war, he spent a good deal of time in Washington as a member of a British Mission to the United States.

In spite of all these war activities, Rutherford still managed to do fundamental experiments in physics and to begin his famous experiments on the artificial disintegration of nitrogen by alpha-particle bombardment, which we discuss below. He completed these experiments by 1919 and published his epoch-making results in four papers in the 37th volume of *Philosophical Magazine*. This research was the last scientific work that Rutherford did at Manchester. He then accepted the post of Cavendish Professor at Cambridge when J. J. Thomson resigned to devote his full time to directing Trinity College as its master. When Rutherford came to Cambridge in 1919, he found that he had also been appointed Director of the Cavendish Laboratory and a Fellow of Trinity College.

In 1920 Rutherford gave the Bakerian lecture to the Royal Society for the second time, presenting a complete theory of radioactivity. During the presentation he proposed two new ideas that were very prophetic but about 12 years before their time. He suggested, first, the possibility of a mass 2 isotope of hydrogen, subsequently discovered by Harold Urey, and, second, the existence of a neutral particle of unit mass. This particle, the neutron, was later discovered by James Chadwick. The same year, at a meeting of the British Association, Rutherford suggested that the nucleus of the hydrogen atom be called the proton. This name was quickly adopted by the scientific community.

This was a period of great lecture activity for Rutherford; he was in demand everywhere and had to refuse most of these invitations. At the same time, he was receiving honors from all over the world: an honorary degree from Copenhagen; foreign member of the Royal Academy of Science at Amsterdam; Professor of Natural Philosophy at the Royal Institution, the Order of Merit, and so forth. In spite of all this acclaim, Rutherford went right along with his scientific work as though the scientist part of him acted quite independently of his more mundane part. He continued—with Chadwick as collaborator—to bombard nuclei of all kinds with alpha particles and was trying, even then, to induce artificial radioactivity by this bombardment. They succeeded in disintegrating many kinds of nuclei. At the same time, he tried to find, without success, the neutron whose existence he had already postulated. In 1925 he was elected President of the Royal Society and held this office for five years. He was so elated by this election that his exuberance was communicated to everyone around him and acted as a stimulant to his students. In 1931 he was created a Baron and became Baron Rutherford of Nelson. He immediately sent a cable to his mother, who had visited him the year before: "Now Lord Rutherford, more your honor than mine, Ernest." But these happy events were also mixed with tragedy. In 1930 his daughter and only child, Eileen, who had married the mathematical physicist R. H. Fowler, died soon after giving birth to her fourth child.

The years from 1930 until his death, from a strangulated hernia on October 19, 1937—he was then 66 years old—were very exciting and productive ones both for himself and his Cavendish Laboratory. Scientific discoveries of the greatest importance were being made all over the world. In 1932 Chadwick discovered the neutron. Shortly after that Cockroft and Walton induced artificial radioactivity decay with proton bombardment. And in 1933 Carl Anderson discovered the positron. These events were soon followed by Enrico Fermi's probing of nuclei with slow neutrons, and the world was suddenly thrust into the nuclear age. Although Rutherford had done as much as, if not more than, any single man to usher in this age, he did not live to see either its destructive or constructive qualities although he certainly had a fairly good idea of the new discoveries that were to come.

Rutherford was a simple man in the sense that Einstein was simple, for they both saw the universe as a very orderly place whose secrets could be discovered by a direct and uncomplicated approach. In a tribute to Rutherford in 1932 Bohr described him as follows: "If a single word could be used to describe so vigorous and many-sided a personality, it would certainly be 'simplicity.' Indeed, all aspects of his life are characterized by a simplicity of a similar kind to that which he has claimed of nature, which he is able to discover, where others before were not able to discover, where others before were not able to see it." He was certainly the greatest experimental physicist of his era; his achievements were so important that most of the developments in physics today stem from the work that he did.

After his basic analysis of the experiments on the scattering of alpha particles by heavy nuclei, which finally led to the nuclear model of the atom and to Bohr's theory of atomic spectra, Rutherford began a series of experiments in 1915 on the scattering of alpha particles by light atoms. These experiments, designed to obtain more detailed information about the nuclear atom, resulted in the first artificially induced nuclear transformation of an element. The description of this remarkable phenomenon, which occurred in 1919, is contained in the fourth of a series of papers that appeared in the *Philosophical Magazine.*

Rutherford began the discussion of his particle collision experiments by noting that radium, in the form of a thin coating on the metal, produced scintillations on a zinc-sulfide screen. But in this case the screen was so far away that these scintillations could not be due to the alpha particles emitted by the radium since such particles were absorbed in relatively short distances by the gas between the radium source and the screen. He then pointed out that these scintillation-producing particles had all the properties of fast protons. Still, he did not know whether they were knocked out of the radium C atoms by the alpha particles or whether they were hydrogen atoms that were attached to the metal surface and then kicked out by the alpha particles. A hydrogen ion—the atom without its electron—is, of course, a proton.

Since protons knocked out of radium C would represent an entirely new physical phenomenon, Rutherford decided to study these "natural" scintillations in more detail. To do this he decided to vary the amount of gas between his scintillation screen and the radium C source of alpha particles and also to use various thicknesses in absorbing foils. In this way he could determine the energy (the range) of the scintillation-producing particles and thus get some idea of their character. He began by first evacuating his apparatus as completely as possible and then putting in various quantities of dry oxygen and dry carbon dioxide. In both cases he found, as was expected, that the number of scintillations decreased with increasing concentrations of the two gases. This result simply verified that the proton-like particles were more effectively absorbed (that is, were robbed of their kinetic energies) if more gas atoms were introduced. More gas atoms meant more collisions and more collisions meant a greater slowing down of the particle.

Rutherford then repreated the experiment using dry air rather than pure oxygen or carbon dioxide and obtained a "surprising effect." He found that the introduction of dry air did not diminish the number of scintillations but rather increased this number. The word "surprising" was something of an understatement, for Rutherford was certainly aware of the full significance of these results. He had already shown that oxygen alone decreased the scintillation count so that the increase could only be due to the nitrogen in the dry air. This conclusion meant that the alpha particles, in passing through the gas, had collided with nitrogen atoms and had knocked protons out of these atoms.

To check this point Rutherford considered all other possible sources of protons that could have given rise to the observed scintillations. He demonstrated conclusively that they could have come only from the nuclei of the nitrogen atoms. He still did not know whether these particles were protons or atomic nuclei of mass 2, but in either case he knew that this meant that "the nitrogen atom [nucleus] was disintegrated under the intense forces developed in a close collision with a swift alpha particle, and that the hydrogen atom which is liberated formed a constituent part of the nitrogen nucleus."

Rutherford pointed out that this conclusion was in agreement with the observed range of the nitrogen atoms after they have been hit by alpha particles. Since they are less massive than oxygen atoms, nitrogen atoms should travel 19 percent farther than oxygen atoms suffering similar collisions. But this is not the case; the nitrogen atoms move about as far in air as the oxygen atoms. As Rutherford noted, this is exactly what one would expect if the collision energy is divided between the ejected proton and the residual nucleus.

The importance of Rutherford's results for the future of atomic physics was twofold. First, he demonstrated experimentally that nuclei of atoms contain individual protons. This idea had already been surmised, but no one had proved it empirically. Second, his experiment showed that nuclei could be disrupted and changed into other nuclei; this demonstration was the first example of the artificial transmutation of chemical elements.

Although Rutherford had no idea of the nature of nuclear forces, he set up a simple model (which is close to our modern concept) of the nitrogen nucleus to account for the observations. He suggested that the 14 particles inside the nitrogen nucleus arranged themselves into three alpha particles and two other particles, one of which is a proton. He pictured these two particles moving in the outer regions of the nucleus where they were easily knocked out by an alpha particle collision. Since Rutherford knew nothing about neutrons in 1919, he spoke of the two outer particles as "either two hydrogen nuclei or one of mass 2." We now know that one of these particles is a proton and the other is a neutron.

The Diversity of Atoms: Francis William Aston (1877–1945)

Frederick Soddy's discovery of isotopes through research in the radiochemical elements about the year 1910, and their exploration in the next three years, must have suggested to more than one investigator that such atoms should be sought in the stable elements. However, the way such a search might be started was by no means clear. It remained for J. J. Thomson with his positive ray apparatus to open the doors for such investigations in 1912. As we have already seen, Thomson analyzed a sample of very pure neon, and his apparatus showed that atoms of mass 20 and 22 were present in the gas. The possibility that the atoms of mass 22 found in the positive rays might be NeH_2 prompted Thomson to put F. W. Aston, his assistant, to work on the diffusion separation of the gas. After the most laborious efforts, Aston achieved what appeared to be a significant separation of fractions of differing mass, but the advent of World War I prevented further researches on the problem.

During the war Aston was assigned to the Royal Aircraft Establishment at Farnborough. Toward the latter part of the war he conceived the idea of extending the isotope study by developing a mass spectrograph as an improvement of Thomson's parabola method. The spectrograph that he designed was built at Cambridge in 1919 following his return to the laboratory. The details were presented in his 1920 letter to *Nature*, entitled "Positive Rays and Isotopes," together with photographs of the mass spectra obtained with its use. Neon was the first element examined and its isotopy was immediately revealed; atomic masses of 20 and 22 (on the scale oxygen = 16) were clearly present, the precision of the mass numbers being estimated at 1/10th percent.

This spectrograph was used to analyze some 50 elements in the following six years, revealing the almost universal existence of isotopes. Exceptions appeared in the elements of odd atomic number, the great majority of these having no isotopes; that is, atoms of one mass only were present. Aston assumed that the most important result of the research was the discovery that the measured mass of all

Francis William Aston (1877–1945)

the atoms, except hydrogen, was integral on the basis oxygen = 16. This is called the whole number rule. But a second and, subsequently, a third mass spectrograph of increasing precision showed that this was not substantiated by more precise measurements. Aston's research thus showed that Prout's hypothesis, made more than a hundred years previously and intermittently revived, was untenable. The addition of masses, equal to that of hydrogen, did not give the masses of the succeeding elements in the periodic table. It was also clear that the elements were not built up, as the whole number rule suggested, of mass units equal to 1/16th of the oxygen atom.

On the other hand, to Prout's credit, the existence of the nuclear atom made it appear that all the elements were made up of protons and electrons, the elemental building material of hydrogen atoms. But even from this point of view it was evident that the masses of the elements were not arrived at by the simple addition of the masses of the protons and electrons of which they were apparently composed. The elemental masses as measured by the spectrograph showed a mass defect as compared to the sum of the masses of their free constituent particles. This defect in mass was explained, on the basis of Einstein's mass-energy equation $E = mc^2$, as representing the energy that was radiated away in the formation of the nucleus. It was therefore the binding energy of the particles in the nucleus; the greater the mass defect, the more stable the nucleus. The further clarification of this point of view had to await the advent of experimental nuclear physics in the 1930s and the discovery of the neutron. These subjects are presented in later chapters.

It is convenient at this point to summarize the nomenclature brought into physics by these advances in the measurement of atomic masses. The atomic number was defined as the number of (integral) elementary nuclear positive charges; the mass number was taken as the nearest whole number to the mass of the isotope considered (later it was taken to be the number of protons and neutrons in the nucleus). The mass of the isotope was its mass on the scale oxygen = 16, although this subsequently had to be changed owing to the discovery that oxygen itself had isotopes. The reference atom was then changed to a comparison of masses on the basis that the most prevalent isotope of oxygen was of mass 16. In order to convey an idea of the stability of nuclei, Aston later introduced a quantity called the packing fraction, defined as the ratio of the mass defect to the mass number. Before closing this commentary it must be pointed out that the importance of determining the isotopic constitution of the elements and the individual isotopic masses drew many able investigators to this field and resulted in the invention of several different mass spectrometers. For details regarding the outstanding investigators and their instruments the reader should consult the indexes of the various physical journals.

Until he came to Cambridge, Aston had a rather varied career. The son of a metal merchant, he was born in a suburb of Birmingham, England, and as a youth displayed a strong bent for experimenting with mechanical devices and with chemicals. He graduated from secondary school with highest honors in mathematics and science and entered Mason College, later the University of Birmingham, in 1893. His chemistry teachers at Birmingham were Tilden and Frankland and he studied physics under J. H. Poynting. With Frankland he worked on problems in organic chemistry, publishing his first paper in 1901. By no means was all of his work done in university laboratories because in 1896 he fitted up a loft in his home as a private laboratory and workshop. Here he carried on private research for many years. In 1900 he entered the employ of a brewing company and continued in industrial work until 1905 when he returned to university work on a scholarship. In 1908, on the death of his father, he took a year off to make a trip around the world. On his return he continued working on glass discharge tubes and on vacuum apparatus to exhaust them, including an automatic Toepler-type pump of his own construction. His outstanding skill in this field led in 1910 to his association with J. J. Thomson at Cambridge as a research assistant. He was soon engaged with Thomson on the development of the latter's positive-ray apparatus and on the diffusion separation of neon in 1913.

After service in World War I, he returned to Cambridge to work independently on his mass spectrograph. He was made a fellow of Trinity College, Cambridge, in 1919 and maintained his residence in the college for the remainder of his life. By 1922 his isotope work was deemed to be so outstanding that he was awarded the Nobel Prize in chemistry. As time went on, other able investigators entered this field, but Aston continued his investigations, increasing the precision of his work.

In his mature years Aston was a quiet, reserved man; above average height, he had an erect carriage. He was a strong individualist and maintained from his youth the conservative views characteristic of a middle-class upbringing in the late Victorian period. He was fond of travel and sports, and his interest in and knowledge of music was so extensive that he served for many years as the music critic of the *Cambridge Review*. In finance, he was also unusually able, so that toward the end of his life he had amassed considerable means. His biographer for the Royal Society summed up his life, perhaps most aptly, with the phrase: "Aston's life was a chain of uninterrupted success."

References

Bohr, Niels, "On the Spectrum of Hydrogen," address before the Physical Society, Copenhagen, 1913.

Fleck, Alexander, "Memoir on Frederick Soddy," *Biographical Memoirs of Fellows of the Royal Society*, vol. 3, 1957, p. 205.

Howorth, Muriel, *The Life Story of Frederick Soddy*. London: New World Publications, 1958.

Nielson, J. Rud, "Memories of Niels Bohr." *Physics Today*, vol. 6. , 1963, p. 23.

Soddy, Frederick, "The Internal Energy of Elements," address to the March Meeting of the Institution of Electrical Engineers, Glascow Local Section, 1311, 1906.

Weisskopf, Victor, "Niels Bohr (A Memorial Tribute)." *Physics Today*, vol. 16. , 1963, p. 59.

Wheeler, John A., *Physics Today*, vol. 16, 1963, p. 44.

10 X Rays and Their Contribution to the Riddle of Matter

Every why hath a wherefore.

—William Shakespeare, *The Comedy of Errors*

Interference Phenomena: Max von Laue (1879–1960)

Max von Laue, like Max Born, was one of the last of the great physicists whose lives spanned the period from the inception of the quantum and relativity theory up to the modern era. Like Max Born, Laue contributed extensively to the development of modern physics, but he was not as well known as some of his more famous countrymen such as Max Planck, Albert Einstein, and Werner Heisenberg.

Laue was born on October 9, 1879, in the small village of Pfaffendorf b. Koblenz (Rhein) and spent his childhood and early adult years there. After completing his gymnasium studies, he entered the University of Strassburg to study mathematics and physics and then went on to Göttingen, Munich, and finally Berlin, where he obtained his Ph.D. in theoretical physics in 1903. Immediately after obtaining his doctorate, he began one of the most productive research careers in the history of modern physics. Although his contributions to physics are to be found in all branches of the subject, he devoted most of his creative efforts to physical optics, thermodynamics, and relativity theory. He was one of the first to recognize the great importance of Einstein's work, and his book on the theory of relativity is still one of the best on this subject.

His great interest in physical optics led him to the general problem of the interference of light. When X rays were discovered, he quite naturally turned his attention to the interference of X rays and immediately became involved in controversy as to the nature of these rays. It was clear to him that if X rays were, indeed, electromagnetic waves, they would have to be of much shorter wavelength than visible light, since X rays could not be refracted or diffracted the same way as visible light. He therefore reasoned that if one could obtain a diffraction grating with the lines spaced much closer together than in an ordinary optical grating, one should observe X-ray diffraction. This observation led him to the idea of using the atoms forming the lattice structure of a crystal as a diffraction grating. After working out the theory for this type of diffraction, Laue had Walter Friedrich and Paul Knipping perform the experiment—with results that completely supported the theory. Laue received the Nobel Prize in physics for this work in 1914.

Max von Laue (1879–1960)

Aside from his work on the diffraction of X rays, Laue made important con-tributions to the fields of physical optics, crystallography, electromagnetic theory, relativity theory, superconductivity, solid state theory, and atomic theory. His more than 130 original papers appeared in the most important physical journals in the world. In addition to his research papers, he wrote numerous articles on the his-torical aspects of physics, with special emphasis on the roles of individual physicists in the development of physics.

While he was doing his trailblazing work on the theory of X-ray diffraction, Laue was assistant professor of theoretical physics at the University of Munich. He left Munich in 1914 to take the chair of theoretical physics at the University of Frankfurt, where he remained until 1919, when he was invited to become professor of theoretical physics at the University of Berlin. At the same time, he was director of the Kaiser Wilhelm Institute of Physics in Berlin, where he remained until 1943. He then went on to the Max Planck Institute of Physics in Göttingen and, in 1957, became director of the Kaiser Wilhelm Institute for Physics, Chemistry, and Electrochemistry. In 1953 he became director of the Fritz Haber Institute of the Max Planck Society of Berlin-Dahlem.

After X rays had been discovered by Wilhelm Roentgen in 1895, a considerable controversy arose as to their nature. That they were not deflected in an electric or magnetic field was no clear proof that they were not particles, for they might very well be neutral particles of some sort and thus unresponsive to electromagnetic forces. It was never evident that they were electromagnetic waves, since it was very difficult to observe their wave properties; as we now know, their wavelengths are very short. Nevertheless, by 1911 more and more evidence had been accumulated in favor of the electromagnetic-wave thesis. Thus, the experiments of C. G. Barkla in 1911 demonstrated that X rays are scattered from small particles in suspension (droplets of water in a cloud, or dust particles) the way light is scattered in a dusty medium. Moreover, Barkla also showed that X rays can excite atoms in a body to emit fluorescence radiation.

Both of these phenomena convinced Max von Laue that X rays are electromagnetic waves, but of such short wavelength that their wave character would be very difficult to establish by the usual interference and diffraction experiments. That the X rays are electromagnetic in character was clear to Laue because they can set atoms, which consist of electrical charges, vibrating. Here then was a challenge to Laue to devise some method of demonstrating the interference of X-ray waves.

To appreciate the difficulty of the problem, we must understand that observable interference between waves can occur only if the dimensions of the particles, apertures, or other sources of the interference are not very much larger than the wavelength of the waves themselves. Light waves can be made to interfere and give observable interference patterns by having them pass through two small holes that are very close together. In the same way, light reflected from a surface of very many closely spaced parallel lines—the diffraction grating—gives an interference or a diffraction pattern. Anyone can easily see such a pattern by allowing light to glance off a long-playing record, since the surface grooves are close together. Since the wavelength of ordinary white light is about 1/50,000 of an inch, one must have a few thousand lines per inch on a diffraction grating to obtain good results. But such a diffraction grating is much too coarse for X rays, which have wavelengths 100 to 1000 times smaller than ordinary light.

Laue, therefore, introduced a very ingenious idea: to use the regular array of the atoms in the lattice structure of a crystal as the grating. This idea goes back to Bravais, who suggested in 1850 that a crystal is a regular lattice structure with the atoms occupying its lattice points. The difference, however, between a crystal lattice structure and an ordinary surface grating is that the former is a space lattice, and hence three-dimensional, whereas the latter is a two-dimensional pattern. But, aside from the difference in complexity, the problem of the interference of waves scattered from these two types of gratings is the same. Moreover, since the distance between two neighboring atoms in a crystal is about 1/100,000,000 cm, the geometrical situation is desirable for X rays, which have wavelengths of the same order of magnitude.

The theoretical analysis of the interference of X rays scattered from a crystal, as given by Laue, is in essence quite simple and straightforward. Since the atoms in a crystal are arranged symmetrically, one can picture the crystal as consisting of elementary patterns or cells that are repeated over and over again like the pattern in wallpaper or floor coverings. Such a fundamental cell is defined by three vectors starting from a point of reference. The position of any atom in the crystal, relative to this reference point, can be calculated by doubling, tripling, quadrupling, and so on, these vectors (that is, laying off these vectors one after the other) until we arrive at the atom in question. This process is like reaching any corner in a city in which the streets are laid out in a regular pattern. We can always get to any corner by following one street after another. This is the content of the first equation in Laue's famous 1912 paper, "Interference Phenomena for X rays."

Suppose now that an atom in a crystal is set vibrating by incident X rays. This atom, then, sends out waves of its own in various directions. What does a wave of this sort look like, at a great distance from the atom? There is first the amplitude of vibration of the wave which Laue called in the second equation in his paper. But this amplitude must be diminished by the factor r (which appears in the denominator of his second equation) because the wave gets weaker and weaker as we go farther and farther away from the source, that is, the atom. The reason for this is that the wave spreads out like the surface of a spherical balloon that is being blown up. Finally, a factor must be included to show that the amplitude of the wave at a given point oscillates with a definite frequency, or, put differently, that if we move a distance λ (the wavelength) along the wave (assuming for the moment that the wave is fixed in space), the amplitude changes from a maximum to a minimum and back again to a maximum. This is the exponential factor e^{-ikr} where i is the square root of -1 and k is proportional to the reciprocal of the wavelength.

If we put these three factors together, we get Laue's second equation which gives the wave emitted by a single atom. To find the total effect, at a point, of all the atoms that are oscillating and emitting waves together, we must add all these contributions, taking into account that each atom is at a different distance from this point. However, before doing this Laue took one more thing into account: At any point there is present not only the electromagnetic field of the waves coming from the excited atoms, but also the field of the waves that excite the atoms. In other words, while waves emitted by the crystal are moving away from the crystal, there are at the same time waves moving toward the crystal; both of these contribute to the electromagnetic field intensity. The field conditions at a point are thus the result of outgoing spherical waves and an incoming plane wave which Laue represented at the point r by an exponential factor which gives the direction of propagation of the incoming waves as well as the x, y, z coordinates of the point. This factor must therefore also be included, and then the sum of this expression must be taken over all atoms in the crystal. This conclusion forms the basis for the third equation in Laue's paper.

To evaluate this sum Laue made the simple assumption that the distance between any two atoms in the crystal is small compared to the distance of any atom from the distant point r. He obtained an expression that is the crucial result of the paper. It gives the field intensity at a particular point and shows (because of the sine functions in its denominator, which vary periodically from 0 to 1) that in a particular plane perpendicular to the direction of the incident ray, and at the distance r from the crystal, the intensity varies from point to point. A photographic plate made to coincide with this plane would therefore show a pattern of dark spots if X rays are indeed waves.

This pattern is called an X-ray diffraction pattern; from its geometry one can not only calculate the wavelength of the X rays but also find out a great deal about the structure of the crystal.

The second part of Laue's paper was devoted to the experimental verification of his analysis for which he won the Nobel Prize in 1914. Laue suggested to Friedrich and Knipping that they try to obtain a diffraction pattern by actually scattering X rays from specific crystals. Following this suggestion, Friedrich and Knipping carried out the experiment which involved an extremely simple arrangement. The rays from an X-ray tube were limited to a thin bundle by a series of small diaphragms arranged along a line, and the crystal that was to be investigated was set up on a rotatable table. The orientations of the crystal with respect to the X rays were carefully measured with a telescope and photographic plates were placed around the crystal to intercept the scattered X rays.

Just as Laue had predicted, Friedrich and Knipping found regular diffraction patterns on the photographic plates, indicating, first, that X rays have wave properties and, second, that crystals do have a regular lattice structure. Laue's theory, and the work of Friedrich and Knipping in verifying it, opened up vast new areas of research both on the structure of matter and on the properties of short-wavelength electromagnetic waves. A direct outgrowth was the trailblazing work of Sir William Henry Bragg and his son in the use of X-ray diffraction patterns in the analysis of crystal structure.

Bragg's Law: William Henry Bragg (1862–1942) and William Lawrence Bragg (1890–1971)

Laue's suggestion that a crystal should act as a diffraction grating for X rays and thus supply proof that they are electromagnetic waves of very short wavelength was, as we have seen, amply verified in 1912 by the experiment of Friedrich and Knipping. In this experiment a very thin pencil of X rays struck perpendicularly on the face of a thin zinc sulfide crystal. Placed a few centimeters behind the crystal was a photographic plate with its surface also perpendicular to the direction of the X ray beam. A pattern of spots formed on the photographic plate after exposing the crystal to X rays. Laue accounted for this with the hypothesis that the X rays were composed of a small number of short wavelengths whose diffraction by the atoms of the crystal resulted in the pattern.

This analysis was immediately examined by W. L. Bragg, then a young research student with J. J. Thomson at Cambridge. Bragg proceeded to show that the Laue pattern could be analyzed in a much simpler way, that it was unnecessary to make the restricting assumption of the presence of only a few wavelengths in the X-ray beam. Bragg adopted the view that the incident X rays consisted of a continuous spectrum extending over a wide range of wavelengths. To simplify Laue's treatment he supposed that the atoms of the exposed crystal act as diffraction centers and radiate secondary waves. When a wave pulse falls on a plane it is reflected. If it falls on a number of atoms situated in a plane, they act as centers of disturbance. The secondary waves from the atoms build up a wave front exactly as if the pulse had been reflected from the plane according to the manner of Christian Huygens's construction for reflection. The intensity of the reflected wave depends, of course, on the number of atoms present in the crystal plane, the intensity being greater as the density of atoms in the plane is increased.

In a crystal the atoms are arranged regularly in space so that if a plane is passed through one set of atoms, say the set of atoms composing a crystal face, then a parallel plane may be passed through the next set of atoms lying underneath the surface layer. The secondary waves from the atoms in the lower plane, when reflected in the same way, may or may not be in phase with the reflected waves from the top layer. Whether or not they are depends upon the perpendicular distance d between the atom planes, the wavelength λ of the incoming X rays, and finally the angle of incidence θ between the ray and the surface of the plane. The relationship between these variables is easily derived from geometric considerations and, as shown by Bragg (and repeated in most books on elementary modern physics), is given by $n\lambda = 2d\sin\theta$, an equation that is now known as Bragg's law. Here n represents what is called the order of the spectrum, a term that has the same meaning for optical gratings where several sets of complete spectra of the source may be produced by increasing angles of incidence to the grating.

Sir William Henry Bragg (1862–1942)

At the time Bragg proposed the above relation, the generation of X rays was believed to take place in the sudden deceleration of electrons within the X-ray tube as they struck the anticathode, usually a heavy metal. Classical electromagnetic theory required that such decelerations must produce radiation, so it was assumed that an electromagnetic pulse was radiated when each electron was stopped by collision with the target of the tube. It was shown by G. J. Stoney in 1898 that such a stream of pulses could be analyzed as wave trains, giving the appearance of a continuous spectrum. On the other hand, C. G. Barkla had shown that when X rays strike a second substance, this substance can emit a characteristic radiation, or radiations, as well as scatter the original X rays. All this knowledge was very fragmentary and the nature of the spectra of X-ray sources remained to be explored. Added to this was the whole field of crystal structure of which only the rudiments were known.

In order to investigate the spectrum of an X-ray source, a spectrometer, similar in many respects to an optical spectrometer, was devised by the elder Bragg. To get approximately parallel rays the investigators used a collimator consisting of a lead block pierced by a hole whose size could be regulated by a slit. The X-ray beam, so limited, emerged as a thin ribbon of radiation directed at the analyzing crystal placed upon a rotatable table whose vertical axis was at the center of the instrument. The reflecting face of the crystal was arranged so that it contained the axis of the table. The X-ray beam reflected from the crystal face was then intercepted by an ionization chamber, the current in the chamber being proportional to the intensity of the reflected X rays. Relative intensities were thus determined by the ratios of the ion-chamber currents.

The X rays in this experiment were produced by a tube containing a platinum anticathode. From considerations outlined above, it might be expected that a continuous spectrum would be present and, on the basis of Barkla's results, perhaps a characteristic radiation as well. Using rock salt as the reflecting crystal, and slowly varying the angle of incidence θ, the intensity of the reflected radiation was found

Sir William Lawrence Bragg (1890–1971)

to follow a curve having three peaks or three monochromatic radiations which were repeated as a second-order spectrum and, probably, as a third order at the largest angles shown. The general appearance of these lines was found to be the same if a different cleavage face of the crystal was used or if other crystals were used, such as zinc blende. Thus it was discovered that the X-ray emission spectrum of the anticathode was composed of a continuous spectrum on which were superimposed definite monochromatic lines. W. H. Bragg later showed that these lines, which became known as the K emission lines, were indeed characteristic of the target emitting the rays. H. G. J. Moseley soon demonstrated that characteristic K lines are generated for all chemical elements used as the anticathode in an X-ray tube and that the frequency of these lines increases as the atomic weight of the anticathode increases, with some exceptions. These exceptions were very important, for they shed a new light on the periodic table of the elements, as is discussed in the section on Moseley and his measurement of the high-frequency spectrum of the elements.

In order to find the actual wavelength of the X-ray lines by the use of the formula $n\lambda = 2d \sin \theta$, Bragg had to know the value of d appropriate to the orientation of the crystal being used. In order to specify the orientation (and therefore the reflecting surface presented to the X rays), Bragg made use of what are called Miller indices. The atom planes appropriate to the Miller indices 100, 110, and 111 for a cubic crystal, such as rock salt, are referred to by Bragg in his 1913 paper, "The Reflection of X rays by Crystals," which he coauthored with his son. The problem in this investigation was to make the selection for d that corresponded to the actual physical situation in the experiment. The Braggs discussed this problem and made a calculation for the wavelength of the B line, using the group 4NaCl as "the smallest complete unit of the crystal pattern." (This corresponds to a simple cube with Na and Cl atoms at successive corners as one goes around the cube.) The value found was 1.78Å, which is too large; a more acceptable value is about 1.10Å.

The importance of this research was twofold. First, it showed that the X-ray emission spectrum of an element is characteristic of that element. Second, it showed that X rays can be used as a powerful and precise means of crystal analysis. Both of these discoveries opened up large fields of investigation. Although X-ray emission spectra were fully mapped out in the course of a few decades, X rays continue to be widely used as an analytical tool in the problems of the structure of matter.

William Henry Bragg, the son of a retired officer of the merchant navy, was born on a farm near Carlisle in the north of England on July 2, 1862. At seven he was sent away to school, and in 1875, at the age of 13, to King William College on the Isle of Man. Here he stood high in his studies, especially in mathematics, but, as he himself often said, his life was far from happy. He was a quiet boy, not socially minded, and did not mix well. The religious atmosphere of the school depressed him, and the Bible readings, emphasizing punishments for the sins of this life, caused him "acute fear and misery."

In 1881, at the age of 19, he entered Trinity College, Cambridge, still a shy and lonely young man, with a liking for boating and tennis, but with a main outlet in study. He entered Cambridge on a small scholarship and earned a larger one for the following years. He took his B.A. in 1884, placing as third wrangler in the first part of the mathematical tripos. This success was entirely unexpected; writing about it many years later he said, "I was lifted into a new world. I had new confidence, I was extraordinarily happy. I can still feel the joy of it." As an added fillip to the achievement, the great philosopher-to-be Alfred North Whitehead, who had taken the examinations a year before and again with Bragg's class, came to him and shook his hand saying "May a Fourth Wrangler congratulate a Third?"

Bragg continued at Cambridge for another year, then quite by chance heard that Horace Lamb had resigned his professorship in mathematics and physics at the University of Adelaide in Australia. Bragg applied for the position; after being interviewed by J. J. Thomson and Lamb, who had come to England, he was soon awarded the appointment. Bragg had studied only a minimum of physics, but the electors, guided by Thomson's and Lamb's recommendations, attached little importance to this drawback. At any rate, he read the subject on his passage out to the island continent. In his new environment he felt completely at home. His shyness vanished; he liked teaching, and through perseverance he became an interesting and polished speaker. In 1889 he married Gwendoline Todd, the daughter of the postmaster general and government astronomer of South Australia.

Until 1904 Bragg had done nothing that could be called research. But in that year he was called upon to give the presidential address to the Australian Association for the Advancement of Science at Dunedin and chose for his discourse a review of radioactivity. He became so interested in the subject that he began experiments to measure the range of alpha particles and the ionization they produced along their paths. Thus, at the age of 41, he was just beginning research in a new country with little support, far from the centers of active research and the stimulation that comes from associating with able workers in a difficult and rapidly moving field. Any assessment of his probable success as a contributing scientist would have been bleak indeed. It is surprising, therefore, that as a consequence of his work in the ensuing few years, Rutherford proposed Bragg for fellowship in the Royal Society—an honor accorded him in 1907. In 1908 he accepted the chair of physics at the University of Leeds and thus returned to the European mainstream of physics. At Leeds he resumed his early interest in X rays and investigated the ionizing effect of X rays arising from the secondary electrons released by the rays. In 1912, when Friedrich, Knipping, and Laue announced their discovery of the diffraction of X rays, Bragg was ready to apply these results to wider research. His discovery of the X-ray spectrum of the elements and the application of the rays to crystal analysis were the basis for his being awarded (with his son) the 1915 Nobel Prize in physics. The elder Bragg was then 53 and William Lawrence, his son, was a mere 25. Both father and son were active in scientific work at that time, supporting England's participation in World War I. The elder Bragg was appointed Quain professor at University College, London, in 1915, an appointment which he held until 1923, when he was made head of the Royal Institution following Sir James Dewar. He remained at that post until his death on March 12, 1942.

Among his later honors were a knighthood in 1920, the Copley medal of the Royal Society in 1930, the Order of Merit in 1931, and the presidency of the Royal Society in 1935 at the age of 73. He also gave a series of highly successful Christmas lectures intended for a juvenile audience; the topics included "The World of Sound," "Concerning the Nature of Things," "Old Trades and New Knowledge," and the "Universe of Light." All of these lectures were published and are masterpieces of sprightly scientific prose. Bragg will always be remembered by those who were fortunate enough to be his auditors. Combined with a quiet, warm dignity, he spoke with a deceptive ease and naturalness. He was a profound and sincere man full of enthusiasm for science and for new ideas. Considering his late start in physics, his achievements are a great tribute to his energy and profound ability.

William Lawrence Bragg, his son, was born at Adelaide, Australia, on March 31, 1890. He was educated at the University of Adelaide, graduating with first honors in mathematics in 1908. The following year he entered Trinity College, Cambridge, finishing in 1912 with first honors in the natural science tripos. In November of that year he published his first paper in the *Proceedings of the Cambridge Philosophical Society*, simplifying Laue's analysis of X-ray diffraction. It was this paper and the one written jointly with his father mentioned above that led to their being awarded the Nobel Prize in physics in 1915. The joint studies of father and son from 1912 to 1915 on crystal analysis were published in their book, *X-Rays and Crystal Structure*, which appeared in 1915. In 1919, after war service, young Bragg was appointed Langworthy professor of physics at the University of Manchester. He was then only 29 years old. He remained at Manchester until 1937, when he accepted a post as director of the National Physical Laboratory. In 1938 he was called to Cambridge to be Cavendish professor of physics, succeeding Lord Rutherford who died in that year. He continued as Cavendish professor until 1953, when he became Fullerian professor of chemistry and scientific director of the Royal Institution of Great Britain. He retired in 1965. Like his father, William Lawrence enjoyed giving science lectures to young people. He was knighted in 1941.

Atomic Number: Antonius van der Broek (1870–1926)

Many ideas in physics have come into focus as the result of simultaneous or nearly simultaneous developments in different branches of the subject. The development of the concept of atomic number seems to have been a phenomenon of this kind. In this instance the simultaneous developments were the Rutherford alpha-particle scattering theory and the X-ray scattering measurements of C. G. Barkla. Each investigation led its author to conclude that the number of electrons per atom is equal to half the atomic weight of the atom, with the restriction in Barkla's case that the generalization was made for the light elements only. Thus, the X-ray scattering data for carbon (substituted in J. J. Thomson's scattering theory) gave the number of electrons in this atom as six, just half the atomic weight of 12.

The development of Rutherford's nuclear atom model again raised the question of the possible number of elements. In 1911, the year of Rutherford's proposal and Barkla's measurements, only highly conjectural answers could be given. Nevertheless, an attempt to cast some light on this problem was made by the Dutch physicist Antonius van der Broek. He suggested in a short note to *Nature* in 1911 that, because the atomic weight of uranium (the heaviest element then known) was about 240, the proposals of Rutherford and Barkla indicated a total of 120 elements in the periodic table. By using a cubic construction of Mendeleev's for the periodic system of the elements, van der Broek found a mean difference between consecutive atomic weights equal to 2. Thus, an atomic weight of 240 for uranium gave 120 possible elements. He therefore suggested that if Mendeleev's cubic model should prove correct, then each element could be characterized by the number of possible permanent charges of both signs. Applied to uranium, one might conclude that the metal would have 120 permanent positive charges and 120 permanent negative charges. The location of these charges was not specified.

Two years later, in a second letter to *Nature*, van der Broek again returned to the idea of an atomic numbering scheme, stating that his earlier suggestion had been that "to each possible intra-atomic charge corresponds a possible element." But, in the meantime, research had shown that for uranium, the last element in the table, the number of [nuclear?] charges was "not . . . even approximately . . . half the atomic weight," if the experimental results were reasonably accurate. Rutherford's nuclear atom theory, however, showed that for a constant incidence of alpha particles on a scattering foil, and at a constant angle of scattering, the quantity (N/ntZ^2) should be constant. Here N is the number of alpha particles scattered at a fixed angle θ, n is the number of atoms per cubic centimeter of the foil, t is its thickness, and Z is the number of positive charges in the nucleus. Stated another way, the number of scattered alpha particles per atom of scatterer divided by the square of the nuclear charge should be constant. Van der Broek made this observation in his article and noted that, because the nuclear charge was unknown for the elements used in the foils, Geiger and Marsden, in order to test the constancy of the ratio, had used as a proportional number the atomic weight, A. He then gave two sets of data from Geiger and Marsden's paper showing that N/ntA^2 is not constant but shows a significant decrease as the atomic weight A increases for the elements copper, silver, tin, platinum, and gold. On the other hand, if A is replaced by M, the number of the place each element occupies in the periodic table, a constant number (within small limits) does result. Essentially, this evidence constitutes the basis for the idea of atomic number, the discovery of which is attributed to van der Broek. It should be emphasized that the preponderant evidence for the validity of the atomic number idea comes from the very powerful support given to it by Moseley's research, which is discussed below.

Before closing this commentary, a word of explanation should be given regarding van der Broek's final sentence in his 1913 note in *Nature*: "Should thus the mass of the atom consist for by far the greatest part of alpha particles, then the nucleus too must contain electrons to compensate this extra charge." The only atomic particles known at that time were electrons and protons. In accounting for atomic weights, the electrons can be neglected to a first approximation as their mass is of the order of 1/2000th of the proton. For example, if we try to reconcile the mass of an atom of copper with its atomic number, then some uncertainty ensues. The atomic number of copper is 29; on a naïve basis let us assume that these 29 positive charges are produced by 29 protons. Because the atomic weight of copper is 63.57, there is a large mass deficit. It was therefore assumed that additional positive charges in the form of protons or alpha particles were present in the nucleus to supply the requisite mass, but that the charge on these additional particles was annulled by the negative charge that was carried by electrons, also assumed to be present in the nucleus.

From these few remarks it is evident that, of necessity, only the crudest guesses about nuclear structure could be made at that time. This was a problem that became much clearer after nuclear masses became amenable to precise measurement, isotopes were recognized, and the neutron was discovered. As a point in favor of assuming the presence of electrons in the nucleus, the emission of beta rays by radioactive nuclei was cited by Frederick Soddy in his 1913 paper, "Intra-Atomic Charge," which appeared in *Nature*. Present evidence indicates that nuclei are composed of neutrons and protons. By rather elementary arguments involving the wave nature of particles, it can be shown that electrons cannot be localized in a space of the dimensions of the nucleus. Such a requirement could be fulfilled only if the electrons were to move at speeds considerably in excess of that of light, a condition forbidden by the theory of relativity.

Antonius van der Broek was born in Zoeterwoude in the Netherlands on May 5, 1870, and lived most of his life in Deventer and Noordwijk. He was trained for the law, but he was also an amateur scientist who published a number of papers, of which only the few dealing with the idea of atomic number are notable. With respect to his first statement of the atomic number idea, published in 1911 in the *Physikalische Zeitschrift*, it is reported that Rutherford expressed considerable annoyance that a layman should publish "a lot of guesses for fun without sufficient foundation."

Van der Broek appears to have been something of an eccentric. His philosophy of life made him a vegetarian, and his attire was marked, in a more conforming era, by the peculiarity of wearing sandals. Although he was deeply interested in science, he did not cultivate scientific acquaintances or engage in the activities of organized science. On the contrary, it appears that he shunned the company of professional scientists. He died in Bilthoven at the age of 56 on October 26, 1926.

11 *Atomic Theory Develops*

Gods are born and die, but the atom endures.

—Alexander Chase, *Perspectives*

Atomic Number: Henry G. J. Moseley (1887–1915)

The Bohr-Rutherford model of the atom was not immediately accepted, even though Bohr had succeeded in deriving the formula for the Balmer lines of hydrogen, for there appeared to be some difficulties in correlating the Bohr theory with the periodic table of Mendeleev. It had occurred to Mendeleev that, if he arranged the chemical elements according to their atomic weights (their masses), he might discern some kind of regularity in their chemical properties. He then discovered that elements with similar chemical properties did occur in the table periodically according to their weights so that one could arrange them into definite families. This periodicity seemed to imply that the mass of an atom is the decisive physical parameter that determines its structure and chemical behavior. However, this view was in conflict with the Bohr-Rutherford model, as H. G. J. Moseley showed. In this model, it is not the mass of the atom that is important for its chemical behavior but, rather, the charge on the nucleus—the atomic number—determined by the number of protons in the nucleus.

According to the Bohr-Rutherford model, a nucleus is composed of protons. As one passes from one element to the next highest element, the number of protons increases by one, so that the charge on the nucleus, and hence the atomic number, increases by one. We should therefore expect to find the chemical properties arranging themselves in a pattern according to the atomic number. Such is the direction that Moseley imparted to atomic physics as a result of his epochal work on the high-frequency spectra of the elements.

Moseley's investigations were concerned with the spectral lines emitted by the heavy elements in the X-ray part of the electromagnetic spectrum. To obtain an X-ray spectrum, Moseley had to excite heavy atoms. Only in such elements as iron, tin, and silver are the electrons close to the nucleus bound with sufficient energy to emit X rays during excitation. In the case of light atoms, such as hydrogen and helium, the most energetic radiation is in the ultraviolet wavelengths, which fall far short of the X-ray region Moseley was interested in investigating. To obtain an X-ray spectrum one subjects a target composed of the heavy element to a stream of very energetic electrons. On colliding with the target, these electrons bring about

Henry G. J. Moseley (1887–1915)

some rearrangement of the deep-lying electrons of the target atoms. When a target atom readjusts itself and returns to its normal state, energy is emitted in the form of X rays.

Moseley thought that if the electrons were moving in orbits according to the Bohr theory, the X-ray spectrum of a heavy element should exhibit a line structure similar to that of the hydrogen spectrum. Moreover, it occurred to Moseley that the X-ray spectrum would change in a regular way according to the charge on the nucleus (the atomic number) as one went from one heavy atom to the next, if the Bohr-Rutherford model of the atom were correct.

In conducting his experiment, Moseley first had to deal with the continuous part of the X-ray spectrum that is emitted by the bombarding electrons as they slow down in hitting the target. This part of the X-ray spectrum has nothing to do with the structure of the target atom and, therefore, must be eliminated. Moseley accomplished this by properly arranging his crystal analyzer relative to the slit from which the X rays were emitted. By allowing the X rays to strike the face of the crystal at large angles of incidence, he was able to reduce the reflection of the continuous radiation to a very small intensity. He was then able to pick out any particular X-ray frequency coming from the slit and reflected from the crystal face by applying the Laue-Bragg law of diffraction to these reflected X-ray beams. According to this law, each particular wavelength is reflected from the face of the crystal at a definite angle that is characteristic of that wavelength and no other.

An examination of Moseley's work shows the simplicity and straightforwardness of his approach. His experimental arrangement for investigating and photographing X-ray spectra involved no complex devices. It was built around a crystal mounted in a simple way on a prism table so that it could be rotated to give any desired angle of incidence for the X rays. A measurement of this angle then gave the wavelength of the X rays striking the photographic plate.

To understand the results obtained by Moseley in terms of the Bohr theory we must remember that the electrons inside an atom occupy orbits (K, L, M, and so on) at increasing distances from the nucleus; as we go from one element to the next in the atomic table, these orbits lie closer to the nucleus because the charge on the nucleus increases each time we go to the next higher element. This statement means that the K electrons in the element with atomic number $Z + 1$ are more tightly bound to the nucleus than are the K electrons in the element with atomic number Z.

If we now consider the X-ray lines resulting from the excitation of the K electrons, we should expect to find the frequencies of these lines increasing by a definite (although not always the same) amount as we go from one element to the next heavier one. This trend should also be true of the lines arising from the L and M electrons, and so on. This change in frequencies is precisely what Moseley found from his investigations, in which he dealt with two distinct lines. The photographs of these lines for brass, copper, nickel, and cobalt clearly indicate the step-by-step change in the frequencies of these lines as one progresses from the lighter to the heavier elements. The major significance of these findings is the discovery that the change in frequency depends not on the change in mass but on the increase in the charge of the nucleus, the atomic number.

Moseley stated this relationship in a simple empirical formula, which he was able to match with the experimental data. This formula may be expressed as follows: The frequency of a particular line in the X-ray spectrum of an atom is equal to a universal constant multiplied by the square of the charge on the nucleus reduced by one. This formula is really equivalent to the Bohr formula for the frequencies of the lines in the hydrogen optical spectrum. It shows that, even when we are dealing with heavy atoms, the electrons arrange themselves in discrete orbits around the nucleus of the atom.

As already mentioned, Moseley's formula is important because it shows that the atomic number and not the atomic weight is the decisive quantity in the arrangement of the elements in the periodic table. But another interesting fact of great importance emerged from this research; it should be mentioned in connection with Moseley's experimental proof of Van der Broek's idea of atomic number. The fact that the chemical elements could be arranged in a step-by-step sequence, changing by one unit at a time, carried the implication that where the numerical sequence was unbroken no unsuspected elements could be present. Thus, Moseley's research not only vindicated the idea of atomic number and pointed out the places of undiscovered elements in the periodic table, but it also showed that no other elements could exist. Nature had made the elements by adding positive charges, one at a time, to the nucleus. Since it was shown by Robert Millikan that fractional electrical charges do not exist, it was clear from these X-ray experiments that the known and predicted elements in the periodic table form a closed system.

When Bohr presented his quantum model of the hydrogen atom, from which he correctly deduced the frequencies of the known Balmer lines and the value of the Rydberg constant, most of the physicists at the time were very skeptical of its validity in general. That it gave a fairly correct picture of how the Balmer lines were produced by the electron in the hydrogen atom was not a sufficiently strong reason for accepting Bohr's atomic model. In any case, Bohr's concept of stationary states (discrete electronic orbits) and his "arbitrary rules" that "guide" the electron in its emission and absorption of photons as it jumps from one allowed orbit to another were highly repugnant to the traditional physicist. Moreover, even if the

Bohr model accounted for the Balmer lines of hydrogen, why should one believe that it could explain the very high frequency spectral lines of heavy atoms, arising from electrons close to the nuclei of such atoms? The difficulty associated with the investigation of such lines is that their freqencies are so large that the electrons, deep in the heavy atoms that produce these lines, can be excited only with very energetic X rays—and few experimental physicists in 1913 knew much about X rays. Moseley had worked with Rutherford on radioactivity and high energy gamma rays. He thus had the kind of experience and experimental techniques that were needed to check the applicability of the Bohr atomic model to heavy atoms.

Henry Gwyn Jeffreys Moseley was born in Weymouth, Dorsetshire, England, on November 23, 1887. His male forbearers were excellent scientists: his paternal grandfather was the first professor of natural philosophy of King's College, London, and an international authority on naval architecture, his maternal grandfather was the dean of England's conchologists, and his father was an outstanding zoologist.

From his early childhood on, with his older sister as his constant companion, Henry studied the flora and fauna of his country neighborhood and would probably have become a great naturalist if he had not turned to physics. At the age of nine Henry entered Summer Fields, an Eton preparatory school and, five years later, in 1901, began attending Eton on a King's Scholarship. Under the supervision of T. C. Porter, an X-ray physicist, Henry acquired his love of experimentation. The subject of his experiments hardly mattered; he wanted only a laboratory in which he could dabble to his heart's content. But he was more than a dabbler or a dilettante because his experiments, which he had begun in Eton, were substantive and enhanced his experimental techniques as well as sharpened his natural insight into physical phenomena. In 1906 Moseley entered Trinity College, Oxford, where he studied physics. He obtained his degree in physics there, but he was very unhappy with its bookish, scholastic atmosphere, which provided him with very little opportunity to do the kind of independent experimentation he loved. Moseley was happy to leave Oxford in 1910 when he was offered a demonstrationship in the physics department at the University of Manchester where Rutherford ruled. Moseley was happy to work long hours each day on radioactivity experiments proposed by Rutherford. In 1912 Moseley received a fellowship at Manchester. This freed him of teaching duties and permitted him to devote all his time to research and to pursue his own ideas without interference from Rutherford, who always wanted to control every aspect of experimental research in his laboratories.

At about this time, important X-ray experiments, conducted by Max von Laue, William H. Bragg, and others, were in progress. Moseley saw that he could use X rays to probe the electrons in heavy atoms. These experiments completely confirmed the Bohr theory for electrons moving in the lowest orbits, close to the nuclei of the heaviest elements. Moseley developed this technique to a point where he could use it to test the periodic table of elements for completeness and show where the gaps in the table were located. He then predicted the atomic weights and numbers of the elements that were needed to fill these gaps. All of these elements were subsequently discovered.

In 1914, while on a vacation trip in Australia with his mother, Moseley learned that World War I had begun. He rushed back to England and applied for a commission, against the advice of friends, colleagues, and the government itself. He persisted, however, and was sent to the Dardanelles, where he was killed in a minor battle on August 10, 1915.

The Quantum Theory of Radiation and Atomic Processes: Albert Einstein (1879–1955)

Although the Bohr theory of atomic spectra and Moseley's work on the X-ray spectra of heavy atoms had placed the nuclear atom in an unassailable position, one more step had yet to be taken to complete the picture and to give the nuclear atom a logically satisfying structure. It was necessary to show that the Planck radiation formula can be derived from the Bohr theory and its stationary states. Einstein achieved this result in 1917 when he wrote the last in his series of monumental papers on the quantum nature of radiation. In some respects this particular paper was the most penetrating of the group. In it Einstein clearly established that the interaction between a radiation field and atoms (or resonators, as he called them) occurs via the emission and absorption of quanta or photons. In addition, he derived the Planck radiation formula from the most general principles, without making any assumptions about the nature of the resonators. At the same time, Einstein clarified the phenomena of emission and absorption of radiation by atoms and showed that these processes are intimately related to the probabilities of an atom passing from one state to another. Finally, he emphasized the importance of taking into account the momentum of a quantum and, in a sense, foresaw the Compton effect. This idea is indicated in Einstein's statement that the emission and absorption of radiation by atoms are "fully directed events" and not spherically symmetrical ones.

We recall that Planck derived his radiation formula by introducing harmonic oscillators (the Planck resonators) as the constituents of matter in equilibrium with the radiation. Bohr later used the quantum theory to explain the line spectrum of the elements by introducing his postulates about the absorption and emission of photons by individual atoms. However, up to that time (1913) nobody had shown that Planck's radiation law was independent of the type of resonators that one assumes and that it could be derived without making any assumptions about the specific nature of the atomic systems. Einstein accomplished this task in his 1917 paper on the quantum theory of radiation. He used general statistical arguments in which the only assumptions were that the atoms (or molecules, as Einstein called them) are in dynamic equilibrium with the radiation field and that each atom can be in only one of a set of discrete atomic states. The atom was then pictured as as interacting with the radiation field by absorbing photons from and emitting photons into the field.

To make the analysis as simple as possible, Einstein considered only two of the possible quantum states of the atom and investigated the condition for statistical equilibrium between these two states and the radiation field surrounding the atoms. He then considered three possible processes involving the emission and absorption of radiation by the atoms and assumed that these processes proceeded at such a rate that the total number of atoms in each of the two quantum states did not change with time (the condition of statistical equilibrium). Two of the processes involved the emission of photons by the atoms and one involved absorption. The mere enumeration of these processes constituted a departure from previous ideas. In considering the emission of radiation, Einstein assumed not only that an atom could radiate a photon spontaneously, and thus pass from a higher to a lower state of energy, but also that the atom in the higher state could be induced or stimulated to emit a photon more readily than it would spontaneously. This result could be achieved in the presence of electromagnetic radiation or photons of the same frequency as that of the emitted photons. Einstein called this phenomenon "induced emission of radiation."

Einstein then introduced probability coefficients (the famous Einstein A and B coefficients) for each of these three processes, spontaneous emission, stimulated emission, and absorption. He assumed that the number of each of these processes occurring per unit time was proportional to the appropriate Einstein coefficient times the number of atoms in the given state. The spontaneous emission process is proportional only to the probability coefficient times this number of atoms, but the process of stimulated emission and the process of absorption are proportional to this number of atoms multiplied by the product of the appropriate coefficient for that process and the density of radiation of frequency corresponding to the energy difference between the two quantum states.

Since statistical equilibrium requires that the number of processes per unit time involving transitions of atoms from the higher to the lower state (emission of photons) equal the number of processes per unit time involving transitions of atoms from the lower to the higher state (absorption), Einstein set up an equation expressing this condition. This equation involved the three Einstein coefficients, the energy density of the radiation, and the numbers of atoms in the two quantum states at any moment.

To derive the Planck formula Einstein then set up two relationships among his three probability coefficients and used the general statistical law (the well-known Boltzmann law) for the distribution of atoms among possible atomic states to eliminate from his equation the number of atoms in each state. This last relationship introduced the temperature of the equilibrium ensemble of atoms and radiation as well as the energy difference between the two states. To obtain the two relations among his three probability coefficients Einstein considered, first, the limiting case of the temperature going to infinity and, second, the limiting case for which the Wien law is valid. In this way he eliminated the probability coefficients from his equilibrium equation and obtained the Planck formula for the energy density of black-body radiation. At the same time he derived, as a necessary condition, Bohr's hypothesis that, in going from one state to another, an atom always emits one photon whose frequency is given by the energy difference between the two states divided by Planck's constant.

One of the remarkable things in Einstein's derivation of Planck's formula was the appearance of the stimulated emission process, which seemed to play a rather mysterious role and which was considered, until recently, to be a rather curious theoretical feature of the problem without an apparent practical application. During the 1950s and 1960s, however, Einstein's hypothesis of stimulated emission of radiation received its most dramatic experimental verification in the development of the maser and the laser by Charles Townes and his collaborators. The word maser is derived from the first letters of the phrase "microwave amplification by the stimulated emission of radiation" and the word laser from the phrase "light amplification by the stimulated emission of radiation."

In both of these devices radiation of a given frequency is greatly amplified by allowing the photons of the radiation to stimulate excited atoms to emit additional photons of the same kind. The remarkable property of the amplified beam thus obtained is its very high coherence. It consists of photons that are all of very nearly the same frequency. Thus, there is almost no spreading of the beam. This amplification means that very energetic, monochromatic beams of radiation can be obtained and sent over great distances without attenuation. The basis for this process was given very early in Einstein's paper when he pointed out that the emission and absorption of radiation is accompanied by a change in the momentum of the atom. When an atom is stimulated by a passing photon to emit a photon, the atom has momentum transferred to it whose direction is opposite to that of the passing photon. This means that the atom must emit another photon in the same direction as that of the passing photon. Einstein stated this as follows: "If as a result of incident radiation the process $Z_n \rightarrow Z_m$ (absorption) occurs, then an amount of momentum $(E_n - E_m)/c$ is transferred to the molecule in the direction of propagation of the radiation. If we have the process $Z_m \rightarrow Z_n$ for the case of incident radiation (stimulated emission), the magnitude of the transferred momentum is the same, but it is in the opposite direction."

Einstein pointed out two other important consequences of his derivation of the Planck formula. The first concerns the direction of photons emitted. Whereas, according to the classical wave theory of radiation, the radiation leaves the atom in the form of a spherical wave (the emission occurs symmetrically in all directions), Planck's formula can be derived only if one assumes that the emission process is a directed one. This statement, as Einstein pointed out, must be true whether the emission is due to stimulation by a beam of incident radiation or whether it is spontaneous.

The second consequence concerns the momentum of radiation. As Einstein stated in the last paragraph of his paper, almost all of the previous work on the interaction of atoms and radiation dealt with the energy exchange between the atom and the radiation field but neglected the momentum exchange. The reason for this is that the momentum of the radiation is very small, so that the momentum transferred to the atom is negligible as compared to the energy. Einstein correctly concluded, however, that for theoretical considerations this momentum transfer is of extreme importance and must be taken into account just like energy interchange: "But for theoretical considerations this small effect [momentum transfer] is on an equal footing with the energy transferred because momentum and energy are very intimately related to each other; a theory may therefore be considered correct only if it can be shown that the momentum transferred according to it from the radiation to the matter leads to the kind of motion that is demanded by thermodynamics." This initial insight into the importance of the momentum of radiation was extremely prophetic because it contained the germ of the Compton effect. In 1923, when Compton analyzed his experiments on the scattering of X rays by electrons, he found that he could understand the results only if he took into account the conservation of both energy and momentum in the interaction between the electron and the X-ray photon.

The Compton Effect: Arthur H. Compton (1892–1962)

Although the Bohr theory had been remarkably successful early on in explaining the gross features of atomic spectra and in providing a model of the atom which appealed to one's sense of the unity in nature, it became clear in the 1920s that the more refined spectral features lay outside its domain. Most of the evidence for the inadequacy of the Bohr theory was derived from experimental work with atomic spectra. But other investigations involving scattering experiments also suggested that fundamental changes would have to be made in the Bohr theory before it could account for all observations. Of the many relevant experiments carried out during this period, the Compton experiment on the scattering of X rays by electrons did most to reveal the shortcomings of the Bohr theory and was the most instructive as to the direction in which atomic theory would have to move.

Compton's experiment was a very simple one in principle. A beam of X rays of known frequency emanating from a molybdenum anticathode was allowed to strike electrons that were weakly bound in atoms such as carbon; the X rays leaving the atoms in a given direction were then carefully analyzed. In Compton's experiment, the primary X-ray beam was directed against a graphite target, and the rays coming away from the target at right angles to the direction of the primary incident beam were studied.

As Compton pointed out in the introductory paragraphs of his paper on the scattering of X rays, which appeared in the *Physical Review* in 1923, his results were in sharp disagreement with J. J. Thomson's classical theory of X-ray scattering and could be accounted for only by a drastic departure from classical theory. What was essential in the experimental results was that the X-ray beam, after it was scattered by the electrons, suffered a definite reduction in frequency. In contrast, the Thomson theory predicted that electrons are set vibrating by the incident X rays with exactly the same frequency as the rays themselves and should, therefore, reradiate X rays of the same frequency.

Compton first attempted to explain his results by assuming that the electrons had a certain size and that X rays emitted from different parts of the electron interfered with each other in such a way as to give the observed results. But he soon realized that this assumption was untenable because it necessitated introducing an electron size that depended on the wavelength of the X rays that were employed in the scattering. He discarded this idea as too "difficult to defend," then proceeded to analyze the scattering data by means of the quantum hypothesis.

He first assumed that the incident beam of electrons consisted of quanta, each with momentum $h\nu/c$ where ν is the frequency of the quantum, h is Planck's

Arthur H. Compton (1892–1962)

constant of action, and c is the speed of light. Compton then analyzed the scattering of each quantum by an electron as though the quantum and the electron were billiard balls colliding with each other. By applying the principles of conservation of momentum and energy to such a collision, and using the appropriate relativistic relationships, he showed that the scattered quantum would move off in some direction making an angle theta with the direction of the incident beam of X-ray quanta, and that the electron that did the scattering (the so-called "recoil electron") would move off at the appropriate angle given by the conservation principles. Compton demonstrated that the frequency of the scattered quantum was reduced by an amount in agreement with the Planck expression for the energy of a photon and the relativistic expression for the kinetic energy of the recoil electron. In other words, the energy of the photon, as given by its frequency, is reduced by the same amount that the kinetic energy of the recoil electron is increased.

Compton's analysis was completely confirmed by his experimental data. These results established the validity of the quantum structure of radiation more firmly than ever before. As Compton stated in his discussion of the experiment, "The experimental support of the theory indicates very convincingly that a radiation quantum carries with it directed momentum as well as energy." This was a striking verification of Einstein's analysis of the emission and absorption of radiation by an atom. It led him to the concept of the photon as a momentum-carrying corpuscle that can transfer its momentum in a given direction to an atom.

Compton's remarkable findings, however, had consequences that went well beyond the concept of the photon; they became the experimental basis for Heisenberg's uncertainty principle. To determine the position of an electron accurately, we must irradiate it with photons of high frequency because a short wavelength (which is equivalent to high frequency) is necessary to minimize the error in measuring this position. One of these photons must then enter our microscope after hitting the electron, if we are to observe it. Because of the Compton effect, the electron receives a recoil of the order of magnitude of Planck's constant of action divided by the wavelength of the photon. But the full recoil of the electron cannot be exactly determined since we know only the approximate direction of the scattered electron. The photon can enter the objective of our observing microscope over a range of directions because of the comparatively broad aperture of the microscope lens. Because of the phenomenon known as the Compton effect, therefore, we can know neither the precise recoil of the electron nor, as a consequence, its precise momentum, if we attempt an accurate determination of its position.

The Compton effect also implies that the electron must be treated as a wave and not a particle. Let us consider the electron moving in one of the Bohr orbits inside an atom. To observe it in its orbit, we must again irradiate the electron with photons of a certain frequency; their wavelength should be about equal to, or smaller than, the radius of the orbit of the electron. However, the electron receives a Compton recoil which may throw it into one of a series of higher orbits. Since we do not know the full recoil, we must allow this electron a whole series of orbits and treat it as though it were associated with all of them. Because each of these orbits has a definite frequency, we must discuss the electron as though it were characterized by an array of frequencies, not just the single one of the orbit, which the Bohr theory would assign. In other words, we must treat the electron not as a particle but as a kind of wave packet.

There is still another consequence of the Compton effect whose full significance is not yet comprehended. It would seem that the Compton effect should lead not only to the uncertainty principle, as expressed by Heisenberg, but also to an uncertainty in the position of any single particle. No matter how small the wavelength of the initial, incident photon is in Compton's experiment, the wavelength of the scattered photon for a stationary electron is never smaller than Planck's constant h divided by mc, where m is the mass of the electron, and c is the speed of light. This remarkable quantity, h/mc, is known as the Compton wavelength of the electron. Today we believe that the position of the electron cannot be determined without allowing for an error of at least the size of this quantity.

It appears from this information that there is some kind of inner structure of the electron that begins to manifest itself at distances from the center of the electron equal to the Compton wavelength. Since physicists are more and more inclined to the belief that the electron has some complex structure, they may obtain insight into this presumed complexity by a more thorough analysis of the Compton effect.

The concept of the photon, as developed by Einstein on the basis of Planck's quantum theory, was never entirely accepted by most of Einstein's contemporaries; even Planck, as late as 1912, considered Einstein's insistence on the reality of the photon a blemish on his superb record and ascribed it to youthful enthusiasm and immaturity. But Einstein persisted and, in the last of his great papers on radiation theory in 1917, he noted that the momentum of the photon, though small, in general, plays a very important role in the interaction of electrons and radiation. This phenomenon is the basis of today's laser technology. In spite of Einstein's very strong support, the photon concept was not generally accepted until Compton, in one of the most famous experiments in modern physics, showed that the photon concept is essential for a thorough understanding of the scattering of light by electrons.

Compton was well equipped intellectually and by his background and schooling to do the kind of exacting experimental research that revealed the effect named after him (the Compton effect) and for which he received the Nobel Prize in physics in 1927. His father was a professor of philosophy, his older brother, Karl, was a physicist and a renowned academic, his brother, Wilson, was a distinguished economist, and his sister, Mary, and her husband were educators. Born on September 10, 1892, Compton was greatly influenced by his mother, Otelia Augspurger, a Mennonite, who guided all the children in their early years. Arthur attended his father's college, the College of Wooster, from which he graduated in 1913. In 1914 he earned his master's degree in physics from Princeton University; he received his Ph.D. there in 1916. The academic atmosphere of Princeton's graduate school delighted Compton. There he was guided especially by the great astronomer Henry Norris Russell and the physicist O. Richardson. On completing his graduate studies, he married a Wooster College classmate, Betty Charity McClosky, with whom he had two sons.

Compton's thesis, which dealt with X-ray diffraction and scattering, contained the germ of the idea that finally led to his important discovery. He developed these ideas further during the years from 1917 to 1919 when he continued his X-ray experiments at Westinghouse. There he conceived of the concept of a basic length, the Compton wavelength, that was to be assigned to the electron and all other elementary particles. Shortly after World War I, Compton spent a year at the Cavendish Laboratory at Cambridge as a National Research Council Fellow. There he met physicists from all parts of the British Empire. This was a most inspiring year, for Compton worked closely with J. J. Thomson and Ernest Rutherford.

Rather than working with high-energy X rays, as he had hoped to do, Compton had to work with gamma rays since the high voltage necessary to produce the X rays of the required frequency was unavailable. The gamma-ray scattering experiments already indicated that the classical picture of a wave interacting electromagnetically with an electron could not explain the observations. The answers to the questions raised by these experiments came to him in 1920 when he began his tenure as the Wayman Crow professor of physics and chairman of the physics department at Washington University in St. Louis, Missouri. The basic discovery—that the frequency of a beam of X rays drops (the wavelength increases) when the beam is scattered from electrons—can only be explained if the beam of X rays consists of individual photons, each of which has energy and momentum. This Compton proposed. He then treated the scattering process as one in which the interaction between an electron and photon is an ordinary collision between two spheres. During this collision momentum and energy are conserved, with the photon moving with its appropriate momentum, that is, its energy divided by velocity (the speed of light). Compton treated the scattering process in this way, using relativistic dynamics, owing to the high energies involved, and obtained excellent agreement with his experimental results. He thus placed the photon on a sound footing, and from that time on the photon concept has been fully accepted by physicists.

In 1923 Compton accepted a position as professor of physics at the University of Chicago, where he remained 22 years. He was an outstanding teacher and an excellent research director and administrator. During World War II he organized and directed the work of the Metallurgical Laboratory of the Manhattan Project which produced the first atomic bombs. He recruited such physicists as Enrico Fermi, Eugene Wigner, and Harold Urey for the project which successfully tested the first uranium nuclear chain reactor on December 2, 1942. After the war Compton accepted the chancellorship of Washington University. He continued his educational activities until his death in Berkeley, California on March 15, 1962, while on a lecture tour.

Space Quantization: Otto Stern (1888–1969)

The planetary theory of the atom as it had been first proposed by Bohr, with its discrete circular orbits for the electrons, was much too simple a model to account for all the observed details of the spectra of atoms. It soon became clear that the theory itself would have to be refined and improved. The direction for improvement was first indicated by Arnold Sommerfeld, who showed that not only must the energy of the electron in the atom be quantized, as Bohr required, but also its angular momentum. Both the sizes of the orbits and their shapes must be discrete. This new type of quantization enabled Sommerfeld to explain the fine structure of spectral lines. But the theory was still not complete because, even with Sommerfeld's improvement, it could not account for the Zeeman pattern of the spectral lines that appeared when an atom was placed in a magnetic field. A new kind of quantization, called space quantization, was needed. The following description will explain what is meant by this term.

One of the most remarkable concepts of the quantum theory is that the angular momentum (the spin of the system) is space-quantized. This means that if an arbitrary direction in space is chosen, let us say the z-direction of a rectangular coordinate system, we find experimentally that the angle between the direction of the angular momentum vector and this z-direction cannot assume continuous values ranging from zero degrees to 180 degrees, as one would expect from classical theory, but can have only a discrete number of values. Put somewhat differently, we find that if we measure the component of the angular momentum vector along the z-direction, this component cannot take on all values between zero and the numerical value of the angular momentum but only certain ones.

This restriction, originally suggested by Pauli, was demonstrated as a reality by the experiments of Otto Stern and Walter Gerlach. A few years later, when quantum mechanics was formulated, it was found to be a direct consequence of the Schrödinger wave equation and also of the Heisenberg matrix mechanics, developments discussed in subsequent sections of this book. From the classical point of view, space quantization is a baffling phenomenon. We need to see why, from a quantum mechanical point of view, it is not only reasonable but necessary. We can show how space quantization arises as a direct consequence of the procedure for measuring the angular momentum. However, we must introduce some way of disturbing the system whose angular momentum we wish to measure; this disturbance must involve the angular momentum. In other words, we can see how the angular momentum interacts with some field that we can impose on the system. This interaction between the spin and the imposed field will, of course, change the energy of the system. By observing this change in energy we should be able to measure the angular momentum of the system and its quantized states.

We can easily guess what field we must impose for our purposes because we note that our spinning system (in this case, an atom) is equivalent to a spinning magnet in that it consists of charges (electrons) that are moving around a nucleus. Now such revolving charges are equivalent to loops of electric current, and such loops carry a magnetic field. In general, there is associated with each atom a spin and a magnetic field. Hence, the atom behaves like a spinning magnet. We should then be able to observe the angular momentum of the atom by subjecting it to an

Otto Stern (1888–1969)

external magnetic field because such a field will disturb these magnetic loops in some way.

Suppose now that we place our atom in a magnetic field pointing in a given direction, let us say the z-direction in a rectangular coordinate system. By "pointing in a given direction," we simply mean that the lines of force of the magnetic field are parallel to the chosen direction. The introduction of such a magnetic field introduces a preferred direction—the direction of the field—and thus eliminates the implied paradox of having the angular momentum adjust itself to any arbitrary direction. Since the magnetic field is necessary for measuring the direction of the angular momentum (spin), the very process that we use for our measurement establishes a direction for us.

Now what happens when we turn on such a magnetic field? The field imparts to the atom additional energy, which—and this is a crucial point—depends on the product of the strength of the magnetic field (if it is not too large) and the magnitude of the projection of the spin along the field. There is another factor, which depends on the electric charge and the mass of the system, but it is not significant for our discussion. The important thing is that the energy depends on the spin component along the field. We have seen, however, that the spin can have only a discrete number of projections, in fact, just $2j + 1$ if j is the spin along the field. We can look upon each of these quantities as representing a different state of the atom, but if no magnetic field is present, all of these atomic states are equal. They coincide. In a magnetic field, however, the energy of each of these states is slightly different. This discrepancy made it possible for Stern and Gerlach to carry out their experiment and demonstrate space quantization, since they were able to use this energy difference to separate one state from another. We shall elaborate on this point in a moment.

We note that the introduction of a magnetic field removes the seemingly paradoxical quality of space quantization because of the energy the field imparts to the atom. We know from the work of Planck, Einstein, and Bohr that energy in an atomic system is quantized; it therefore follows that the magnetic energy is

Walter Gerlach (1889–)

also quantized. But the magnetic energy, as we have just seen, is proportional to the component of the spin along the field. Hence this component must be quantized and we have, in a sense, solved the mystery.

We may enquire into the nature of the magnetic energy. There is, of course, potential energy, which is measured by the work we would have to do to remove the atom from the magnetic field. But there is also kinetic energy, which arises as follows. When the magnetic field is turned on, it tries to turn the atom so that the atom's magnet is lined up with the field just as it would tend to turn an ordinary bar magnet. Compare, for example, the way a compass needle turns toward the north in the earth's magnetic field. The "little magnet" associated with an atom has an effective pole strength and length; the product of these two quantities is called the magnetic moment of the atom. The magnetic moment is a vector that is parallel to and proportional to the spin of the atom. We may therefore say that the magnetic field, when turned on, tries to turn the magnetic moment of the atom parallel to the field. But we must remember that the atom is spinning and therefore resists being turned, just the way a gyroscope does. Rather than turning parallel to the magnetic field, the spin vector of the atom precesses around the magnetic field the way the axis of a spinning top precesses around the vertical. The atom thus acquires additional kinetic energy—the energy of precession—in the field.

We can see now why space quantization must exist. It is a consequence of the general rule that all periodic motion must be quantized, that action is discontinuous. Since the precession of the atom is periodic, it must be quantized. This means that only a discrete number of precessional states is allowed and that each of these states has a slightly different amount of energy. This is the basis of the Stern-Gerlach experiment which we now discuss.

To take a specific example, suppose we have an atom whose total spin, j, is $3/2$. There are then four spin states corresponding to the values $3/2$, $1/2$, $-1/2$, $-3/2$ of the z-component of the spin, where z is any direction in space. If a constant magnetic field is imposed, we obtain four different energy states, whose energies are proportional to $3/2\ B$, $1/2\ B$, $-1/2\ B$, $-3/2\ B$, where B is the strength or intensity of the magnetic field. This simply means that if a system of such atoms is placed in a magnetic field, these atoms are distributed randomly among these states, or almost so, since the energies of the four states are almost, but not quite, equal. In other words, there are approximately equal numbers of atoms precessing at four different angles about the magnetic field.

The purpose of the Stern-Gerlach experiment was in effect to show that the spin vectors of the atoms do, indeed, distribute themselves among these four states only (along four different directions relative to the magnetic field). In addition, it showed that the energy difference between any two of these states could serve to demonstrate this phenomenon. Now the energy of the atom in a magnetic field depends on the intensity of the field. If the magnetic field is inhomogeneous, the atom will experience a force. Depending on how the magnetic moment, or the spin, of the atom aligns itself with respect to this field, the atom is pulled toward the strong part of the field or pushed toward the weak part. In our example the atoms in the spin state $3/2$ are pulled most strongly toward the strongest part of the field; the atoms in spin state $1/2$ are pulled in the same direction but not so strongly; the atoms in the two negative spin states are pushed away from the strong part of the field toward the weak part. Hence, if a beam of such atoms is allowed to pass through an inhomogeneous magnetic field, it will be split into four different beams owing to the four different forces experienced by the atoms in the different spin states.

In 1922, Stern and Gerlach devised an experiment to demonstrate this phenomenon with silver atoms. They obtained a beam of silver atoms by first evaporating the metal in a hot oven, then passing these gaseous atoms through a series of small holes arranged in a straight line. This beam passed between the two poles of a magnet designed to give a strongly inhomogeneous field at right angles to the motion of the atoms in the beam. Finally, after moving through the magnetic field, the atoms were collected on a photographic plate.

When these atoms left the oven they were distributed randomly along their various possible spin states. If Stern and Gerlach had used a homogeneous magnetic field, the atoms would have been subjected to no forces by the magnetic field. The beam of atoms would thus have passed right through the field unchanged. But instead, they used a magnet of which one pole was a flat surface, the other a sharp edge parallel to the flat surface and to the atomic beam. This setup gave an inhomogeneous field in which the magnetic lines of force were spread out at the flat pole but highly concentrated at the pole with the sharp edge. Consequently, atoms in the beam experienced strong forces perpendicular to their direction of motion. Had the spins of the silver atoms behaved classically, the beam would have been spread out uniformly since all possible spin orientations are allowed classically and each orientation would have resulted in a different force.

But Stern and Gerlach did not obtain this result; they found just two spots on their photographic plate, showing that the magnetic field had split, not spread, the initial beam. This result clearly demonstrated space quantization and showed that the spin, or angular momentum, of the silver atom is just $1/2$ unit. This value means that the magnetic moment of the silver atom can align itself either parallel to the direction of the magnetic field or opposite to its direction. In other words, when the silver atoms passed between the poles of the Stern-Gerlach magnet, their spins either pointed toward the sharp-edged pole or toward the flat pole, and in about equal numbers. The atoms in the first group were pulled toward the flat pole and the others were pulled toward the sharp edge, so that two beams were formed, as predicted by space quantization.

Otto Stern was trained in theoretical physics but went on to do even more significant work in experimental physics. Born in Sorau, Upper Silesia, Germany, on February 17, 1888, he moved to Breslau with his parents in 1892, where he completed his elementary and secondary schooling. After graduating from the gymnasium there in 1906, he entered the University of Breslau to study physical chemistry. In 1912 he received his doctorate and came under the influence of Einstein, whom he joined at the University of Prague that same year; he later followed Einstein to the University of Zurich. In 1913 he was appointed privatdozent (unpaid lecturer) of physical chemistry at the Eidgenössische Technische Hochschule of Zurich.

During this early period in his scientific career, Stern published important theoretical papers in statistical mechanics, thermodynamics, and quantum theory. In 1913, he derived the correct quantum theoretical expression for the entropy of a gas, using a step-by-step analysis of the changes taking place in a substance when it passes from the crystalline to the gaseous phase. This derivation required Stern to apply Einstein's method of quantizing the vibrations of the crystal to introduce Planck's constant into the formula for the entropy. It was one of the earliest applications of the quantum theory to physical systems other than radiation and occurred just about the time when Bohr was applying the quantum theory to the dynamics of the atom.

In 1914, Stern went to the University of Frankfurt am Main as lecturer in theoretical physics; there his attention turned more and more to problems of experimental physics. In 1919, he became particularly interested in experimental methods of proving the theoretical deductions of the kinetic theory of gases and began to devise procedures for measuring the velocities of molecules in a gaseous phase. He finally hit on the method of molecular beams, which is one of the most powerful techniques in experimental physics for studying the behavior of individual molecules, atoms, and atomic nuclei. Stern began his experimental work by applying the molecular beam technique to verify Maxwell's law of the distribution of molecular velocities in a gas.

It soon occurred to Stern that his molecular beam method was ideally suited for studying the magnetic properties of molecules and atoms; it was a simple matter to impose a magnetic field, of whatever nature might be desired, onto the molecular beam and examine the results. In 1922, working with Gerlach, he passed beams of silver atoms through an inhomogeneous magnetic field and proved experimentally—for the first time—that the direction of the magnetic moment of an atom is quantized in a magnetic field. At that time, Stern was associate professor of theoretical physics at the University of Rostock, to which he had come in 1921. As a result of his molecular beam experiments, he was invited to become professor of physical chemistry and director of the laboratory at the University of Hamburg. He remained at Hamburg until Hitler came to power in 1933, then left Germany to become research professor of physics at the Carnegie Institute of Technology in Pittsburgh, Pennsylvania.

Just before he left Germany, Stern made one of the most important applications of the molecular beam method. In collaboration with Estermann and Frisch, he passed a beam of neutral hydrogen molecules through an inhomogeneous magnetic field and measured the magnetic moment of the proton, which, like the electron, is a small spinning magnet. The results of this experiment were quite surprising because they showed that the magnetic moment of the proton is $2^{1}/_{2}$ times as large as one would expect if the proton were a simple particle governed, as is the electron, by the Dirac wave equation. This experiment indicated that the proton is some kind of composite particle, a finding that was borne out by the important experiments of Robert Hofstadter.

In 1943, Stern received the Nobel Prize in physics for developing the molecular beam method and for his measurement of the magnetic moment of the proton. Stern remained at the Carnegie Institute of Technology until 1945, when he retired as professor emeritus. He moved to Berkeley, California, where he remained until his death on August 17, 1969.

Electron Spin: Samuel A. Goudsmit (1902–1978) and George E. Uhlenbeck (1900–)

When Niels Bohr applied Planck's quantum hypothesis to the structure of the hydrogen atom and succeeded in deriving the wavelengths of the Balmer lines in the hydrogen spectrum, he did so with the very simplest possible model. According to this model, the electron in the atom was pictured as moving around the proton in circular orbits of discrete radii. With this model Bohr could compute the frequencies of the various spectral lines by assigning to each orbit a very definite energy. As we know, he pictured the electron as emitting or absorbing a quantum of energy (a photon) every time it changed orbits. Since the energy of the electron in a circular orbit is determined solely by the radius of the orbit, and the radius, in Bohr's quantum picture, is determined by one of the integers (the lowest orbit is assigned the integer 1, and so on), it follows that the energy of the electron depends on the value of the integer that is associated with its orbit. This integer is called the principal quantum number of the orbit.

It soon became clear, however, that for the electron to jump from one orbit to any other, the orbits could not be circles. If the orbits were circular, the electron could jump from a given orbit only to the adjacent orbit lying either directly above or below according to certain rules—selection rules—that had to be introduced. Thus, it was necessary to replace the circular orbits by elliptical orbits to allow for all possible jumps. However, we know from planetary theory that an orbit having a definite semimajor axis is associated with an infinitude of elliptical shapes ranging all the way from a circle to a straight line, and that an electron in any of these orbits, regardless of shape, has the same energy. One might have been inclined therefore to picture an infinitude of possible elliptical orbits associated with an electron moving around a proton, all having the same size, that is, the same principal quantum number n and, hence, the same energy. This is not the case in the Bohr picture of the atom since the shape of the orbit is itself quantized (associated with integers). Only certain shapes, equal in number to n, are permissible. If the principal quantum number of the orbit of the electron is 3, for instance, only three elliptical orbits are permissible.

When the elliptical orbits associated with a given principal quantum number were introduced into the theory, it was first thought that they should all be assigned the same energy. Hence, an electron jumping from any one of these orbits to any elliptical orbit of a set associated with some other principal quantum number would emit or absorb a photon of exactly the same frequency. In other words, it was thought that as far as the observed spectral lines of hydrogen were concerned, it made no difference whether the orbits were circular or elliptical, except that in the latter case each spectral line was to be associated not with just two orbits, but with two sets of orbits. These sets of orbits were thought to be a convenient mathematical fiction without any bearing on the physics.

This simple picture was soon discarded. It was quickly shown by Arnold Sommerfeld that the different elliptical orbits associated with a given principal quantum number do not have the same energy. This discrepancy arises from the relativistic variation in the mass of the electron as it moves about in an elliptical orbit, which causes the entire orbit to precess. This means that the electron will have slightly different energies if it is in different elliptical orbits of a given set. The energy of an electron, then, is determined by two quantum numbers: the previously discussed principal quantum number n and another, let us say l, called the azimuthal quantum number. The latter gives the shape and therefore the angular momentum of the electron in the given elliptical orbit. This state of affairs means that rather than one spectral line associated with the jump of an electron when it changes its principal quantum number, there is a group of lines all lying very close together.

Each of the Balmer lines (the lines as predicted by the first Bohr theory) has a fine structure, so that each line is a multiplet. In fact, each Balmer line consists of three closely spaced components, which is exactly what the spectroscopists discovered.

However, very complex analysis of the spectral lines shows that they are even more complicated than the Sommerfeld relativistic theory indicated. Whereas the Sommerfeld theory predicted that each Balmer line should consist of three almost identical members, actual observations prove that each Balmer line consists of five close components. We shall come back to this point in connection with the idea of electron spin as introduced by Goudsmit and Uhlenbeck.

Not only did the Bohr theory run into difficulty in trying to explain the fine structure of the hydrogen spectral lines under ordinary conditions, but there was also trouble (even with Sommerfeld's relativistic refinement superimposed) in connection with the multiplet structure of the lines when the atom emitting the radiation was placed in a magnetic field. We have already noted, with relativity taken into account, that two quantum numbers—the principal quantum number, giving the size of the orbit of the electron, and the azimuthal quantum number, giving the shape of the orbit—are required to specify the energy of the electron. However, it is clear that at least one other quantum number is also needed, since the orientation of the orbit in space must still be specified. As long as no magnetic field is present, the spatial orientation of the orbit has no influence on the energy of the orbiting electron; as already noted in our discussion of space quantization, the situation changes when the atom is in a magnetic field.

We must keep in mind that the electron circling around the nucleus behaves like a small magnet. As such, it will precess around a magnetic field with a frequency dependent on the orientation of the plane of its orbit to the field. Since this magnetic precession contributes to the energy of the electron, the orientation of the atom must also be specified by a quantum number. Such a quantum number, the magnetic quantum number m, was introduced into the theory by Stern and Gerlach. To some extent it helped explain the complexity of the spectral lines when the atom is in a magnetic field.

That a magnetic field should bring about a splitting of the spectral lines was recognized as early as 1896 when the Dutch physicist Zeeman observed that each spectral line of an atom in a magnetic field breaks up into three lines. This is known as the normal Zeeman effect and was explained by Lorentz with the aid of his classical theory of the electron. Lorentz pictured an atom as a small magnet that can precess in a magnetic field. Since the magnetic precession (the so-called Larmor precession) can either be added to or subtracted from the electron's ordinary motion, we obtain two additional lines in the spectrum. This same result can be obtained from the Bohr theory since the introduction of the magnetic quantum number gives rise to an additional term, which depends on the magnetic quantum number in the expression for the energy of the electron in any one of the Bohr orbits. When the electron jumps from one orbit to another the magnetic quantum number may either remain unaltered or may change by becoming one unit smaller or larger. Thus, there is a possibility for three lines arising in place of each ordinary line when no field is present.

But it was not long before spectroscopists, with their ever-increasing refinements, observed that the Zeeman effect is considerably more complicated than had previously been imagined. Rather than consistent sets of three lines, a magnetic field induces a splitting into three lines in some cases but more than three in others. At first attempts were made to explain these spectral complexities by picturing the atom as a core (the nucleus plus the inner electrons) with an external electron, the core possessing certain magnetic properties. Pauli, however, showed that a model of this sort runs into other serious difficulties. He then suggested that many features of the multiple spectral lines and the Zeeman effect could be explained if one pictured the electron with an additional degree of freedom, a fourth quantum number.

As Pauli demonstrated, the assignment of a fourth quantum number to the electron fell very nicely into the relativistic scheme. According to relativity theory four numbers—three space coordinates and a time coordinate—are required to describe the electron. Pauli, however, did not suggest just how this fourth quantum number was to be assigned to the electron nor what its physical significance was to be. This is where Goudsmit and Uhlenbeck came into the picture. In a note to *Naturwissenschaften* in November 1925 and in a letter to *Nature* the following month, they proposed a scheme to account for the fine structure in the lines of hydrogen-like atoms and, at the same time, for the complexities of the Zeeman spectral pattern. The essential feature of their scheme was a spinning electron, that is, an electron possessing an intrinsic angular momentum and, hence, a magnetic moment. Since a spinning electron precesses when placed in a magnetic field and this precession can be quantized, the spinning hypothesis automatically introduced another quantum number.

To explain the multiplicity of the fine structure of the lines, Goudsmit and Uhlenbeck found it necessary to suppose that the angular momentum of the spinning electron was not an integral multiple of $h/2\pi$, the unit of angular momentum in quantum mechanics, but only half of this quantity. This means that the spin of the electron can only be added to the angular momentum of the orbital motion or subtracted from it. There are thus possible suborbits associated with an orbit of given angular momentum, depending upon whether the electron spin adds to or subtracts from the orbital angular momentum.

Before the introduction of the spinning electron, the electron was pictured in its orbit as equivalent to a small magnet because of its motion around the nucleus. But if the electron is spinning, it must, by virtue of this spin, behave like a little magnet even if it is not going around in an orbit. Thus, two magnets may be imagined in association with the motion of the electron in its orbit around a nucleus. These two magnets can either align themselves so that they are parallel to each other (north pole next to north pole and south pole next to south pole) or align themselves so that they are antiparallel. Since these two alignments correspond to states of different energy, two suborbits are associated with each orbit. This gives rise to a doubling of the spectral lines for a single electron moving around a nucleus (a hydrogen-like atom).

Although the introduction of an intrinsic magnetic moment, that is, a spin, eliminated the spectroscopic difficulties, the idea of a spinning electron was repugnant to many prominent physicists. Indeed, Pauli himself presented apparently insurmountable objections to such an idea, and Ehrenfest, the great Dutch theoretical physicist, rejected it because, he argued, the surface of the electron would have to be moving with a speed greater than the speed of light if the electron possessed an angular momentum equal to $1/2(h/2\pi)$. Other objections dealt with the fact that the electron has to be treated as a point from relativity considerations, and it is difficult to see how a point can be spinning. All of these oddities in connection

Samuel A. Goudsmit (1902–1978)

with the spin of the electron show that it cannot be treated as one would treat ordinary classical angular momentum.

In spite of all the objections raised against the spin of the electron, it finally had to be fully accepted in light of the experimental evidence. A particularly strong point in its favor was the famous Stern-Gerlach experiment. This experiment demonstrated that a beam of atoms, which should show no magnetic effects if the electron had no spin, breaks up into two beams when passing through an inhomogeneous magnetic field, as though the electron were a small magnet of the sort postulated by Goudsmit and Uhlenbeck. The final triumph of the spinning electron came with Dirac's discovery that the spin itself can be derived from relativity theory.

The year 1924 to 1925 was one of the most remarkable in the entire history of physics: in that period Pauli published his paper on the exclusion principle, de Broglie announced his discovery of the wave nature of electrons, Heisenberg discovered the matrix mechanics, Dirac published his version of quantum mechanics, Schrödinger was writing his paper on the wave mechanics, and Goudsmit and Uhlenbeck proposed their hypothesis of a spinning electron. It is interesting that this concept of spin should have originated in the Leiden school of physics since it was in the tradition of Lorentzian electron theory and, at first sight, somewhat out of step with the other rapidly occurring developments. Indeed, Pauli, whose exclusion principle, as applied to the arrangement of electrons inside the atom, requires a fourth quantum number—which the spin provides—initially doubted the validity of the idea because of what he called its "classical-mechanical" character. But the spin was needed to clear up the difficulty with the Zeeman effect, and Zeeman's great influence on Dutch physics must have played its role in the Goudsmit-Uhlenbeck discovery.

When they wrote their first paper on the spinning electron, Samuel A. Goudsmit and George E. Uhlenbeck were young graduate students at Leiden, still two years away from their Ph.D. degrees. Goudsmit was born in The Hague, Holland, on July 11, 1902; after attending elementary and secondary school there, he began his

George E. Uhlenbeck (1900–)

career in physics as an assistant at Amsterdam in 1923. After collaborating with Uhlenbeck at Leiden in 1925, Goudsmit spent a year at Tübingen as a fellow of the Institute of International Education and returned to Leiden in 1927 to receive his Ph.D.

At this time American universities were beginning to build up their science facilities, particularly in the field of theoretical physics, and they were looking for promising European physicists who were willing to come to the United States. The University of Michigan was in the forefront of this effort and invited both Goudsmit and Uhlenbeck to come there as instructors in physics. Both men, newly married, came to Michigan in 1927 with their wives. After one year as an instructor, Goudsmit was promoted to an associate professorship in 1928; in 1932, at the age of 30, he became professor of physics. He remained on the Michigan faculty until 1946, devoting his time to research and teaching. In 1938 he spent a year in Rome and Paris as a Guggenheim Fellow, and in 1941 he was a visiting lecturer at Harvard.

When the United States entered World War II, Goudsmit, like most of the other influential physicists, participated in military scientific research. He became a member of the staff of the Radiation Laboratory at the Massachusetts Institute of Technology and was also a consultant for the War Department. At the end of the war in 1946, he went to Northwestern University as professor of physics where he remained until 1948 when he became senior scientist at Brookhaven National Laboratories, Upton, Long Island. In 1952 he was appointed chairman of the physics department there and held this post for the remainder of his career. In 1951 Goudsmit became editor of *Physical Review*, the outstanding journal of American physics. He was also editor of *Physical Review Letters*, which is issued weekly to permit physicists to publish new and important discoveries quickly. In addition to a book on the theory of atomic spectra, published jointly with Linus Pauling, and a book on German nuclear physics during the war, Goudsmit published many papers on atomic structure, the theory of atomic spectra, nuclear spin, and statistical mechanics. For his contributions in physics he received the Order of the British Empire and the award of the Research Corporation. He was a member of the National Academy of Sciences, a fellow of the Physical Society, and a correspondent of the Royal Netherland Academy of Sciences. He died on December 4, 1978, at Reno, Nevada.

That George Uhlenbeck should have collaborated with Goudsmit in the discovery of the spin of the electron was quite natural in view of their similar backgrounds. Although Uhlenbeck was born in Batavia, Java, on December 6, 1900—two years before Goudsmit—his training in physics paralleled that of Goudsmit. Like Goudsmit, he was a product of the Leiden school of physics, receiving his Ph.D. there in 1927.

From 1925 to 1927 he was an assistant in physics at Leiden but then went to Michigan with his wife to become an instructor at the university. A year later he was appointed assistant professor in the department of physics and, in 1930, associate professor. In 1935 he returned to Holland as professor of physics at the University of Utrecht but remained there only until 1939 when he returned to the University of Michigan as the Henry S. Carhart professor of physics, retaining his post until 1960.

In the decade before World War II, the summer session of the University of Michigan became famous as a center of modern physics, graced, as it was for many summers, by the presence of Enrico Fermi. The emergence in the United States of the first international meeting ground of physicists—for it attracted such men as Heisenberg, Dirac, and Pauli—in large measure was due to the efforts of Uhlenbeck and Goudsmit. Unfortunately, the advent of war suspended this remarkable conclave. During the war, Uhlenbeck carried on military research as a member of the staff of the Radiation Laboratory at the Massachusetts Institute of Technology. Immediately after the war he returned to Michigan to resume his teaching duties and his own research. He left the University of Michigan in 1960 to become professor of physics and member of the Rockefeller Institute (now Rockefeller University) in New York.

Uhlenbeck received many honors for his contributions to physics. He was a member of the National Academy of Sciences and received honorary degrees of Doctor of Science from the University of Notre Dame and from Case. He served as president of the American Physical Society in 1959 and was a member of the Dutch Physical Society.

Although Uhlenbeck wrote numerous papers on atomic structure, quantum mechanics, and nuclear physics, his greatest love was statistical mechanics, of which he was one of the great masters. This proclivity stemmed from the early influence of Paul Ehrenfest at Leiden, who insisted on having his students pursue rigorous training in thermodynamics and statistical mechanics. Moreover, Uhlenbeck learned how to impart his knowledge to others; he was one of the great teachers of his time. Like Fermi, Uhlenbeck made the most difficult and obscure topic clear and understandable.

The Exclusion Principle: Wolfgang Pauli (1900–1958)

In the early 1920s, just before the development of quantum mechanics, a serious difficulty arose in the interpretation of the spectra of complex atoms. Here the Bohr theory appeared inadequate. This difficulty concerned the so-called anomalous Zeeman effect, which, as Pauli stated in his Nobel address, "was hardly understandable from the standpoint of the mechanical model of the atom, since classical theory, as well as quantum theory always led to the same triplet." To understand the nature of the difficulty and how its solution led Pauli to the discovery of one of the most profound and mysterious principles in nature, we must first consider a refinement of the Bohr theory introduced by Sommerfeld.

When Bohr propounded his theory, he made the simplest possible assumption about the orbits of the electron, namely, that they were circles. According to this theory, the electrons could have only one degree of freedom in their motion and, hence, just one periodicity. The energy corresponding to this motion would therefore depend on a single quantum number, the integer n, which determines the radius of the orbit. As we have already noted, this integer is called the principal quantum number of the electron. However, if one solves the general problem of an electron moving in the electric field of a positively charged nucleus, using the classical Newtonian laws of motion, one obtains elliptical Keplerian orbits; when an electron moves in such an orbit, a new degree of freedom and a new periodicity of motion are introduced, as Sommerfeld suggested. He pointed out that because of the rapid motion of the electron in its orbit, its mass while in motion is not equal to its rest mass: it increases because of the relativistic change of mass with speed. If the orbit of the electron were circular, this increase in mass would be of constant size because in a circular orbit the speed is constant. Hence, the somewhat larger, but still constant, mass of the electron would not affect its regular motion.

For an elliptical orbit, however, the situation is quite different; the mass of the electron changes continuously. Because of this continuous change in mass, the electron cannot move in a closed ellipse—it describes a kind of rosette figure. This phenomenon can best be imagined by picturing the electron as moving in a closed ellipse, which is itself rotating with a definite period in the same direction as the electron is moving in its ellipse. Since the period of this precessional motion is independent of the period of the motion of the electron in the ellipse, we have two degrees of freedom of motion of the electron and thus have two energy modes associated with the orbit of the electron. Sommerfeld suggested that to describe this motion in accordance with the quantum theory a second quantum number, the orbital or azimuthal quantum number, be introduced.

This number, which is designated by the letter l, is also an integer, but its value is governed by the value of the principal quantum number n. The azimuthal quantum number can take on all integral values (starting with zero) up to and including $n - 1$. We see from this fact that associated with each Bohr orbit there are n suborbits corresponding to the n orbital quantum numbers l. These suborbits give rise to spectral lines that lie close together, so that each of the spectral lines, as predicted by the original Bohr theory, is really a combination of closely packed lines. Consequently, each of the Balmer lines should show a "fine structure." This structure is precisely what spectroscopists found. In fact, they discovered a somewhat more complex structure than the Sommerfeld theory predicted.

We should consider the physical significance of the new quantum number before we examine Pauli's discovery of the exclusion principle. The principal quantum number n is a measure of the size of the ellipse in which the electron moves and, hence, the total energy of the electron in its orbit. But we know that for an ellipse of a given size we can have different shapes from the roundest, a circle, to the narrowest, a straight line with the electron oscillating back and forth along it. The new quantum number l is related to the shape of the orbit; but since, as we know from planetary theory, the shape of the orbit determines the

Wolfgang Pauli (1900–1958)

angular momentum of a planet in its orbit, the azimuthal quantum number really is a measure of the angular momentum of the electron.

We see that both the energy of the electron and the shape of its orbit (its angular momentum) is quantized. We may illustrate this concept by considering an electron in the third Bohr orbit so that its principal quantum number is 3. We thus have the three azimuthal quantum numbers 0, 1, and 2, which determine the eccentricity—the flatness—of the suborbits. The angular momentum is 0 for value 0 of the azimuthal quantum number (when the orbital ellipse most resembles a straight line, or is flattest) and is largest for value 2 of the azimuthal quantum number (when the orbit is circular). The electron has a slightly different energy in each of these states because of the precession of the orbit. Hence, the third Bohr level actually corresponds to a combination of three closely spaced sublevels. If the energy of the electron in each of these sublevels were the same, the three energy sublevels would coincide to form a single level; we would then say that this state or this level is degenerate. We may note that the unit of angular momentum in this theory is $h/2\pi$ and that the quantization rule always involves the product of the quantum number, in this case l, and Planck's constant divided by 2π.

In connection with the angular momentum, we must take up one final point. The azimuthal quantum number is a measure only of the magnitude of the angular momentum, which is a vector quantity. Hence, we must also introduce some way of taking into account the direction of this vector. We may understand this quantity if we consider the electron moving in one of the elliptical suborbits discussed in the previous paragraph. If we want to specify this orbit completely, we should give not only its size and shape but also its orientation in space. The question that naturally arises is whether the specification of the orientation in space involves a quantum number the way the size and shape do, or whether the axis of the electron's orbit can take on any direction and change continuously from one direction to another. As we have seen in our discussion of the Stern-Gerlach experiment, we do, indeed, have to assign a quantum number to the direction of the angular momentum. An additional type of quantization, space quantization, must be introduced. This phenomenon, first proved experimentally by Stern and Gerlach, plays an important role in the Zeeman effect, which so puzzled Pauli and finally led him to the discovery of the exclusion principle.

We may infer that a third quantum number has to be introduced in the description of the motion of the electron because each of the other two quantum numbers is associated with one of the degrees of freedom of the motion of the electron. Since the electron has three degrees of freedom, it should have three quantum numbers attached to it. The orientation of the orbit gives us this third degree of freedom, but it is a bit puzzling to see just how a quantum number is to be assigned to this degree of freedom because it is not clear how we are to associate periodicity or energy with orientation. If there were no energy assigned to this degree of freedom (in other words, if there were no orbital motion of any kind related to the orientation of the axis of the orbit of the electron) there could be no sublevels associated with it. Therefore, we would be dealing with a case of degeneracy.

We can assign a quantum number, and hence energy, to the orientation of the angular momentum vector only if certain directions in space are physically more meaningful than any other directions. At first it may seem that this concept is untenable, that all directions in space are equivalent. But all directions are not equivalent if there is a physical phenomenon in the neighborhood of the electron that establishes some preferred direction. It is precisely in establishing such a preferred direction that we can assign energy to this third degree of freedom and thus eliminate the degeneracy. For instance, the spectral lines can be made to take on an additional pattern of complexity, the Zeeman effect, if a preferred direction is obtained by means of a magnetic field.

We can easily see why a magnetic field, in some particular direction in the neighborhood of the electron, introduces additional states of energy. As the electron moves around in its elliptical orbit, it is tantamount to an electric current. Therefore, it is accompanied by a magnetic field whose direction is parallel to the direction of the angular momentum of the electron. In other words, the electron behaves like a very small magnet. If we now have an external magnetic field pointing in some fixed direction, the angular momentum vector of the electron is not free to point in any direction whatsoever but has to adjust itself to this field.

If the atom were no more than a small magnet, and there were no angular momentum associated with it, its axis (or, as it is called in physics, the axis of the magnetic moment) would line up in the direction of the magnetic field. We still could assign neither energy nor a quantum number to this direction. But we must recall that, in addition to behaving like a small magnet, the atom behaves like a spinning top because the electron is revolving in its orbit. A spinning top has one very important property that gives it additional motion and therefore an additional sublevel of energy. We need only consider an ordinary top that is set spinning on the ground. If its axis of spin is perfectly upright, the only motion of the top is the one of spinning. But if the axis is tilted, not only does the top continue to spin, but its axis points in changing directions: it precesses around the vertical direction. This motion of precession is an additional state of energy of the top; we have a similar situation when a spinning magnet is placed in a magnetic field.

The rate at which this magnet precesses was first calculated on the basis of classical electrodynamic theory by Larmor. Thus, this frequency of precession is called the Larmor frequency. It is equal to the magnetic moment of the magnet (the length of the magnet multiplied by the pole strength of the magnet) multiplied by the strength of the magnetic field, with this product then divided by the angular momentum of the spinning magnet.

We can apply this calculation to the motion of the electron inside an atom, such as the one in hydrogen. As the electron revolves in its orbit, it gives the atom the properties of a magnet and also makes the atom behave like a top. This means that the atom must precess in a magnetic field; the Larmor frequency of this precession is obtained by multiplying the electric charge on the electron by the strength of the magnetic field and by dividing this quantity by 4π times the mass of the electron and the speed of light. Hence, when an atom is in a magnetic field, the energy levels are not the same as in the absence of a field, since the Larmor precession takes place. Additional energy is associated with the additional periodicity of the Larmor precession. We must therefore introduce a third quantum number, since the periodicity of the Larmor precession is quite unrelated to the periodicities of the other two notions we have already discussed. This third number is called the magnetic quantum number and is designated by the letter m.

The magnetic quantum number m cannot be chosen arbitrarily but must be related to the azimuthal quantum number in a definite way; it is an integer that can range in unitary steps from -1 to $+1$. In the case of an electron in the third Bohr orbit, for example, with principal quantum number 3, and with three possible azimuthal quantum numbers 0, 1, 2, there are nine possible magnetic quantum numbers. Thus for $l = 0$, m can only be 0; for $l = 1$, m can take on the three values, -1, 0, $+1$; for $l = 2$, m can have the values -2, -1, 0, $+1$, $+2$.

The significance of these magnetic quantum numbers is that the electron orbit (or the angular momentum vector) cannot orient itself in any direction with respect to the magnetic field but is limited to a finite number of directions. The elliptical orbit for which the azimuthal quantum number in the above example is 2 has only five possible orientations in a magnetic field. Of course, if no magnetic field is present, there is no restriction as to the way in which the plane of the orbit of the electron can be tilted in space.

Even before Bohr had introduced his theory of the atom, it was known that the energy of an atom in a magnetic field is different from what it is in empty space. The spectroscope gave evidence of this difference in the form of the Zeeman effect: when an atom is placed in a magnetic field, the usual lines in its spectrum are no longer simple but break up into multiplets. In the case of the hydrogen atom (and also of the atoms of the alkali metals like lithium and sodium), each line is replaced by a triplet. The central component has the same frequency as when no magnetic field is present; the frequency of the other components differs from that of the central component by the amount expressed by the Larmor frequency. This result had already been predicted by Lorentz on the basis of his classical theory of the electron. His argument was very simple because it was clear to him that a vibrating electron in a magnetic field would suffer a perturbation that would give rise to new vibration frequencies. The electron would vibrate not only with its original frequency but also with two new frequencies: one obtained by adding the Larmor precession frequency to the original frequency of the electron and the other obtained by subtracting the Larmor precession frequency from the original frequency. In other words, when an atom is in a magnetic field, each of the spectral lines should break up into three lines; this phenomenon is called the normal Zeeman effect.

Although the Lorentz theory gives the correct result for the Zeeman effect of the hydrogen atom, it does not account for the much more complicated magnetic splitting of spectral lines for the heavier atoms, the so-called anomalous Zeeman effect. If we now use the Bohr theory of the atom, we find that it gives us exactly the same results for the normal Zeeman effect as the classical Lorentz theory; but the Bohr theory can also explain the anomalous Zeeman effect for the heavier atoms.

At the time these investigations into the Zeeman effect were being pursued, another complication arose—a multiplicity of spectral lines was observed when the atom was not subjected to external magnetism. This fine structure appeared in addition to the relativistic effect explained by Sommerfeld as the result of the orbital precession. The phenomenon emphasized the need for still another quantum number. Sommerfeld attempted to provide it; he demonstrated that he could account for the observed multiplicities by introducing a quantum number that could have but two values.

To justify the introduction of the new quantum number, Landè suggested the core + electron model of the atom. In this model the electron that is responsible for the spectral lines is pictured as moving alone in an external orbit. At the center of the orbit is the core of the atom, which consists of the nucleus surrounded by the inner electrons. If this core is visualized as spinning, we are able to postulate the generation of a magnetic field. What follows is a kind of internal Zeeman effect that in turn gives rise to an internal, magnetic quantum number. Although this model appeared promising upon its introduction, it soon was discarded. Pauli demonstrated its inaccuracy for heavy atoms if one took into account the relativistic change in mass of the core electrons.

In his Nobel address, Pauli noted that he became interested in the anomalous Zeeman effect in 1922 and soon was convinced of the necessity of still another approach. But there was a twofold difficulty in attempting to discover a new scheme. On the one hand, he needed to find a suitable mechanical or classical model for heavy atoms; on the other hand, it was necessary to incorporate such a model into the quantum picture. Pauli was well aware that the Bohr model and its correspondence principle were inadequate for his purposes.

Pauli was also aware that the anomalous Zeeman effect was associated with the problem of the closed electron shells. In analyzing this phase of the problem, Pauli rejected as incorrect the core + electron model of the atom. He developed the concept of a new, nonclassical, twofold property of the electron—which actually amounted to assigning another quantum number to it as Sommerfeld had attempted to do. Consequently, the electron was pictured as having a fourth quantum number, in addition to the three already discussed. This fourth quantum number, as Sommerfeld suggested, can take on only two values, which is what Pauli meant when he spoke of the "classically non-describable two-valuedness" of the electron. By "classically non-describable" Pauli simply meant that there was no classical model he could adopt to explain this fourth quantum number.

We have seen that the quantum number carries with it the idea of a classical periodic motion of the electron. Landè, in his core + electron model, had tried to incorporate this additional periodic motion by assigning it to the core of the atom. By contrast, Pauli assigned the additional periodic motion to the outer electrons and, in fact, to every electron. The most difficult aspect of the conundrum, as far as Pauli was concerned, was determining just what classical periodic motion corresponded to this fourth degree of freedom, or fourth quantum number. This difficulty was partially overcome by Goudsmit and Uhlenbeck's concept of electron spin. However, Pauli was reluctant at first to accept without reservations the spinning electron model on the grounds that it was too mechanical an explanation for what he considered to be an essentially nonclassical, or nonmechanical effect. However, he finally accepted the spin phenomenon when Bohr pointed out that a spinning electron may differ considerably from classical analogs.

Having concluded that there are four, not three, quantum numbers associated with the motion of each electron, Pauli was led to the exclusion principle by his study of the classification of spectral lines of an atom in a strong magnetic field. In addition, he was influenced by E. C. Stoner, who had pointed out that in a magnetic field the electron most distant from the nucleus of an atom like lithium, sodium, or potassium—the so-called optical electron—has as many energy states as there are electrons in the closed shell of the first inert gas (neon, argon, and so on) that follows this particular alkali metal in the atomic periodic chart.

Pauli's interpretation of this phenomenon, which is the heart of the exclusion principle, takes the following course. Picture the optical electron moving alone in the second Bohr orbit of the lithium atom. If this atom is in a magnetic field, the electron is subject to the four independent periodic motions associated with its four quantum numbers, since the magnetic field removes all the degeneracies of the electronic motion. The actual energy of the electron corresponds to some particular combination of all of these periodic motions. Stoner stated that in the case of lithium the total number of possible combinations of motions for this electron is equal to the number of electrons in the closed shell of neon. The total number of such electrons turns out to be eight, and therefore, according to Stoner, there are eight energy levels for this optical electron.

Pauli proceeded by suggesting that each combination of the four periodic motions, or of the quantum numbers, defined a sublevel of the electron. He stated that each sublevel of any particular Bohr level can accommodate but one electron, so that every sublevel is closed as soon as it contains one electron. Within a given atom no two electrons can have identical full sets of quantum numbers. If we assign one set of quantum numbers to one electron and another set to another electron, the two sets must differ by at least one number.

Let us consider the second Bohr level, for which the principal quantum number is $n = 2$; we have in association with n the two azimuthal quantum numbers 0 and 1. Associated with the azimuthal quantum number 0 there is only one magnetic quantum number 0; in addition, the azimuthal quantum number 1 carries with it the three magnetic quantum numbers -1, 0, and $+1$. By this scheme, any electron within the second Bohr orbit is assigned just four sets of combinations of azimuthal and magnetic quantum numbers: $(0,0)$, $(1,-1)$, $(1,0)$, and $(1,1)$. According to Pauli, in addition to these three quantum numbers, the electron has a fourth degree of freedom because of its spin. This fourth degree of freedom, or quantum number, can have only two distinct values (regardless of what combination of the other three quantum numbers is assigned to the electron). Therefore, we can have twice the number of distinct quantum-number combinations than we have listed above, or, all together, eight, the number of electrons in the closed shell of neon. It can be shown that if the principal quantum number is n, then the shell having this principal quantum number is closed when it is occupied by exactly $2n^2$ electrons, since there exists just this number of distinct combinations of the four quantum numbers.

Conversely, by using the same factors, we may arrive at an understanding of the freedom of action of lithium's single optical electron. A simple analogy may help. One person in a room has a wide range of possible activities; likewise, the optical electron's solitude makes lithium very active chemically. In the case of the element neon, the room is full, and there is little freedom of activity. Hence, with eight electrons completely filling the shell, the gas is inert.

The importance of the Pauli exclusion principle extends far beyond its application to atomic spectra and to the distribution of electrons in the various closed shells of complex atoms. The full implications of this principle became evident, as Pauli pointed out, only after the introduction of wave mechanics to describe the properties of particles. Since the description of a particular state of an atom is given by its wave function (a mathematical expression involving the positions and the spins of the electrons in the atoms), one must look to this wave function for information about the atom. The wave functions that are found in nature have definite properties of symmetry. One finds wave functions that are either completely symmetrical (if two particles are interchanged the wave function is unaffected) or wave functions that are antisymmetrical (if two particles are interchanged the wave function changes its sign, becoming negative if originally positive, or vice versa). We find that a given type of particle is described by either symmetrical or antisymmetrical wave functions and that these two symmetries are never mixed.

Because electrons are described only by antisymmetrical wave functions, they are governed by the exclusion principle. Electrons, protons, and neutrons, in fact all particles that have a spin, or intrinsic angular momentum equal to that of the electron, are described only by antisymmetrical wave functions that change sign when two particles are interchanged. All such particles, whether they are inside atoms or not, must obey the Pauli exclusion principle. Particles such as the photons, and certain mesons that are described by symmetrical wave functions, are not governed by this principle. As Pauli pointed out in his Nobel address, this phenomenon leads to two types of statistical mechanics to explain the complexity of spectral lines.

In the investigation that finally led to the formulation of the exclusion principle, Pauli first suggested that the complexity of spectral lines (the so-called hyperfine structure) might be caused by the spinning of the nucleus of the atom. Although this hypothesis was unable to account for the fine structure of the spectral lines, it led Goudsmit and Uhlenbeck to the hypothesis of the spinning electron. Subsequently, Pauli's picture of a spinning nucleus was found to be correct, and it was shown that the hyperfine structure of the spectral lines could be explained by this phenomenon.

The numerical value of the spin of a nucleus—and hence its symmetry class and the kind of statistics that it obeys—can be found by studying the molecular spectra of molecules composed of two similar atoms such as H_2 or O_2. Such molecules rotate the way a dumbbell does when it is thrown through the air; the transition of the molecule from one such rotational state to another gives rise to a line in the spectrum of the molecule. Since the rotational energy levels of such a molecule lie very close together, all the lines that arise from a given set of transitions are so closely packed that they form a bright band in the spectrum of the molecule. By studying the intensities of bands arising from transitions involving only even or odd rotational quantum numbers we can identify the spin of the nucleus of the atom in the two-atom molecule.

It was in connection with the symmetry properties of the rotational energy levels of molecules that the concept of the parity of the wave function was introduced. Transitions in any system occur either between states in which the wave function is symmetrical or between states in which the function is antisymmetrical. Transitions seldom occur between a state defined by a symmetrical wave function and one described by an antisymmetrical wave function. (In principle they never occur if one neglects the effect of the spin of the nuclei on their motions within the molecule.) The states described by symmetrical wave functions are called states of even parity and the antisymmetrical ones, states of odd parity. It was assumed, as a general principle throughout nature, that only those processes can occur for which parity is conserved. This assumption, which is called the principle of the conservation of parity, proved to be incorrect under certain conditions, as was shown in 1957 by Lee and Yang.

In his Nobel address, Pauli pointed out how important the exclusion principle is for the determination of the spin properties of nuclei and for such particles as the proton, the neutron, and the deuteron. From studies of the spectral bands emitted by the ortho and para types of hydrogen, the spins of the proton and electron were found to be the same. Since the spin of the proton is $1/2$ (the same as that of the electron), protons are also governed by the exclusion principle. It follows that a nucleus cannot contain electrons in addition to protons because the symmetry properties of any system composed of a number of particles is determined by the number of these particles that obey the exclusion principle. If this is an odd number, the system must obey the exclusion principle and must be governed by the antisymmetrical type of statistics. But if this number is even, the system is described by symmetrical wave functions and it obeys the same type of statistics that governs the photon.

We know from experiment that the nitrogen nucleus, for example, does not obey the exclusion principle. But if there were electrons in this nucleus, the total number of antisymmetrical type particles in it (electrons and protons together) would be 21, an odd number. Therefore, it would have to obey the exclusion principle. Hence, it follows that there can be no electrons in this or any other nucleus. The number 21 for nitrogen is arrived at as follows. The nucleus is known to have an atomic weight of 14 so that it must consist of 14 particles, each with the mass of the proton (14 units of mass). If all of these particles were protons, there would have to be 7 electrons present as well, in order to compensate for the 7 positive electric charges and to leave a residue of 7 positive charges in agreement with the observed charge on the nitrogen nucleus. We would thus have a total of 21 particles (7 electrons and 14 protons) each of spin $1/2$ in the nucleus. This same analysis can be applied to all nuclei. With the discovery of the neutron the mystery of the particles in the nuclei of atoms was cleared up. If one assigns a spin $1/2$ to the neutron, imparting to it the same statistical properties as those of the electron (it is governed by the exclusion principle), one can explain all the spin and symmetry properties of the nuclei. The importance of this concept is indicated by the absence in nature of a nucleus consisting of two protons only. Such a nucleus would be a helium nucleus having the same weight as deuterium. However, because hydrogen obeys the exclusion principle, two protons, which are identical particles, cannot combine with their spins parallel to form a nucleus as a neutron and a proton can.

The second part of Pauli's Nobel lecture was devoted to questions relating to the classical and quantum-mechanical properties of fields arising from particles and how the exclusion principle ultimately determines the way in which these fields are to be quantized. If one is dealing with classical fields, such as those obtained from the solution of Maxwell's equations, where quantization has no part, the exclusion principle plays no role. But all fields are governed by quantum principles (for example, the uncertainty principle in the case of the electromagnetic field), and therefore they should be treated from a quantum-mechanical point of view. Then the fields break up into two kinds: those fields, such as the electromagnetic field, that are generated by particles like photons, which do not obey the exclusion principle; those fields, such as the de Broglie fields, that are generated by particles like electrons, which do obey the exclusion principle. In these two cases, the algebras that govern the behavior of the fields are different, as expressed in the form of the commutation relations for these fields.

Pauli ended his Nobel lecture with a statement of his strong dislike for the current theories of elementary particles that endow the vacuum with what he considered artificial properties. Such a criticism could be leveled at the Dirac theory of the electron, which is really not a theory of a single particle but rather of an infinite number of particles. In Dirac's theory we have, in addition to the "perceived" electron, an infinitude of electrons in negative-energy states in the vacuum. This hypothesis is what Pauli meant when he referred to the "infinite zero charge," since these electrons contribute nothing to the electric field surrounding the electron being studied. He finally expressed his opinion that such theories must be faulty because they use "mathematical tricks" and "mathematical fictions" "to formulate the correct interpretation of the actual physical world."

Although Pauli was one of the ablest mathematicians in the family of physicists, he was always highly critical and suspicious of a purely formalistic approach to physics. He, more than any other physicist, was able to draw the greatest content from the mathematics of current theories. By means of very keen mathematical analysis, he was able to subject these theories to the most severe tests. In connection with this exacting method, it may be noted that his article on the theory of relativity, which he wrote in 1921 at the age of 20, is still the most penetrating analysis of that theory available.

Wolfgang Pauli was one of the giants of 20th-century theoretical physics. He was born in Vienna on April 25, 1900, the very year that Planck discovered the quantum of action, which was to play so important a role in Pauli's scientific life. It was the theory to which he, in turn, was to contribute so much.

Pauli was the son of a physician, who himself was a university professor of chemistry. Young Pauli was raised in a scientific atmosphere, and this household certainly must have contributed to his unusually rapid intellectual development. By the time he was 19 he had already mastered enough physics and mathematics to be invited by Arnold Sommerfeld, his professor of theoretical physics at the University of Munich, to write the article on the theory of relativity for the *Encyclopädie der Mathematischen Wissenschaft*. When Einstein read this article, which is still the leading text on relativity, it is said that he felt that perhaps Pauli knew more about relativity than he himself did. In reviewing this article, Einstein wrote: "Whoever studies this mature, and grandly conceived work, can hardly believe that the author is a 21-year-old man. One hardly knows what to admire and wonder at most, the psychological understanding of the development of the ideas, the sureness of the mathematical deductions, the deep physical insight, the capacity for a systematic, clear exposition, the knowledge of the literature, the completeness of the treatment, the sureness of the critical approach" (Einstein, 184).

Pauli's article on relativity was the first in a long series of scientific papers, each of which maintained the same high level of excellence set in the relativity article. All of these papers have been issued in two volumes under the joint editorship of R. Kronig and V. F. Weisskopf.

After graduating from a local gymnasium in 1918, Pauli enrolled in the University of Munich where he concentrated on theoretical physics for six semesters; here he was greatly influenced by Sommerfeld and by W. Lenz. Although he began his scientific career in the field of relativity, he soon shifted to quantum theory and to quantum mechanics, which was just beginning its magnificent growth. Indeed, Pauli's scientific life may be said to recapitulate or mirror the developments in this subject, for he made essential contributions to every phase of it. He was numbered in the famous group of physicists, directed from Niels Bohr's headquarters at Copenhagen, who created quantum mechanics.

Although all of Pauli's contributions were of the first magnitude, his greatest achievement was the discovery of the exclusion principle for which he received the Nobel Prize in 1945; it is one of the pillars on which modern physics and the quantum mechanical understanding of the structure of matter rest. Not content with having discovered the exclusion principle, Pauli sought continuously to explain it in terms of more basic principles. He finally succeeded in showing that the exclusion principle is the result of the symmetry properties of nature.

After writing his Ph.D. dissertation in 1921 (a paper on a special type of molecular model), Pauli went to the University of Göttingen where he served as Max Born's assistant. From there he went to Copenhagen to spend a year with Niels Bohr and then to the University of Hamburg to accept an assistant professorship. Finally, in 1928, he accepted the chair of theoretical physics at the Federal Institute of Technology in Zurich, where he remained until he died, a victim of cancer, on December 14, 1958.

Pauli influenced the development of physics not only by his papers but by his critical discussions and his voluminous correspondence with physicists all over the world.

> Pauli stimulated the thoughts and ideas of many leading theorists by letters and critical discussions, by suggestions and inspiration. He was a profuse letter writer. His correspondence reflects the development of modern physics in the most intimate way, and will some day serve as a fertile source for the historians of science. His sharp criticism by mouth and pen is part of the legend of modern physics. He could not tolerate vagueness or half-truth and his biting wit found devastating formulations for his contempt of ideas that did not measure up to his high standards.

Many considered him the conscience of theoretical physics. His death left a gap which still remains to be filled.

Secondary Radiation:
Chandrasekhara Venkata Raman (1888–1970)

The scattering or diffusion of light is a well-known phenomenon that has been observed and studied for hundreds of years. But only within a relatively short

Chandrasekhara Venkata Raman (1888–1970)

time—since the advent of the quantum mechanics—has it been fully understood and certain of its strange features, such as the Raman effect, been explained. The simplest example of this phenomenon is the effect produced in the path of a beam of light by extremely fine dust. The eye cannot perceive light unless the light enters it; nevertheless, we can see a beam of light traversing a dust-filled room even though the beam itself is not moving toward the eye. The reason for this effect is that each particle of dust scatters the light, striking it so that the dust particle behaves as though it itself were a center of radiation, sending out rays in all directions. Thus, observers standing anywhere in relation to the beam receive scattered light.

A brief discussion of how a dust particle does this will be useful for what we shall state later. We know that light is an oscillatory phenomenon consisting of rapidly oscillating electric and magnetic fields. As the light passes through the dust-laden air, each dust particle is pulled back and forth by the vibrating electric field of the light and, as a result, emits electromagnetic radiation in all directions, thus scattering the incident light. This scattering happens because the incident electric field pulls the negative electric charges in the dust particle to one side and the positive charges to the other side, forming a small electric dipole. As this dipole vibrates, it emits radiation. But it vibrates with the same frequency as that of the incident light so that the scattered light is of the same frequency as the incident light, at least according to this simple analysis. The color of the scattered light is thus the same as that of the incident light.

Here one more point must be mentioned. The response of a dust particle to light striking it is not the same for all colors. The shorter the wavelengths (the bluer the light), the more violently the dust particle responds and the more intense is the scattering. If white light (a fairly uniform mixture of all colors) passes through a dusty medium, the blue and green colors are scattered more readily than the red and yellow, and dust, such as smoke in a room, appears bluish. This is known as the Tyndall effect.

The question naturally arises as to whether molecules themselves scatter light. At first it may seem that they do not, since pure air appears perfectly clear. But a little thought and more careful observation show us that molecules do, indeed, scatter light. The best evidence for this phenomenon is the blueness of the sky, first investigated by the British physicist Lord Rayleigh, who suggested that it was caused by molecular scattering. When we look at the sky in a direction away from the sun, our eyes receive the scattered rays of sunlight; since these are mostly blue in color, the sky appears blue. When we look toward the sun at sunset, our eyes receive the rays of light coming through the atmosphere—the unscattered rays— and these are red. Therefore, the sun appears red.

That molecules should also scatter light follows, in principle, from the same general analysis. The molecule in the electromagnetic field of light becomes a small electric dipole which vibrates with the oscillating electromagnetic field and emits light in all directions. Since the molecule vibrates with the same frequency as the incident light, we should expect to find the scattered light to be of the same frequency as the incident light. This similarity is, indeed, what one generally finds. But a careful investigation of the scattered light shows that, in addition to light of the same frequency as the incident light, weak components, different in frequency from that of the incident light, are present in the scattered radiation. In 1928 Raman discovered this surprising and completely unexpected effect, which is now known as the Raman effect.

To study this phenomenon more thoroughly, Raman used incident light of a definite frequency rather than white light. He isolated the light of a definite spectral line from a mercury arc lamp and used this line as the primary radiation. When the light was sent into a homogeneous medium, Raman found that the scattered light contained not only this primary spectral line but also other faint lines on either side of the given line. In other words, the scattered light contained frequencies equal to, smaller than, and larger than the frequency of primary light (of the mercury line). Raman found that this effect was present no matter which mercury line was used for the incident light. In other words, a Raman spectrum is present in the scattered radiation of any spectral line and the lines of this Raman spectrum in the scattered light lie symmetrically on either side of the incident line.

The following problem is associated with the Raman spectrum. Because of the discrete orbits permitted to an electron in an atom or the discrete states that a molecule can have, an atom or a molecule can absorb or emit photons of a definite frequency. But the lines in the Raman spectrum of scattered light are not those found either in the emission or absorption spectrum of the atom or the molecule. How, then, do the Raman lines arise?

We may get an idea of how this phenomenon happens by first considering a simple classical argument. Consider a molecule in an electric field. This field displaces the negative and positive charges with respect to each other so that the molecule becomes a dipole. Now this dipole has its own natural frequencies. If it were set vibrating, it would vibrate with its natural frequencies, like the strings of a violin, and emit light of these frequencies. If light of a given frequency—different from any one of the natural frequencies of the dipole—strikes the molecule, it oscillates under the influence of the oscillating electric field of this light. One can now show mathematically that the molecule, that is, our dipole, oscillates not only with its own natural frequencies but with additional frequencies obtained by adding and subtracting the frequency of the incident light from the natural frequency. In other words, if ν_n is one of the natural frequencies of the molecule and ν is the frequency of the incident light, then the light scattered by the molecule contains not only the frequency ν but also neighboring frequencies $\nu + \nu_n$ and $\nu - \nu_n$, lying at equal intervals to the right of the incident line.

Although this simple classical analysis shows why we can expect to find frequencies in the scattered light different from those of the incident radiation, we can obtain a complete understanding of the Raman effect only by using the quantum theory. Consider the discrete energy states of a molecule. Under ordinary conditions, the only photons that the molecule can absorb or emit are those associated with transitions of the molecule from one of these discrete states to another. But the spectral lines arising from all such transitions are not present because there are certain rules, called selection rules, that prohibit certain transitions. The spectral lines associated with these transitions do not occur in the spectrum of the molecule. However, these lines and others may appear in the Raman spectrum under the following conditions.

Consider a photon that has a frequency different from that which the atom or molecule can emit under ordinary conditions. Such a photon will, in general, not be absorbed by the molecule because the absorption would bring the molecule to a state of energy that does not correspond to any of its possible energy levels. But suppose the molecule absorbs this photon and immediately re-emits another one whose frequency is such (let us say smaller than that of the incident photon) that the molecule ends up in one of its possible excited states above the state from which it started. The energy of the emitted photon is clearly the difference between the energy of the incident photon and the energy of a photon that the molecules would have to absorb to bring it from its initial to its final state. A line is thus present in the Raman spectrum of frequency smaller than that of the incident photon.

Suppose, now, that immediately on absorbing the incident photon, the molecule emits a photon of greater frequency than that of the one it absorbed. It can do this provided it ends up in a state lower in energy than the one from which it started. The emitted photon in this case has an energy equal to the energy of the incident photon, plus the energy of the photon that the molecule would emit if it jumped directly from its initial to this final state. A line is thus present in the Raman spectrum whose frequency is higher than that of the incident photon.

For Raman lines to appear both to the left and to the right of the incident light (for larger and smaller frequencies to appear in the scattered light), the molecule does not have to be in its ground state. If it were always in its ground state, only frequencies lower than the frequency of the incident photon could appear since the end state of the molecule would then have to lie above the initial state (because the initial state in this case is the ground state and, hence, the lowest possible state). Under ordinary conditions, the thermal motions of the molecules bring about collisions between them, causing the molecules to spin and vibrate. Hence, the molecules are initially in rotational and vibrational states that lie above the ground state. Thus, the final state of the molecule can be lower than the initial state.

We would therefore expect to find a difference between the intensity of the Raman lines below (in frequency) the incident line and the intensity of the lines above, depending on the temperature of the medium. As the temperature gets lower and lower, the number of molecules in excited rotational or vibrational states gets smaller and smaller. There are thus fewer and fewer molecules available to emit photons of frequency higher than that of the incident photon. This effect is observed.

One more point is of interest in connection with the Raman spectrum. The lines in this spectrum are associated with transitions of the molecule between rotational (or, to a lesser extent, vibrational) energy levels. The energies associated with such transitions, excited by the thermal motion, are quite small and correspond to photons in the infrared part of the molecular spectrum. In a sense, the Raman spectrum gives a picture of the infrared lines of the molecule.

Raman initially examined only the Raman effect in gases. But since this early work was done, the Raman spectrum in crystals has also been studied extensively. Since the fundamental vibrations of a crystal are difficult to analyze under ordinary conditions, the Raman lines arising when a crystal is excited by photons of a definite frequency are a source of information for the structure of the crystal.

Chandrasekhara Venkata Raman, whose father was a lecturer in mathematics and physics, was raised in an academic environment. From the time of his birth at Trichonopoly in southern India on November 7, 1888, Raman lived in a world of ideas and quickly absorbed advanced concepts of mathematics and physics. He was so precocious and showed such a grasp of science that he was admitted to the Presidency College, Madras, in 1902 at the age of 14 and passed his B.A. examinations two years later. He placed first in this examination and won the gold medal in physics. In 1907 he received his M.A. degree with highest distinction.

From the time he began his graduate studies, Raman devoted himself to optics and acoustics. In the early years of his career, he had to do research on a part-time basis because he took a job in the Indian Finance Department to support himself. He carried on his scientific work at every spare moment, using the laboratories of the Indian Association for the Cultivation of Science in Calcutta.

Raman was very successful in his part-time research and soon acquired a reputation as an excellent experimentalist. When a newly endowed Palit chair of physics was established at Calcutta University in 1917, it was offered to Raman, who accepted it. During his stay at Calcutta, he performed the experiments that led to his discovery in 1928 of what is now called the Raman effect, for which he received the Nobel Prize in 1930. Raman's curiosity about the blue opalescence of the sea, which he had observed in the Mediterranean on a trip to Europe in the summer of 1921, started the chain of events that led to his final discovery. In thinking about this strange glow of the sea, Raman stated in his Nobel address that "[i]t seemed not unlikely that the phenomenon owed its origins to the scattering of sunlight by the molecules of the water."

Raman decided to test this hypothesis by investigating the laws that govern the diffusion of light in liquids; he started experiments immediately on his return to Calcutta in September 1921. This work entailed a much greater effort than he had originally anticipated since it very quickly broadened into a general investigation of the molecular scattering of light. A critical review of the entire subject and the results of Raman's initial experiments were published as an essay by the Calcutta University Press in February 1922. From 1922 until 1927, Raman, with various collaborators, investigated as many facets of optical scattering as possible. In one of these experiments in April 1923, a collaborator, Ramanathan, discovered that, in addition to the usual type of scattered light, which is similar in wavelength to the incident light, there was present a very weak secondary radiation differing in wavelength from that of the incidental light.

Raman became greatly interested in this secondary radiation and devised various experiments to increase its intensity so that it might be studied spectroscopically. Finally, on February 18, 1928, he found that if he used a mercury arc lamp as his incident light source, the spectra of the scattered light from a variety of liquids and solids included a number of lines and bands not present in the spectrum of the mercury arc light. The production of these lines by the scattering molecules is called the Raman effect.

After receiving the Nobel Prize for this discovery, Raman became professor of physics at the Indian Institute of Science at Bangalore where he continued his work in optics. He remained there until 1948 when he became director of the Raman Institute of Research at Bangalore, which he himself established and endowed. Raman remained very active in research. He also inspired many of his countrymen to go into physics and did a great deal to further physical research in India. In 1926 he founded the *Indian Journal of Physics*, of which he served as editor, and initiated the establishment of the Indian Academy of Sciences, serving as its first president. He also began publishing the *Proceedings* of that academy and was president of the Current Science Association of Bangalore, which publishes *Current Science (India)*.

Besides his research into the scattering of light, Raman did a considerable amount of study in acoustical vibrations, X-ray diffraction by crystals, crystal dynamics, crystal structure, optics of colloids, electric and magnetic anisotropy, and the physiology of human vision. His interests in acoustics led him to study the manner in which the violin maintains its vibrations, and he published on the subject of string and other musical instruments.

Raman received his share of laurels, among them many honorary doctorates and honorary memberships in scientific societies in several countries. He was elected a fellow of the Royal Society in 1924 and was knighted in 1929. He died in Bangalore on November 21, 1970.

Statistical Mechanics: S. N. Bose (1894–1974)

One of the most powerful techniques the physicist has for the solution of problems dealing with ensembles, such as groups of electrons, molecules, or photons, is statistical mechanics. It had its beginnings in the kinetic theory of gases as formulated by Maxwell and Boltzmann. This discipline was later developed more fully and applied to the equilibrium of chemical systems by the well-known American physical chemist, Willard Gibbs. In fact, the form in which Gibbs developed this branch of physics is what came to be known as statistical mechanics. The procedures of Maxwell and Boltzmann are generally referred to as kinetic theory.

The real power of statistical mechanics did not become fully apparent until Einstein redeveloped it and applied it to all types of phenomena from Brownian motion to the behavior of photons. In all of these applications of statistical mechanics, the type of statistics used is called Boltzmann statistics or classical statistics. No one doubted its validity since there was then no reason to question the validity of classical physics. But, as the quantum theory developed from its early stages to its more complex phases, the question of the validity of Boltzmann statistics began to intrude into the discussions of physicists.

To understand the problem involved in the application of statistical mechanics to an ensemble of identical particles (e.g., the molecules of a gas), we must first define the state of a system. One way is to consider the microscopic measurable properties of the ensemble. The state of the gas can thus be represented macroscopically by specifying its pressure, its volume, and its temperature. Actually, we need specify only two of these three quantities since an equation exists that connects all three quantities; therefore, we can obtain the third quantity if we know any two. This equation is called the equation of state of the gas. This macroscopic representation of the state of an ensemble gives us no insight into the way this state depends on the motion of the individual particles in the system. It is clear that the gross or macroscopic condition of the ensemble is determined by how the individual particles are moving about and how they happen to be distributed throughout the volume. We should thus be able to define the macroscopic state of the ensemble by specifying the distribution of the particles in the ensemble and giving their motions. We may refer to this as the microscopic specification of the state. Statistical mechanics is the branch of physics that deals with the microscopic representation of states of a system.

To define the microscopic state of the ensemble we first consider a single particle. As we follow this particle, we note that its position and its motion, that is, its velocity or its momentum, change continuously. At each moment, therefore, we can, on the basis of classical physics, assign to this particle a position and a velocity. Suppose we continued this assignment of values for every particle in our ensemble. We would then, at each moment, have a detailed microscopic picture of the state of the system—the position and velocity of each of its particles.

If we think about this idea for a moment, we realize that we do not have to go into such detail to define the state of the ensemble. It is not really necessary to know where each particle is and how it is moving at each moment because, since all the particles are identical, it does not matter—insofar as the state of the system goes—whether a given particle or some other one is at a particular point, moving with a particular velocity. It is clear that the only thing that counts here is how many particles in a given neighborhood are moving in a given way.

In considering groups of particles instead of individual particles to describe the state of our system, we encounter a certain arbitrariness that becomes apparent when we speak of the positions of the particles in the group. We clearly cannot speak of the particles in a group as occupying the same place at the same time. We must therefore consider a small region of space in our ensemble and ask how many particles at any given moment are in this small region. Clearly, the smallness of this region is arbitrary, but we can select a region so small that all points in it may be considered identical insofar as defining the state of the system is concerned. How small this volume must be cannot be precisely specified in classical physics. All that can be said is that it must be so small that any change of position of a particle within this restricted volume leaves the state of the ensemble unaltered.

We can now picture the entire volume of the ensemble broken up into these small volume elements and consider the number of particles in each. But the knowledge of the number of particles in each such element of volume is not enough to give us the microscopic state of the ensemble. We must also know in what manner the particles in each volume element are moving. To see what this means, we single out a given element of volume that, though small, contains many particles moving with various velocities. We now separate these particles into groups such that all the particles in a given group have the same velocities (the same speed in the same direction). Here again we must introduce an arbitrary feature because it is impossible to group particles with precisely the same velocities. We must consider a particular velocity and then assign all particles to the same group if they have velocities close enough to the chosen velocity. The arbitrariness arises because the expression "close enough" is not precisely defined. All that can be said about this concept in classical physics is that no appreciable change in the state of the system is said to occur if a particle changes its velocity by an amount that keeps it within the range of velocity defined as close enough.

We have now laid the basis for defining the microscopic state of an ensemble, at least classically. Consider all the particles in our ensemble—let us say N in number—and suppose that at a certain moment we distribute them so that a definite number—let us say N_1—lie in a given volume element and have velocities lying in the neighborhood of some given velocity, and another group—say N_2— lies in another element volume with velocities lying close to some other velocity, and so on, until we have exhausted all N molecules. Notice that in doing this we differentiate between two groups if either the volume element or the velocity value (or both) differs for the two groups.

A distribution of the N particles of our ensemble into the subgroups N_1, N_2, and so on, defines the microscopic state of the system. But this same microscopic state has a corresponding macroscopic state, that is, a description in terms of pressure, volume, and temperature. What alterations in the distribution of particles can change either or both of these pictures? We first note that there are many more ways of changing the microscopic than the macroscopic state. The macroscopic state is affected only if we change the number of particles within groups. For example, the macroscopic state is changed if the number of particles in N_1 is decreased or increased (which implies an increase or loss in one or more of the other groups of particles since the total number of gas molecules in the system is conserved). But the microscopic state is altered if we merely interchange particles between groups, neither adding to nor subtracting from the numbers within the groups. In other words, if a particle from one group is interchanged with a particle from another, the macroscopic state is unaffected; the microscopic state, on the other hand, does change, even though this change is not measurable in terms of the volume, pressure, and temperature of the gas.

A simple numerical example illustrates this idea. Suppose our ensemble consists of ten molecules of a gas distributed in five groups. A possible macroscopic state might be defined by the distribution of these particles into groups of 3, 1, 3, 1, and 2 particles. Another macroscopic state would be given by the groups 1, 3, 0, 4, 2, another by 1, 1, 1, 1, 6, and so on, until the various combinations of 10 particles arranged in five groups are exhausted.

Corresponding to the macroscopic state that consists of the groups 3, 1, 3, 1, and 2 of particles, there are many related states at the microscopic level because of the numerous ways of assigning the ten molecules to places within these groups. We may choose any of the ten molecules to occupy position one in the first group. After filling this first opening, we have nine remaining particles from which to assign position two. Similarly, after this choice is made, we are left with eight particles from which to assign the third and final position of this group. In assigning particles to this first group with its three places, we have had $10 \times 9 \times 8$ possible orders of selection altogether. Not all of these orders denote different microscopic states because we may discount the order in which we arrange particles within a group. Since interchanging particles within a group has no effect even on the microscopic state, we divide the product $10 \times 9 \times 8$ by the possible combinations of arranging the three particles among themselves within this group, in this case $3 \times 2 \times 1$ or

6. We thus arrive at the number of microscopic states arising from the assignment of particles to the first group: $(10 \times 9 \times 8)/(3 \times 2 \times 1)$. We now carry out the same calculation for the remaining groups 1, 3, 1 and 2, and get a similar product for each of them. In carrying out this procedure for each remaining group, we must, in each instance, substract from the total number of particles the number we have already arranged in groups since these are no longer available for forming others. In the example being considered, only seven particles are left to form the remaining groups because three have been used for the first group.

The total number of microscopic states corresponding to the first macroscopic state is the product of all these choices or $(10 \times 9 \times 8 \times 7 \times 6 \times 5 \times 4 \times 3 \times 2 \times 1)/$ $[(3 \times 2 \times 1)(1)(3 \times 2 \times 1)(1)(2 \times 1)]$. It is convenient to introduce a shorthand notation here. The product of successively decreasing numbers like $4 \times 3 \times 2 \times 1$ is written as 4! and, in general, $N(N-1)(N-2) \ldots$ is written as $N!$ Thus, the number of ways we can distribute the N particles of an ensemble so that there are N_1 particles in the first group, N_2 in the second, N_3 in the third, and so on, until all N particles have been exhausted is $N!/N_1!N_2!N_3! \ldots$. This is the number of microscopic states that corresponds to the macroscopic state given by the distribution N_1, N_2, $N_3 \ldots$. How does one use this combinatorial procedure in the analysis of the ensemble? In other words, how can this approach lead to the laws that govern the ensemble? We note, first, that the more microscopic states there are that give a particular macroscopic state, the greater is the chance for that macroscopic state to occur. The probability for finding our ensemble in a particular macroscopic state is proportional to the expression we wrote above for the number of microscopic states that correspond to the given macroscopic state. Second, we note that if the ensemble is left to itself, it will ultimately reach that particular macroscopic state with the maximum probability of occurrence. Put differently, the populations N_1, N_2, N_3 of the various groups will ultimately be such that the expression $N!/N_1! N_2! N_3!$ is a maximum.

We now see how to go about solving our problem. We examine all possible values of the group populations $N_1, N_2, N_3 \ldots$ and find that particular set for which the expression for the number of microscopic states is a maximum. This expression gives us the distribution law for the group populations, and from these populations the macroscopic properties of the ensemble can be derived. However, we must observe two conditions. First, in considering all possible group populations, we must make sure that the sum of the group populations is always equal to N, the total number of particles. Clearly, playing this combinatorial game, we must abide by the rule that the total number of particles is a constant; particles can neither be created nor destroyed. Second, we must impose the restriction that the total energy of our ensemble remains constant. In other words, if we add the energies of all the particles in all the groups, no matter how we vary the composition of the groups, the answer must always be the same. Any macroscopic state for which this statement is not true (any set of group populations for which this condition does not hold) must not be considered in seeking the maximum probability.

One can then apply well-known mathematical techniques for finding the maximum probability, or the most probable macroscopic state. This procedure leads to what is known as the Boltzmann statistics, or the Boltzmann distribution. When we apply this procedure to a gas, we obtain the Maxwell-Boltzmann distribution of velocities, discussed in our commentary on Maxwell, which, in turn, leads to the equation of state of a perfect gas—the law of Charles and Gay-Lussac.

Since the entire discussion above is based on classical physics, we must reexamine the concepts of the macroscopic and microscopic states of a system in the light of the quantum laws. These laws introduce two very important changes in our outlook. To begin with, the arbitrariness associated with choosing an element of volume and the values of the velocity near a given velocity is removed. This result follows from the uncertainty principle which says that the product of the positions and the momentum of a particle cannot be known with an accuracy greater than a value related to Planck's constant of action. This means that we may not make the volume elements of our ensemble, multiplied by the momenta of the particles, smaller than this value.

The second modification introduced by the quantum theory is related to the identity of the particles in our ensemble. One of the basic tenets of quantum mechanics is that we must not introduce into our theory any concept that cannot be verified experimentally. Consider now, in the light of this statement, a system of identical particles. Since there is no observational way of distinguishing one particle from another, we must not differentiate in our theory between two microscopic states differing simply by the interchange of two particles since there is no way, in practice, of telling the difference between such distributions. We now see that this fact alters the relationship between microscopic and macroscopic states introduced in our discussion of the Boltzmann statistics. We saw that for a given distribution of our N particles into groups N_1, N_2, and so on, which defines a macroscopic state, there are as many microscopic states as there are ways of interchanging particles among the groups. But this is not true in quantum statistics. All of these microscopic states are identical. Thus, there is just one microscopic state associated with the given macroscopic state. In fact, a microscopic state in the quantum theory coincides with the classical macroscopic state.

But if this is the case, how are we to calculate the most probable macroscopic state? It appears from what we have just stated that all macroscopic states are equally probable since just one microscopic state is associated with each of them. How are we to proceed? We must alter our definition of a macroscopic state in such a way that we obtain macroscopic states of different probability. The easiest way to do this is to arrange the particles into groups according to the energies of the particles. Suppose that N_1 is the number of particles in our ensemble, regardless of where in the ensemble they may be, with energies ranging from 0 to a certain small value. Suppose that N_2 is the number of particles with energies lying between this small value and twice this value, N_3 is the number with energies between twice and three times this value, and so on. We see that such a grouping, which we now define as a macroscopic state, consists of many of our previously defined macroscopic states and, hence, many quantum-mechanical microscopic states.

To calculate the most probable distribution now we must do one more thing. The way in which we have grouped the particles of our ensemble still does not describe the state of the ensemble completely because in quantum mechanics there are discrete energy levels. In assigning N_1 particles to the energy range from 0 to some small value, we must take into account the presence within this energy range of a certain number, let us say Z_1, of energy levels. Any two distributions of the N_1 particles among these Z_1 levels must be treated as distinct microscopic states in the quantum sense if they differ in the actual numbers of particles assigned to the different Z_1 levels. In other words, two distributions of the N_1 particles among the Z_1 energy levels are to be counted as one if they differ only by an interchange of particles, with the numbers remaining the same.

Thus, the difference between classical and quantum statistics lies in the definition of microscopic states: classically, these states are defined without taking into account the identity of the particles; the quantum definition takes identity into account. S. N. Bose was the first physicist to consider this difference and to set up what we now call the Bose-Einstein statistics. He used this tool to derive Planck's formula for black-body radiation without reference to any classical ideas or the introduction of oscillators in equilibrium with the radiation. He simply pictured an ensemble of photons of all frequencies in a state of equilibrium at a given temperature T and then determined the most probable distribution of these photons among all energy states—that is, the number of photons of each frequency—by using the combinatorial analysis outlined above, in which the exact identity of all photons is taken into account.

When Bose did this work in 1924 and communicated his paper to Einstein, who translated it into German and submitted it to the *Zeitschrift für Physik*, quantum mechanics had not yet been invented. The only justification for what Bose did lay in the correctness of his results. Even without quantum mechanics Einstein saw at once the importance of Bose's work. Indeed, as he remarked in the translator's note at the end of the Bose paper, he was, himself, at that very time, applying the Bose-type statistics to a gas of molecules. The importance of Einstein's work is that it showed clearly that the Bose-Einstein statistics is not limited to photons but can be applied to material particles (molecules) as well. It leads to a distribution different from the Maxwell-Boltzmann distribution and hence to a different equation of state for a perfect gas. Boyle's law does not hold for such a gas and the departure from Boyle's law becomes greater and greater as the temperature decreases.

Shortly after Bose did his work, Fermi presented another type of quantum statistics, which we discuss in some detail in our commentary on Fermi. However, it is worth noting here the distinction between Fermi's and Bose's statistics. Both of these statistics start from the assumption that, because of the identity of the particles, no distinction must be made between two group distributions of the particles that differ only in an interchange of two particles. But the statistics differ from one another in counting the number of microscopic states that must be assigned to a macroscopic state. The difference arises for the following reason. In the Bose statistics no restriction is placed on the number of particles that can be in a given group, that is, be in approximately the same position and have the same momentum (the same velocity in magnitude and direction). In other words, the numbers N_1, N_2, N_3, and so on, in the Bose statistics are not restricted. This is not the case in the Fermi statistics because Fermi assumed that his particles obeyed the Pauli exclusion principle. He therefore introduced the restriction that no two particles can occupy approximately the same positions and have the same momentum. This restriction leads to Fermi's statistics, later discovered independently by Dirac, and now known as the Fermi-Dirac statistics.

The differences between the ways microscopic states are counted in the Boltzmann, Bose, and Fermi statistics can best be illustrated by a simple example. Suppose we have two identical particles that must be distributed among three energy levels, represented as three boxes. Since, in the Boltzmann statistics, each particle is considered as a distinct individual, we label the two particles a and b. However, in the Bose and in the Fermi counting of states we label each particle *a* since they are not to be distinguished. Let us first distribute the particles among the three energy levels on the basis of the numerical possibilities for these levels. We would then have the following distributions: (2,0,0), (0,2,0), (0,1,1), (1,0,1), and (1,1,0). Each of these groupings corresponds to a macroscopic state according to the Boltzmann reckoning.

While there are nine microscopic states in the Boltzmann statistics, there are just six in the Bose and three in the Fermi statistics. Thus, there are two microscopic states in the Boltzmann statistics for the grouping (0,1,1) but only one for the Bose statistics and only one for the Fermi statistics. There is no microscopic state at all in the Fermi case for the first groups since they have two identical particles in the same energy box and that is forbidden by the Pauli exclusion principle.

To complete this discussion, we must give some criterion for distinguishing between particles that obey the Bose statistics and those that obey the Fermi statistics. From what we have said above, we see that this is the same as distinguishing between particles that are governed by the Pauli exclusion principle and those that are not. The general rule is that particles such as electrons, protons, neutrons, neutrinos, and so on, which have a half unit of spin, obey the Pauli principle and hence are governed by the Fermi-Dirac statistics. Particles that have zero spin, such as alpha particles, or an integral number of spin units, are not governed by the Pauli principle and hence obey the Bose statistics.

One other criterion can be used, based upon wave mechanics, the subject of Chapter 12. The state of an ensemble of particles is given by a particular wave function. Those particles for which the wave function is antisymmetrical (its sign changes when any two particles are interchanged) are governed by the Pauli principle. Those particles for which the wave function is symmetrical (with no change of sign when two particles are interchanged) obey the Bose statistics. Thus, the two different kinds of quantum statistics are a direct consequence of the wave nature of particles and the laws of quantum mechanics.

Although the great schools of physics, during the early decades of the 20th century, grew up around the dominant European physicists, a number of important discoveries were made by men and women who were outside the mainstream of events and, in a sense, were self-taught. This was true of the physicists of India, who, although to some extent under the influence of British physics, developed an excellent school of their own. This was primarily owing to the work of such men as Raman, M. N. Saha, who derived the equations of thermal ionization, and S. N. Bose, who developed one branch of quantum statistics.

Bose was born in 1894 and studied physics at Calcutta University where he received his M.Sc. degree just at the time when the Bohr theory of the atom was enjoying its greatest triumphs and European physicists were busily applying this theory to all phases of atomic dynamics. It was partly because of this feverish concern with the consequences of the Bohr theory, particularly in its relation to spectroscopy, that a physicist like Bose, rather than one of the Europeans, discovered the relationship between quantum theory and statistical mechanics.

Bose was well acquainted with European scientific literature and in 1924, arriving in Germany, fortunately came under the influence of Einstein, who was primarily concerned with the general concepts of physics rather than the specific consequences of the Bohr theory. Einstein was one of the few physicists who, early in the development of the quantum theory, insisted on the importance of applying quantum principles to all branches of physics, and not just to phenomena involving the emission and absorption of radiation. It is curious that Einstein, at the first Solvay Conference of Physics in 1911, stood alone in his insistence on the reality of the photon as a particle; at that time Planck himself, the founder of the quantum theory, was reluctant to accept this particular consequence. Einstein had already applied quantum ideas to the vibrations of atoms in crystals and had, thereby, cleared up the difficulty associated with the specific heats of solids; he thus had ample evidence of the power of quantum concepts.

When Bose came to Germany, Einstein was greatly interested in obtaining as general a derivation of the Planck black-body radiation formula as possible, and as free of classical ideas. He had already taken a big step in that direction in his 1917 paper in which he had used only the Bohr concepts of discrete states and the emission and absorption of photons to derive this formula. Since, in both Einstein's 1917 paper and in his work on specific heats, the approach was a statistical one, he undoubtedly was convinced that all of statistical mechanics should be reformulated in terms of quantum ideas.

In addition to the stimulation derived from his contacts with Einstein, three other developments influenced Bose. The first was Nernst's discovery of the third law of thermodynamics, which states that the entropy of a system goes to zero when the absolute temperature goes to zero. Nernst showed that this entropy principle is a direct consequence of the application of quantum theory to thermodynamics. The second was Stern's derivation of the expression for the entropy of a gas, again using quantum principles. The third was Planck's suggestion, in 1921, that the size of a cell in phase space must be placed equal to the cube of Planck's constant. (Phase space is a fictitious six-dimensional space constructed by combining the three dimensions of ordinary space and the three components of momentum; it plays an important role in statistical mechanics.) This was one of the decisive points in Bose's derivation of his quantum statistics.

It is interesting that Bose derived his quantum statistics just when De Broglie was discovering the wave properties of particles. Since one can show that the Bose quantum statistics can be derived as a consequence of the wave nature of particles, it may well be that Bose would have discovered electron waves had not De Broglie done so first.

From 1924 to 1927 Bose was reader in physics at Dacca University. He became professor and head of the physics department and remained at Dacca until 1946. He was then appointed Khaira professor of physics at Calcutta University and held this position until 1956. In addition to his academic posts, he was chairman of the National Institute of Science of India from 1948 to 1950. In 1958 he was elected a fellow of the Royal Society of London and was also appointed professor and vice-chancellor of Visva-Bharati University. He died on February 4, 1974 in Calcutta.

References

Einstein, Albert, *Die Naturwissenschaften*, 10 (1922), p. 184.

Kronig, R. and Weisskopf, V. F., *Pauli's Collected Works*. New York: Interscience, 1964.

12 *Wave Mechanics*

Beware that you do not lose the substance by grasping at the shadow.

—Aesop, *Fables*

The Principle of Least Action:
William Rowan Hamilton (1805–1865)

To understand how Louis de Broglie was led to the wave theory of matter, the subject of the next chapter, we must begin by considering the behavior of light from two different points of view. Suppose that a beam of light is moving between two points in a homogeneous medium in which the index of refraction and, therefore, the speed of light are constant. The light travels in a straight line between the two points. This fact, borne out by experiment, was the starting point of an important principle developed by the famous 17th-century French mathematician Pierre de Fermat; it describes the propagation of a beam of light when the index of refraction of the medium is not constant. If the light is traveling in an inhomogeneous medium, its speed changes from point to point because of the variation of the index of refraction. It does not travel in a straight line.

Fermat described the motion of the light correctly by introducing the principle that light always travels between any two points in the shortest possible time. In other words, among all the possible paths in the medium connecting the two points, the light chooses that particular path along which it has to spend the least time. This is called Fermat's principle of least or minimum time. It leads to what we now call geometrical optics since it treats light as though it were composed of geometrical rays that travel along geometrical lines. The path along which the light travels may be geometrically longer than another path connecting the two points, but the time along the chosen path is always the shortest.

Fermat's principle of least time accounts for the gross behavior of light, including such phenomena as reflection and refraction (the bending of a ray of light when it passes from a rarer medium such as air to a denser medium such as glass). This principle can be used to explain the formation of images by lenses and mirrors, but it cannot explain the refined details of the behavior of light, such as interference and diffraction (the bending of light around corners). To explain these properties we must use physical optics, which takes into account the wave structure of light.

As long as we are dealing with phenomena in which the wavelength of light is not important, geometrical optics (Fermat's principle of least time) gives an

William Rowan Hamilton (1805–1865)

adequate description of optical events. But as soon as the wavelength of the light becomes an important factor, geometrical optics gives wrong results and physical optics must be used. This is the case when light rays pass through tiny openings or around very small bodies. Since such openings and bodies are comparable in size to the wavelength of the light, its wave properties play an essential role in the phenomena. We may state this differently by saying that the laws of geometrical optics approach those of physical optics when the wavelength of the light approaches zero or when it is very small. In general, therefore, geometrical optics gives a much better description of ordinary light than it does of radio waves.

There is another important distinction between the two points of view that has a bearing on the corpuscular theory of light and on the trajectories of particles. Since Newtonian corpuscles must be described in terms of well-defined trajectories, and since well-defined ray trajectories are the hallmark of geometrical optics, it follows that geometrical optics is really a corpuscular theory of light, and physical optics is a wave theory. As the wavelength of light gets smaller and smaller, the corpuscular theory becomes more applicable. But the wave theory becomes more important as the wavelength of light gets larger. This fact suggests an important analogy between the behavior of light of very short wavelength and the behavior of ordinary particles of matter, such as atoms, molecules, and electrons. We know, according to Newton's laws of motion, that such particles travel along geometrical trajectories. The analogy between geometrical optics and the trajectories of particles of matter is further strengthened when we consider that the classical laws of mechanics can also be formulated in terms of a minimum principle (referred to as a stationary or variational principle by physicists) similar, in its mathematical form, to Fermat's principle.

The great Irish mathematician and astronomer Sir William Rowan Hamilton was the first to recognize the importance of such a formulation of the laws of mechanics; he developed the necessary mathematical scheme for handling problems in Newtonian mechanics in a manner similar to that of geometrical optics. He first redeveloped the laws of geometrical optics, starting with Fermat's principle. He showed that the path of light rays can be computed from a single mathematical quantity, the characteristic function, whose properties are quite similar to those of the action function that had been used to study the dynamics of a particle. In this formulation of geometrical optics, with its emphasis on the trajectories of rays, one sees the parallel to mechanics.

This similarity between the ray trajectories of geometrical optics and the Newtonian paths of particles suggested to Hamilton the possibility of expressing the laws of mechanics in a mathematical form closely resembling that of geometrical optics. He was further encouraged in this project by his knowledge that the Newtonian laws of motion of a particle can be derived from a minimum principle similar to Fermat's.

As Hamilton saw it, one merely had to start from this minimum principle in mechanics and transform it in such a way that it resembled the Fermat principle as closely as possible; the analogy between mechanics and optics would then be complete, at least as far as the formal aspects of these two branches of physics are concerned. Hamilton therefore started from the principle of least action that was discovered and introduced into mechanics by Maupertuis. This principle states that a particle, or any system of particles in a force field, for example a gravitational field, moves between any two points so that the total action of the system taken along its path between the two points is a minimum.

To understand this relation more fully, consider the momentum of a particle—its velocity multiplied by its mass. As the particle moves along its path under the action of the force, its momentum changes continuously. Now consider the particle for a very short time interval during which its momentum does not change appreciably. We can always make the time interval short enough for this to be true. During this interval the particle describes a small section of its path. If we multiply the length of this element of path by the momentum, we obtain what Maupertuis called the element of action. We can now formulate Maupertuis's principle of least action by stating that of all possible paths a particle can take in going from point A to point B, the path that it actually takes is the one for which the sum of all the elements of action along the path is a minimum.

This statement does not describe the behavior of a mechanical system at all times, although Maupertuis was so convinced of its universal validity that he considered it to manifest divine efficiency in the guidance of the world: "Here then is this principle, so wise, so worthy of the Supreme Being: Whenever any change takes place in Nature, the amount of action expended in this change is always the smallest possible."

Hamilton started from Maupertuis's principle in his great work on dynamics and generalized it into what is called the principle of stationary action: The path chosen by a particle or a system of particles in going from one point to another is such that the action, measured along any path that differs only infinitesimally from the actual path, does not differ from the action measured along the actual path. This means that the action is stationary in terms of its near neighbors: either a maximum or a minimum, but not necessarily the minimum imposed by Maupertuis.

In his paper entitled "On a General Method of Expressing the Paths of Light, and of the Planets, by the Coefficients of a Characteristic Function," Hamilton offered the following commentary.

> In mathematical language, the integral called action, instead of being always a minimum, is often a maximum; and often it is neither one nor the other; though it has always

a certain stationary property.... We cannot therefore suppose the economy of this quantity to have been designed in the divine idea of the universe; though a simplicity of some kind may be believed to be included in that idea. And though we may retain the name of action to denote the stationary integral to which it has become appropriate— which we may do without adopting either the metaphysical or (in optics) the physical opinions that first suggested the name—yet we ought not (I think) to retain the epithet *least*: but rather to adopt the alteration proposed above, and to speak, in mechanics and in optics, of the Law of Stationary Action.

Hamilton did not introduce the action directly but used a related quantity called the Lagrangian, or the kinetic potential, of the system. This quantity is obtained by subtracting the potential energy from the kinetic energy of a system, such as that of a particle, at each point along its path. If we observe a particle for a very short time interval and multiply this interval by the momentary value of the particle's kinetic energy minus its potential energy, we obtain the element of Hamiltonian action, which is related to, but not always the same as, the element of action defined by Maupertuis. The Hamiltonian action can be obtained from the Maupertuis action by subtracting from the latter the total energy (the sum of the kinetic and potential energies) multiplied by the small time interval during which we are observing the system. We may restate the Hamiltonian principle of action as follows: If a particle is allowed to move from point A to point B in a field of force, and if the mechanical energy of the particle is always conserved, or remains constant, then the path actually followed by the particle is the one for which its total action is stationary. Consider a large number of possible paths connecting points A and B, and compute the total Hamiltonian action along each one of these paths. If these paths do not differ very much from one another, all the Hamiltonian actions lie close to the same value. But there is one path among the many for which there is practically no change in action as we go to a very nearby path on either side of it. This path is called the stationary path.

By using this principle of least or stationary action, Hamilton was able to express the Newtonian dynamical equations in a form appropriate for describing the motions of a particle or a collection of individual particles as well as the motion of rigid bodies. He achieved this with the following reasoning. To describe in complete detail the motion of a system of particles rigidly connected to each other, it is not necessary to consider the position and the motion of each particle in the system separately. Since they are all connected, it is evident that the complete description of the system can be given by introducing six quantities (six generalized coordinates) to give the position of the system and the six velocities associated with the generalized coordinates. This procedure of working with generalized coordinates was first introduced by Lagrange. Hamilton took over the Lagrangian idea of generalized coordinates and, in addition, introduced generalized momenta corresponding to them. These momenta can be obtained from the Lagrangian, that is, the kinetic potential, by introducing the generalized coordinates and the generalized velocities, then differentiating the Lagrangian with respect to each of the generalized velocities in turn.

By using generalized coordinates and generalized momenta Hamilton was able to transform the Newtonian equations of motion for any complex system into a set of simple and symmetrical equations—one set for each pair of generalized coordinates and momenta. A still greater simplification was achieved by Jacobi, who was able to show that one could replace the system of Hamilton's equations by a single partial differential equation for a single mathematical quantity, the total action of the system. This function enables one to determine completely the trajectory of the particle and thus to solve the dynamical problem. The procedure is analogous to the treatment of geometrical optics by Hamilton. He derived the paths of rays by solving an equation, similar to the one introduced by Jacobi, for a single optical quantity called the characteristic function. According to this point of view the dynamical problem is solved by considering the motion of a system as though it were a gradual unfolding of a series of states, each one derived from the preceding one by an infinitesimal transformation similar to that of a ray of light advancing from one wave front to the next.

By considering the similarity between Fermat's principle for geometrical optics and the Maupertuis principle of least action for dynamical systems, Hamilton hypothesized that the behavior of particles might be described by a kind of wave mechanics. He showed that the trajectories of particles having the same total amount of energy are identical with the paths of rays of light in a medium having the proper index of refraction. In other words, it is possible to find an index of refraction such that the trajectory of any particle can be described by the path of some ray of light in a medium having the given index. Since, however, rays are only an approximation (which becomes more accurate as the wavelength gets smaller) to the correct wave description of light, it follows that the Newtonian trajectories are only an approximation to a wave description of the motion of particles. Just as in optics the rays of light are orthogonal to the wave fronts (which are surfaces of equal phase), so in mechanics the trajectories of particles are orthogonal to another kind of wave front (the surfaces of equal action). In other words, in particle wave mechanics the action plays the role of the phase in optics. This Hamiltonian formalism is just what was needed to go from the classical mechanics of a particle to the quantum and wave mechanics; it was taken over bodily by Erwin Schrödinger.

Hamilton's great contribution to classical dynamics enables us to describe the path of a particle as though it were a ray of light moving through an optical medium with an index of refraction related in a definite way to the force field through which the particle is moving. We may surmise from this (which Hamilton did not but Schrödinger did) that just as ray optics, that is, geometrical optics, gives only an approximate description of light, so classical dynamics gives only an approximate description of the motion of a particle. And just as ray optics has its wave-optical supplement, so particle dynamics has its wave-dynamical aspect.

It is not often that a great mathematician leaves his abstract world to contribute to another field. Mathematics is so vast, and one problem leads so naturally to another, that the mathematician is often unaware of the mathematical needs of other disciplines. But this does happen every now and then, and the results are quite remarkable. It seems that only the very greatest minds can create important mathematics at the same time that they master enough of another field, like physics, to do creative work there also. In this select group we have men like Joseph Lagrange, Pierre Laplace, Daniel Bernoulli, Carl Friedrich Gauss; more recently, David Hilbert, John von Neumann, and Hermann Weyl. Near the very top of the first group is William Rowan Hamilton, who reshaped Newtonian mechanics into a mathematical form from which its extension to quantum mechanics is a very simple and natural step. This was just one of the many domains of physics to which Hamilton contributed. One volume of his collected works, consisting of 460 pages, deals only with his work in geometrical optics. We have here, indeed, one of the mathematical giants of the world of physics.

Hamilton was born in August 1805 in Dublin, Ireland, and displayed his precocity and unusual powers at a very early age. His father therefore decided to let his brother, the Reverend James Hamilton, a member of the Royal Irish Academy, direct young William's education. From the age of three until Hamilton entered Trinity College in 1823, his uncle carefully nurtured the youngster and introduced him to subjects that were calculated to stimulate his intellectual powers to their fullest capacity.

At the age of 13, five years before entering college, Hamilton had already begun to think very deeply about optics, and at the age of 17 he began to write on this subject. While carrying on his first-year college studies, at the age of 19, he completed his first paper, "On Caustics" (a branch of geometrical optics) which he submitted for publication to the Royal Irish Academy.

The committee to whom the paper was referred did not quite know what to do with it, for it was in such an abstract mathematical form that they did not really understand it. They therefore asked that Hamilton explain his presentation more fully. While still an undergraduate, he completed the project and presented it to the Academy in 1827 as a paper entitled "Theory of Systems of Rays." It is one of the great classics in theoretical physics and has remained the basis of most treatises on geometrical optics. Moreover, it contains the germ of the idea which was to lead Hamilton to his famous formulation of dynamics. In this paper Hamilton introduced the characteristic function of a system of rays from which all the properties of the system can be deduced by simple mathematical operations such as differentiation. The function is directly related to the action integral, which plays the dominant role in Hamiltonian dynamics and which was the starting point of Schrödinger's derivation of his wave equation.

The importance of the characteristic function for optics is that it can be used as a starting point to develop either a corpuscular theory of propagation or a wave theory. In the corpuscular theory the characteristic function is related to the principle of least action. In the wave theory the characteristic function is related to Huygens's wave front.

Hamilton surmised that the concept of the characteristic function would also be of great importance in dynamics, since dynamics can be derived from a principle of least action. He indicated this extension in his essay on the relationship between the path of light and the path of a planet. With the publication of his paper on the theory of rays Hamilton's genius was universally recognized; his reputation was further enhanced when the phenomenon of conical refraction, which he had discovered theoretically, was observed experimentally.

When Dr. Brinkley was promoted to the bishopric of Cloyne in 1827 the professorship of astronomy at Trinity College became vacant and Hamilton, then only 21 and still an undergraduate, was offered the post—perhaps the most remarkable example of foresight in the history of university administration. Hamilton accepted the chair and devoted himself not only to his researches into mathematics, mathematical physics, and astronomy, but also to the practical problems of teaching and supervising the observatory.

In 1834 Hamilton was awarded the Cunningham medal of the Royal Irish Academy and the Royal medal of the Royal Society for his discovery of conical refraction. This was considered one of the wonders of mathematical physics and a most remarkable prediction, attesting to the great usefulness of mathematical analysis.

On the death of the president of the Irish Academy in 1837 Hamilton was elected to that position, which he held for eight years; this was one of his most productive periods during which he developed his dynamical theories in a famous paper "General Methods in Dynamics." In this paper he applied the concept of the characteristic function to the motion of a system of particles in a force field and developed the equations of motion in the form now universally known as Hamilton's equations. These equations are based on the concept of the Hamiltonian function, so useful today in quantum mechanics.

The basic idea of the Hamiltonian function is as follows. Suppose we have a system, say a particle, moving in a field of force. We may express the total energy of this system (kinetic plus potential) in terms of the components of the momentum and the coordinates. This expression is called the Hamiltonian of the system. The equation describing the motion of the system can be obtained from this quantity by certain mathematical operations. The advantage of such a formulation of dynamics is that it can be applied to very complex systems by a judicious choice of coordinates.

While developing his theory of dynamics Hamilton also did extensive work in many branches of mathematics. He published papers on algebra, theory of functions, calculus of probability, theory of equations, and so on. During the last 22 years of his life he invented and extensively developed the calculus of quaternions, an extension of vector analysis.

In spite of his great skill in abstract analysis and profound intuition, Hamilton never shied away from extended and tedious calculations that he felt were necessary to prove a point. Often he would reach a result by a series of involved computations and then, having established the truth of a theorem for a special case, generalize it in its most elegant mathematical form. As the Very Reverend Charles Graves, president of the Irish Academy, stated in his eulogy on Hamilton's death: "He engaged in exercises of this kind [extensive numerical calculations] sometimes from a wish to strengthen his intellectual hold in general propositions by scrutinizing results obtained by applying them in a number of particular cases; and sometimes from a wish to mature and keep in exercise those powers of calculation upon the exactitude and prompt operation of which so much depends in difficult mathematical investigations."

In spite of his deep involvement in the most abstract problems, Hamilton was very much aware of the people about him and was as much concerned with the least learned as with the most. He was a kind and patient man who could spend two hours convincing some "half-crazed squarer of the circle that his proposed construction was inaccurate" or answering with gentle good nature the most elementary questions of students and visitors. Hamilton was also a deeply religious man and devoted much time to establishing a metaphysical basis for his discoveries. For example, his paper "Algebra Considered as a Science of Pure Time" was a metaphysical attempt to derive algebra from the elements of space and time and establish it as a science rather than as a branch of mathematics. He undoubtedly was attracted to the principle of least action by his predisposition to metaphysics, which he studied extensively. He considered himself a disciple of Bishop Berkeley's and studied Berkeley's writings avidly, although he was also devoted to Immanuel Kant, whose works he had mastered when he learned German. Finally, he was greatly influenced by René Descartes as well as Francis Bacon and Isaac Newton. He indicated his intellectual and spiritual preferences in assigning religion to the highest place in human activities, followed by metaphysics. Next in order was mathematics and then poetry, physics, and, finally, general literature.

Early in his life Hamilton began to master various languages and was soon reading the classics in Greek and Latin. He was generally familiar with the literature of the world and was on friendly terms with William Wordsworth and Samuel Coleridge, as well as others. He might, indeed, have become a great poet had science not captured him; on a few occasions he won prizes for English poetry. To him mathematics and poetry were closely related for they both expressed the essence of beauty in different forms.

Hamilton was very much a humanist. Like John Donne he felt that "no man is an island entire in itself," but that all humanity is interrelated and every person is diminished when any single individual is diminished. At the time of his death on September 2, 1865, he was Andrews professor of astronomy and Astronomer Royal of Ireland and had completed 60 volumes of carefully written manuscripts, all chronologically arranged and carefully dated, showing the day-to-day progress of his work. In addition to the unpublished manuscripts there are more than 200 of his notebooks deposited in the Library of Trinity College, Dublin.

Hamilton was undoubtedly the most outstanding man of science Ireland has ever produced. His equations brought the science of dynamics to the peak of its development.

The Wavelengths of Particles: Prince Louis de Broglie (1892–1987)

Although the Bohr model of the atom proved successful in accounting for the spectral series in hydrogen and helium, it was clear from its inception that it was something of a hybrid and not based upon logically sound axioms. Bohr himself in his fundamental paper had already indicated that the theory was a mixture of classical Newtonian dynamics and nonclassical electrodynamics. The motion of the electron in its equilibrium orbit, that is, in a stationary state, was described by Newtonian dynamics, but the transition of the electron from one stationary state to the other was described by a new electrodynamics, which allowed the electron to emit and absorb radiation in bundles (quanta) rather than continuously.

As the Bohr model was applied to more and more atomic processes it became increasingly evident that it suffered from basic inconsistencies and gave wrong answers, or none at all. It was unable to account for the differential intensities of the spectral lines of atoms or for the absence of certain lines that should have been present on the basis of the model.

By the beginning of the 1920s, so many difficulties faced the Bohr theory that every physicist was aware that changes of a fundamental nature in all physical concepts would have to be made to produce a consistent theory. But where was this new theory to come from? The solution to the problem came from three separate investigations, each beginning from a different point but ending with essentially the same result.

Prince Louis de Broglie was the first to see where the difficulty lay in the Bohr theory and what had to be done to establish the foundations of a correct theory. de Broglie began his study of the problem in 1920; within a three-year period he had

Prince Louis de Broglie (1892–1987)

formulated the basis of our contemporary understanding of the behavior of matter. De Broglie first considered the fundamental duality of nature (matter and energy), assuming that these two aspects must have some common meeting ground.

Until 1900, physicists believed that matter and energy were two distinct realms in nature governed by different laws. Pure energy, manifesting itself as radiation, was thought of in terms of waves. Matter was considered to be corpuscular in nature. The first important departure from this concept occurred in 1900 when Max Planck showed that radiation cannot be described entirely in terms of waves; it also possesses corpuscular properties. The final destruction of the classical notion of energy as waves and matter as corpuscles was accomplished by de Broglie, whose chain of reasoning started from Einstein's famous mass-energy relationship.

De Broglie's reasoning was based upon the concept of symmetry in nature and the special theory of relativity. He was prompted in his discovery of the wave nature of particles by what at first appeared to be a serious contradiction in the quantum picture of radiation; namely, that one must speak of a radiant corpuscle—a photon—as having a frequency and a wavelength, as did Planck when he introduced the quantum of energy. Since the assumption of these dual properties—wave and corpuscle—was necessary for the correct description of radiation, why, then, on the basis of symmetry, should not material particles also have wave properties?

Once de Broglie had convinced himself that there was nothing wrong, however strange, in assigning a frequency and a wavelength to particles—even material ones like electrons and protons—he hit upon the fundamental equations of relativity for determining the relationship between the mechanical and wave properties of all particles. These equations already contained a relationship between radiant energy and matter, and therefore between matter and frequency, a property of radiation.

In his procedure, de Broglie introduced concepts that require some prior explanation. His idea of the proper time of a particle is derived from the special theory of relativity. In the theory of relativity one has to distinguish between time measured by a clock moving with a particle and time measured by a clock relative to which the particle is moving. The proper time of the particle is measured by the clock that is moving along with the particle. It is related to the time given by the clock relative to which the particle is moving by the simple transformation formula that de Broglie gave in his Nobel lecture. This is the factor that gives the amount by which a moving clock slows down.

De Broglie also found it necessary to introduce the concepts of group velocity and wave or phase velocity. Group velocity must be related to the speed of the particle; de Broglie succeeded in accomplishing this and thereby established the wave properties of electrons. His approach owed much to the investigations of Lord Rayleigh, who demonstrated that the speed of a group of waves—each of which varies slightly in wavelength and frequency from its neighbors—is not just equal to the speeds of the component, individual waves. If we had just a single wave in the group—a wave extending infinitely far in both directions—the group and phase velocities would be identical and would have the value obtained by multiplying the wavelength by the frequency of the wave. But if many waves are present, each with its own phase velocity, the overall velocity, or group velocity, results from the variations among the component waves. One can actually see this with water waves by dropping a pebble into the water. The main wave is seen to advance with a definite speed while tiny wavelets move up to the main wave front and then disappear.

In his analysis of the electron, the concept of group velocity was very helpful to de Broglie. He realized that if he assigned a wavelength and a frequency to an electron, using the equations of relativity theory, the number obtained by multiplying these two quantities would not equal the velocity of the particle. Indeed, this number is always larger than the speed of light, since it is the phase velocity and shows that the particle is really compounded of a group of waves.

In a sense the work of de Broglie merged the special relativity and quantum theories by relating Planck's constant to the momentum-energy concept of the relativity theory. That a relationship between waves and particles should exist on the basis of the theory of relativity is clear from a consideration of the invariant properties of waves and particles. One of the most important concepts derived from the theory of relativity is that of invariance: physical systems are characterized by certain quantities that appear as constants to observers moving with uniform speed with respect to one another. For example, consider a train of waves of radiation moving in a certain direction and suppose that we have only a single frequency and wavelength associated with this train. What is the invariant property of this train? What physical quantity associated with this wave train would all observers moving with uniform speed with respect to one another find the same?

The constant quantity is, of course, the number of waves that the observers would tally, starting from a given wave and counting for the same period of time. This can be shown to be equal to what is called the phase of the wave so that the phase is invariant when one goes from one coordinate system to another. The phase of a wave is obtained as follows. Divide the distance a wave has moved in a given time interval by its wavelength. Multiply the frequency of the wave by this time interval and subtract this quantity from the previous one. We may picture this invariant of a wave somewhat differently. Since the distance a wave moves and the time during which it is moving may be defined as the space-time interval of the wave, we can say that the invariant of the wave is obtained by properly combining (usually multiplying) the space-time interval of the wave and its wavelength-frequency interval.

Let us now look at the invariant properties of a moving particle such as an electron. A particle has momentum and energy; if one properly combines the space-time interval of the particle with its momentum-energy one also gets an invariant property of the particle. It thus appears from considerations of symmetry that the momentum-energy of a particle plays a role in its motion similar to the role played by the wavelength-frequency in the motion of a wave. It is, therefore, quite natural to seek a correspondence between momentum-energy and wavelength-frequency. Planck had already established the fact that frequency is related to energy for waves, and de Broglie completed the work by showing that wavelength is related to momentum. This means that wherever energy is present there is a frequency and wherever momentum is present there is a wavelength, and vice versa. It follows that waves must have momentum and, therefore, particle properties and particles must have wavelengths and, therefore, wave characteristics.

De Broglie considered one more important point—the relationship between his wave theory of the electron and the discrete orbits introduced by Bohr. Why does an electron move in special orbits inside an atom? He answered this question simply by writing down the expression for the wavelength of a particle which is Planck's constant divided by the particle's momentum. In other words, the faster a particle moves, the shorter is the wavelength that must be assigned to it. If a particle moves in a closed orbit (always retracing the same path like the electron inside a hydrogen atom) the orbit can be closed only if the waves in the wave packet associated with the circling electron do not destroy each other as they move around. Such waves move with the phase velocity described above, whereas the electron moves with the group velocity; thus the waves move faster than the electron does and catch up with the electron time and time again.

Consider a wave that has left the electron, gone around the orbit, and caught up with the electron again. At that moment another wave will be leaving the electron. If the motion of the electron is stable, the crest of the wave that has caught up with the electron must just match the crest of the wave that is leaving; otherwise these two waves would cancel and the motion would be disturbed. Therefore, the circumference of the path of the electron must be equal in length to an integral number of wavelengths. But this turns out to be equivalent, as de Broglie showed, to the condition that Bohr imposed on the motion of the electron.

In developing his theory, de Broglie made another significant contribution that clarified a puzzling point concerning the quantum picture. Classical physics is based on the concept of continuity, fundamental for an understanding of the properties of wave motion. Quantum theory departs radically from classical physics in introducing discrete discontinuous processes. However, it still retains the ideas of frequency and wavelength that were derived from the continuous wave picture. How is it possible to describe something as discrete and as continuous at the same time?

De Broglie pointed out that the notion of discreteness is not foreign even to the classical wave theory. When two waves meet, an interference pattern is formed that can be described only by a discrete set of numbers, since one sees a discrete set of interference fringes (light and dark spots). In other words, whenever one is dealing with collections of waves, a discrete pattern becomes evident. The same thing happens when a membrane vibrates. If one first scatters particles of sand on a membrane and then sets the membrane vibrating, the sand rearranges itself in definite patterns. These patterns may be compared to the stationary states of the electron in the Bohr theory of the atom.

Compare the de Broglie picture with the wave mechanics developed by Hamilton. Hamilton had shown that a formal similarity exists between the laws of classical mechanics and the laws of geometrical optics; both sets of laws can be obtained from minimum, or stationary, principles. In a sense we might say that Hamilton had foreshadowed de Broglie. But this is not really the case because Hamilton's work never went beyond pure formal analogy. Although the Hamiltonian formulation is of great importance in modern wave mechanics, at no point in his work did he consider the possibility that particles might have wave characteristics associated with them. de Broglie used the formal relationship between geometrical optics and Newtonian mechanics to establish a relationship between wave optics and a new kind of mechanics, that of waves. Since geometrical optics is wholly inadequate to describe such optical phenomena as interference and diffraction— phenomena in which the wavelength of the light becomes apparent—so, too, the Newtonian mechanics is wholly inadequate to describe the behavior of particles in atoms where phenomena involving discrete processes and whole numbers occur (in this respect similar to diffraction and interference in optics). As de Broglie said, "This suggests the idea that the older Mechanics too may be no more than an approximation as compared with a more comprehensive mechanics of an undulatory character."

It should be noted that the waves associated with particles in de Broglie's theory are to be considered real, not merely as a fictitious, useful device for describing an electron. As he pointed out, the dualism of waves and particles is real and the wavelength associated with a particle is as concrete a physical quantity as the mass of the particle. It is upon the dualism in nature between waves and corpuscles, expressed in more or less abstract form, that the entire recent development of theoretical physics has been built and where its immediate future development is likely to be erected.

In 1911, when Louis de Broglie was still a young student, his brother Maurice, a well-known physicist, attended the first of the famous Solvay conferences on physics. This must have been a very exciting conference, indeed, for the quantum theory was just beginning to make itself felt, primarily because of Einstein's remarkable success in applying it to various phenomena; Einstein's relativity theory had, to some extent, been accepted. H. A. Lorentz, recognized by all as the leading physicist, presided, and Planck, Einstein, Nernst, and other top scientists were present. Most of the discussion was devoted to the photon; Einstein insisted that it be recognized as a real physical particle, but the majority accepted it only as an artificial device for clarifying certain radiation phenomena. Maurice de Broglie carried a description of these events back to Louis, who was deeply stirred by the account and decided that he, too, was going to participate in this exciting search for the laws of nature.

Except for the fact that his brother was a physicist, no one would have thought that Louis would become a physicist and win the Nobel Prize for his remarkable synthesis of waves and particles into a wave-particle dualism. De Broglie was born a prince, the son of Victor, Duc de Broglie, and Paulini d'Armaillé, and very few of the royalty adopted science as a career. Indeed, Louis was almost lost to science because he first devoted himself to literary studies and took a degree in history in 1910, a year after he had graduated from the Lycée Janson of Sailly. Born on August 15, 1892, at Dieppe, de Broglie lived in an ancestral manor and, raised with an emphasis on tradition and duty to family, must have experienced little contact with science. Nevertheless, after completing his history degree, he went over completely to science and took his degree in physics in 1913.

After World War I erupted, de Broglie was conscripted for military service and spent 1914 to 1918 with the wireless section of the army at the Eiffel Tower. Although the wireless technical problems were of some interest, he could not keep his mind off the problem of the photon. Immediately after the war he threw himself into this phase of physics with great vigor and passion. Experimental physics interested him to some extent, but theoretical physics dominated his life.

In 1920 de Broglie became interested in the apparent contradictions that seemed to stem from the quantum theory of radiation. It appeared that if the corpuscular theory (Einstein's photons) were accepted, one could not explain interference and diffraction, which are essentially wave phenomena. On the other hand, if the wave picture were accepted, there would be no way to account for black-body radiation. After thinking about this problem for many days, de Broglie finally saw the solution. There was no need to accept either the wave theory or the photon theory exclusively because both theories were correct and light consists of both waves and corpuscles. Indeed, Planck's basic concept, that a photon (or quantum) has an amount of energy proportional to its frequency, shows that particles and waves go together—one cannot have one without the other. (This involves the famous $E = h\nu$ relation, where E is energy, h is Planck's constant and ν is frequency.) In his Nobel address, de Broglie stated, "I thus arrived at the following overall concept which guided my studies: For both matter and radiations, light in particular, it is necessary to introduce the corpuscle concept and the wave concept at the same time. In other words, the existence of corpuscles accompanied by waves has to be assumed in all cases." Here one sees de Broglie's generalization: If corpuscles (photons) are associated with waves, then waves must be associated with corpuscles. From this assumption the derivation of de Broglie's equation relating the wavelength of a particle to its momentum and Planck's constant was a matter of straightforward algebra applied to the relativistic kinematics of a particle. Thus began wave mechanics, which has so drastically altered our concepts of matter.

De Broglie presented his theory in a thesis entitled "Recherches sur la Théorie des Quanta" to the faculty of sciences at Paris University in 1924. It aroused considerable astonishment at first because of the strangeness of its ideas and was not fully accepted until the experiments of Davisson and Germer proved it to be correct.

Soon after presenting his thesis, de Broglie began teaching, first as a free lecturer at the Sorbonne and then as professor of theoretical physics at the new Henri Poincaré Institute. In 1932 he occupied the science faculty's chair of theoretical physics at the University of Paris. De Broglie spent much of the remainder of his career trying to obtain a causal interpretation of the wave mechanics to replace the probabilistic theories of Born, Bohr and Heisenberg. However, he was never very successful in this endeavor. The uncertainty principle and wave-particle dualism still dominate modern physics. He received the Nobel Prize for his work in 1929.

A Wave Equation for Particles: Erwin Schrödinger (1887–1961)

Almost at the very moment that Heisenberg and his collaborators were developing the matrix phase of quantum mechanics and, with its aid, deriving the frequencies of the lines of the normal hydrogen spectrum, an entirely different approach to the same problem was being taken by Erwin Schrödinger, a Viennese physicist at the University of Zurich.

Schrödinger's starting point was de Broglie's discovery of the wave nature of particles. He believed that it should be possible to describe this wave pattern by a wave equation, just as it is possible to describe the propagation of light by a wave equation. This conviction was greatly strengthened by his knowledge that under certain conditions the motion of a system described by a wave equation (for example, vibrating strings or a vibrating metal plate) can be represented by integers. Here, then, was the kernel of an exciting idea and, possibly, the source of a new theory of the atom. Waves are associated with particles and described by wave equations, the solutions of which involve integers. Thus, it should be possible to introduce quantum numbers quite naturally into the description of the motion of an electron inside an atom by using a wave equation. With this approach, the solution of atomic problems would become fairly simple, as opposed to the complex processes of matrix mechanics.

To obtain his wave equation, Schrödinger went back to the work of Hamilton, aware that this great mathematician had established an analogy between the Newtonian mechanics of a particle and the geometrical optics of a ray. Hamilton had shown that we can duplicate the path of a particle in a given force field (for example, a freely falling body in a gravitational field) by the path of a ray of light in a given medium—provided we choose the correct index of refraction of the medium.

Erwin Schrödinger (1887–1961)

We remember that the equations of wave optics become identical to those of geometrical optics if the wavelength of the former is placed equal to zero. Schrödinger suggested that, since classical geometrical optics is the limiting case of wave optics when the wavelength is infinitesimally small, classical Newtonian mechanics is nothing more than the limiting case of a more general wave mechanics, which invests particles with wavelengths and wave properties. If this were so, one should be able to obtain a wave equation that would describe the motion of a particle and obtain it by a method similar to the one used in going from geometrical optics to wave optics. This was Schrödinger's plan.

In 1928 Schrödinger gave a series of four lectures shortly after he had discovered his wave equation for the electron and used it to obtain the correct energy levels of an electron inside an atom (the frequencies of the spectral lines) and to explain the motion of the electron in these levels. He began his first lecture by restating the Hamiltonian theory in the form of a variational principle involving the kinetic energy, or the difference between the total and potential energies, of a particle moving in a force field. He formulated an equation to express this idea.

To understand this equation, consider a body such as a planet falling freely in a gravitational field. If this planet is to move from point A to point B, it is clear that, with a given velocity (a given speed and direction of motion at point A), it can reach point B along only one possible path. (This path is, of course, a segment of the actual orbit of the planet.) What distinguishes this path from all other lines that can be drawn between A and B? Hamilton's principle, as contained in Schrödinger's equation, tells us the following: If we divide the actual path of the particle into infinitesimal segments, we note that in each of these segments the particle has a kinetic, a potential, and a total energy—the last being the sum of the first two. If for any segment we represent the kinetic energy as the difference between the total energy E of the particle and its potential energy V, multiply this by twice the mass of the particle, take the square root of the result, and multiply this quantity by the length of the segment of the path, we obtain an element of Maupertuis action. If we now add all of these elements of action along the true path of the particle from point A to point B by integrating along the path, the sum we obtain is smaller than for any nearby—but, of course, not permissible—path connecting these two points; or, more generally, it differs only slightly from that of any neighboring path.

The analogy with a ray of light in geometric optics is obtained by considering Fermat's principle: A ray of light in an inhomogeneous medium moves from A to B along a path that requires the shortest time. The two situations are similar because in each case the actual path is described in terms of a minimum or stationary principle. The analogy is even more striking if we divide the path of the ray into infinitesimal segments, as we did that of the particle. If we then divide the length of each of these segments by the speed of the ray along that segment, we obtain a series of infinitesimal time intervals; the sum of these intervals is smaller along the actual path of the ray than it is along any nearby path.

If we compare these two minimum principles, we see that the particle moves in the force field as though it were a ray of light moving in a special optical medium, in which the speed depends on the wavelength or frequency of the light (on the energy of the particle) so that the medium is dispersive.

With this important first step, Schrödinger obtained a wave equation for a particle such as an electron moving in the electric field of a proton, as in the case of the hydrogen atom. He started from the usual equation for the propagation of a wave disturbance in some medium. Since the equation contains the speed of propagation of this disturbance, called the phase velocity of the wave, Schrödinger saw that it should be possible to obtain a wave equation for a particle by replacing the phase velocity in one equation with the expression from another equation for the phase velocity of a particle. He had already obtained such an expression from his analogy between geometrical optics and the motion of a particle. To complete the derivation, one more step was necessary: the introduction of Planck's fundamental hypothesis. Therefore, Schrödinger, like de Broglie, assumed that the frequency of a particle-wave is equal to the total energy of the particle divided by the Planck constant of action.

When we consider the derivation of the Schrödinger equation, we realize that few of Schrödinger's procedures were based on anything more than analogy. Nothing in Schrödinger's analysis can be taken as a proof for any step in his reasoning, and yet his inexact and rather amorphous method led to one of the most powerful equations in all of physics. It must be noted that when Schrödinger introduced his wave equation, he had no idea of the physical significance of the wave function that it governed. He remarked that "in the case of waves which are to replace in our thought the motion of the electron, there must also be some quantity p subject to a wave equation . . . though we cannot yet tell the physical meaning of p." When we consider that the correct interpretation of the wave function brought into consideration some of the most subtle concepts of modern physics, it is all the more remarkable that this reasoning by analogy guided Schrödinger to a correct wave equation. Furthermore, the complex nature of the wave function (consisting of both a real and an imaginary part) must at first have been somewhat puzzling, if not disturbing.

Schrödinger's derivation of his wave equation must have looked like pulling rabbits out of a hat to the physicists of his day. It is doubtful that his equation would have been accepted had it not come at a time when events demanded a radical break with the past. Even so, physicists found it extremely difficult to reconcile the Schrödinger equation with their understanding of quantum theory. It was known that quantum phenomena were essentially discontinuous and discrete, whereas his equation described the behavior of a continuously propagated wave. The wave-particle dualism (Bohr's complementarity) was then unknown.

Most physicists continued to believe that the only correct approach to the quantum mechanics of the atom was the use of the Heisenberg matrix mechanics. Matrices, being arrays of discrete numbers, seemed perfectly suited for the description of the motion of an electron inside an atom with its discrete orbits. They therefore accepted the Schrödinger picture only with the greatest reservations, as a sort of mathematical trick that could be used to solve problems easily but had no true physical content. It remained for Dirac, Born and Jordan to show that the Schrödinger equation and wave function are related to the probability of finding an electron at a given point and that the Schrödinger wave mechanics and the matrix mechanics are two sides of the same coin. Schrödinger himself had already shown how wave mechanics and matrix mechanics were to be reconciled in terms of operators.

Erwin Schrödinger's personality was much more complex than those of the other physicists who created the quantum mechanics; except for Born, he approached the subject with deeper roots in classical philosophy. Moreover, with the exception of Born, he was a good deal older than the others and less inclined to accept revolutionary solutions to problems. Indeed, we shall see that the very revolutionary creation of his own—the famous Schrödinger wave equation—was, to some extent, an attempt to escape from the discontinuity of quantum physics as represented by Bohr's quantum jumps.

Schrödinger was born August 12, 1887, in Vienna, the only child of Rudolf Schrödinger and the daughter of Alexander Bauer, professor of chemistry at the Vienna Institute of Technology. Erwin spent his childhood in the delightful atmosphere of the academic circles of late 19th-century Vienna; he was raised in a household devoted to learning and culture. His father, who came from a Bavarian family that had settled in Vienna generations before Erwin was born, was an intellectually gifted man with a broad educational background, who studied chemistry at the Vienna Institute under Professor Bauer. He did not become a professional chemist but devoted himself to Italian painting and, later, to botany. This work led him to write a series of papers on plant phylogeny. There is no doubt that these activities of his father greatly influenced Erwin who had a profound interest in the arts and in the origin and nature of life.

Schrödinger attended the Gymnasium at Vienna and there developed a wide range of interests in addition to his scientific work. He was particularly fond of the grammatical structures of ancient languages and was devoted to literature in general and to poetry in particular. Like most creative people he was repelled by rote learning and the need to follow a rigid curriculum.

In 1906 Schrödinger entered the University of Vienna and began his studies of classical physics under Fritz Hasenöhrl, who had succeeded Boltzmann. He steered Schrödinger into the work that ultimately led him to the idea of a wave equation for the atom. This work dealt with the physics of continuous media, which, like a smooth jelly, are assumed to have no granular structure. The vibrations of such media are classified as eigenvalue problems; Schrödinger became a master of these problems.

We may illustrate an eigenvalue problem in a simple way by considering a violin string. Its vibrations can be described by a differential equation that contains a constant, which can be chosen at will. However, only those vibrations of the string are actually observed (are possible) for which this constant is an integer. These integers are called the eigenvalues of the problem. Although the string is continuous, its vibrations are thus governed by integers. We shall see presently how Schrödinger used this idea.

After receiving his degree in physics in 1910, Schrödinger remained at the university as an assistant in physics, setting up elementary experiments from which, as he said, he never really learned about experimental work. His academic career was interrupted by World War I during which he served as an artillery officer.

In 1920 he became assistant to Willy Wien, then went to the University of Stuttgart as assistant professor and next to Breslau as full professor. When Max von Laue left the University of Zurich in the early 1920s, Schrödinger was invited to replace him as full professor of physics. The six years he spent there were his most fruitful and happy. He had married Annemarie Bertel in 1920 and life was very pleasant and exciting. It was during this period that he developed the wave equation.

In the early years of his residence at Zurich, Schrödinger published theoretical papers in the fields of the specific heats of solids, thermodynamics, statistical mechanics, and atomic spectra. He also investigated various problems in the physiology of color.

The idea of a wave equation that would describe the motion of an electron inside an atom and would, at the same time, give the spectral lines, was very appealing to him, for he was unhappy about, and unreconciled to, the idea of quantum jumps. He therefore sought to return to some kind of continuous classical description by treating the spectrum as the solution of an eigenvalue problem. He reasoned that if the discrete modes of vibration of a classical system like a violin string could be obtained as the solution of an eigenvalue problem, so could the Bohr stationary states. One could thus eliminate, so he thought, quantum jumps and replace them by transitions from one mode of vibration (eigenvalue) to another.

Schrödinger did not quite see how to do this before 1926, but toward the end of 1925 he learned about de Broglie's wave theory of the electron through some words of praise by Einstein. This was all that Schrödinger needed. If de Broglie was right and electrons were waves, this meant vibrations, and vibrations in turn meant eigenvalue problems, and eigenvalues meant integers, and these, finally, meant the discrete spectral lines like the Balmer series. In a masterful display of creativity, Schrödinger grafted de Broglie's wave ideas onto Hamilton's formulation of Newtonian dynamics. In a few months he developed a wave equation that answered all the questions about the atom. He presented it to the scientific world without revealing how he had come to this remarkable invention. Only in later papers did he outline his derivation and show how his wave equation followed naturally from the work of de Broglie and Hamilton. He shared the 1933 Nobel Prize in physics with Dirac for his discovery.

The Schrödinger wave equation was and still is the wonder of atomic physics. It presents a well-known and easily handled mathematical procedure for solving any imaginable problem without using the cumbersome matrix algebra of Born and his group. After discovering the wave equation, Schrödinger took the first step in unifying all the different approaches to quantum mechanics—Heisenberg's matrices, Dirac's noncommuting algebra, and his own wave mechanics. He showed that if one solved the Schrödinger wave equation, the solution itself, which is called the Schrödinger wave function, could be used to obtain the Heisenberg matrices. The unification of quantum mechanics was later generalized and stated in its complete form by Dirac.

Schrödinger parted company with most physicists (particularly Born) when it came to the interpretation of the Schrödinger wave function. He proposed to treat the electron not as a particle but as a real wave spread throughout all of space in varying concentrations. He rejected wave-particle dualism. In this view he differed radically from Born's probability interpretation. This divergence led to many private and public debates between Born and Schrödinger. In one letter to Born, Schrödinger wrote, "You, Maxel, you know that I am very fond of you, and nothing can change that, but I must once and for all give you a basic scolding . . ." and then followed an admonition taking Born to task for the "boldness" with which he constantly asserted that the "Copenhagen interpretation" of the quantum mechanics was generally accepted, when he knew that Einstein, Planck, de Broglie, von Laue, and Schrödinger himself were not satisfied with it. When Born replied, giving a list of good physicists who did not agree with the group mentioned by Schrödinger, he in turn answered, "Since when, moreover, is a scientific thesis determined by a majority? (You could perhaps answer: at least since Newton's time)." And so it went between these two warm friends and intellectual opponents.

After Schrödinger's great success with his wave equation, he was invited to Berlin as Planck's successor in 1927 and remained there until 1933, taking an active part in the vigorous exchange of ideas among the great physicists that occurred in the weekly colloquia. When Hitler came to power in 1933, Schrödinger, racially acceptable to the Nazis, who would have showered him with wealth and honors had he collaborated with them, decided to leave immediately. As Born, who was forced to leave Germany because of the racial laws, put it, "He did it, and we admired him. For it is no small matter to be uprooted in one's middle age and to live in a foreign land. But he would have it no other way."

Schrödinger left Berlin to hold a fellowship at Oxford, but in 1936 he accepted a position at the University of Graz, Austria. He did this only after a great deal of thought, for he was well aware of the Nazi danger. But his love for Austria outweighed his caution. In 1938, when Austria was annexed, Schrödinger escaped to Italy and from there he came to Princeton. de Valera then founded the Institute of Advanced Studies in Dublin, primarily for Schrödinger, who became the director of its School for Theoretical Physics. He remained in Dublin until his retirement in 1955.

During this period, Schrödinger did active research in numerous fields and published many papers on topics ranging from the unification of gravity and electromagnetism to the problems of life. In 1944 he published a small book, *What is Life?*, a question which continued to interest him until his death. In an article "Are There Quantum Jumps?," published in two parts in the *British Journal for the Philosophy of Science* in 1952, he challenged the whole statistical concept of the Born school. A public debate between him and Born, who was then at Edinburgh, was arranged, but Schrödinger could not attend because of illness. After his retirement, Schrödinger returned to Vienna as a highly honored citizen. He died there on January 4, 1961.

More than any of his contemporaries, Schrödinger was a lone worker, finding it very difficult to collaborate with anyone else. He was reserved in his personal life and shared his experiences with only a few of his very intimate friends. Wanting to be free to wander off at a moment's notice, he carried his belongings in his rucksack. He usually was last to arrive at conferences, for he always walked to the hotel from the station, lost in his own thoughts.

The name of this many-sided genius, who contributed so profoundly to our understanding of nature, will never be forgotten. Just as we cannot today speak of the motion of an electron without mentioning the Schrödinger wave equation or the Schrödinger wave function so will it be for ages to come.

Statistics and Waves: Max Born (1882–1970)

As we have seen, the Schrödinger wave mechanics gave physicists a complete mathematical procedure (the Schrödinger wave equation) for solving the dynamic problems associated with the motion of electrons in the atom. Moreover, de Broglie and Schrödinger had succeeded in relating the quantum conditions introduced by Bohr to the wave properties of electrons; the various stationary states were to be considered as similar to the nodes in the vibration of a metal plate. With his wave equation, Schrödinger solved the problem of deriving from the de Broglie wave picture the Bohr stationary states, or energy levels of the electron. One had only

Max Born (1882–1970)

to solve the appropriate Schrödinger wave equation, using the proper boundary conditions, and the energy levels of the electron came out automatically.

In spite of this great achievement there were still some difficulties in the interpretation of the Schrödinger wave mechanics that gave the entire theory an appearance of unreality. The Schrödinger equation gave the correct energy levels of the electrons inside an atom. But it was difficult to see just what physical meaning one could give to the waves that the Schrödinger equation assigned to the motion of an electron.

De Broglie considered his waves as real properties of the electron, representing physical vibrations of the electron itself. However, Schrödinger's development of the wave equation cast some serious doubts on the reality of these waves. The amplitudes of the waves obtained from a solution of the Schrödinger equation are complex mathematical quantities composed of real and imaginary parts. Although complex quantities often appear in the mathematical description of physical systems, the only physical interpretation that can have any meaning must involve only the real parts of these mathematical expressions. In all theories up to the time of Schrödinger's, all physical quantities such as the amplitudes of physical waves (sound waves, waves of light, etc.) always appeared as real mathematical quantities even though complex and imaginary mathematical quantities (to simplify the mathematics) were used to obtain these physical quantities. In the case of the Schrödinger theory, however, the amplitudes of the waves describing the electrons are not real quantities. Incorrect results occur if one works only with the real parts of these waves.

Just like de Broglie, Schrödinger believed in the physical reality of his waves and assumed that electrons were not particles but vibrating clouds extending in all directions in space. He thought that the particle picture of the electron was completely untenable and that physical reality existed only in the waves. He pictured the radiation from an electron as resulting from the vibrations associated with its waves.

Aside from the purely formal difficulty of describing real things with imaginary mathematical quantities, a more serious objection to the Schrödinger interpretation was pointed out by Max Born, who finally gave the correct interpretation of the wave functions associated with the motion of an electron. As Born noted in his book *Atomic Physics*, the wave packet interpretation of the electron (the notion that the electron is a vibrating cloud of charge) is incorrect. Such a cloud, on the basis of the elementary properties of wave packets, could not retain its structure but would spread until it became thoroughly dissipated. To overcome these difficulties, Born proposed to interpret wave function in terms of probabilities: "According to this view, the whole course of events is determined by the laws of probability; to a state in space there corresponds a definite probability, which is given by the de Broglie wave associated with the state." This means that waves do not represent physical vibrations but rather the unfolding of the probabilities of future events from a given initial state. In arriving at this interpretation, Born was guided by the behavior of waves and particles that are scattered by some body or field of force. In scattering processes involving light, the intensity of the scattered beam is determined by the square of the amplitude of the wave associated with the light. Therefore this must be a measure of the number of photons present in the scattered wave. But the number of photons in a scattered beam is a measure of the probability of finding a photon in that beam. In other words, the probability for finding a particular state for scattering, or a beam scattered in a given direction, is determined by the amplitude of the scattered wave.

Born assumed that the de Broglie wave associated with a scattered beam of electrons would be a measure of the probability for that particular state of scattering. In general, therefore, the square of the amplitude of the de Broglie wave of an electron determines the probability of finding the electron in a given region of space. We may illustrate this point by drawing an analogy between the de Broglie waves of a collection of particles and the water waves on the ocean. If we watch the ocean, we observe that the surface is in constant agitation. The disturbances range from small ripples to large waves. The greater the amplitude of the wave (the higher above the surface of the ocean is the crest of the wave), the more intense is the disturbance of the water at that point. Where there is no disturbance at all, the water is perfectly smooth and the amplitude of the wave is zero. We now picture the space surrounding an electron or a swarm of electrons as similar to the surface of the ocean. But the waves with which we are now concerned are probability waves, a measure of the probability for finding the electron, or groups of electrons, at particular points in space. The greater the intensity of the de Broglie wave at a point, the greater the probability for finding the electron there; should the amplitude of the de Broglie wave vanish at any point, the probability for finding the electron there is zero.

Suppose we consider an electron in a region of space in which the de Broglie wave is spread out uniformly so that the amplitude of the wave is everywhere the same. Thus there is an equal probability of finding the electron at any point in this region. What happens, then, if we perform an experiment in which we locate the electron at a given point in this region? For example, we can place a fluorescent screen in some position and if the electron happens to strike it, we observe a scintillation at that point. Since we know that the electron is precisely at the point of scintillation on the screen, the probability of finding the electron there is one and the probability of finding the electron elsewhere at this time is zero. This means that we may think of the de Broglie wave describing the electron, which originally was spread out uniformly, as now concentrated in a tiny packet. By performing an experiment we have altered the nature of the wave describing the electron and therefore the future unfolding of the probability.

In this connection, it is interesting to consider Einstein's penetrating objection to the hypothesis that the de Broglie waves are real. If a beam of light is allowed to fall on a surface such as glass, part of the beam is reflected and part is transmitted. If we attribute this fact to the wave character of light, the same thing must then be true of the electron wave packet. If an electron, represented by a de Broglie wave packet, impinges upon a surface or passes into an electric field, which is the same thing, part of the packet should pass through the surface and part should be reflected. If we place a fluorescent screen to intercept these two packets, a scintillation occurs where one, but not both, strikes the screen. This is because the electron is not divisible. If we look for it, we find it in either one or the other of the paths but never in both. Of course, if we do not look for the electron, we must consider the electron as being simultaneously in both the transmitted and reflected beams; then the probability of finding the electron in either beam is given by the square of the amplitude of the de Broglie wave of the respective beam. However, by looking for the electron and finding it in a particular beam we immediately destroy the wave packet in the other beam. A mere increase in our knowledge of the electron's position has altered the structure of the wave describing the electron after its interaction with the reflecting surface.

The remarkable part about this phenomenon is that by finding the electron in one of the beams, we destroy the wave packet in the other beam regardless of how far apart in space the two packets were located. This occurs instantaneously and, according to Einstein, means that the waves defining the packets cannot be physical quantities; if they were, we should be faced with a situation in which a causal action is transmitted from one packet to the other with a speed greater than the speed of light.

In spite of Einstein's important observation, one cannot get away entirely from the physical reality of the waves. We may understand the nature of this difficulty better if we consider the interference of waves. We know that if we allow a beam of light to fall on an opaque surface in which two small holes have been punched fairly close together, two trains of waves leave the holes on the other side of the screen. If we now permit these two trains of waves to fall upon another screen, we discover that the light is not distributed uniformly in two small regions where the two wave trains hit, but rather that the light is distributed in a series of light and dark bands over a single region of the screen. In other words, the two wave trains interfere with each other, giving rise to bands of maximum intensity alternating with bands of zero intensity. The same is true if a stream of electrons falls on a surface with two small holes in it. When the electrons leave the other side of the surface, there are two interfering de Broglie wave trains. In accordance with the principles of the interference of such wave trains, a scintillating screen shows that the electrons are not distributed uniformly over two small regions but instead are found in the regions of maximum intensity of the interfering waves, which are spread over a large area.

But now we come to one of the strange paradoxes that beset this theory. If we allow just one electron at a time to hit the surface, we find that the pattern left on the scintillating screen after many such electrons have passed through the two holes is precisely the same as when the electrons move together in a stream. What does this mean? It is clear that the interference pattern could not have been caused by the interference of two or more electrons, since each electron arrived at the scintillating screen alone. How then do we account for the interference pattern? Only by assuming that each electron interferes with itself by passing through both holes! Or, put differently, that the wave train representing the electron passes through both holes, and the resulting two wave trains representing the electron after it has passed through the holes interfere with each other. This implies that both wave trains have a physical reality because there could be no interference otherwise.

According to the views of Werner Heisenberg and Niels Bohr, we must assume that the electron is simultaneously in both beams and, in a sense, has passed through both holes simultaneously. It is in this way that a particle, like the electron, loses its particle properties and becomes a wave diffused over an extended region of space. As Bohr pointed out, the wave and the particle pictures are complementary aspects of the same physical entity; one exists at the expense of the other. The electron has this diffused wave character so long as we do not try to observe it. As soon as we find it as a particle, we immediately condense it into a small wave packet concentrated in a small region of space. The more accurately we observe its position, the less diffuse its wavelike character becomes and the more do its corpuscular properties manifest themselves. This duality is intimately related to the Heisenberg uncertainty principle, which is considered in greater detail below.

In the year 1901, Max Born, studying mathematics at the University of Breslau, his place of birth, was determined to become a professional mathematician. Attending the algebra lectures given by Rosanes, a student of the famous algebraist Frobenius, he was introduced to the theory of matrices. At that time, he could not know that this course in algebra, with its emphasis on matrices, would give him the clue to the meaning of Heisenberg's strange rule for multiplying the coordinates and momenta of a particle and lead to the quantum mechanics. Still intent on becoming a mathematician, Born continued his studies at Heidelberg and at Zurich under Hurwitz, whose lectures on higher analysis were a great delight to him.

Probably the greatest influence on Born's future development came at Göttingen where he attended the lectures of the outstanding mathematicians of the time—Felix Klein, David Hilbert, and Hermann Minkowski. These men must have directed his attention to physical problems, for their mathematics had many applications to physics. Undoubtedly, Bohr's contact with Minkowski, who had developed the elegant four-dimensional mathematics for treating the theory of relativity, gave Born his abiding interest in this theory; no one can read Minkowski's work without experiencing a sense of great wonder and beauty. At that time Born was also studying with Runge, one of the pioneers in modern numerical methods, and pursuing courses in astronomy with Karl Schwarzschild, who, more than a decade later, was to give the first solution of Einstein's famous gravitational field equations. Born's budding interest in physics was further stimulated by work with Voigt, an expert in the elastic properties of matter. This work in physics must have made a very great impression on Born; his first venture into physics dealt with the elastic properties of wires and tapes and he later became an authority on crystal structure. In 1906 he was awarded the prize of the philosophical faculty of the University of Göttingen for his work on the stability of elastic wires and tapes. From that time on his devotion to physics never faltered.

Like many other great physicists in our roster, Born came from a highly cultured family, steeped in the finest traditions of German humanism. He was born on December 11, 1882, to Professor Gustav Born, an anatomist and embryologist, and Margarete Kauffmann, a member of a Silesian family of industrialists. With this background it was quite natural for Born to retain his interest in the humanities and literature while pursuing his research in physics; one finds in all of his writings a recognition and deep understanding of the unity of all knowledge. In particular, Born always wrote with a profound sense of the history of physics and the evolution of modern concepts from classical physics.

Born subsequently studied at Cambridge under Larmor and J. J. Thomson and, still later, worked with Lummer and Pringsheim at Breslau. There he acquired his ever-abiding interest in optics, which years later led him to write outstanding optics textbooks. During this period he also studied relativity theory intensively and wrote his first paper on one phase of the theory; as a consequence, Minkowski invited Born to Göttingen. Minkowski died shortly after Born returned to Göttingen, but Born remained to prepare some of Minkowski's unfinished papers for publication. He later accepted a position as lecturer there and published several papers on the relativity theory of the electron. Albert Michelson, then the outstanding physicist in the United States, recognized the importance of this work and in 1912 invited Born to lecture on relativity at the University of Chicago.

Although appointed to assist Planck at the University of Berlin in 1915, Born entered the German Army where he worked on military technical problems. During this period he did considerable work on the theory of crystals and published his first book, *The Dynamics of Crystal Lattices*. This was the first in a series of papers, books, and monographs on crystal structure that spanned all of Born's creative life.

After the war, in 1919, Born accepted a professorship at the University of Frankfurt am Main, where his assistant was Otto Stern, later to win the Nobel Prize for the famous Stern-Gerlach experiment that measured the spin and magnetic moment of an atom. In fact, Stern began these experiments under Born's direction.

By 1921 Born had attained a maturity and mastery of physics that clearly gave promise of great creative work. At this stage of his life he was contributing to almost every phase of physics; he had the ability to handle diverse problems with equal skill. In addition to his remarkable grasp of physics, he was an excellent mathematician who could apply the most powerful type of mathematical analysis to physical problems. With such an array of intellectual skills it is small wonder that, in 1921, he was offered the chair of theoretical physics at Göttingen, one of the most coveted posts in the German university system.

Born came to Göttingen at the same time as James Franck; these top-ranking physicists made the university the outstanding center of atomic research in Germany, surpassed in worldwide renown only by Bohr's Copenhagen Institute. So powerful was Born's influence that no serious student of physics considered his education complete without spending at least a year at Göttingen; among some of the outstanding physicists who worked with him were Pauli, Heisenberg, Jordan, Fermi, Von Neumann, Wigner, Weisskopf, Oppenheimer, Dirac, and Maria Goeppert-Mayer.

From 1921 to 1933, when Born was forced to leave Germany, he did his greatest work. With Heisenberg and Jordan he discovered quantum mechanics. At the same time he extended and improved his theory of crystal structure and modernized his early book on the subject. He also wrote two books on atomic theory: one in the early 1920s, based on the Bohr theory; and a later one with Jordan, using the newly discovered quantum mechanics. This was the period during which Born found the correct interpretation of the wave function and assigned to quantum mechanics the statistical interpretation that has been its principal feature ever since that time. It was for this work that he received the Nobel Prize in 1954, about fifteen years later than it should have been awarded. He also found the correct mathematical interpretation of Heisenberg's multiplication rule for physical quantities, which was the beginning of matrix mechanics.

This was a remarkable period in the history of physics. While Born, Heisenberg, and Jordan were developing matrix mechanics, Paul Dirac was publishing his paper on noncommuting dynamic operators, and Schrödinger was preparing his famous paper on wave mechanics. Born declared in his Nobel address

> The result [his collaboration with Heisenberg and Jordan] was a three-author paper which brought the formal side of the investigation to a definite conclusion. Before this paper appeared, came the first dramatic surprise: Paul Dirac's paper on the same subject. . . . he did not resort to the known matrix theory of the mathematicians, but discovered the tool for himself and worked out the theory of such non-commuting symbols. . . .
>
> What this formalism really signified, was, however, by no means clear. Mathematics, as often happens, was cleverer than interpretative thought. While we were still discussing this point, there came the second dramatic surprise, the appearance of Schrödinger's famous paper. He took up quite a different line of thought which originated from Louis de Broglie. (Born, 266)

Interestingly enough, in 1925 Born received a letter from Davisson about the peculiar results that he had found in the reflection of electrons from metallic surfaces. Both Born and Franck suspected that Davisson's patterns were diffraction patterns of the de Broglie electron waves. On investigation they found that Davisson's results did, indeed, confirm de Broglie's theory and that de Broglie's formula for the wavelength of an electron moving with a particular speed agreed perfectly with the wavelength as measured from the diffraction pattern.

The variety of work in quantum mechanics at that time produced considerable confusion about the significance of many of the results. In particular, nobody knew the exact physical meaning of Schrödinger's wave function. Schrödinger proposed that the particle concept be entirely discarded and his concept of wave function be given all the physical reality, which meant that the electron was to be pictured as spread out continuously throughout space. But the Göttingen group, led by Born, refused to accept this interpretation. They could not reconcile such a concept with distinct particle counts in Geiger counters, cloud chambers, and so on. It was at this point that Born discovered the correct statistical interpretation of the wave function, which has been used in all atomic processes ever since.

After leaving Germany in 1933, Born went to Cambridge, where he remained for three years as Stokes lecturer. During this period he proposed, in a series of papers published with L. Infeld, a new type of electrodynamics to replace Maxwell's theory. His idea was to incorporate the structure of the electron into the electromagnetic field. Although this theory has many attractive features, it has not been generally accepted because of its mathematical complexity.

After spending some time at the Indian Institute of Science in Bangalore, in 1936 Born went to the University of Edinburgh as Tait professor of natural philosophy, where he remained until his retirement. During his career, Born received numerous awards, medals, prizes, fellowships, and honorary doctorates from many countries and universities, such as Bristol, Bordeaux, Oxford, Edinburgh, Oslo, and Brussels, to name but a few. He was awarded the Stokes medal of Cambridge, the Max Planck medal of Germany, the Hughes medal of the Royal Society of London, and the Hugo Grotius medal for international law. In 1953 he was made Honorary Citizen of Göttingen and in 1959 he was awarded the Grand Cross of Merit with Star of the Order of Merit of the German Federal Republic.

Born was not only a great scientist who made significant contributions to all aspects of physics. He was also a great human being who was as much concerned with humanity as he was with science. For many years he was at the forefront of those persons who, like Einstein, have sought and are still seeking to make this a peaceful, more beautiful world. He never divorced his science from the broad field of human knowledge and philosophy but attempted to relate all recent discoveries in physics to basic philosophical principles.

The Uncertainty Principle: Werner Karl Heisenberg (1901–1976)

The serious discrepancies between observation and the Bohr model of the atom did not give way immediately to a single unified theory leading to an unambiguous picture of the atom and the electron inside the atom. Instead, the new developments came from different approaches that at first seemed quite irreconcilable, if not actually contradictory. On the one hand, de Broglie's theoretical discoveries suggested that the Bohr theory gave an inadequate picture of the physical nature of the electron. Rather than treating the electron as a particle, de Broglie considered its wave characteristics in order to understand its behavior. This approach, which dealt primarily with the nature of the electron and not with the relationship of the observer to the electron, reached its final development in the Schrödinger wave equation.

Against these discoveries, which indicated that the electron could no longer be treated as a particle, Heisenberg, Born, and Jordan propounded a new theory in 1925. This theory, called matrix mechanics, was the first complete form of quantum mechanics presented as a technique for solving atomic problems. The starting point was an investigation carried out by Heisenberg and the Dutch physicist H. A. Kramers into the nature of the dispersion of light passing through a gas. Since the Bohr theory gave incorrect results in this analysis, it was clear that a new approach was necessary. Rather than changing the concept of the particle nature of the electron, Heisenberg rejected the classical concept of the relationship of the observer to the phenomenon being observed. Just as Einstein had been led to a revolutionary picture of space and time by analyzing the manner in which the physicist performs measurements of time and distance, so Heisenberg was now led to a radically different view of the atom. He subjected accepted concepts, such as the orbit of an electron, the position of an electron inside an atom, and the momentum of an electron, to profound criticism.

At first sight, it appears highly improbable that an approach of this sort could lead to results identical with those obtained by ascribing wavelike properties to the electron, and, indeed, to all particles. Yet this is exactly what happened. Soon after the discovery of the Schrödinger wave equation, Schrödinger, Dirac, and others

Werner Karl Heisenberg (1901–1976)

demonstrated that the two apparently unrelated prescriptions for discovering what is going on inside an atom were different ways of looking at the same thing.

To physicists, this result was especially puzzling since the two theories appeared to be contraries. One theory (Schrödinger-de Broglie) introduced a revolutionary picture of the intrinsic nature of the electron. The other (Born, Heisenberg, and Jordan) dealt with the way we make measurements and records of material things and events. But the basic agreement between the two theories should not have been altogether surprising, because a precedent had been established in the theory of relativity. Einstein's revision of the methods of measuring space and time intervals led to a change in our concepts of the very nature of matter.

Heisenberg's great contribution was to show that in analyzing physical systems it is not permissible to use concepts that cannot be measured directly. Only those quantities that have meaning in terms of experiment—that is, which have been measured—should enter into our calculations. By analyzing this requirement, Heisenberg discovered the uncertainty principle—the foundation of quantum mechanics—which forms a bridge between the wave picture and matrix mechanics.

There is, however, a fundamental difference between the relativity approach to the measurement of space and time and the Heisenberg approach to the measurement of physical quantities. The theory of relativity starts by placing an upper limit on a physical quantity, namely, the speed of a body. This limitation arises from the observation that the speed of light is the same for all observers moving with uniform speed with respect to each other, regardless of how fast they may be moving. Thus, the point of departure of the theory of relativity is a statement about the intrinsic nature of a physical quantity. The quantum mechanics of Heisenberg also begins by introducing a limit—not on the value of an intrinsic physical quantity but on the accuracy with which we can measure certain pairs of quantities simultaneously.

As we have noted, the matrix form of quantum mechanics grew out of an attempt by Kramers and Heisenberg to calculate the dispersion of polarized light by atoms using the Bohr quantum theory. In this investigation, which appeared as a paper in the *Zeitschrift für Physik* (1925), the authors applied the Bohr theory to the interaction of radiation with an atom. They assumed the atom to be in a particular energy state, or the electron in the atom to be in a particular Bohr orbit as described by the concepts of classical mechanics applied to an electron moving under the action of a Coulomb force. To perform a calculation of this sort, it was necessary to use what Bohr called the correspondence principle, since the Bohr theory without this empirical device lacks completeness. To explain this idea we must consider briefly how the classical theories and that of Bohr describe the emission of radiation by an electron circling the nucleus in an atom.

The classical approach is quite straightforward. Imagine an electron moving with periodic motion in some orbit inside an atom. To describe its position inside the atom at any time we draw a vector from the nucleus to the electron. The length and direction of this arrow at any moment gives us the position of the electron. If the motion of the electron (like that of the earth around the sun) is periodic, then this vector has certain sets of regularly recurring lengths and directions. The same set recurs at intervals equal to the fundamental period of the motion, just as the earth comes back to its same position relative to the sun and the stars every year. Mathematically speaking, if $q(t)$ represents the position of an electron at time t, the $q(t)$ will always be the same for values of t that differ from each other by some integer multiplied by t_0, where t_0 is the period of the motion. We may express the relation as $q(t) = q(t + nt_0)$ where n is an integer.

Although the motion of the electron is periodic, it is usually quite complicated, just the way a sound can be a complex mixture of simple tones even though it is constantly repeated, like a bird call. Consider the difference between the sound of a pure note and that of the same note obtained by sounding some complex instrument. We can get the pure note by striking a tuning fork that is allowed to vibrate only in its fundamental tone or by plucking a string that vibrates only as a whole. Such vibrations are said to be simple harmonic vibrations. Their frequency is also the frequency or pitch of the note that is produced. Such vibrations are the simplest kind known and are represented mathematically by sine and cosine formulas.

If we now produce the same note in an impure form, not by a simple harmonic vibration of the string but by complex vibrations so that the string oscillates both in its fundamental mode and in various higher modes, the resulting sound is complex. If this complex sound is analyzed with a harmonic analyzer, we find that in addition to the fundamental frequency of the pure note, corresponding to the simple harmonic vibration, many higher frequencies are present. All of these higher frequencies are simple integer multiples of the fundamental frequency.

Assume that we know at each instant of time the position of each point of the vibrating string, or, as the mathematician would express it, that we know the shape of the vibrating string as a function of the time. We can then do mathematically what the harmonic analyzer does—analyze the vibration of the string into its component simple harmonic vibrations.

This mathematical analysis was discovered by the 19th-century French mathematician Jean Fourier and is known as a Fourier analysis. It can be applied to any periodic function of the time and leads to a representation of the function as an infinite series of sine and cosine terms, each one with a different frequency.

But only those frequencies appear that are integral multiples of the fundamental frequency of the vibrating string.

If the fundamental frequency is ν, the function representing the distortion of the string at any moment is given as a sum of terms of the form $\sin 2\pi n\nu t$ and $\cos 2\pi n\nu t$, where n is an integer that takes on all the values 1, 2, 3, and so on. Each term in this series appears multiplied by an appropriate coefficient, which is a measure of the amplitude of that particular harmonic in the complete vibration of the string. Or, put differently, if the sound issuing from the vibrating string is analyzed mechanically into its component frequencies—or into its various harmonics—the intensity of a harmonic having a particular frequency is given by the coefficient multiplying the sine or cosine term of that frequency in the Fourier expansion of the function representing the periodic distortion of the vibrating string.

We may draw an analogy between this concept and Ptolemy's description of the motion of a planet in the pre-Copernican geocentric theory of the solar system. The observed motion of a planet, with the earth assumed to be fixed, is so complex that it cannot be represented by a single circular orbit around the earth. Ptolemy therefore assigned an entire series of circular orbits of smaller and smaller radii to each planet. Each succeeding circle had its center on the circumference of the preceding one. These small circles were called epicycles. It is easy to see that if one uses enough epicycles, one can describe the most complicated type of periodic motion. The epicycles are equivalent to a Fourier analysis of pure vibrations.

What has this to do with the matrix mechanics of Heisenberg? The relationship becomes apparent if we consider from the classical point of view the problem of the radiation of a vibrating electron inside an atom. According to this point of view, an electron moves around the nucleus in an orbit similar to that of a planet around the sun; therefore, the motion is periodic and, in a sense, similar to that of our vibrating string. Since the position of the electron in its orbit can be represented by a vector from the nucleus to the electron, this quantity, which we have called $q(t)$, can be represented by a Fourier series.

Note that the orbit of the electron is, in general, quite complicated, even though it is periodic. Thus various overtones are present. In other words, the Fourier analysis of the orbit shows that the electron is really vibrating around the nucleus with a whole series of frequencies, each of which is an integral multiple of some fundamental frequency.

According to classical theory an electron that moves in this fashion must radiate energy consisting of light with all the frequencies that are present in the Fourier analysis of the motion of the electron. But it is precisely here that the classical theory disagrees with observation. A spectral breakdown or harmonic analysis of the light emitted by an atom shows that the frequencies of the component radiation, the lines in the spectrum, are not associated with a single fundamental frequency of an orbit and its overtones. Rather, the frequency of any spectral line is expressed in terms of the differences of two numbers, as though two different orbits were associated with each line in the spectrum. This is known as the Ritz combination principle.

We have seen how Bohr treated this difficulty. He discarded the association of classical frequencies with classical orbits of the electron. Although he retained the idea of a classical orbit, he did not connect the classical frequencies obtained by a Fourier analysis of this orbit with the frequencies of the spectral lines. Nevertheless, it was still necessary for Bohr to retain the Fourier analysis to complete his picture of the atom. His theory of stationary jumps, with the emission of radiation from the electron occurring during jumps from a higher stationary state, or orbit, to a lower one, predicts which lines will be found in the spectrum of an atom. But it does not predict the intensities of these lines. Indeed, it predicts that all the lines associated with all the possible jumps of the electron should be present with equal intensity. This is not the case. In fact, certain lines are not present at all, as though the electron were forbidden to make the jumps that would produce these lines.

Bohr managed to get around this difficulty by introducing his famous correspondence principle, the subject of our present discussion. He observed that the classical theory and his quantum theory of the atom merge if the electron is moving in an orbit that is at a great distance from the nucleus—that is, if the principal quantum number n of the orbit is very large. The reason for this is that for orbits of this sort the energy differences are extremely small. In going from one such orbit to another the electron does not behave as though it were jumping but as though it were moving continuously—in other words, classically. Bohr assumed that for these outer orbits the classical Fourier analysis gives the correct picture of the spectral lines and their frequencies. If a particular term in the Fourier analysis were not present, the particular spectral line associated with that frequency would not be present in the spectrum.

The correspondence principle involves the extension of this idea to all the orbits of the electron, not only to the outer orbits with their very large quantum numbers. Consequently, one must use the Fourier analysis of the motion of an electron not to determine the frequencies of the spectral lines in the radiation but to determine which lines are present in the spectrum. This means that the Fourier analysis gives incorrect answers it used to determine the frequencies of the spectral lines but correct answers in the determination of the intensities of various spectral lines. It is this use of the classical Fourier analytical procedure that is the essence of the correspondence principle. That it works for the intensities, even though it does not give the correct frequencies of the spectral lines, is a tribute to the great genius of Bohr and an indication of the depth of his insight.

We now come to the work of Heisenberg and Kramers, who used the correspondence principle and the Bohr theory to analyze the dispersion of light by an atom. Consider an atom with its optically active electron in a particular Bohr orbit. By analyzing the motion of the electron into a Fourier series and then applying the rules for the interaction of a monochromatic beam of radiation with an electron moving in this way, we can solve the problem of the scattering of a monochromatic beam of light (a beam of definite frequency). We thus obtain the classical formula for the scattering of light by an electron moving in a bound orbit. It involves the fundamental frequency and the various harmonics of the electron's orbit. Since this formula is based on classical dynamics, it is incorrect. However, Heisenberg and Kramers found that by using the correspondence principle it is possible to start from this formula and obtain one that is in agreement with the quantum theory.

Their first step was to replace the Fourier frequencies which refer to the classical orbit, with proper quantum theoretical frequencies as given by the Planck relationship. When this is done, each frequency that appears in the revised Fourier expansions refers to two electronic states and not to one particular orbit. In mathematical terms, the frequencies appear in the form v_{ik} where i and k refer to the states i and k of the electron. These are just the observed frequencies of the spectral lines. (The frequency of the scattered radiation can be represented as a square array since both i and k can take on all the integer values.) The next and final step in the application of the correspondence principle is to replace the coefficients in the Fourier expansion by coefficients that have quantum theoretical significance. This requires the Bohr quantization rule relating the motion of the orbiting electron to Planck's constant.

When the classical dispersion formula was thus modified, it led Heisenberg to a profound deduction that was the first step in the development of his matrix mechanics and led finally to today's quantum mechanics. To follow Heisenberg's reasoning, consider an electron moving in a classical Bohr orbit and note how it would scatter light if it interacted with radiation according to the classical theory. The scattered radiation would contain as many different frequencies, all multiples of a fundamental frequency, as there are terms in the particular Fourier series that represents the position of the electron with respect to the nucleus. The intensity of each such component in the scattered radiation would be given by the coefficient of that particular term in the Fourier analysis. According to classical theory, then, scattered radiation, in its frequencies and amplitude, refers to one particular orbit of the electron. In other words, it is the result of the electron's moving in a particular orbit.

But what does the correspondence principle tell us? First, that the component harmonic frequencies are not associated with a definite orbit but with two orbits, or suborbits. Moreover, they are not associated with the frequencies of the electrons in these orbits but with their energies. Second—and this is the importance of the Heisenberg-Kramers calculations—the amplitudes of the various components of the radiation do not refer to one particular orbit; instead, each amplitude is associated with the quantum numbers of two orbits, or suborbits. The amplitudes, and therefore the quantities that describe the motion of an electron (its position at any moment and its momentum or velocity), appear as square arrays. These quantities always have two numbers associated with them (the quantum numbers) that can take on integer values.

We return to the electron moving in its Bohr orbit and consider its position $q(t)$ at any moment. Since the motion of the electron is periodic, we may, from the classical point of view, replace $q(t)$ by its Fourier analysis. We can then say that the orbit of the electron is represented by the complete collection of coefficients in the Fourier expansion. This means that if these Fourier coefficients are A_0, A_1, A_2, and so on, we can picture the electron as moving simultaneously in a collection of circles with radii A_0, A_1, A_2, and so forth. All of these circles together give the effect of the actual orbit of the electron. This is purely a mathematical fiction; from the classical point of view the orbit of the electron is the only real thing.

What happens when we pass over to the quantum theory using the correspondence principle? The quantities A_0, A_1, A_2, and so on, are now replaced by quantities referring to two orbits, so that we must write them as $q_{1,2}, q_{4,5}, q_{2,3}$, and so forth. That is, we must write them as q_{ik}, where i and k refer to orbits i and k and can run through all the integers. If we want to replace $q(t)$ in quantum theory by the equivalent of a Fourier series, using the correspondence principle, we must use a square array of numbers, each of which refers to two different orbits. But what is more important, and what struck Heisenberg so forcefully, is that when such a square array is used in studying some atomic process (the scattering of light, for example), all its numbers must be taken into consideration and not just those that refer to two special orbits.

To Heisenberg this fact meant that it is not permissible to treat the electron inside the atom as though it were moving in a definite orbit. In this denial of the possibility of assigning anything like a Bohr orbit to an electron, Heisenberg departed from the Bohr theory. Bohr had developed his quantum theory of the atom by discarding the idea of a classical frequency associated with the orbit of an electron, but he still retained the concept of the classical orbit. Heisenberg went one step further and discarded the concept of the orbit ifself. Rather than the classical idea of the position and the motion, or momentum, of the electron at each instant of time, Heisenberg introduced his square arrays or matrices, which depict the electron as existing simultaneously in all possible Bohr orbits. After Heisenberg's discovery, the classical concept of the electron as a particle was no longer tenable.

Heisenberg was led to these revolutionary ideas by his insistence on utilizing only those quantities in a theory that are directly observable. Since the orbit of an electron is not observable, it can have no place in a theory. Only the spectral lines are observed, and, since these involve pairs of orbits, all quantities that are used to describe the electron inside the atom should be associated with such pairs.

Such thinking led directly to Heisenberg's matrices. Once these matrices were introduced to represent the position and the momentum of an electron, it was necessary to use a different kind of algebra to study these quantities. If we have two such arrays to be multiplied, how are we to perform the multiplication? Certainly not in the way we usually multiply two numbers since we are here dealing with arrays containing many ordinary numbers.

Heisenberg discovered the correct way of multiplying such arrays to obtain results that agreed with observation. Born then pointed out that this multiplication rule had been known to mathematicians for many years and was, indeed, the rule for the multiplication of matrices. One of the important features of this rule is that it is not commutative. If the array representing the position of an electron is q and the array representing its momentum is p, then the product pq is not the same as the product qp.

This showed Heisenberg that the uncertainty relationship is purely an algebraic consequence of his matrix theory. Picture the product pq as representing a measurement of the position of the electron followed by a measurement of its momentum; qp, on the other hand, represents the measurement of the momentum of a particle followed by the measurement of its position. That these two sets of measurements give different results simply means that the measurement of the momentum of a particle destroys our knowledge of its position, and vice versa. It follows that it is impossible to obtain or to have precise knowledge of the position and the momentum of a particle simultaneously; this is the essence of the uncertainty principle.

Its significance for the structure of the atom (or our knowledge of this structure) is that we have no way of determining the orbit of an electron inside the atom observationally. As Heisenberg pointed out in his analysis of the Copenhagen interpretation of quantum theory, an electron can be observed inside an atom only with a gamma-ray microscope which, because of the short wavelength of gamma rays, has a high resolving power. This microscope shows us where the electron is at any moment, but at least one gamma-ray photon must be reflected from the electron. In this very process the electron is knocked out of the atom. It is senseless then to speak of its orbit.

We can see now the connection between the de Broglie-Schrödinger picture, which assigns wave properties to particles, and the Heisenberg picture with its emphasis on the measurement process that alters the behavior of the electron in the process of observing it. To say that a particle has wave properties implies that its motion and its position cannot be known simultaneously. We can, indeed, locate the particle-wave as accurately as we wish but only if, in so doing, we forego the opportunity to discern its wave character. On the other hand, this same particle-wave entity will lose all of its particle-like properties if our measurement gives us precise information about its motion.

Although the uncertainty relations can be derived mathematically from the formalism of the theory, it is much more instructive to derive them from the physical picture. This method shows clearly the interrelationship between the wave and the particle. In fact, it is clear from Heisenberg's analysis that wave and particle are complementary aspects, as are position and momentum. It was from considerations such as these that Bohr developed his theory of complementarity, which is essential for an understanding of modern atomic theories.

The uncertainty relations completely change our ideas of causality. If we cannot determine the position and the momentum of a particle simultaneously to any desired degree of accuracy, we cannot determine its future course. We can solve equations for the motion of the particle. But these solutions can tell us its future history only if at some moment in the past or at the present instant we know its position and momentum. The farther we try to peer into the future, the less accurate our predictions become because our present uncertainty, however small, leads to greater deviations from the predicted pattern of the motion as the time increases. We can understand this situation by considering the lunar missile probes carried out by the United States and the Soviet Union. To hit a target as far away as the moon involves extreme accuracy in aiming the rocket and giving it the correct initial momentum; if we wish to hit targets at greater distances, our accuracy will have to be increased considerably because the further the distance, the greater the multiplication of any initial error.

Today we use the term quantum mechanics (the term that was then applied to Heisenberg's work) for the entire mathematical scheme (i.e., the algebra of noncommutative operators) that is used to treat problems in atomic, nuclear, elementary-particle, and field physics. The mathematics of quantum mechanics stems directly from Heisenberg's matrix mechanics and is a consequence of his uncertainty principle.

Werner Karl Heisenberg was born in Würzburg, Germany, on December 5, 1901, and grew up in academic surroundings, in a household devoted to the humanities and steeped in culture. His father was a professor at the University of Munich and undoubtedly greatly influenced young Werner, who was a student at the Maximilian Gymnasium. This influence was apparent in Werner's intense interest in the classics, especially in early Greek science to which he devoted himself in his high-school years. During his free time, while at the gymnasium,

Werner could usually be found deeply engrossed in Plato's dialogue "Timaeus," in the atomic theory of Democritus, in the writings of Thales of Miletus, or in some other ancient philosopher, all in the original Greek. This interest in the philosophy and history of science remained with Heisenberg throughout his life.

When not engrossed in physics and Greek philosophy, young Heisenberg divided his time between music and tennis, as well as climbing a mountain whenever one was available. He probably could have gone far in music because he had an excellent ear and became a fairly good pianist, but his passion for theoretical physics dominated his life. When he graduated from the gymnasium early in 1920 he immediately presented himself to Professor Arnold Sommerfeld (the dean of German atomic physicists and the greatest teacher and developer of new talent) and asked to be allowed to study theoretical physics with him. Sommerfeld, somewhat taken aback by so bold an approach by so young a man, proposed that Heisenberg first pursue the standard course of study in physics, including laboratory work, before thinking of specializing in theoretical physics. Nevertheless, Sommerfeld, who himself was a great theortical physicist, allowed Heisenberg to attend his seminars during Heisenberg's first semester at the University of Munich. In 1923, six semesters later, and not yet 23 years old, Heisenberg took his doctoral examination and received his Ph.D. Like many of the brilliant young physicists of that period he decided to go to the University of Göttingen, which, under Max Born's direction, was rapidly becoming the center of theoretical research in physics. He became Born's assistant in 1923 and two years later, in 1925, placed a revolutionary manuscript on Born's desk; this paper marked the beginning of quantum mechanics. Heisenberg had then been appointed a lecturer at the university.

Born quickly recognized the significance of Heisenberg's work and sent the manuscript to the *Zeitschrift für Physik*, one of the two top German journals of physics. After the paper was published, Heisenberg's ideas were developed more fully in a series of papers published jointly with Born and Pascual Jordan, Born's other brilliant assistant.

Heisenberg's discovery stemmed from work he had done previously with the Dutch that physicist H. A. Kramers on the dispersion of light by gaseous atoms, work which showed clearly that the Bohr theory breaks down when pushed too far. When Heisenberg looked for the cause of this breakdown, he realized that the concepts introduced in that Bohr theory, such as the orbits of an electron and the electron's position and motion inside the atom, cannot be observed directly and therefore have no place in a self-consistent theory. He therefore decided to reformulate the theory solely in terms of observables and found that this could be done only by using noncommutative algebra.

This insistence on freeing atomic theory from unobservable quantities finally led Heisenberg to his famous uncertainty principle. It was clear to him that if the theory was to deal only with physical quantities that could be observed directly, it would have to be cast in a form in which it could not speak of the exact position and momentum of a particle simultaneously, since measuring either one of these quantities would affect the other and hence our knowledge of it.

Although matrix mechanics as we know it today evolved from Heisenberg's theory, it was not present as such in Heisenberg's first paper. There Heisenberg had simply introduced a multiplication rule for the momentum and the position of an electron, indicating that a special kind of algebra would have to be used in dealing with these quantities.

When Born saw this type of multiplication in Heisenberg's paper, certain memories stirred in him. He described this "remembrance of things past" in his Nobel address: "Heisenberg's multiplication rule left no peace, and after eight days of intense reflection and probing, I suddenly recalled an algebraic theory that I learned from my teacher in Breslau. I applied this quadratic, algebraic scheme, called matrices, to Heisenberg's quantum condition." So the actual mathematical scheme of matrix mechanics came from Born.

One may conclude from this episode that even outstanding theoretical physicists in the early days were not too well equipped in advanced mathematics, particularly if such mathematics did not have an immediate bearing on physics. Today, no theoretical physicist who wishes to contribute significantly to physics can afford to be ignorant of any of the branches of modern mathematics.

After Heisenberg had spent three years at Copenhagen, he was offered the chair of theoretical physics at the University of Leipzig, where he remained until 1941. This was a period of great productivity. He not only published many papers on the technical aspects of the new quantum mechanics but also applied it to diverse problems. He collaborated with Pauli to lay the foundation of quantum electrodymanics and quantum field theory in general, and he also applied quantum mechanics to the study of magnetic phenomena. During this period he became interested in nuclear physics and introduced one of its most useful ideas, isotopic spin, which treats the neutron and proton as two different energy states of the same basic particle, the nucleon. While at Liepzig, Heisenberg also made fundamental contributions to cosmic ray theory. In 1934 he was awarded the Nobel Prize in physics for having discovered the quantum mechanics.

After World War II began, Heisenberg left Leipzig. From 1942 to 1945 he was director of the Max Planck Institute of Physics at Berlin. There is no doubt that this was a very unhappy period for him since he was strongly opposed to the Nazi regime and remained in his academic post only because he felt that someone had to hold up the "torch of culture" against the barbarians. Before the war he had strongly defended his Jewish assistant, Guido Beck, against scurrilous racist attacks.

After the war Heisenberg continued with his research as director of the Max Planck Institute at Göttingen. He then became interested in high-energy particle physics and introduced the technique of what is now called the S-matrix theory (the so-called scattering matrix). This technique attempts to describe events only in terms of what one sees initially before two systems are brought together (two nuclei, for example, or two protons) and what one sees finally, after they have interacted and separated. It does not try to describe the actual mechanism of the interaction (since this cannot be observed) but only the products of it. This S-matrix procedure became very fashionable in particle physics in the 1960s and 1970s.

Heisenberg spent the final years of his career trying to derive the properties of such elementary particles as electrons, protons, and so on, from a departure from quantum field theory by having the field itself construct its own particles. Unfortunately, this approach led to a very complex mathematical formulation which spoiled the great beauty of quantum mechanics.

In addition to his research Heisenberg was also greatly interested in the exposition and the philosophy of physics. His books such as *The Principles of the Quantum Theory*, *Cosmic Rays*, and *Nuclear Physics* are models of clarity and excellent introductions to these subjects, and his *Physics and Philosophy* remains one of the best books on the philosophy of modern physics.

The Barrier around the Nucleus: George Gamow (1904–1968)

Shortly after the discovery of quantum mechanics and the wave properties of particles, the new physical ideas these theories produced were further supported and extended to larger domains by George Gamow's brilliant application of wave mechanics to the analysis of the radioactivity of nuclei. Wave mechanics had been remarkably successful in accounting for the behavior of electrons both inside and outside atoms and in explaining correctly the origin of atomic spectra. But it was not known, before the work of Gamow (and of Condon and Gurney, who did a similar analysis independently of Gamow), that wave mechanics is applicable to alpha particles and in regions such as the interior of the nucleus. Gamow's work represented an important step in the development of a complete quantum-mechanical picture of nature.

Gamow's prime purpose was to explain the spontaneous emission of alpha particles (alpha decay) from radioactive nuclei, such as those of uranium, and to deduce decay lifetimes theoretically (which he did very successfully). But he also showed that quantum mechanics is valid inside nuclei and that alpha particles obey the same kind of wave equation (the Schrödinger equation) as electrons. In addition, he demonstrated the correctness of Born's concept that the Schrödinger wave function is the probability amplitude for finding a particle in a given small neighborhood of space. Gamow's work also threw into sharp relief the essential difference between the classical dynamics of Newton and the modern wave picture of matter.

Until the work of Gamow and of Condon and Gurney the radioactivity of nuclei, resulting in the emission of alpha particles, was a complete mystery, senseless in terms of classical physics. To see why this is so, consider the kinetic energy with which alpha particles are emitted from radioactive nuclei. This energy is considerably less than what it would be if an alpha particle were placed at the surface of a nucleus, such as that of uranium, and allowed to move away under the action of the Coulomb repulsion resulting from the positive charges on the nucleus and the alpha particle. Coulomb's law of electrostatic force between charged particles is valid for distances the size of the nucleus. Since we know the size of the uranium nucleus, we can easily compute the kinetic energy the alpha particle would acquire in this way and compare it with the kinetic energy of observed alpha particles. We find that the kinetic energy of these particles is much smaller than would be expected from this classical picture. The observed alpha particles behave as though they came from a point quite distant from the nucleus, rather than from within it. Classical theory is thus in complete disagreement with observation.

To account for the observations and, at the same time, to obtain an expression for the lifetime of a radioactive nucleus—that is, how long on the average a nucleus remains intact—Gamow applied quantum mechanics (more specifically, the Schrödinger wave equation) to spontaneous alpha-particle emission. In 1928, little was known about nuclear forces (and we do not know very much even now). As a result, the prospects for a wave-mechanical treatment of the problem did not seem hopeful. But because of the very short range of the attractive nuclear forces (the forces that oppose the Coulomb repulsive forces and thus keep the nucleus together), it was not necessary to have a detailed picture of those forces to obtain a quantum mechanical model of alpha-particle emission. Gamow sensed that this was the case and went boldly ahead with his analysis. He constructed

George Gamow (1904–1968)

an approximate force model of the nucleus that was good enough to allow him to solve the problem quantum-mechanically. This model consisted of a repulsive barrier (the Coulomb repulsive force or, more precisely, the Coulomb repulsive potential) that surrounded the nucleus like a hill and prevented positively charged particles from coming close to the nucleus unless they were moving fast enough (with enough kinetic energy) to break through this barrier. In addition, the model included a deep well in which the particles in the nucleus moved about.

Classically, according to this picture, no particle (e.g., alpha particle) could leave the well, that is, the nucleus, unless its kinetic energy were greater than the potential energy represented by the height of the Coulomb barrier. But if that were the case, the particle would have to leave the nucleus with kinetic energy much greater than that observed. For a particle to leave the well with the observed kinetic energy, it would have to tunnel through the Coulomb barrier. Classically this feat is impossible because while it is in the tunnel, its kinetic energy (its total energy less its potential energy) is negative, a meaningless concept in classical physics. However, this concept offers no difficulty in quantum mechanics, which, in considering what is possible in nature, is only concerned with what one can observe at the end of a process. Quantum mechanics allows a particle to pass through an intermediate state with negative kinetic energy so long as the final state of the particle is in agreement with the correct principles.

If we keep this fact in mind the quantum-mechanical analysis of alpha-particle emission is quite straightforward. It falls into the class of potential barrier problems, which may be described briefly in terms of an optical model. Consider a particle that is moving along a straight line and impinges on a barrier of a certain height. Classically it would bounce off this barrier unless it had enough energy to rise above it and roll over. But we must now take into account the wave properties of the particle and note that a wave spreads out into all regions and can penetrate barriers. Thus, if a wave of light, moving along a straight line, strikes a barrier such as a glass surface, part of the light is reflected from the barrier and part of it passes through. The amount of light that is reflected and the amount that is transmitted depend on the nature of the barrier and can be expressed in terms of the index of refraction of the barrier. The latter depends on the density of the barrier and how the molecules in it are arranged.

In the case of a charged particle such as an electron or an alpha particle, the barriers are electric and magnetic fields, for which one can introduce the equivalent of the index of refraction. This index depends on the field strengths that determine the height of the barrier. Just as in the case of an optical wave, the material de Broglie wave suffers both reflection and refraction when its strikes the barrier. Part of the wave passes through the barrier, even though the particle counterpart of the wave does not have enough energy to get over the barrier.

By solving Schrödinger's wave equation for a particle surrounded by a potential barrier of finite height we can show that the wave that describes the particle has two parts. One part lies inside the barrier. The other lies outside the barrier even if the particle is moving inside the barrier with a kinetic energy less than the barrier's height. This is the significant difference between classical and quantum physics.

Born's interpretation of the Schrödinger wave function supports this analysis. According to Born, the probability of finding a particle in a small region of space is reached by multiplying the volume of the region by the square of the absolute value of the particle's wave function in this region. The procedure tells us at once that wherever the wave function of the particle is not zero, there is a finite probability for finding the particle. Thus, in the case of the particle inside the potential barrier, the existence of a part of its wave function outside the barrier tells us that the particle has a finite probability of penetrating the barrier and being found outside it.

Gamow applied these very subtle ideas to the emission of alpha particles from radioactive nuclei. An alpha particle inside the nucleus is described by a wave (the solution of the Schrödinger equation for the particle); part of the wave lies inside the nucleus but part extends into the outside world. Thus there is a finite probability for finding the alpha particle on the outward side of the barrier — outside the nucleus. In other words, as the alpha particle oscillates back and forth inside the nucleus, it has a finite chance of penetrating the barrier every time it comes near the surface of the nucleus. This probability becomes greater the larger the amplitude of its wave outside the nucleus.

We can now calculate the lifetime of a radioactive nucleus as opposed to its decay by alpha-particle emission. We first solve the Schrödinger wave equation for an alpha particle in the nucleus and then seek among all the possible solutions that particular wave function that describes an alpha particle having the observed energy when it is emitted from the nucleus. The sum of the squares of the absolute values of this function taken over all regions of space outside the nucleus gives the total probability for finding the alpha particle outside the nucleus. If this amount is multiplied by the number of times per second that the alpha particle oscillates back and forth inside the nucleus, we obtain the probability per unit time that the alpha particle will escape from the nucleus. The reciprocal of this probability is the lifetime of the nucleus. Gamow carried through these calculations and showed that the theory agrees with observations.

Gamow's theory had important applications far beyond those associated with radioactive decay. The same analysis that he used can be reversed to analyze the capture of alpha particles, protons, and other particles by nuclei. This is of great importance in astrophysics, for we know today that the energy released by stars like the sun comes from the fusion of nuclei. The growth, evolution, and death of stars are governed by thermonuclear fusion in their deep interiors. To understand such nuclear processes we must consider the penetration of potential barriers by interacting nuclei. The analysis of such thermonuclear processes is carried out by using Gamow's results for the penetration of the potential barriers that surround all nuclei.

George Gamow was born in Odessa, Russia, on March 4, 1904, and received his doctorate from the University of Leningrad in 1928. This was a fortunate time for him to have completed his graduate studies because quantum mechanics had just been discovered and many exciting problems were on the verge of solution. Gamow had the further good fortune of receiving a fellowship to spend a year at the Niels Bohr Theoretical Institute of Physics in Copenhagen, just when Max Born had proposed his statistical interpretation of the Schrödinger wave function.

Before Gamow tackled the problem of the emission of alpha particles from radioactive nuclei, the only way of testing Born's interpretation of the wave function was by analyzing collision phenomena. Gamow's paper on the structure of nuclei, published in 1928, showed that the Born statistical interpretation of the wave function gave a correct picture of radioactive decay processes. This paper was also important because it demonstrated clearly that quantum mechanics is valid inside the nucleus of an atom.

From Copenhagen, Gamow went to the University of Cambridge from 1929 to 1930 as a fellow of the Rockefeller Foundation. He then returned to Copenhagen for another year. In 1930 he returned to Leningrad as professor of theoretical physics and remained there until 1933. He left Leningrad to spend a year as visiting professor at the University of Paris.

In 1934 Gamow came to the United States as professor of theoretical physics at George Washington University in Washington, D.C., and remained at this post until 1956. During this period, his research interests shifted from pure physics to astrophysics and cosmology. While still continuing his research work in nuclear physics, he began a series of investigations into the evolution of stars. It was natural for him to become involved in the problem of stellar interiors because the theoretical technique he used in analyzing the emission of alpha particles by radioactive nuclei is just what was needed to study thermonuclear reactions in stars. In fact, one of the factors that enters into the formula for energy generation in stars is referred to as the Gamow reflection factor.

In one of his investigations during this period, Gamow discovered a process that can play a very important role in the life history of a star. This is the rapid emission of neutrinos, which allows a star to collapse to very high densities. In collaboration with Edward Teller, Gamow also studied the thermonuclear reactions in the early stages of a star's life.

After these investigations into stellar structure, Gamow became interested in cosmological problems and proposed the big-bang theory of the origin of the universe. He argued that all the heavy elements now present in the universe were built up from protons and neutrons in the first half hour after the big bang. This view of the formation of the heavy elements is no longer accepted, since we now know that they are built up in the very hot interiors of stars during certain stages of their evolution.

As Gamow became less active in research, he became more active as a science writer. Besides his book on nuclear structure, which is a technical treatise, he wrote numerous popular books, beginning with his famous *The Life and Death of a Sun*, which went through many printings. All of his books were brilliantly written, with emphasis on clarity, simplicity, and reader interest.

In 1956 Gamow became a professor of physics at the University of Colorado in Boulder where he remained until his retirement.

Electron Waves: Clinton J. Davisson (1881–1958) and George Paget Thomson (1982–1975)

When de Broglie first published his wave theory of the electron (and of matter in general) there was no experimental evidence to support his bold hypothesis. No one had any reason to think that electrons might have wave in addition to corpuscular properties. De Broglie reasoned that since light exhibits both wave and corpuscular properties, this dualism might also extend to electrons.

De Broglie did more than present the basic ideas of his wave theory. He also clothed it in a complete and consistent mathematical formulation. From it, one can predict the behavior of the waves and, in particular, the wavelength of an electron moving with a given velocity. It is easy to find the wavelength of an electron moving with speed v or with momentum mv, where m is its mass, by analogy with a photon. The energy of a photon equals $h\nu$, where h is Planck's constant of action and ν is the frequency of the photon. Now one can show from the theory of relativity, as Einstein pointed out, that the momentum of a photon equals $h\nu/c$. But $c = \lambda\nu$ where λ is the wavelength of the photon. Thus we find that the momentum of the photon equals Planck's constant divided by the wavelength, h/λ. It follows that wavelength $= h/$momentum. This is the expression for the wavelength of an electron as derived by de Broglie.

We see that the wavelength depends on the momentum and hence on the speed of the electron. Therefore, one can obtain various wavelengths for the same electron by altering its energy. This is important for experimental purposes: For electron speeds well below the speed of light (let us say less than one-tenth of the speed of light) we may place the momentum equal to mv; for speeds closer to that of light the expression must be replaced by $mv/[1 - (v^2/c^2)]^{1/2}$. From this we see that even for speeds of only a few thousand kilometers per second, the wavelength of an electron is about equal to that of X rays. As the speed goes up, the wavelength decreases. This fact was demonstrated by the experiments of G. P. Thomson and of Davisson and Germer. Just as crystals reflect and diffract X rays, they should also reflect and diffract electron waves, if indeed electrons have wave properties.

Before describing the work of Thomson, and Davisson and Germer, we must discuss the classical electron particle. If we suppose that the electric charge e of an electron is spread over the surface of a small sphere of radius a (which we may take as the classical radius of the electron) we obtain the electrostatic energy e^2/a. This gives us a particle picture that is very much at variance with the wave picture, since it indicates that the electron is an extremely tiny particle concentrated in a very small volume. To illustrate this point, we note that the energy e^2/a must equal the energy mc^2 (obtained from Einstein's mass-energy formula). This means that the classical radius a of the electron is of the order of e^2/mc^2, which is about one ten-trillionth of a centimeter.

We may now consider the experiments of Davisson, who, with the assistance of Germer, first observed the wave properties of electrons. If electrons were particles

Clinton J. Davisson (1881–1958)

in the classical sense (as described above), there would be no regular pattern in the reflection of electrons from the surfaces of crystals because an electron—pictured as a classical particle—is thousands of times smaller than the spacings between the atoms in a crystal. Hence, its behavior when it enters the crystal should depend only on whether or not it happens to pass close to some atom, which, in general, would not be very probable. Of course, as the electron pushes into the crystal, it is bound to meet some atom, but its behavior depends only on this single interaction, since all the other atoms are so far away that they do not affect the electron.

But this is not what Davisson and Germer found in 1928, when they scattered electrons from a nickel surface. Their discovery came from a happy accident. Although they had designed an experiment to study the scattering of slowly moving electrons from smooth metal surfaces, they had no idea of looking for a wave pattern in the scattered electrons. Indeed, they were unaware of de Broglie's work, which had been done three years earlier. In any case, their experiment as first set up would not have revealed the wave structure of the electron since, under ordinary conditions, the atoms in a metal are not arranged in an ordered scheme but haphazardly in small crystalline groups. In other words, the nickel strip that Davisson and Germer first used was not one large crystal. Therefore it could not have scattered the incident electrons in a regular pattern. No diffraction patterns, the hallmark of a wave, should have been observable.

But the vacuum tube in which they had placed the nickel strip accidentally burst. They found it necessary to heat the nickel, then to cool it slowly before it could be put into another tube for the resumption of the experiments. The heating and slow cooling caused the atoms to arrange themselves into a regular crystalline pattern, which produced the diffraction they observed when they sent the electron beam against the nickel surface. Rather than finding electrons reflected at random from the face of the nickel, they saw a regular pattern, similar in every respect to that of a reflected wave. This is remarkable because such a pattern can arise only if the reflection of the electron is caused by its simultaneous interaction with many atoms, rather than with one. The reflection of each electron seemed to be a collective phenomenon, precisely what would be found if the electron were accompanied by a wave.

When a wave, an X ray for example, strikes a crystal surface, it is reflected not only from the face of the crystal but also from different layers within the crystal. We must keep in mind that the atoms in a crystal are arranged in a definite formation, row after row, starting from the face of the crystal and going toward the interior. We may picture these rows as parallel to the face of the crystal, spaced at equal distances. A wave striking the face of the crystal at a given angle will, in part, be reflected from the first row of atoms. But parts of it will also be reflected from the second, third, and fourth rows, and so on, each in succession contributing less to the final reflected wave. The intensity of the reflected wave will depend on how all these partial waves, reflected in a given direction from the various layers, combine. If the reflected waves from the various layers are in phase when they leave the surface of the crystal (if all their crests coincide), the intensity of the reflected wave will be at a maximum. If they are out of phase (crest coinciding with trough), the intensity of the reflected wave will be zero.

Using their crystalline nickel surface, Davisson and Germer found that if the speed of electrons in a beam is constant, the intensity of the reflected beam is a maximum for only one angle of incidence; if the speed is varied, however, this angle of maximum intensity changes inversely with the speed of the electrons, just as it does with the wavelength of a reflected wave. They found that the relationship between the angle of maximum intensity, the speed of the electrons, and the lattice spacing in the crystal is the same as that for a wave, provided one assigns to the electrons a wavelength given by the de Broglie formula. This observation definitely established the wave-particle dualism of the electron and set the study of wave mechanics on a sound experimental basis.

Clinton J. Davisson began his experimental work on the scattering of electrons by crystals and metallic surfaces at a most opportune time, just before de Broglie and Schrödinger had introduced their wave picture of the electron. By the time the de Broglie-Schrödinger wave theory had been introduced, Davisson was ready to perform the experiments that were to establish the wave theory on a sound basis. Yet he had not undertaken his study to show that electrons have wave properties; he began in 1919 simply to inquire how electrons are scattered in various directions from metallic surfaces.

Davisson exhibited proficiency in physics and mathematics while still in high school, which he attended at a somewhat older age than is normal, graduating at 21. His father, Joseph, was an artisan and a descendant of the early Dutch and French settlers of Virginia; his mother was Mary Calvert, a Pennsylvania schoolteacher of English and Scottish parentage. After attending Bloomington, Illinois, grammar and high schools, he was given a scholarship to the University of Chicago for his high-school achievements in mathematics and physics. While at the university he came under the influence of Robert Millikan. After a year, he left the university for lack of funds and went to work for the telephone company in Bloomington. In 1904, at the recommendation of Millikan, he was appointed an assistant in physics at Purdue University, although still without an undergraduate degree. From Purdue, again with Millikan's help, he went to Princeton University as a part-time instructor in physics. While holding this post he studied under Professors James Jeans and O. W. Richardson, returning to the University of Chicago to attend summer sessions in order to complete his undergraduate studies; in 1908 he received a B.S. degree.

Princeton awarded him a fellowship in physics for the year 1910 to 1911, which allowed him to devote all of his time to his graduate studies and to complete his requirements for the Ph.D. He wrote his doctoral thesis, under Richardson, on *Thermal Emission of Positive Ions from Alkaline Earth Salts*. Richardson himself was to discover the basic laws of thermionic phenomena, for which he received the Nobel Prize in 1929. From Princeton, Davisson went to the Carnegie Institute of Technology. For a short time he also worked for J. J. Thomson at the Cavendish Laboratory. During World War I he joined the engineering division of the Western Electric Company and after the war continued as a member of the Bell Telephone Laboratory research staff.

The research which Davisson began in 1919 had, at first sight, no unusual facets or hidden meanings. Its purpose was simply to find how the energies of the electron reflected from a metallic surface are related to the energies of the incident electrons. But, on obtaining some unexpected results, Davisson investigated the scattering from crystals placed at various angles to the incident beam and found, "purely by accident," he said, that the intensity of the scattered beam depended on the orientation of the crystal. This was the beginning of the crucial experiment, for Davisson sent his results to Born, asking for his interpretation. Born and Franck, on the basis of calculations made by Elsasser, one of their students, suggested that Davisson's results stemmed from the de Broglie wave character of the electrons and advised him to perform an experiment designed specifically to look for de Broglie diffraction patterns. This experiment, begun in 1926, completely verified the de Broglie-Schrödinger theory. In 1937, Davisson shared the Nobel Prize in physics with G. P. Thomson for his work.

Davisson continued to study electron scattering after completing his basic research, but he became more interested in electron optics than in fundamental questions relating to the structure of matter. During World War II he contributed to the development of electronic devices and to crystal physics. After retiring from Bell Telephone Laboratories in 1946, he became visiting professor of physics at the University of Virginia in Charlottesville, where he died on February 1, 1958, at the age of 76. In his lifetime he received honorary degrees from various universities as well as numerous medals and prizes.

George Paget Thomson was born on May 3, 1892, at Cambridge, England. The only son of J. J. Thomson, he was educated at Trinity College, Cambridge, where he achieved the distinction of winning first class honors in the mathematical and natural sciences tripos in 1913. In World War I, young Thomson was commissioned in the Queen's Regiment and served in France from 1914 to 1915; he was then assigned to the Royal Flying Corps for experimental work on aerodynamic problems. In 1918, he came to the United States as a member of the British War Mission.

After the war, Thomson returned to Cambridge and carried out various researches in the Cavendish Laboratory. In 1922, he became professor of natural philosophy at the University of Aberdeen, where he carried out independent experiments on the diffraction of electrons in thin films, receiving the Nobel Prize, for this work along with Davisson, in 1937. In 1930, he was appointed professor of physics at the Imperial College of Science and Technology, University of London,

Sir George Paget Thomson (1892–1975)

a chair which he held until 1952 when he was elected master of Corpus Christi College, Cambridge. He retired from this position in 1962. During World War II, Thomson was chairman of the British Atomic Energy Commission (1940–1941) and served as scientific advisor to the Air Ministry (1943–1944), during which period he received a knighthood. In 1946 and 1947, he was scientific adviser to the British Delegation on Atomic Energy to the United Nations Commission. He was elected a fellow of the Royal Society of London in 1929 and was honored by numerous universities and scientific societies during his lifetime.

Thomson's demonstration of the wave nature of electrons was executed in a manner significantly different from the experiments of Davisson and Germer. In his first experiment, with A. Reid, Thomson directed a beam of cathode rays at a thin film of celluloid and examined the effect of the transmitted rays on a photographic plate placed behind this film. On developing the plate, a central spot made by the undeflected rays was found to be surrounded by diffuse rings. By means of densitometer measurements of the exposed plate, the rings were discovered to be clearly defined. Using 13,000-volt electrons, Thomson found two rings inside the one apparent on visual inspection. These represented diffraction halos in the same way that Debye-Sherrer halos were obtained with X rays reflected from powdered crystals. The rings arising from the diffraction of waves disclosed the wave properties of the electron. The fact that the rings showed a radius roughly inversely proportional to the velocity of the electrons was evidence to support the de Broglie equation $\lambda = h/mv$, relating the wavelength of the electron to the velocity of the electrons themselves. Subsequent experiments in which thin platinum films replaced the celluloid completely confirmed the de Broglie relation.

Since the original experiments of Thomson and of Davisson and Germer, numerous others have been performed to demonstrate the interference of electron waves. All such experiments have shown that electron waves interfere with one another in accordance with the de Broglie theory.

The Electron and Relativity:
Paul Adrian Maurice Dirac (1902–1983)

The quantum mechanics, as we have already noted, seemed to develop along two divergent lines—one ending with the matrix mechanics of Born, Heisenberg, and Jordan, the other with the wave mechanics of de Broglie and Schrödinger. For some time it appeared that the two theories, although giving the same results, were unrelated and irreconcilable. For how could the Heisenberg infinite arrays or matrices—representing the positions and momenta of electrons and identified with all possible electromagnetic radiation that an atom can emit during transitions of the electron—be equivalent to the position and the momentum of a wave packet? Dirac was one of the first to show that both the Schrödinger wave picture and the Heisenberg matrix structure are two aspects of a much more general theory of operators.

It occurred to Dirac, in a moment of deep insight, that Heisenberg's p and q matrices, representing the electron's momentum and position, are just special cases of a general group of quantities characteristic of quantum mechanics, with no counterpart in classical physics. Dirac pointed out that when we pass over from classical physics to quantum mechanics, the ordinary algebraic rules do not apply to all the quantities with which we are forced to deal. There are, indeed, two kinds of quantities in nature: the ordinary numbers of arithmetic, which Dirac called c-numbers, and the observable quantities that describe the behavior of a physical system. Dirac called these observables q-numbers and stated that they were governed by an algebra different from that which governed ordinary c-numbers.

Dirac generalized Heisenberg's results in the following way. For the Heisenberg p's and q's (the momenta and positions of a moving particle), the product pq is different from qp because these q's and p's are matrices in Heisenberg's scheme. More generally, we may say that pq is different from qp not because they are matrices but because we are dealing with quantities governed by a different kind of

Paul Adrian Maurice Dirac (1902–1983)

algebra, a noncommutative algebra.* That Heisenberg represented these quantities as matrices was the result of a very special way of looking at them. But this method of observation does not really capture the essence of the quantities.

This argument was Dirac's starting point. He went on to show that the quantities that Heisenberg had represented as matrices, the p's and q's, were special cases of a general class of q-numbers, all governed by a noncommutative algebra. Mathematicians for years had been dealing with noncommutative algebras in which the product of two factors depended on their order. Consequently, Dirac had little trouble in applying these mathematical ideas to his q-numbers and developing the algebra that governs them. He demonstrated that in dealing with the observable phenomena of quantum mechanics (the q-numbers) it is not necessary to represent them as matrices, or in any other formal way, to understand the processes within the atom or the behavior of the fundamental particles. All one need know about the observable q-numbers in order to understand their physics is the algebra that governs them. Representing them in one way or another is merely a matter of mathematical convenience dictated by the particular problem with which one is dealing.

We may illustrate this point by considering the Schrödinger and the Heisenberg pictures of the momentum of a particle. In the latter, momentum is represented as a particular kind of infinite matrix. But the Schrödinger equation for a particle is obtained only if momentum is represented as a mathematical operator of a specific kind, a so-called differential operator. Both the Heisenberg matrices and the Schrödinger waves are correct. They are, however, dependent on special representations of q-numbers, themselves mere abstract, algebraic representations of the operators whose values are sought in making laboratory measurements of physical systems.

Dirac expanded the idea of an operator to include all physical quantities such as the energy and angular momentum of a system, as well as the p's and q's, the linear momentum, and the position. In fact, any quantity represented in terms of the p's and q's can be thought of as an operator. To obtain the operator corresponding to this quantity (the quantity is called a function of the p's and q's) we replace each q or p that appears in the quantity by the suitable operator.

As an example, consider the expression for the energy of a system such as an electron in a hydrogen atom. This mathematical expression can be expressed on the basis of classical physics, in terms of the position and the momentum of the electron, that is, the p's and q's of the electron. To obtain the operator that represents the energy of the electron in quantum mechanics we must replace each p and q by its own appropriate operator. If the Heisenberg representation for these operators is used, the energy becomes an infinite matrix. But if the Schrödinger

---

*Noncommutative means that the product of two factors is different if the order of the two factors is changed. This kind of algebra is quite different from ordinary arithmetic where 3×4 is the same as 4×3. The concept of noncommutativity can be understood in terms of ordinary experiences. Every chemist and cook knows that the final product that one gets depends on the order in which things are done. Beating the egg whites first and then adding them to a cake is not the same as first adding them and then beating the mixture. Similarly, pulling the trigger of a gun and then putting the bullet in the gun does not give the same result as first inserting the bullet and then pulling the trigger. Thus these operations are noncommutative.

representation is used, the energy becomes a differential operator and the quantity on which it operates is the Schrödinger wave function.

To complete this theory Dirac had to show how to apply these ideas to any general situation involving a variety of measurable, observed quantities. For example, suppose we are dealing with an electron whose position at some instant we know with infinite accuracy. Now we must use noncommutative algebra because the theory tells us that two quantities that do not commute cannot be measured with infinite accuracy simultaneously. We can have no knowledge of the momentum of the electron at the instant we know its position, nor can we know the energy of the electron since the energy involves the momentum. Indeed, the precise knowledge of the position of the electron, because of the noncommutative nature of the quantities, precludes the further knowledge of measurable quantities that can be constructed from knowledge of the momentum.

Dirac described this situation by supposing that the electron is in a particular quantum state into which it has been forced by the measurement of its position. This measurement has destroyed our knowledge of any observable quantity whose algebraic expression does not commute with the position operator q. It has taken the electron from a state about which we had some precise information (for instance, the energy) into another state about which our measurement has given us some other precise information but, at the same time, destroyed our original information.

By introducing the concept of the state of a system Dirac generalized the uncertainty relationship in the following way. We know from classical physics that whenever we are dealing with a physical system we require the knowledge of more than one quantity to describe the behavior of the system in its entirety. This need arises because the system, in general, has more than one degree of freedom. Thus, in the case of an electron moving freely in space, we have three degrees of freedom to consider—(the three independent directions in space along which the electron moves). The measurements of three distinct quantities must be made before the state of the electron is fully known. Dirac pointed out that these three measurements must be of a sort to give simultaneous, infinitely precise information about these three quantities. It follows, then, because a noncommutative algebra is equivalent to an uncertainty principle, that all operators must commute if they represent observables whose values must be precisely known to define the state of a system completely.

Now the operators that represent the observables of a system have their classical counterparts in certain algebraic expressions involving the p's and q's of the electrons in the system. It is not always easy to determine whether these operators commute. Of course, since these operators are composed of p's and q's, the commutative properties of the p's and q's should enable one to determine the commutation properties of these operators. This is a difficult procedure. It is convenient, therefore, to introduce a way of determining whether operators commute without having recourse to the p's and q's directly. Dirac did this by carrying over into quantum mechanics certain mathematical expressions which had been introduced into classical mechanics by Poisson. Poisson had shown that the equations of motion in classical mechanics can be expressed in terms of certain quantities known as Poisson brackets. There are definite classical rules for calculating these brackets for any two algebraic quantities that depend on the p's and q's which describe the mechanical system with which we are dealing. Although these rules have nothing to do with quantum mechanics or noncommutative algebra, Dirac showed that if F and G are any two observables, then the value of the commutator $[FG - GF]$ in quantum mechanics is, except for a numerical constant, equal to the classical Poisson bracket for these two quantities multiplied by Planck's constant. If the Poisson bracket vanishes the two quantities commute.

In his book on quantum mechanics Dirac stressed the essential difference between the concept of the state of a system and the same concept in classical physics. The difference, as Dirac aptly pointed out, lay in the principle of superposition of states, which has no analog in classical physics. It is "one of the most fundamental and most drastic" of a new set of accurate laws required to set up the quantum mechanics. Essentially, this principle of superposition says that a system in quantum mechanics must be looked upon as being simultaneously in a whole set of states rather than in some particular state, as is the case in classical physics.

Dirac illustrated this principle by considering the polarization of a beam of light consisting of photons and also the interference of photons. Regardless of whether we are dealing with polarization or with interference, we must consider the photon as being simultaneously in two different states. In the case of the interference phenomenon we are forced to picture each photon as being "partly in each of the two components into which the incident beam is split," even though the photon must be regarded as an indivisible particle. Of course, the photon can never actually be found in both beams simultaneously because in the very process of looking for the photon we force it into one or the other of the partial beams.

To illustrate this point and show how quantum physics differs from classical physics, consider a person moving in a direction that is 30° north of east. This motion can be represented by an arrow of appropriate length pointing in this direction. We may now say that this person is moving simultaneously in an eastward direction and in a northward direction. Of course, the eastward and northward motion are not equal. In classical physics there is no doubt about the person's state of motion. Saying that he is moving partly in an eastward and partly in a northward direction is a convenient way of representing the fact that the vector describing the actual motion can be pictured as consisting of a vector lying in the eastward direction and another lying in the northward direction (the two being of unequal length). It is clear, however, that in classical physics this is merely a way of talking. If one wants to find the direction in which the man is walking, one must look in the direction that is 30° north of east.

The quantum-mechanical way of describing the motion of the person would be to say that the person is really moving partly along the east-west line and partly along the north-south line; the probability of finding him moving entirely along either of these directions is determined by the cosine of the angle which his actual motion makes with each of these directions. Moreover, in the process of looking for the person along one direction or the other we would force him to move entirely in the direction in which we found him to be moving. Of course, it is clear that such considerations cannot apply to a person or any other massive object, but this is merely because we disturb a massive object only very minutely when we look at it. In the case of an electron this is not so. The disturbance we introduce when we make any measurement on it forces it to change suddenly from one state to another.

Dirac's idea of states connected the Schrödinger wave function with the probability concept ascribed to the Schrödinger wave function by Born. Consider an electron about which we wish to obtain some information. It is necessary for us to perform some measurement to get this information. Before the measurement, the electron is in some definite state but not a pure state in the sense that we have no definite information about it. We must therefore consider it as a combination of many different pure states, each one entering into the composition of the actual state of the electron with a different weight. When we perform a measurement and obtain definite information about the electron, we force it into one of the pure states. The probability of our finding it in one of these pure states when we carry out our meaurement is given by the weight with which that particular pure state enters into the composition of the original state of the electron. As long as we perform no measurement on the electron, the state of the electron (the wave function that describes it) changes continuously and this continuous change is described by the Schrödinger wave equation. Any measurement, however, causes a sudden change.

According to Dirac's theory, at any instant of time we can describe the behavior of a system by a mathematical quantity that represents what he has called the state of the system. If the system is left to itself, this mathematical quantity (function) changes continuously according to a definite mathematical prescription, the Schrödinger equation. However, whenever a measurement is made on the system, its state changes suddenly, since in performing our measurement we have disturbed it and thus forced it into a new state. After the measurement the system continues to change its state continuously according to the Schrödinger wave equation.

The importance of the Schrödinger wave function here is that it enables us to pass from one pure state to another by means of a mathematical transformation in which the Schrödinger function is involved. Suppose that at some instant of time we definitely knew the position of the electron. We would then have no knowledge at that moment of the energy of the electron. Let us now measure the energy of the electron and thus force it into a definite energy state. This means that we have learned about the energy of the electron but know nothing about its position. When we have done this measurement, it is equivalent to solving the Schrödinger equation. The Schrödinger wave function may thus be considered as the quantity that enables us to pass from the state of the system in which we know the position precisely, at some moment, to one in which we know the energy precisely, at the same moment. Thus, Born's probability interpretation of the Schrödinger wave function means that if the electron is in a state in which its energy is precisely known, the value of the Schrödinger wave function, at a particular point of space, when multiplied by the value of its complex conjugate, gives the probability for finding the electron at this point in space.

While matrix mechanics was being developed by the Göttingen group under Born, and wave mechanics was being proposed by Schrödinger in his famous wave equation, the form of quantum mechanics described above, more general than either, was being developed in England by Paul Dirac, who must be ranked as one of the giants of modern physics. Dirac's success in deriving his form of quantum mechanics, which is now universally used, stemmed from his general philosophy that mathematics is not just the handmaiden of physics but an integral part of it. He believed that one should not peer too deeply behind the mathematics to find the physical reality because they are too closely interrelated. This view led him to seek the most general and elegant mathematical methods of presenting a physical theory because he felt that the most refined representation would reveal physical realities that would be otherwise obscured, as is the case in approximate or nongeneral mathematical procedures. Indeed, he felt that every mathematical theory has its physical counterpart. This philosophy also led him to follow the physical consequences of the mathematics of any theory to their ultimate conclusion, however strange and contrary to our experience they might appear. Dirac's great genius lay in his ability to follow unbeaten paths and to use unorthodox mathematical methods whenever necessary to develop his physical theories. It is quite understandable that he should have taken this approach because his graduate studies were in mathematics; in a sense, he entered physics through the backdoors of mathematics and engineering.

Paul Dirac was born on August 8, 1902, in Bristol, England, of a Swiss father and an English mother. After attending elementary and secondary schools in Bristol, Dirac entered Bristol University where he studied electrical engineering, receiving his B.S. degree in 1921. At this point, a definite change occurred in his concept of the knowledge required to become a good physicist; as a result, he studied mathematics at Bristol for two years after receiving his undergraduate degree. From there he went on to St. John's College, Cambridge, as a research student in mathematics and received his Ph.D. in mathematics in 1926. He remained at St. John's College as a fellow and, finally, was named Lucasian professor of mathematics at Cambridge. Dirac received his degrees in mathematics just when the world of physics was exploding with activity; this period is still considered the golden age of physics. The years from 1926 to 1932 were his most creative and his contributions to physics cover a wide range of subjects.

The year 1926 was particularly fruitful. Not only did Dirac present his own general version of the quantum mechanics, but he also wrote his famous papers on quantum electrodynamics (the quantum mechanics of the radiation field) and developed the quantum statistics of particles that obey the Pauli exclusion principle. The quantum statistic is now known as the Fermi-Dirac statistics since Fermi made the discovery shortly before, and independently of, Dirac.

Dirac's first important contribution to physics was his formulation of quantum mechanics in which neither the Schrödinger wave equation nor the Born-Heisenberg matrices were introduced. Instead, he used a general noncommutative algebra for all physical observables so that the physical properties of a system were then contained in the algebraic rules governing the multiplication of any two physical observables. Dirac showed that the Born-Heisenberg and Schrödinger formulations are special cases of the general theory.

Dirac's next great contribution was to demonstrate the relationship of the various formulations of quantum mechanics to one another and to show how to make the transition from one to the other. This work constituted what is now called the transformation theory of quantum mechanics.

Almost at the same time Dirac was developing a quantum electrodynamics in which the radiation field itself was seen as quantized. The absorption of a photon was pictured as a quantum process in which the quantum state of the radiation field changed through the disappearance (or destruction) of a photon, and the emission of radiation was shown as a quantum change of the field resulting in the creation of a photon. Dirac pictured all radiation processes as describable in terms of creation and destruction of photons and obtained the general quantum laws of radiation as a consequence of the algebra of his creation and destruction operators. Following closely upon this work, he developed the special quantum statistical mechanics that bears his and Fermi's names.

Each of these developments was a great achievement in itself. For one man to have done all of this while still very young and in the span of a few years was a mark of tremendous genius. Dirac was immediately recognized as one of the giants of the modern era on an equal footing with Heisenberg, Pauli, Schrödinger, and the other originators of the quantum mechanics.

But Dirac's greatest creation was yet to come—his relativistic theory of the electron for which he shared the Nobel Prize with Schrödinger in 1933. Dirac's approach to this problem showed his great physical intuition and his unusually ingenious utilization of mathematical and physical knowledge. The problem that he faced was to derive a wave equation that satisfied the demands of the theory of relativity and, at the same time, retained the de Broglie-Schrödinger wave picture of the electron. Dirac achieved this by using a special mathematical technique (his famous spin matrices) in which the spin of the electrons enters automatically, without requiring separate introduction. Dirac thus showed that the spin is a consequence of the relativity theory.

He also demonstrated that the negative energy solutions of his wave equation must have their physical counterpart, although negative energy has no meaning in classical physics. Dirac's interpretation of negative energy states led to his theory of holes, namely that empty negative energy states (holes) are positively charged particles. These holes are the positrons that were later discovered by Carl Anderson. This theory of holes—stemming from his relativistic wave equation—opened up vast new fields in physics, and, as Dirac himself first suggested, the possibility, now completely verified, that every particle in nature had its antiparticle (hole) counterpart.

After discovering the relativistic wave equation, Dirac spent many years improving the mathematical formulation of quantum mechanics. At the same time, he tried to eliminate certain undesirable intrinsic features of all quantum theories which gave infinite answers to simple questions regarding the manner of the absorption and emission of radiation by the electron. These investigations were the starting point from which Tomonaga, Schwinger, Feynman, and other physicists developed methods that could be used in conjunction with the Dirac theory to give reasonable (noninfinite) answers to reasonable questions.

Quantum mechanics, introduced in its wave form by Schrödinger and in the form of a matrix mechanics by Born, Jordan, and Heisenberg, was expressed in its more general form by Dirac's transformation theory. His interpretation was remarkably successful in solving problems involving the interactions of fundamental particles. In spite of these successes it became increasingly evident that in its original form the Schrödinger theory was incomplete. This is evident from a consideration of the fine structure of spectral lines. There is nothing in the Schrödinger wave equation that is the counterpart of the spin of the electron, introduced by Goudsmit and Uhlenbeck to explain the fine structure. It remained for Dirac to remove this last inadequacy and to derive an improved wave equation, which automatically introduced spin into the theory. But Dirac's new wave equation produced difficulties of a profound nature that were finally eliminated only by a radical revision of our concept of the vacuum. To understand the nature of this new concept, we must examine how Dirac altered the Schrödinger wave equation in order to introduce the spin of the electron.

The Schrödinger wave equation is incomplete because the Schrödinger wave function for the electron contains only the three quantum numbers already introduced by Sommerfeld and others in the extended Bohr theory of the atom. To describe the spin of an electron, however, a fourth quantum number was needed. Pauli had already demonstrated this need on the basis of the relativity theory, which demands that all descriptions of particles be in terms of four coordinates, or four degrees of freedom, not three. For this reason Pauli suggested that a fourth degree of freedom (a fourth quantum number) be assigned to the electron to explain the fine structure of spectral lines. To account for the doubling of the spectral lines, Pauli restricted the fourth quantum number to only two values; this restriction fitted neatly into the spin concept of Goudsmit and Uhlenbeck, who postulated that in a magnetic field the spin of the electron could align itself either in the direction of the field (parallel alignment) or opposite to the field (antiparallel alignment).

To describe an electron interacting in the most general way with an electromagnetic field, Pauli replaced the single Schrödinger wave equation with two similar equations, each containing an added term representing the interaction of the magnetic moment of the spinning electron with the magnetic field. One equation represented the parallel arrangement of spin and field, the other the antiparallel arrangement. Although the use of two simultaneous equations to describe the motion of the electron seems, at first glance, to destroy the idea of a wave associated with the electron, this is not the case. Something similar happens in the electromagnetic picture of light. There, too, one has to introduce a group of equations to describe the propagation of a light wave because it involves vectors, the electric and magnetic field strengths, each consisting of three components.

By introducing a matrix with two rows and two columns to represent the spin of the electron, Pauli was able to combine his two equations into one. It bears a formal resemblance to the Schrödinger equation, with two exceptions: an additional term represents the interaction of the spin with the magnetic field; and the wave function consists of two components that represent the two possible spin orientations of the electron.

In his well-known paper on the relativistic theory of the electron, Dirac derived its spin properties by combining the wave mechanics of Schrödinger with the special theory of relativity. From what we have already said it is clear that the Schrödinger equation is deficient since it is based entirely on nonrelativistic considerations. A number of physicists, including Schrödinger himself, had attempted to obtain a relativistic wave equation before Dirac solved the problem. But they failed because in relativity theory it is not the energy that is directly related to the square of the momentum but the square of the energy; therefore, the energy itself has to be expressed as the square root of a quantity involving the momentum. Since the momentum must be taken as an operator, one cannot work with the square root. The square root of a sum of operators is not defined and there is no way to interpret it physically.

To overcome this difficulty, physicists, such as Klein, Gordon, and Schrödinger, set up a relativistic wave equation in terms of the square of the energy rather than in terms of the energy itself. Their wave equation, which is, indeed, in accord with the requirements of relativity theory, contains the operator that corresponds to the energy, the quantity that appears in the nonrelativistic Schrödinger equation and in the Puali spin equations. The mathematical operator in quantum mechanics that corresponds to the energy in classical mechanics is the time rate of change, that is, the derivative with respect to the time. Thus, the square of the energy must correspond to the operator representing the time rate of change of the time rate of change, or the second time rate of change. However, the substitution of the second time rate of change for the time rate of change itself leads to a serious difficulty. It is impossible to correlate the wave function with the probability for finding an electron at a given point of space-time, as was done by Born for the nonrelativistic wave function. Indeed, it is clear that this so-called second-order wave equation named for the second time rate of change operator cannot represent the motion of a single particle, since it leads to probabilities that can be understood only in terms of the motions of many particles having both negative and positive electric charges.

Dirac overcame this difficulty by discovering a wave equation that was relativistically correct yet contained only the first time rate of change operator. He took the expression for the square of the energy in its correct relativistic form and factored it, splitting it into a product of two equivalent terms. He used only one of these terms to obtain the wave equation. It turned out, however, that the factoring process automatically introduced matrices into the mathematics, structures similar in nature to those Pauli had introduced in an ad hoc fashion. The Dirac matrices thus entered into the theory quite naturally because of the requirement that the final wave equation involve only the first time rate of change operator. This was all to the good since, as we have noted, the introduction of Pauli-like matrices must lead to particles having spin.

There is, however, one important difference between the matrices that appear in the Dirac equation of an electron and those of Pauli. Whereas the latter are just three matrices, each having two rows and two columns, the former constitute a group of four matrices, each having four rows and four columns. In fact, three of the Dirac matrices can be obtained by taking each Pauli spin matrix twice and arranging these into a four-rowed structure. The significance of four-rowed matrices in the wave equation is that the equation becomes fourfold, with the wave function composed of four components. Dirac showed that his equation gives the same results that Pauli had obtained by arbitrarily introducing spin into the Schrödinger mathematics. But Dirac's work went further because it introduced a four-component wave function, governed by four simultaneous wave equations. The sudden appearance of this fourfold wave function gave rise to a grave and apparently insurmountable difficulty.

The relativistic Dirac theory shows that the electron is a far more complicated structure than was ever dreamed of before, that it possesses properties even beyond those of spin. This is apparent from the four-component wave function itself, since we know from Pauli's work that to describe the spin alone only a two-component wave function is required. If only two of the components of the Dirac wave function are needed to describe a spinning electron, what is the role of the other two components? When Dirac examined these other two components, he was inclined to dismiss them as nonsensical. They describe a spinning electron with less than zero energy—that is, an electron with negative energy and hence, in terms of our ordinary understanding of things, an impossibility. Dirac suggested that we disregard these other two components of his wave function and use only the two "sensible" components.

But if one tries to describe an electron using only the two positive energy components of the Dirac wave function, the description does not tally with observations. If one uses only the two "sensible" components to describe an electron in a box, one suddenly discovers that this description allows the electron to leave the box (seemingly without passing through the walls) and appear outside it. In other words, all four wave components must be employed to describe an electron properly. Here the full dilemma of the situation is revealed. If one pictures the negative-energy states of the electron as having physical reality, how can ordinary matter occupying the states of positive energy (all matter) not jump down into the unoccupied negative-energy states, thus bringing about a cataclysmic annihilation of all matter? It was to show a way out of this difficulty that Dirac published his paper on electrons and protons in 1930.

Dirac solved the problem in a most ingenious fashion—a tribute to his great imaginative powers—endowing the theory with a new dimension that had not been dreamed of before. He imparted a new meaning to the vacuum. Since the negative-energy states could not be eliminated from the theory without destroying it, and since such states, if left empty, would lead to the complete collapse of the physical world, Dirac hypothesized that each negative-energy state was already occupied by an electron. This assumption saved the entire theory because the Pauli exclusion principle stated that only one electron could occupy a particular energy state at any time. By filling the entire vacuum with an infinitude of electrons in states of negative energy (electrons in negative-energy states cannot be observed), we can use the Dirac theory to describe ordinary electrons in positive-energy states.

There was one error in Dirac's paper which was later corrected by J. R. Oppenheimer and by Dirac himself. As Dirac pointed out, a negative-energy state that for some reason or other is lacking an electron behaves like a hole in a sea of negative-energy states, or like a bubble in an ocean of water. Hence it resembles a positive charge with positive mass. (The absence of negative charge and of negative energy or mass is equivalent to the presence of positive charge and positive energy.) This analogy led Dirac to his idea that protons are these holes in the vacuum. We know now that this interpretation was incorrect. The holes that behave like positive charges are, indeed, particles in their own right, the so-called positrons. In 1934 they were discovered in cosmic ray photographs by Carl Anderson. Since positrons can annihilate electrons, it is customary to consider them as antielectrons. With the more recent discovery by Segré and Chamberlain of antiprotons and antineutrons, there is good reason to suppose that our universe is composed of equal quantities of matter and antimatter.

Dirac's paper was an excellent example of his unorthodox approach to physical problems and his boldness in postulating ideas that did such violence to the traditional ideas of space, time, and matter. In this respect, Dirac was closer to Einstein than were any of the other physicists of the modern quantum-mechanical era. Dirac was able to extract from the formalism of a theory everything that was of physical significance. For him a particular mathematical expression or relationship went far beyond the correlation that it might establish between two or more entities. Its very form and the branch of mathematics from which it stemmed were of deep physical significance.

Another feature of Dirac's approach to theoretical physics was his insistence on making the mathematical formulation of a theory as beautiful as possible. He felt that a theory that was expressed in an inelegant and cumbersome mathematical form could not be correct and must be incomplete in some respect. One example is the formulation of the relativistic wave equation as compared to that of the nonrelativistic Schrödinger equation. In one of his articles, Dirac pointed out that the former is much more beautiful than the latter and that, if Schrödinger had insisted on beauty, he might well have arrived at the correct relativistic equation.

Dirac's contribution to physics went beyond quantum mechanics and the theory of the wave electron. He also did provocative work in cosmology. Based on very general principles, he proposed the theory that the constants of nature, such as Newton's gravitational constant, do not remain the same but change with time because of the expansion of the universe. Dirac also introduced methods of quantizing Einstein's gravitational field to obtain a quantum theory of gravity. Although he was not alone in this endeavor, his procedure bore the usual Dirac stamp of individuality and ingenuity.

Dirac lectured extensively, gave courses in quantum mechanics and quantum field theory in universities throughout the world, and was a visiting professor at the Institute for Advanced Studies at Princeton. His lectures reflected, as one would expect from a man of his stature, his remarkable grasp of physics. He presented the subject matter with great simplicity and clarity; he always emphasized the physical concepts and relegated formalism to a minor role. His book, *The Principles of Quantum Mechanics*, was a model of textbook writing and one of the best introductions to the operator approach to quantum mechanics.

"Holes" in the Dirac Theory: J. Robert Oppenheimer (1904–1967)

When Dirac proposed his relativistic theory of the electron, it was overburdened with difficulties since it predicted the existence of an infinite continuum of negative energy states. According to this theory, the universe consists of two parts divided by the line of zero energy. Above this line are the states of positive energy, most of which are unoccupied since the amount of matter in the universe is negligible compared to the amount of empty space; below the line are the states of negative energy, whose very existence poses a dilemma. Quantum mechanics predicts that ordinary electrons have a finite probability of jumping down into these negative-energy states with the emission of radiation. Thus, if all of these states were empty, no particles, such as ordinary electrons, could exist; there would be a finite probability for all of them to fall into negative-energy states.

To overcome this difficulty Dirac proposed that all but a few of these negative-energy states were filled with electrons with negative energy. Of course, an electron with negative energy has no real meaning and certainly has no classical counterpart. But the idea makes it possible to work with the Dirac theory of the electron, which accounts beautifully for all the observed phenomena involving these particles. By placing electrons in most of these negative-energy states, Dirac made it impossible for ordinary electrons to jump down, since the Pauli exclusion principle permits only one electron in each state. Dirac proposed further that negative-energy states that do not have electrons represent protons, since such holes in the sea of negative-energy electrons behave like positive charges. In this way Dirac hoped to account both for protons and electrons. He suggested that the difference in mass between the electron and the proton might be accounted for by the interaction of the ordinary electrons with the infinite sea of negative-energy electrons.

However, J. R. Oppenheimer pointed out that to assume that the ordinary protons in the universe are gaps in the sea of negative-energy electrons leads to insurmountable difficulties. The assumption that there are no distinct positive charges requires an infinite distribution of positive charge to compensate for the effect of

J. Robert Oppenheimer (1904–1967)

the infinite distribution of the negative-energy electrons on the scattering of light and on other electromagnetic phenomena. Moreover, if the Dirac holes were protons, the scattering of light by protons would have to be explained by a process involving the interaction of the ordinary electrons with the holes. Consequently, the formula that shows how the scattering depends on the mass of the proton would be wrong. The correct formula requires that the scattering vary inversely as the square of the mass of the scattering particle whereas the hole theory gives a different mass dependence.

Finally, as Oppenheimer noted, no protons could exist in nature (nor could electrons) if the gaps or holes in Dirac's sea of negative-energy electrons were protons. The theory shows that if any hole is present, there is a finite probability for an electron in its neighborhood to jump down into the hole and thus to destroy itself as it fills the hole. Under these conditions, Oppenheimer showed that in an ordinary sample of matter there are so many protons around that, if they were holes, all the electrons in this matter would jump down into the holes in less than a billionth of a second. This means that there can be no empty negative-energy states under ordinary conditions.

Oppenheimer therefore proposed to retain the concept of the electron and the proton as two independent particles of opposite sign and dissimilar mass and to picture all of Dirac's negative-energy states as filled. With all the negative-energy states filled there can be no transitions of ordinary electrons from their positive-energy states to the negative continuum and all the difficulties pointed out by Oppenheimer thereby disappear. At this point Oppenheimer was but one step away from predicting the existence of the positron, discovered in cosmic rays a few years later by Carl Anderson. We now know that it is not necessary for all the negative-energy states to be filled. Under certain conditions, if enough energy is available, an electron in a state of negative energy can be lifted into a state of positive energy so that a pair is created—an ordinary electron and a Dirac hole, which behaves like a positively charged particle with exactly the same mass as the ordinary electron. These short-lived holes are the positrons.

It is curious that despite the technical dominance of the United States in the first two decades of this century and the excellence of its educational system, there was, during this period, no American school of physics comparable to those that sprang up in Europe. There were, of course, excellent experimentalists such as Michelson, Millikan, Compton, and Davisson, but there were scarcely any first-rate theoreticians, without whose stimulus and constant analysis of existing theories in the light of new experimental data no great progress can occur. With the exception of a few isolated cases, such as Willard Gibbs of Yale, who laid the foundations of statistical mechanics in the early part of this century, most of the great ideas that influenced the experimental physicists in the United States came from Europe.

In the 1920s, when the great quantum-mechanical ferment was stirring European physics and, as the great German mathematician David Hilbert put it, Nobel prizes were lying around to be picked up in the streets, American theoretical physics began its spectacular growth, which resulted in its present dominant position. At that time such excellent theoretical physicists as G. Breit (a Russian exile), Van Vleck, Condon, and Slater were beginning to attract more and more brilliant young college students to physics. By the time J. Robert Oppenheimer began to hold sway in California, enough good raw material was around for him to organize what might be called an American school of theoretical physics. An additional impetus was given to American physics in the 1930s by the influx of such European physicists as Wigner, Einstein, Bethe, and Fermi.

Oppenheimer was very well suited for this task by virtue of his innate ability and the excellence of his schooling. He was born in New York City on April 22, 1904, of well-to-do German-Jewish parents, who gave him every encouragement in his intellectual pursuits. From early boyhood he showed a pronounced interest in science and mathematics. While still in high school he mastered calculus. His interest in science extended to all phases that required an orderly analysis of data and a formulation of new concepts. In addition to his abiding interest in science, Oppenheimer was also deeply concerned with the humanities, with the social sciences, and with the arts. Indeed, throughout his life he never allowed his deep dedication to science to rob him of the pleasures that can be found in other pursuits.

After graduating from high school Oppenheimer went to Harvard, where he received his A.B. degree in 1925. During this phase of his education, he so clearly demonstrated his great skill in mathematical physics that it was assumed he would become one of the top theoretical physicists of the future. From Harvard he went on to the University of Göttingen to study with Born, then the leading spirit in the development of the quantum mechanics. At that time the atmosphere in Göttingen was particularly appealing to a young graduate student interested in the new ideas that were rapidly altering the basic concepts of physics. In addition to Born himself, and outstanding experimentalists such as James Franck, it had attracted many of the young Ph.D.'s who were the source of these new ideas— men like Heisenberg and Jordan, Born's assistants, and other young physicists like Kramers, Pauli, Fermi, Dirac, Weisskopf, and Wigner. It seemed as though no promising student of theoretical physics could consider his education complete without spending some time at Göttingen.

Göttingen had something else to offer—its school of mathematicians. Starting with Carl Friedrich Gauss and continuing with such men as Georg Riemann, Klein, Hermann Minkowski, and David Hilbert, Göttingen had become the center of the kind of mathematics that the new physics required. Minkowski had developed the four-dimensional geometry that was needed for a full understanding of the relativity theory and Hilbert had formulated the mathematics required for the full flowering of the quantum mechanics. He had developed the mathematics of an abstract infinite-dimensional function space—called Hilbert space—that, in a sense, occupies the same position in quantum mechanics that Minkowski's four-dimensional space occupies in relativity theory. In addition to the resident professors of mathematics like Hilbert, young mathematicians like John von Neumann, who did much to relate Hilbert's mathematics directly to the quantum mechanics, and Van der Waerden, who formulated the mathematics of spinors—mathematical quantities that play an important role in the Dirac theory of the electron—were constantly visiting Göttingen.

When Oppenheimer came to Göttingen, Heisenberg had just completed his first important paper on the matrix mechanics and Schrödinger had published his first paper on the wave mechanics. Born himself became interested in the interpretation of the Schrödinger wave function and hit upon the idea of analyzing collisions between particles to arrive at the correct interpretation. Oppenheimer was quite naturally brought into this work. In collaboration with Born, he published some basic papers on collision theory. As a result of this, he received his Ph.D. at Göttingen in 1927 at the age of 23.

From Göttingen he went to Zurich and then to Leyden as a fellow of the Institute of International Education. In 1929 he became assistant professor of theoretical physics at both the California Institute of Technology and the University of California. In quick succession he became associate professor in 1931 and full professor in 1936 at both institutions.

During this period, Oppenheimer contributed many important papers to quantum mechanics and to atomic theory. At the same time, he attracted the best of the young American physicists and even some Europeans, so that the University of California became the American center of theoretical physics and remained so until the Institute for Advanced Study at Princeton began to outstrip it.

Oppenheimer's research activities cut across many branches of physics. In addition to his work on the properties of the Dirac electron and on the interpretation of the negative-energy states that are a consequence of the Dirac relativistic wave equation, he made important contributions to the theory of cosmic rays, to nuclear physics, to fundamental particles, and to astrophysics. In his work on cosmic rays he was one of the first to study and analyze the origin of cosmic ray showers. These are large bundles of secondary particles originating from a single high-energy cosmic ray particle (proton). As Oppenheimer demonstrated, the primary proton creates many pairs of electrons and positrons as it passes through the earth's atmosphere; these secondary particles form the showers that accompany the primary cosmic ray particles.

When the neutron was discovered in 1932, and nuclear physics began to occupy the minds of physicists more and more, Oppenheimer began to work in this field. He wrote an important paper, in collaboration with Volkov, on the application of neutron physics to stars. He showed that, under appropriate conditions, a star can contract gravitationally to such an extent that all the protons inside the star are changed into neutrons and a pure neutron star, with a diameter of about 10 miles, results.

With the beginning of World War II, Oppenheimer, like most other leading physicists, became involved in the problem of nuclear fission. When it became clear that a chain reaction was possible, he was chosen (by almost unanimous agreement among the physicists working on the atomic bomb project) to head the work at Los Alamos where the bomb was to be constructed.

Immediately after the war Oppenheimer remained for a year at Los Alamos, one of the first of a number of permanent National Scientific Laboratories established in the wake of the vast scientific-military research activity. In 1947 he left Los Alamos to become director of the Institute for Advanced Studies at Princeton, where he remained until his retirement in 1965. During this period, Oppenheimer devoted himself to his own research and to making the institute the outstanding center of theoretical physics in the world. His research dealt with the theory of fundamental particles and particularly with the nature of mesons—the particles by means of which nucleons (protons and neutrons) attract each other. He was one the first to postulate the existence of a neutral pi-meson in addition to positively and negatively charged pi-mesons.

Oppenheimer received many honors for his contributions to physics and the world of science in general. In addition to receiving the Fermi award he was made an honorary fellow of Christ's College, Cambridge, a member of the National Academy of Sciences, a fellow of the American Philosophical Society and the American Academy of Arts and Sciences, honorary member of the Japanese Academy of Sciences, of the Royal Danish Academy of Sciences, and the Brazilian Academy of Sciences.

Oppenheimer was best known to the general public for his highly publicized hearing before a Personnel Security Board in 1953. This grueling three-week hearing was prompted by the federal government's withdrawal of Oppenheimer's security clearance. The government action seems to have been motivated by an assortment of half-truths and rumors about the political orientations of some of Oppenheimer's former colleagues as well as his own lack of enthusiasm for the hydrogen bomb project. Although Oppenheimer's testimony was elegant and demonstrated that there was little basis for the charges being levied against him, the Atomic Energy Commission upheld the withdrawal of his security clearance.

From 1946 to 1952 Oppenheimer was chairman of the General Advisory Committee of the Atomic Energy Commission, and from 1949 to 1955 he was a member of the Board of Overseers of Harvard. In 1948 American physicists honored him by electing him President of the American Physical Society—the highest honor they can bestow upon a colleague.

Complementarity: Niels Bohr (1885–1962)

Quantum mechanics, the modern view of matter and energy, has evolved from a number of apparently contradictory theories. In 1924, de Broglie, who initiated the new developments, showed clearly that particles such as electrons have wave properties that can be described in terms of wavelengths and frequencies related to their momenta and Planck's constant of action. Schrödinger subsequently developed the wave equation for a particle, the solution of which gives its possible states of energy and allows one to follow the motion of the particle to within certain limits of accuracy. Born then demonstrated that the waves associated with a particle, as obtained from the solution of the Schrödinger equation, must be interpreted as probability amplitudes for finding the particle in a given region of space at any given moment. A particle thus appears to behave like a wave in the sense that one has to associate with its motion a probability amplitude that obeys a wave equation. In spite of this interpretation, one must still speak of a particle as having an actual wave character. This led to apparent paradoxes that persuaded Einstein to question very seriously the entire foundation of quantum mechanics and to doubt that it could give a complete description of nature.

While de Broglie and Schrödinger were developing the wave mechanics, Born, Heisenberg, Dirac, and Jordan were approaching the problem from an entirely different point of view. They showed that the electron's emission and absorption of quanta of energy, or photons, is comprehensible only if the position of an electron and its momentum are treated as operators rather than ordinary numbers. These quantities obey a different kind of algebra from that of ordinary numbers— an algebra, as we have noted, that is noncommutative, in which the product of momentum and position differs from that of position and momentum.

Born and his collaborators proved that the difference between these two products is proportional to Planck's constant of action, h, a discovery that led to matrix mechanics. Matrix mechanics gives correct descriptions of the energy states of an electron inside an atom, as does Schrödinger's wave equation. Although Schrödinger, and later Dirac, showed that these two descriptions are equivalent, with differences attributable only to the varying points of view of the observers, there still appeared to be an inherent contradiction between the theories. On the one hand, the de Broglie-Schrödinger view describes the electron as a wave. Matrix mechanics, however, treats electrons as particles, describing their dynamic behavior with a new kind of algebra. The following questions then arose: Is a particle really a particle or is it a wave? And if it can be both particle and wave, when is it one and when is it the other? The same question arose in connection with the photon theory of radiation; here, too, one can ask whether one is dealing with particles or waves. We shall now see how Bohr approached this problem and resolved the paradox by using Heisenberg's important discovery.

Heisenberg was the first to recognize that the need for noncommutative algebra implied, in Bohr's words, a "peculiar indeterminacy" associated with the momentum and position of a particle. As Heisenberg pointed out, this indeterminacy arises because of the quantum of action; if we wish to determine the position of an electron by detecting it with a photon, the interaction between the electron and the photon must involve an action interchange that cannot be smaller than Planck's constant h; hence, the electron's momentum is uncertain by an amount obtained by dividing h by the uncertainty in the position of the electron.

It is clear from elementary considerations that an uncertainty principle implies certain wave aspects of a particle. If we have a wavelength λ, then the uncertainty in the position of the wave packet cannot be smaller than λ. The uncertainty principle then tells us that there must be a momentum associated with the packet that equals Planck's constant h divided by λ. This is exactly the relationship between the wavelength and the momentum of a particle that de Broglie discovered. Consequently, the uncertainty principle indicates that we may consider a particle as a wave packet.

The wave picture also tells us that there is an uncertainty relationship associated with the energy of a particle and the time required to measure the energy. If we wish to measure the frequency of a wave, the uncertainty in this measurement cannot be less than the reciprocal of the time allotted for the measurement. Using Planck's simple formula relating energy to frequency, this means that the uncertainty in the energy of a particle multiplied by the time duration of the measurement cannot be smaller than Planck's constant of action.

We come now to the nature of the paradox that bedeviled quantum mechanics in its early years and was the subject of many penetrating and profound discussions between Einstein and Bohr. At issue was a problem best illustrated by the behavior of a beam of electrons passing through two holes in a screen and then striking a second screen. If each electron is treated as a plane wave, then when it passes through the two holes, two new spherical waves move away from the screen on the other side (one from each hole). These two waves then interfere with one another to give the typical wave interference pattern on the second screen where the electrons are recorded. This pattern is present regardless of the weakness of the electron beam—even though the electrons hit the recording screen at widely separated time intervals. Eventually, the pattern is enhanced as more electrons hit the screen and the interference pattern becomes more clearly defined. Remarkably, the wave interference pattern is present even though each electron strikes a single definite point on the second screen and is recorded there as a particle.

This state of affairs appears to hold an inherent contradiction. It begins with a set of initial conditions that can be stated quite unambiguously in the usual classical terms: particles (electrons) are emitted by a source (for instance, a cathode) in a given direction, pass through two holes in a screen, and then are recorded as particles on a second screen. In principle, we could determine the initial position and momentum of each electron, with an accuracy governed by the Heisenberg uncertainty principle. And we know the position of each electron at the end of the experiment since it strikes a point on the recording screen. In other words, our description of the experiment at its beginning and end is given in classical terms. But no such description can be given for the behavior of the particles between the beginning and end of the experiment. It is impossible to follow the motion of a particle from its beginning to its end, to say through which hole in the first screen the particle passed, and, at the same time, to obtain an interference pattern. In other words, if we are to obtain an interference pattern, we must give up the possibility of providing a complete description of the motion of the particles.

This seeming paradox led Einstein to reject the ability of quantum mechanics to give a complete description of physical phenomena. Einstein believed that quantum mechanics was a statistical description of nature, that it could refer only to a collection of particles, not to the motion of a single particle. He felt that physicists would in time find some way to describe completely the behavior of a single particle and that such a description would go beyond the quantum-mechanical or wave description.

Einstein presented a number of idealized experiments in order to ascertain whether it might be possible to obtain a more complete description than wave mechanics could give. For example, in the case of electrons passing through two holes in a screen, Einstein argued that a careful analysis of the momentum transferred to the screen by the electron should tell us through which hole the electron had passed and that this information would enable us to predict where on the recording screen the electron would strike. He also proposed weighing a box containing radiation before and after allowing a small amount of the radiant energy to leave the box. In this way he proposed to determine the moment the energy left the box as well as its exact amount (which is more than quantum mechanics can tell us).

A famous paper by Einstein, Podolsky, and Rosen has had an important influence on the thinking of physicists since its publication in 1935. Here again Einstein and his collaborators questioned the completeness of the quantum-mechanical description of nature and analyzed a simple experiment to illustrate their point. They considered two particles that come together, interact, and then separate. According to classical physics one should be able to determine or predict the position and momentum of either particle at any moment by observing the other particle and using the conservation principles. The situation is different, however, in quantum mechanics, because of the uncertainty principle (momentum and position do not commute).

Consider the position and momentum of each particle at any moment after they have collided and separated. Although the uncertainty principle tells us that we cannot measure simultaneously the position and momentum of either particle with infinite accuracy, we can show (using quantum mechanics) that the difference between the positions of the two particles and the sum of their momenta can be measured simultaneously and accurately. If we measure the position of one of the particles we can calculate the position of the other, but this procedure precludes any knowledge about the momentum of the second particle. On the other hand, if we measure the momentum of the first particle we can determine the momentum of the second from our quantum-mechanical knowledge of the sum of the two momenta. In other words, even though the two particles have separated, our knowledge of the second particle is still determined by what measurement we make on the first.

To Einstein this outcome meant that quantum mechanics did not give a complete description of nature. It appeared to introduce a mysterious connection between two particles that were no longer interacting. It is difficult to see how measuring something on one particle can determine what we can measure accurately on the other particle. Bohr analyzed this difficulty, again on the basis of his principle of complementarity.

The concept of the complementarity of physical quantities has played an important role in the development of the philosophy of modern physics and in the theory of measurement. Essentially, it tells us that there are always two complementary and mutually exclusive ways of looking at a physical phenomenon, depending on how we arrange our apparatus to measure it. According to Bohr, when we deal with an electron, we must use both the wave picture and the particle picture; one is complementary to the other in the sense that the more our apparatus is designed to look for the electron as a particle, the less the electron behaves like a wave, and vice versa.

Heisenberg himself gave a few examples to illustrate how the very process of measurement forces us to discard the classical deterministic approach to physics. In the classical sense it is entirely correct to speak of the precise orbit of a particle. But such descriptions must be rejected in quantum mechanics where it is meaningless to speak of an orbit since we can never determine more than one point of an orbit at any time by observation. In the very process of pinpointing a particle by bouncing a gamma-ray photon off it we knock it out of its orbit. Gamma-ray photons are required for accurate position measurements because they have very short wavelengths. But they are very energetic because of their high frequencies and thus strongly disturb the particle. Thus, the orbit of a particle, which is its deterministic description since an orbit would enable us to predict where the particle would be at any time in the future, is complementary to its space-time description, which gives its position at a particular time.

Heisenberg also considered the flow of radiation through two holes in a screen. He pointed out that the wave description leading to the well-known interference pattern formed (when the radiation is collected on a photographic plate behind the two holes) is complementary to the photon description, which would allow one to determine through which of the two holes each photon passes. Thus, a wave picture precludes a detailed description of what happens to each photon between the beginning and the end of the experiment. As long as we wish to obtain an interference pattern, it is meaningless to ask through which of the holes any particular photon passes.

In each of the experiments outlined by Einstein, the challenge was taken up by Bohr, the great champion and spiritual leader of quantum mechanics. To answer Einstein, Bohr developed a remarkable generalization of quantum mechanics, referred to as the principle of complementarity, or the Copenhagen interpretation of the quantum theory. Bohr's analysis of Einstein's experiments occurred at a series of international meetings of world-renowned physicists.

In every case Bohr was able to show that, taking the Heisenberg uncertainty relations into account properly, the conservation principles themselves (conservation of energy and of momentum) made a detailed space-time description of the particle between the beginning and the end of the experiment impossible, if one tried to obtain an accurate picture of the energy-momentum interchange between the particle and the apparatus (the holes in the screen). On the other hand, any attempt to obtain an accurate space-time description destroyed the possibility of obtaining, in Bohr's words, a "closer account as regards the balance of momentum and energy." Bohr pointed out that the complete description of an experiment must tell not only what happens to the particle but also must take into account the interaction of the particle and the apparatus. He said that the description of an experiment must include "the impossibility of any sharp separation between the behavior of atomic objects and the interaction with the measuring instruments which serve to define the conditions, under which the phenomena appear."

References

Born, Max, *Atomic Physics*. New York: Hafner, 1962, chapter 4.

————*Nobel Lecture—Physics: 1942–1962*. Amsterdam: Elsevier Publishing Co., 1964, p. 266.

Graves, Charles, in W. R. Hamilton, *Collected Mathematical Papers*. Cambridge: Cambridge University Press, 1931, vol. 1.

13 *New Particles and Atomic Accelerators*

The universe is transformation; our life is what our thoughts make it.

—Marcus Aurelius Antonias, *Meditations*

The Positive Electron—The First Particle of Antimatter: Carl D. Anderson (1905–)

One of the most remarkable features of Dirac's theory of the electron was the concept of an infinite sea of negative-energy electrons at all points of the vacuum. Although this idea followed directly from the mathematics of the theory, it was very difficult to accept it when the theory was first announced. The very idea of a negative energy particle is "unphysical"—contrary to our feeling that the laws of physics must describe a real observable world.

Because of this difficulty, Dirac himself at first proposed that the negative-energy electron states inherent in his theory be regarded as mathematical fictions and that they be disregarded in the application of his theory to physical phenomena. However, the theory gives a correct description of the behavior of an electron only if the negative-energy states are actually taken into account. To do so is to impart a physical reality to the negative-energy states; one must assign to each of them one electron (thus introducing an infinite number of negative-energy electrons in the vacuum). Otherwise all the positive-energy electrons in the universe would jump down to the negative-energy states and the material universe would disappear instantaneously in a vast burst of energy. The presence of negative-energy electrons in all the negative-energy states prevents such a catastrophe, owing to the Pauli exclusion principle.

When Dirac proposed this infinite sea of negative-energy electrons, he indicated that it would not conflict with observation, since such a sea of negative-energy electrons can never be observed. All that is observable is some departure from the uniform distribution of negative-energy electrons in the vacuum. Dirac showed that such a departure could occur if one of the negative-energy electrons absorbed enough energy (in the form of a photon or by means of some sort of collision) to lift it into a state of positive energy.

If this were to occur, the negative state occupied by the negative-energy electron would be empty and the negative-energy electron itself would appear as a real electron with positive energy. The hole thus created in the sea of negative-energy electrons would now be observable as a positively charged particle with

Carl D. Anderson (1905–)

positive energy. In this way a pair of particles would be created, one an ordinary electron, and the other a positively charged particle with the same mass as the electron. Dirac suggested that the chance of creating this pair in some laboratory on earth would be small because it would require energy equivalent to at least twice the mass of the electron. However, enough energy is present in cosmic radiation to create such a pair as it passes through a sheet of matter. Therefore, one should look at the photographs of cosmic ray tracks to detect them; this procedure is how the positrons were first discerned.

On August 2, 1932, Carl Anderson found tracks of positively charged particles on his cosmic ray photographs. He knew that they were not tracks of protons, the only positively charged particles recognized at the time. To study the energy and the charge on cosmic ray particles one introduces a magnetic field in the cloud chamber through which the particles pass. The magnetic field turns the negatively charged particles in one direction and the positively charged particles in the other. If one knows the mass of the particle, one can then calculate its energy from the curvature of its path in the magnetic field. If the path is only slightly curved, the energy is large; if the path is highly curved, the energy is small. For a particle of given charge and mass, the radius of curvature of the path varies as the square root of the kinetic energy of the particle.

The paths of the particles Anderson observed had large curvatures. He knew that if these particles were protons, they would have had rather low energies. In fact, they would have had hardly enough kinetic energy to go more than a few millimeters in air. The lengths of the actual paths in air were found to be ten or more times greater than would be expected for a proton of such energy. The particles were also discovered to have enough energy to penetrate 6 mm of lead. From this analysis Anderson concluded that these positive particles must have masses of the electron's order of magnitude and must have a unit positive charge.

At first, Anderson did not identify these positive electrons, or positrons, with the Dirac holes in the sea of negative-energy states. The early photographs did not show a pair being created, as demanded by the Dirac theory. However, this oversight was due to faulty observation. Soon afterwards it was found that whenever the track of a positron appeared on a photographic plate, it was accompanied by the track of an electron of opposite curvature. Moreover, the positron tracks ended abruptly, whereas the electron tracks twisted around and died out gradually.

It is easy to see the reason for this on the basis of the Dirac theory. Since a positron is a hole in a sea of negative-energy states, any electron it meets will be annihilated with the emission of energy. The positron thus disappears abruptly at the end of a short path because there are always many free electrons about in the cloud chamber. The electron that is created, on the other hand, is an ordinary particle that loses its kinetic energy slowly through collisions with other ordinary particles in the cloud chamber. Its track is therefore an intricate one. As the electron loses kinetic energy and slows down, the magnetic field in the cloud chamber curves it more and more sharply so that the path may end in a series of small loops.

Anderson's discovery, for which he won the Nobel Prize, established the Dirac theory as one of the most reliable in physics and, for the first time, showed that the complexities of elementary-particle physics were far greater than had been imagined. It also opened up a vast new domain of research, suggesting the existence of negative-energy protons and neutrons. These particles were discovered many years later. We now know that for every type of fundamental particle in the universe there is an antiparticle; the members of any pair are related to one another in the same way the electron is related to the positron. This finding has led to the concept of antimatter and to the speculation that there are parts of our universe (perhaps half of it) that consist of antimatter. When matter and antimatter meet, they destroy one another and give rise to pure energy.

Carl David Anderson was born in New York City of Swedish parents on September 3, 1905. After the typical American boyhood, with its grammar and high-school education, Anderson entered the California Institute of Technology at Pasadena and in 1927 received his B.Sc. degree in physics and engineering. After his graduation, he devoted all his time to physics and in 1930 received his Ph.D. degree in physics from the same institution, where he remained as a research fellow until 1933. He was then appointed assistant professor of physics and, in 1939, full professor, a position which he held for the remainder of his professional career. Anderson's early career was his most productive. In 1932 he discovered the positron, for which he received the Nobel Prize in 1936; two years later, working with S. Neddermeyer, he discovered the mu-meson.

Although Anderson began his research in the field of X rays, and wrote his Ph.D. thesis on the emission of electrons from gases by X rays, under the direction of Robert Millikan, he soon turned to cosmic rays. Once Millikan and Anderson decided to investigate cosmic rays by means of the Wilson cloud chamber, as had been done for the first time in 1927 by the Russian physicist Skobelzyn, it was only a matter of time until the positron was discovered.

By designing his instruments properly and working with powerful magnetic fields, Anderson was able to measure cosmic ray events of much higher energy than had previously been done. After finding tracks of positively charged particles, which he proved could not be those of protons, Anderson suggested that they were the Dirac positrons. This prediction was later confirmed in independent experiments by Blackett and Occhialini. After completing this basic work, Anderson studied various other cosmic ray phenomena, particularly the characteristics of cosmic ray showers. This work led to his discovery of the mu-meson.

The Discovery of the Deuteron: Harold Clayton Urey (1893–1981)

With the discovery of the neutron, discussed below, it immediately became clear to physicists that all the difficulties in trying to construct nuclei with protons and electrons would disappear. The nucleus could be pictured as consisting only of neutrons and protons, and no conflicts with the Heisenberg uncertainty principle, or the known spin-and-statistical properties of nuclei, would arise. For an electron to reside in a nucleus, its de Broglie wavelength would have to be about as small as the nucleus. This would require it to move so fast (the wavelength varies inversely with the speed) that it could not be contained by known nuclear forces. Electrons would also give nuclei of even atomic weight and odd atomic number the wrong spins.

The discovery of the existence of the neutron eliminated these difficulties and revolutionized the theory of the atomic nucleus. Rather than considering the electrostatic interaction between electrons and a nucleus, as in atomic structure, one treats the nuclear interactions between neutrons and protons. In general, however, this is a much more difficult problem than the electron-nucleus problem in atomic structure. In the study of electrons moving about a nucleus, the motion of an individual electron can be treated quite accurately, without considering the presence of any of the other electrons. Protons and neutrons, however, are of about the same mass; therefore, one faces the many-body problem.

The nuclear two-body problem (a nucleus consisting of a single proton and a single neutron) is comparatively easy. Such a nucleus exists in the deuteron, or heavy hydrogen, which is the simplest compound nucleus known. It has been investigated more extensively than any other nuclear structure. Theoreticians quickly realized the usefulness of a thorough analysis of this system for an understanding of nuclear forces, not obtained easily from a study of more complex nuclei.

There are two phases to the study of the two-body, or the two-nucleon, nuclear system. On the one hand, we may investigate how two nucleons (two protons or a proton and a neutron) collide without combining to form a stable, compound system. This is called the two-body scattering problem, on which much experimental and theoretical work has been done. The other phase of the two-body system deals with the stable compound system, the deuteron or the hydrogen 2 nucleus. The problem here is, in principle, the same as that with the hydrogen atom—to find the stationary states or, what is the same thing, the energy levels. Just as Schrödinger set up a wave equation to describe the motion of the electron in the electrostatic field of the proton, one sets up a wave equation to describe the motions of the proton and the neutron in their mutual nuclear field of force. There is, however, one important difference between the hydrogen problem and the deuteron problem. Whereas the mathematical form of the force between the electron and the proton (the Coulomb force) is known, the mathematical formula for the nuclear force is not.

Consequently, various forces were used with the hope that the experimental evidence would demonstrate which one was correct. The idea was to calculate the energy levels (and other nuclear features) with various mathematical expressions for the force and then to check the results with observations made on the two-body system itself. Hence, the existence of a stable two-body nucleus was of great significance.

Fortunately, this particle was discovered by Harold C. Urey in 1932, almost coincidentally with the neutron, so that the theory could be tested directly by comparing the theoretical models with the observed properties of the deuteron. In a 1932 paper in *Physical Review*, Urey, F. G. Brickwedde, and G. M. Murphy outlined and analyzed the techniques employed to isolate the heavy isotope of hydrogen. After presenting various chemical and physical arguments for its existence, the authors considered various methods for separating the isotopes and concluded that fractional distillation offered the best chance for success. Although this procedure had not been very successful in other cases, they argued that it had a much better chance to succeed with hydrogen, since the mass ratios of the different hydrogen isotopes are much larger than for the heavier elements. To see how effective fractional distillation might be for hydrogen, Urey and his collaborators first made an approximate calculation of the vapor pressure of the three molecular isotopes H^1H^1 (ordinary hydrogen), H^1H^2 (hydrogen-deuteron), and H^1H^3 (hydrogen-tritium) in equilibrium with their solid phases. The greater the difference in vapor pressure for these various isotopes, the more easily they can be separated by fractional distillation.

To calculate the vapor pressure of each isotopic gas in equilibrium with its solid phase, Urey and his coworkers wrote down an expression for the free energy of the solid phase and the free energy of the gaseous phase (which contains the vapor pressure as one of its terms) and then equated the two. Since the free energy of a system is a measure of the amount of work it can do, it is clear that if two phases (solid and gaseous, for example) are to be in equilibrium, the free energy of one phase must equal that of the other; if this were not so, one phase would do work on the other, until both free energies were equal.

By this procedure, Urey and his fellow authors found that the calculated vapor pressure for the H^1H^2 molecule should be about a third of that of the ordinary hydrogen molecule (H^1H^1) when the gas is in equilibrium with the solid phase at the triple point temperature of ordinary hydrogen. However, this procedure does not give an accurate description of the experimental situation. One must know the vapor pressure of the gaseous phase in equilibrium with the liquid phase, not the solid phase, because fractionation is carried out from the liquid phase. However, since the theory of liquids was quite unsatisfactory when this experiment was performed, the authors simply argued that the ratio of the water pressure for the gas in equilibrium with the liquid phase close to the solidification temperature should be about the same as for the solid phase. From this argument they concluded that fractional distillation can produce an appreciable concentration, since the highest isotope (ordinary hydrogen) will evaporate more rapidly than the heavier ones.

Three different lots of liquid hydrogen were prepared by Brickwedde for the experiment; one was six liters and the other two were four liters each. The six liter lot was allowed to evaporate at the usual boiling point of liquid hydrogen, 20.4°K, and the gas evaporating from the last few cc's of liquid was collected for analysis. The two four liter lots were evaporated at close to the triple point temperature of ordinary hydrogen, 13.95°K and the gas evaporating from the last few cc's of liquid was collected. Each of the gas samples was then introduced into a discharge tube and the light from the discharge analyzed with the aid of a 21-foot-grating spectrograph. The simple Bohr theory states that for atoms of given atomic number, the line spectrum should be slightly different for those with nuclei of different mass, that is, for different isotopes. Hence the heavy hydrogen atoms should produce a line spectrum with slightly different wavelengths than those characteristic of light hydrogen atoms. When the spectroscopic measurements were made, the gas sample evaporated at 20°K showed nothing but lines appropriate to light hydrogen (disregarding the molecular rotation spectrum), but each of the two samples evaporated close to 13.95°K gave faint lines in the positions expected for hydrogen of mass 2. It was expected that greater fractionation would occur at the lower temperature, so the appearance of H^2 lines for this condition gave added confidence in the result.

The son of a country clergyman, Harold Clayton Urey was born in Walkerton, Indiana, on April 25, 1893. His father died when he was six years old and his mother remarried, again to a clergyman. He obtained his early education in the Indiana public schools and graduated from high school in 1911. Being unable to go directly to college, Urey taught for several years in rural schools before entering the University of Montana in 1914. Three years later he obtained his B.S. in chemistry. Because of World War I, Urey took a position in Philadelphia in a chemical plant engaged in war production. He later described this experience as most fortunate because it convinced him that academic work, not industrial chemistry, was his major interest. The direction of this academic work was to the University of Montana where he became an instructor in chemistry. Five years later, in 1921, he went to the University of California at Berkeley to study for a doctorate in chemistry under Professor Gilbert N. Lewis. Lewis was an inspiring teacher and leader in chemistry and undoubtedly greatly influenced Urey's career.

Following the receipt of his Ph.D. degree, Urey received a fellowship from the American-Scandinavian Foundation which he used for a year's postdoctoral study at Bohr's Institute for Theoretical Physics in Copenhagen. Returning to the United States in 1924, he served for five years as associate in chemistry at the Johns Hopkins University in Baltimore. It was during this time that he wrote with Arthur E. Ruark his widely used book *Atoms, Molecules and Quanta*, published in 1930.

In 1929 Urey accepted a position as associate professor of chemistry at Columbia University. Very shortly thereafter he began the investigation for which he received the Nobel Prize in 1934—the isolation of the hydrogen isotope of mass two, heavy hydrogen or deuterium. This research was carried out with the assistance of Dr. Ferdinand G. Brickwedde, then head of the cryogenic section of the U.S. National Bureau of Standards, and Dr. George M. Murphy at Columbia University.

Research with heavy hydrogen and methods of concentrating stable isotopes occupied Urey during the following years until the United States's entry into World War II. As a consequence of his experience in isotope separation he was appointed to head the group designated as the SAM (Substitute Alloy Materials) Laboratory established at Columbia University to investigate the separation of uranium isotopes 235 and 238 for the purposes of making an atomic bomb. This laboratory was part of the Army's Manhattan District Project and subsequently became the research laboratory that provided the fundamental data necessary for the construction of the large-scale diffusion separation plant built by the Kellex Corporation at Oak Ridge, Tennessee.

At the end of the war, Urey joined the Institute for Nuclear Studies at the University of Chicago and became distinguished service professor of chemistry. His scientific interests now took a new direction. He began research on the origin of the earth and the planetary system, the evolution of life on the earth and the temperatures of the oceans in past geologic ages. His book, *The Planets, Their Origin and Development*, was published by Yale University Press in 1952. In 1958 he left Chicago to assume the chair of professor of chemistry-at-large in the University of California at La Jolla. He remained a resident of La Jolla until his death on January 5, 1981.

The Discovery of the Neutron

In the second decade of the 20th century physicists were concerned primarily with the outer regions of the atom and with the development of the theoretical procedures necessary to understand the behavior of the electrons as they move around the nucleus. The third decade marked the exploration of the nucleus itself. Of course, well before the 1930s Ernest Rutherford had used alpha particles ejected from radioactive nuclei to study the nuclei of other atoms, and in 1929 George Gamow had applied quantum mechanics to an analysis and explanation of the emission of alpha particles from radioactive nuclei. But many difficulties stood in the way of carrying out detailed experimental investigations of the nucleus and no theoretical model had yet been introduced that could account for even the most elementary observed properties of atomic nuclei.

Since the electric charge on the atomic nucleus increases by one unit as we move from one element to the next in the periodic table, very energetic protons or alpha particles are required to penetrate the repulsive Coulomb barrier of the heavier nuclei. Electrons are not massive enough to do much damage to a nucleus. Since only these particles were available to physicists before 1932, the amount of nuclear investigation was necessarily meager. Rutherford succeeded in bringing about nuclear transformations but only on a very small scale, since he had at his disposal only minute quantities of radioactive material. With the development of the cyclotron by E. O. Lawrence, and linear types of accelerators, beams of high-energy protons became available but, again, in rather small quantities and at energies that were suitable only for working with light nuclei.

More serious than the practical difficulties hampering nuclear research were the theoretical contradictions. The essential difficulty lay in trying to account for the mass of a nucleus and its charge in terms of the fundamental particles that were known at the time—the electron and the proton. It was quite natural to suppose that the nucleus of an atom consisted of electrons and protons, since protons were positively charged and electrons were known to be emitted by radioactive nuclei. To illustrate the nature of the difficulty let us consider the nucleus of the helium atom, the alpha particle. We know that its mass is very nearly four times that of a single proton. Hence, our first inclination is to assume that it contains four protons to account for this mass. However, this would give us a nucleus with twice the observed charge. To overcome this inconsistency all we need do, it appears, is to add two electrons to the four protons inside the nucleus. We then have a structure that seems to mirror reality. This was, indeed, the earlier model proposed and it seemed satisfactory until the advent of quantum mechanics and the uncertainty principle. It then became obvious that an electron could not exist inside a nucleus

Sir James Chadwick (1891–1974)

since there was no room for its large wave packet. Trying to put it in the nucleus would be like trying to force an elephant into a dog house.

Now let us consider the application of the uncertainty principle to an electron in a nucleus. Since we can calculate the size of any nucleus, we can also know the position of an electron assumed to lie inside it with enormously high accuracy, of the order of a ten-trillionth of a centimeter, the size of a nucleus. But if we apply the uncertainty principle, it follows that there is a correspondingly high uncertainty in the momentum and, therefore, in the kinetic energy of the electron. In fact, a simple calculation, on the basis of the formula for the de Broglie wavelength, shows that the electron would have to be moving with such high speed inside a nucleus that it could not possibly be held imprisoned by the forces of the order we know are present there. In short, the wave pattern that describes the electron is so distended that it cannot be contained within a nucleus.

There is still another objection to having electrons inside nuclei. This point relates to the spin properties of electrons, protons, and nuclei. Electrons and protons both have the same spin, which we may take as half a unit (actually it is one half of the fundamental unit of angular momentum, which is Planck's constant divided by 2π). If a nucleus consisted of electrons and protons, its spin would be the sum of the spins of the constituent particles. In general these spins arrange themselves so that they cancel out in pairs. This means that the spin of a nucleus should either be zero or $1/2$, depending upon whether the total number of electrons plus protons is even or odd. As an example consider the nucleus of nitrogen, which must contain at least 7 protons to account for its 7-fold positive charge. To justify a mass of 14, the proton-electron model of the nitrogen nucleus would have to assign 7 more protons to the nucleus, as well as 7 electrons, in order to achieve a positive charge equal to 7. But this would leave us with 21 spinning particles altogether inside the nucleus; hence the spin of the nitrogen nucleus would have to be $1/2$ or some odd integral multiple of $1/2$. Experiments, however, show that the spin is zero. Therefore, this proton-electron model is untenable.

This contradiction was removed from nuclear theory in 1932 when James Chadwick proved the existence of a new type of particle, the neutron. In his paper, published that year in the *Proceedings of the Royal Society*, Chadwick carefully examined the experiments of Bothe and Becker, as well as those of Mme. Joliot-Curie and of Webster, involving the bombardment of light elements by alpha particles. Up to that time it was thought that the radiation emitted as a result of these bombardments (particularly in the case of beryllium) was electromagnetic in nature (photons) but extremely energetic, as indicated by its penetrating power. Chadwick ingeniously showed, however, that the protons ejected by this radiation from hydrogen containing material (paraffin wax, for example) do not follow the accepted pattern predicted by well-established theory. Moreover, he pointed out that there was no nuclear mechanism that could give rise to such energetic electromagnetic radiation from nuclei bombarded by alpha particles. As a result of these considerations, Chadwick decided to carry out further experiments on the radiation emitted when beryllium is bombarded by alpha particles.

From these experiments, Chadwick concluded that the only way to account for the effects of beryllium radiation on various atoms was either to "relinquish the application of the conservation of energy and momentum in these collisions or to adopt another hypothesis about the nature of the radiation." As he pointed out, all the difficulties disappeared if the beryllium radiation was not viewed as electromagnetic but as consisting of neutral particles with masses equal to the mass of the proton. Because of the large mass of such a particle it can easily eject atoms from various targets with the velocities that Chadwick and the other investigators had observed when the beryllium radiation impinged upon matter. Furthermore, because these neutral particles of Chadwick's have no charge, and are thus not repelled by the electric fields surrounding nuclei, they are highly penetrating.

In bombarding various substances with these neutrons, as Chadwick called them, he undoubtedly had created artificially radioactive isotopes without realizing it. It was not until a few years later that a systematic investigation of this artificial radioactivity was carried out by the Joliot-Curies and, independently, by Enrico Fermi. A point of interest in connection with the introduction of the neutron is that, as Chadwick noted, the need for such a particle had already been recognized by Rutherford as early as 1920. How vastly different the world might now be in its political, scientific, and geographical division, had Rutherford pursued his idea experimentally and discovered the neutron at that time! In the 1920s and early 1930s, the center of scientific investigation was in Germany. In the United States, fundamental scientific work on a much smaller scale was then being done. Very likely the atomic bomb would have been developed first in Europe, undoubtedly by the Germans.

To see how the neutron swept away all the difficulties that plagued nuclear physicists, consider the nitrogen nucleus again. We can account for the charge and the mass of the nitrogen nucleus by introducing just 7 protons and 7 neutrons; since each of these particles has a half unit of spin (the neutron has the same spin properties as the electron), the spin of the nucleus must either be zero or some integer value. The value turns out to be zero as determined by actual measurement.

In addition, there is sufficient room inside a nucleus to accommodate a neutron. Of course, the uncertainty principle still applies and the uncertainty in the momentum of the neutron in a nucleus is as large as it would be for an electron. But because the mass of the neutron is about 2,000 times larger than that of the electron, its kinetic energy (for the same amount of momentum) is some 2,000 times smaller. Therefore, the neutron is not too energetic to remain bound in the nucleus.

In the second half of his paper Chadwick concluded that the neutron was a proton and an electron in very close combination, since it had no charge and had a mass slightly less than the hydrogen atom but larger than that of the proton. According to this point of view, held by many physicists in the period immediately following the discovery of the neutron, the particle was assumed to be a compound structure. However, principally as the result of the ideas of Heisenberg, the compound picture was soon discarded and the neutron was admitted into the gallery of fundamental particles already occupied by the electron, the proton, the positron, and the photon. We now consider the neutron and the proton to be two different energy states of the same fundamental particle, the nucleon. Since the neutron is more massive than the proton, it represents the nucleon in a higher energy state. If left to itself, the neutron decays into a proton with the emission of an electron and a neutrino.

The most recent experiments show that the neutron has a half-life of about 12 minutes. Under ordinary conditions inside a nucleus the nuclear forces between the proton and the neutron prevent the neutron from decaying into a proton. But in nuclei in which the number of neutrons is much larger than the number of protons, decay occurs and electrons are then emitted. Such nuclei are called beta-radioactive and the electrons are called beta rays.

One of the consequences of this decay is the need to introduce still another fundamental particle, the neutrino (the little neutral one), a name first suggested by Fermi. Such a particle was proposed by Wolfgang Pauli to explain certain discrepancies in the spin and energy balance of nuclei involved in beta radioactivity. When a neutron changes into a proton with the emission of an electron, the total spin of the nucleus no longer balances. We start with one half unit of spin, the neutron, and end with two half units, the electron and the proton. The only way total spin balance can be maintained is by the introduction of another half unit of spin, which is the exact function of the neutrino. It also leads to a correct energy balance, for without the neutrino the principle of the conservation of energy would be violated.

In 1920 Rutherford had proposed a neutral particle, formed by the combination of an electron and proton, as the first step in the formation of a nucleus. On the basis of Rutherford's suggestion, Glasson and Roberts attempted to observe the formation of such a particle by passing electrons through hydrogen, but without success. The first successful attempt to detect this neutral particle, now called the neutron, was made in 1932 by Chadwick as a consequence of experiments performed by the Joliot-Curies.

James Chadwick was well equipped intellectually and educationally to carry out the precise and ingenious experiments that led to the discovery of the neutron. He was born in Cheshire, England, on October 20, 1891, the son of John Joseph Chadwick and Ann Mary Knowles. After attending Manchester High School, he enrolled in the University of Manchester in 1908 and graduated from the Honors School of Physics in 1911. He then went to work with Rutherford at the Manchester Physical Laboratory and received his M.S. degree there in 1913. Simultaneously he received the 1851 Exhibition scholarship and went to work with Professor Hans Geiger in Berlin. Chadwick was in Germany in 1914 when World War I broke out and was interned for the duration.

In 1919 Chadwick returned to England and worked with Rutherford at the Cavendish Laboratory on the artificial transformation of light nuclei by bombardment with alpha particles. This proved his ability as a physicist of top quality. He was elected fellow of Gonville and Caius College in 1921, became assistant director of research of the Cavendish Laboratory in 1923, and was elected fellow of the Royal Society in 1927.

During this period, Chadwick was actively at work with Rutherford in disintegration experiments involving various types of atoms. By 1924 they had demonstrated that the nuclei of all elements ranging up to potassium—except lithium, carbon, and oxygen—could be disintegrated by alpha particles from radium C, which have energies of about 7,000,000 volts. The idea of a neutral particle inside the nucleus, as suggested by Rutherford, was very much in Chadwick's mind, even at that time, for he was thinking of various ways to detect such a particle. As early as September 1924, in a letter to Rutherford, he wrote: "I think we shall have to make a real search for the neutron. I believe I have a scheme which may just work, but I must consult [Francis] Aston first."

Chadwick began his famous neutron experiments in 1931 after Frédéric and Irène Joliot-Curie reported that radiation from beryllium excited by alpha particles could eject protons from paraffin wax or "any other matter containing hydrogen." H. C. Webster of the Cavendish Laboratory had already detected and examined this strange beryllium radiation and had found that it had peculiar features not explained by the usual theories. In particular, the behavior of the beryllium radiation could not be explained if one assumed that it consisted of high-energy photons. Chadwick therefore started from the hypothesis that this radiation consisted of neutral particles. As he said, "I therefore began immediately the study of this new effect using different methods—the counter, the expansion chamber, and the high-pressure ionization chamber." With these instruments, designed to detect material particles rather than photons, Chadwick discovered the neutron in 1932. This discovery was the beginning of the vast growth of nuclear physics that culminated in the fission of uranium and the development of the chain-reacting atomic pile. For this research Chadwick received the Hughes Medal of the Royal Society in 1932 and the Nobel Prize in 1935.

Chadwick left Cambridge in 1935 to accept the Lyon Jones chair of physics at the University of Liverpool. During World War II he worked in the United States as head of the British Mission attached to the Manhattan Project for the development of the atomic bomb. After the war, in 1948, he returned to England and retired from his Liverpool chair and from active physics research in general to become Master of Gonville and Caius College, Cambridge. From 1957 to 1962 he was a member of the United Kingdom Atomic Energy Authority.

In 1945 Chadwick was knighted. In addition to his Nobel Prize, Chadwick received such awards and honors as the Copley Medal in 1950 and the Franklin Medal in 1951, as well as honorary doctorates from the universities of Reading, Dublin, Leeds, Oxford, and McGill, among others; he was also a member of several foreign academies. He died on July 24, 1974, at Cambridge.

Fermi's Contributions: Enrico Fermi (1901–1954)

The science of physics was born in Italy at the close of the 16th century with the investigations of Galileo. By the beginning of the 20th century, the center

Enrico Fermi (1901–1954)

of scientific activity had shifted to the great German universities, which drew researchers from England and the Continent. Physics developed slowly in Italy. During the first two decades of the present century there was no Italian physicist with the international stature of men like Kelvin, Poincaré, Lorentz, or Planck. But just as modern physics entered a golden age in the 1920s, with exciting developments in quantum physics and atomic structure, Enrico Fermi appeared. Singlehandedly, without the benefit of training at European scientific schools, he raised Italian physics to a summit of achievement.

Endowed with creativity, great mathematical gifts, and a veritable genius for implementing theory and experiment, Fermi attracted the best of the young Italian physicists to Rome, where he organized what came to be known as the Roman school of physics. By the time Fermi left Italy in 1938, the year he won the Nobel Prize, his school was sought after by young physicists of many lands, as well as the leading international scientists of the day. From 1934 on, through the efforts of Fermi and his group of nuclear experts, Rome was an important center of nuclear research.

In his first papers, Fermi, at the age of 20, dealt with relativistic electrodynamics, in particular with the electromagnetic mass of charged particles, such as electrons, and with the general problem of the equivalence of gravitational and inertial mass. In the course of this work he proved an important theorem in the absolute differential calculus, the mathematics used in the general relativity theory. In addition, he resolved a certain difficulty that had cropped up in the assumption of the equivalence of mass and energy.

Shortly after his initial work, Fermi began an investigation of certain theoretical problems in classical mechanics governing groups of bodies and became especially interested in statistical mechanics. This interest, which never left him, led him to analyze the statistical mechanics of a gas consisting of particles that obey the Pauli exclusion principle—particles such as electrons with a spin of $1/2$. In this analysis Fermi derived his famous quantum-mechanical statistics of degenerate matter, introduced independently but somewhat later by Dirac and now known as the Fermi-Dirac statistics. Fermi quickly realized the significance of his work for the understanding of very dense matter and the behavior of solids. His subsequent series of papers dealt with the quantum mechanics of metals, in which electrons are treated as a degenerate gas. This work became the foundation of modern solid-state physics and is today the basis for the analysis of white dwarfs and the evolution of stars in their super-dense phase.

Fermi developed his statistical mechanics by applying quantization to the molecules of a gas. He reasoned correctly that, since at very low temperatures both the specific heat and the equation of state of a gas differ from the classical expressions, some type of degeneracy must set in that can be accounted for only by the quantum theory. Einstein had already shown that the correct specific heats of solids at low temperature cannot be obtained without quantizing the vibrations of the molecules. Moreover, Bose and Einstein, separately, had developed a new kind of statistical mechanics for gases. This statistics, based on the quantum theory, leads to an equation of state for a gas different from the classical equation of state. Indeed, Bose introduced the new statistics in order to obtain the Planck radiation formula for a photon gas.

The Bose-Einstein statistics and the Fermi statistics are similar in that they both take into account the quantum properties of matter. In fact, it can be shown that each of these two types of statistics is equivalent to taking into account the wave properties of particles. But they differ in the way that quantization is applied to the gas. The region of agreement between these two statistics, as well as the area in which they part company with the classical Boltzmann statistics, is related to the identity of the molecules composing the gas. Classical statistics is set up on the assumption that a certain state of a gas (its pressure and its temperature for a given volume) is defined by the way the molecules in the gas are distributed in the volume and by their motions. It was noted in the discussion on Bose statistics that each classical state is defined by a particular partition of the molecules in a given volume element according to the possible states of energy and momentum. In the classical analysis, each molecule is assigned its own identity, and it is assumed that one molecule can be distinguished from any other. This leads to the classical or Boltzmann statistics.

Both the Bose statistics and the Fermi statistics are similar to classical statistics in the sense that they, too, define the state of a system in terms of the way the molecules in a given element of volume are distributed among the possible states of energy and momentum. However, in counting the number of different distributions that can lead to a given state of the gas (a given temperature and pressure), both Bose and Fermi state that all molecules are exactly the same. Consequently, two different distributions involving merely the interchange of any two molecules cannot be distinguished from one another and must be counted as identical. This identity of all the molecules (assuming, of course, that our gas consists of just one kind of molecule, such as hydrogen) and the inability to distinguish one from another arise from the wave properties of particles, since there is no way to differentiate between two similar waves.

This method of treating molecules distinguishes quantum statistics from classical statistics and is common to both the Fermi and Bose treatments. They differ from each other, however, in the manner in which molecules in a given small volume element are assigned to different states of energy and momentum. In the Bose statistics there is no restriction on the number of molecules that can be assigned to the same state of energy and momentum. But in the Fermi statistics the situation is quite different. Fermi was guided by Pauli's discovery that the electrons inside an atom arrange themselves in such a way that no two electrons have the same set of four quantum numbers; that is, no two are in the same state of motion. In other words, the electrons influence one another in such a way as to preempt or exclude identical motion in the same volume element. This is the well-known exclusion principle which we discuss elsewhere in this book.

Fermi was prompted to use the exclusion principle to set up the statistics of an ideal gas. He reasoned that if the molecules of a gas had no influence on one another (as one assumes in the case of a classical ideal gas), then the gas would never exhibit degeneracy at low temperatures. He therefore introduced the concept of the influence of one molecule on another. He did not intend this to be the usual type of classical influence that is expressed in terms of forces between the molecules. Insofar as forces between molecules go, the situation here is the same as in the classical case; the new kind of influence between any two molecules is of a quantum-mechanical nature and stems from the fact that no two molecules can be distinguished from each other. Fermi therefore introduced the Pauli exclusion principle and limited the way in which molecules in a gas can move by supposing that in any small element of volume of a gas no two molecules can be in the same state of motion (the same energy and direction of motion) or, as expressed by the Pauli exclusion principle, may not have the same set of quantum numbers.

The Fermi statistics was first elucidated in a paper entitled "On Quantizing an Ideal Monatomic Gas." It later became known as the Fermi-Dirac statistics because of Dirac's independent research, made known in a paper published about six months later. Fermi's paper, one of the masterpieces of reasoning in theoretical physics, is an excellent example of his approach to a problem. It is presented with such simplicity that anyone with a knowledge of elementary calculus can follow the mathematics. Throughout, the emphasis is on the physical content of the theory; the mathematical formalism is used merely to elucidate the physics.

Fermi began his paper with the statement that classical thermodynamics leads to a constant molecular heat for a gas, whereas the experimental data demand that the molecular heat vanish at absolute zero. Therefore, it is necessary to quantize the motions of the molecules of a gas. He then proposed to carry out a new type of quantization by introducing the Pauli exclusion principle. In addition, he represented each molecule as being in an external force field so that its motion was periodic and therefore could be characterized by a set of quantum numbers. In addition he represented each molecule oscillating about the origin as though it were connected to the origin by a perfect spring (simple harmonic motion, the quantum harmonic oscillator). He pointed out, however, that this concept was only an artifice and could be replaced by any number of other types of force fields with the same result.

Fermi then wrote down the set of quantum numbers that defined the motion of each molecule. (These are the well-known quantum numbers of the harmonic oscillator: an integer for each degree of freedom of motion). The motion of each molecule was described by three integers, one for each of the three mutually perpendicular directions in space. He then applied the Pauli exclusion principle by imposing the condition that no two molecules in the gas could have the same set of integers describing their motion. In counting configurations all molecules must be treated as identical so that two configurations differing only by the permutations of molecules are the same. This imposed a definite condition on the way a given amount of energy in the gas could be distributed among the molecules because the energy of a harmonic oscillator (a molecule in this case) is determined by its quantum numbers. Only one molecule can be at rest (zero energy), only three moving with one quantum of energy each, no more than six moving with two quanta of energy, and so on. This use of the exclusion principle enabled Fermi to obtain a general formula for the way the total kinetic energy in a gas can be distributed among the N molecules. From this formula he was then able to write down the formula for the number of molecules having a given energy in a gas at a given absolute temperature as a function of the temperature. This led to the new equation of state of the gas.

The statistics for a degenerate gas thus obtained was used shortly after by Arnold Sommerfeld to explain the properties of metals by treating the electrons in the metal as a degenerate gas. In their now famous work, Hans Bethe and Sommerfeld developed the physics of degenerate systems to a point where it could be used in the analysis of all types of solid-state problems.

Dirac later showed that the Bose-Einstein statistics, which does not limit the number of molecules that can be assigned to the same state of motion, and the Fermi-Dirac statistics, which limits to one the number of molecules in each energy state, are both derivable from quantum mechanics and are special cases of a very general principle of symmetry. Dirac demonstrated that two types of particles occur in nature that have to be described by wave functions with different symmetry. Particles such as electrons, protons, neutrons, and neutrinos, which have a half unit of spin (or an odd multiple of a half unit), are described by wave functions whose signs change when the coordinates of the positions and spin of the particles are replaced by the negative values in the wave functions. These particles obey the Fermi-Dirac statistics. Particles that have an integral multiple of the unit of spin (such as photons) are described by symmetrical wave functions (they do not change sign when the coordinates change sign). These obey the Bose-Einstein statistics.

Another outgrowth of Fermi's work in statistical mechanics was his well-known model of the atom. Fermi reasoned that one may treat the cloud of electrons surrounding the nucleus inside a heavy atom like a degenerate gas obeying the Pauli exclusion principle (the electrons arrange themselves one per atomic energy level). Therefore one may apply the Fermi-Dirac statistics to such an atom. In this way he showed how various atomic problems can be treated statistically to give results that are fairly accurate. Although Fermi was unaware of it, a similar theory had been developed a year earlier by L. H. Thomas. Today this work is referred to as the Thomas-Fermi model of the atom. The model is of basic importance in the study of crystal structure and other solid-state problems.

During this same period, from about 1924 to 1928, Fermi became involved in a wide range of problems in optics, spectroscopy, and molecular structure. He published papers on the quantum mechanics of the interference of light, on the spectroscopy of alkali atoms, on the hyperfine structure of spectral lines and the magnetic moments of nuclei, and on molecular spectra. His work on the quantum mechanics of the interference of light was prompted by his reading of Dirac's papers on the quantum mechanics of the emission and absorption of radiation. After he had mastered Dirac's treatment, he reformulated the theory and recast it into a much simpler and more amenable mathematical form. This work became one of the most popular articles in the *Reviews of Modern Physics* (1932), and for years served all newcomers as the gateway to the quantum electrodynamics. It is a model of beauty and clarity and illustrates Fermi's gifts as a teacher. The content of this paper was contained in a series of courses he gave at the Henri Poincaré Institute in Paris in 1929 and at the 1930 Summer School of Theoretical Physics at the University of Michigan in Ann Arbor. Fermi became a fixture at the Ann Arbor school and was there almost every summer until he finally left Italy. During this period he influenced the trend in American physics and made a summer's stay at the University of Michigan one of the most exciting intellectual adventures in science.

This extremely productive period was followed by Fermi's investigations into nuclear theory. These occupied him for most of his remaining years and led to his discovery of the self-sustaining chain reaction of uranium and to the theory of the atomic pile. Fermi had spent most of his formative years in Italy. Except for a profitless year at Göttingen (part of which was spent at Leyden) he came into very little direct contact with physicists of international reputation. However, as his papers became known to increasing numbers of physicists throughout the world, he was invited to attend international conferences and to present his ideas. By the end of the 1920s his reputation was well established and he was recognized as the outstanding representative of Italian physics. Near the end of the 1920s, many of the international conferences devoted increasing time and publication space to problems of nuclear physics. Although the neutron had not yet been discovered, sufficient experimental data had been collected to present important questions in nuclear physics to scientists attending conferences. Two of the puzzling features about the nucleus at that time concerned the beta decay of nuclei (the radioactivity of heavy nuclei with the emission of electrons) and the composition of nuclei. Since electrons are emitted from nuclei during beta decay, it was thought that nuclei consisted of both protons and electrons in the proper number to give the correct charge and mass. But this picture was in conflict with the principles of wave mechanics—in particular, with the Heisenberg uncertainty principle. Because of the wave structure of the electron, it can be shown that an electron inside a nucleus would have to move about with so much kinetic energy that it could not be held there by the known forces. Since the nucleus is very tiny, the position of the electron inside a nucleus would be so accurately known that its momentum would, according to the uncertainty principle, be extremely large and hence its kinetic energy would be very large.

The other puzzle in connection with beta decay of nuclei involved the energy of the emitted electron. Since all the nuclei of the same radioactive element decay into identical, less massive nuclei with the emission of an electron, the energy given up by each nucleus must always be the same. It is measured by the difference between the mass of the initial nucleus and the mass of the final nucleus. Hence, it was argued, each electron emitted should emerge with the same kinetic energy. But this result does not agree with observation. In beta decay there is a spectrum of energy values for the emitted electrons ranging from zero to a maximum value. Yet the actual energy lost is always equal to the difference in mass between the initial and final nuclei multiplied by the square of the speed of light, regardless of the energy of the emitted electron. This was suprising because at that time there was no direct experimental evidence that any particle other than the electron was emitted during beta decay. Some of the outstanding physicists, particularly Bohr, argued that if electrons were inside a nucleus they would have to be treated outside the framework of quantum mechanics. Moreover, they stated that energy would not be conserved during the beta decay process. Bohr was further prompted to take this position by the difficulty in accounting for the spin and the statistics of nuclei after they had undergone beta decay. Since the electron has a half unit of spin and obeys the Fermi statistics, all nuclei suffering beta decay should change their spins by a half unit and also change their statistics. This is not the case. The spin changes either by one unit or not at all, and the statistics of the nuclei remain unchanged.

Pauli, however, did not agree with Bohr and felt that the principle of the conservation of energy should not be abandoned. He pointed out that one can retain the energy principle and at the same time account for the energy spectrum of the electrons emitted in beta decay, as well as the spin and statistics of the decaying nuclei, by assuming that another particle, in addition to the electron, is emitted during beta decay. Pauli described this hypothetical particle as one with no charge, almost no rest mass, and a half unit of spin (Fermi statistics), but capable of carrying energy. Later, during the conference at which Pauli proposed the new particle, Fermi referred to it as the neutrino and the name stuck. We now know that the neutrino is an actual, fundamental particle; recently, two different kinds of neutrinos have been discovered.

Shortly after the neutrino was named, in 1932, the neutron was discovered and its properties correctly analyzed in England by Chadwick and in France by the Joliot-Curies. In Italy the brilliant theoretical physicist E. Majorana, whose gifts exceeded those of Fermi in some phases of mathematical physics, immediately saw the importance of the neutron as a constituent of the nucleus and developed a nuclear model based on protons and neutrons. Majorana's ideas about nuclear forces are still valid and of great importance, but he never followed up this work because of the severe depression that frequently incapacitated him. Fermi took over the neutron as his own special pet and in his hands it became the dominant tool of nuclear research. He realized that the properties of the neutron (chiefly its charge neutrality) made it especially suitable for penetrating the positively charged nucleus. In 1938, he received the Nobel Prize for "his demonstration of the existence of new radioactive elements produced by neutron irradiation, and for his related discovery of nuclear reaction brought about by slow neutrons." In his Nobel address, he stated that the most natural choice of a bombarding particle to effect nuclear transformations and artificial radioactivity was the neutron, rather than the alpha particle. Fermi reasoned that alpha particles could be successful in such efforts only with the very light elements because the nuclear charge beyond atomic number 15 is so great as to inhibit alpha particle capture by the nucleus. This argument, however, does not apply to neutrons. Therefore, these particles can be used to probe atomic nuclei. He stated his point of view as follows.

> Compared with [alpha] particles, the neutrons have the obvious drawback that the available neutron sources emit only a comparatively small number of neutrons. Indeed neutrons are emitted as products of nuclear reactions, whose yield is only seldom larger than 10^{-4}. This drawback, however, is compensated by the fact that neutrons, having no electric charge, can reach the nuclei of all atoms, without having to overcome the potential barrier, due to the Coulomb field that surrounds the nucleus. Furthermore, since neutrons practically do not interact with electrons, their range is very long, and the probability of a nuclear collision is correspondingly larger than in the case of alpha particle or proton bombardment.

Fermi, using the neutron, investigated all the elements of the atomic table. He conceived the experiments, designed the apparatus, and predicted his results theoretically. Fortunately, his recruitment in earlier years of young Italian physicists now stood him in good stead. He led a group of devoted, brilliant men who collaborated beautifully in the Herculean task he set for them. His idea was to bombard each element in turn with neutrons and to determine the nature of the end products.

In a 1934 paper in the *Proceedings of the Royal Society*, Fermi described how his team began with hydrogen and systematically bombarded all the elements up to uranium. He began the experiments himself with rather crude homemade Geiger counters, using very weak neutron sources. He obtained no results with the elements from hydrogen through oxygen, but this did not deter him. Finally, he met success with fluorine, which showed a strong radioactivity as a result of the bombardment.

During these continuing investigations many problems, both of an experimental and theoretical nature, arose. In each case Fermi found the correct solution. One of the most interesting developments was the great increase in artificial radioactivity resulting from the use of slow neutrons. This discovery was an accidental one, made when elements irradiated on wooden tables proved much more active than those irradiated on marble tables. Fermi quickly understood the significance of this discovery. The difference was due to the effect of the intervening wood, which slowed down the neutrons much more effectively than did the marble. To demonstrate this fact Fermi placed paraffin filters between the neutron beam and the irradiated target and thus increased enormously the resulting artificial radioactivity of the target. Fermi correctly pointed out that the light elements in paraffin, similar to those in wood, particularly hydrogen, slow down the neutrons through collisions with their own atomic nuclei. The slow neturons, spending more time in the neighborhood of an irradiated target nucleus, can then do more damage.

One of the most puzzling historical features of this work concerns the production of elements with atomic number higher than 92, that is, uranium. In the note Fermi sent to *Nature* in 1934 concerning this research he analyzed the possible reactions that could be obtained and concluded that the evidence strongly favored the transuranic (across the boundary of uranium) elements 93 and 94. At this point in their investigations, the Rome group was on the threshold of discovering the fission of uranium. Some five years later Hahn and Strassman made this discovery in Germany and communicated their data to Lise Meitner, then in exile in Sweden, thereby ushering in the Nuclear Age. It is now clear that the fission fragments were present in the earlier Rome uranium-irradiation experiments. Emilio Segré, one of Fermi's outstanding collaborators and disciples, who received the Nobel Prize for discovering the antiproton, believed that the failure to detect fission was due to the use of an aluminum sheet covering the uranium and the thorium, which filtered out all the fission fragments. Although the possibility of fission had been called to Fermi's attention by I. Noddack, who wrote an article predicting it, Fermi dismissed this possibility because the data on the mass defects of nuclei seemed to preclude it. It later turned out that these data were wrong.

Although Fermi was then still engaged in theoretical physics, most of it was in connection with the nuclear experimental work. In a letter written in April 1934, Rutherford welcomed Fermi into the experimental fraternity. He congratulated Fermi on his "successful escape from theoretical physics." Rutherford also remarked that even Dirac was doing some experimental work which all "seems to be a good augury for the future of theoretical physics." Fermi's semideparture from the provinces of theory was marked by another great paper that appeared in the *Zeitschrift für Physik* in 1934. This paper presented a complete theory of beta decay and beta emission; with some minor modifications and extensions, the theory is still the accepted one. Starting from the idea that the nucleus contains no electrons, Fermi argued that the electron and the neutrino that are emitted in beta decay are created at the moment of their emission, as is the photon at the moment of its emission from an electron. Just as the emission of the photon by an electron is due to the transition of the electron from one state of energy to another state, so, too, is beta emission the result of the transition of the nucleus from one energy state to another. In this case, however, Fermi pictured the electron and the neutrino as being emitted by a neutron in the nucleus. Like Heisenberg, he looked upon the proton and the neutron as two different energy states of the same fundamental particle. Transitions from one state to the other occur because of the emission or absorption of electrons and neutrinos. With this knowledge as a basis, Fermi set up the mathematical machinery for describing beta decay.

This was done by carrying out the analogy between beta emission and the emission and absorption of photons. Fermi employed a formula to describe the interaction of a nucleon (neutron or proton) with an electron and neutrino that is similar to the equation used to explain photon emission. Fermi used the same quantum-mechanical technique that Dirac and others utilized to set up a quantum electrodynamics of radiation. This is the so-called technique of second quantization in which one introduces annihilation and creation operators to describe the processes under consideration. Since electrons and neutrinos appear suddenly in beta-decay processes, Fermi described the phenomena in terms of electrons and neutrino creation operators, which he treated in the usual quantum-mechanical fashion.

This theory successfully explained the spectral shape of beta-ray emission, the lifetimes of beta emitters, positron decay, orbital electron capture, and so on. What is more interesting is that the type of interaction that Fermi used at the time is now recognized as a special case of a more general universal Fermi interaction, now used to describe the decay of the muon and its interaction with protons.

Shortly after Fermi came to New York in 1939 to accept a professorship at Columbia University, the fission of the uranium atom was discovered. He then devoted his full time to uranium research. Together with Leo Szilard, he was instrumental in interesting the government in nuclear fission. Since Fermi at that time was the greatest authority on neutrons, it was quite natural that he should become a dominant force in the Manhattan Project. This neutron research finally culminated in the first self-sustaining nuclear chain reaction under the stadium at the University of Chicago, where the nuclear reactor there became critical on December 2, 1942.

Following the war, Fermi returned to his forsaken theoretical work. He began to devote more time to high-energy research and to fundamental particle physics. The properties of such particles as the pi- and mu-mesons interested him greatly. He also began some investigations into the nature and origin of cosmic rays. One of the mechanisms now accepted for the creation of cosmic rays was discovered by Fermi and presented in one of his last great theoretical papers. In the final years of his life he became interested in the problems of galactic structure and, with

S. Chandrasekhar, wrote a basic paper on the importance of galactic magnetic fields in the equilibrium of the spiral arms of galaxies.

A few years before his death in 1954, Fermi was invited to give the Silliman Foundation lectures at Yale University; he chose elementary particles as his topic. These lectures have been collected in book form and are excellent examples of the simplicity of Fermi's lecture technique. To listen to Fermi was a most inspiring experience, not only for the student but also for the hardened worker in physics. He had the knack of all great teachers: he made the student wonder why there had ever been any difficulty in understanding the topic he was presenting.

Enrico Fermi was born in Rome on September 29, 1901, the third child of Alberto Fermi and Ida DeGattis, both of whom originally came from the north Italian town of Piacenza. Like all geniuses, and especially those in physics (with the possible exception of Einstein), Enrico demonstrated his ability and his interest in physics at a very early age. He told Emilio Segré that when he was about ten years old he tried to understand what is represented by the equation $x^2 + y^2 = r^2$.

Fermi spent his early school years in Rome and, while still a high-school student, began to educate himself in physics by reading a Latin book on mathematical physics written in 1840 by a Jesuit. That he mastered its 900 pages is indicated by the numerous notations he made in its margins. He certainly must have been an amazing phenomenon to those who knew him in his early years, for he questioned everybody he met about mathematics and physics. At the age of 13, he met one of his father's colleagues, A. Amidei, an engineer, and asked him whether it was true that there was a branch of mathematics in which important geometric properties were derived without using the "notion of measure." He was referring to projective geometry, just then becoming very popular among mathematicians. Amidei told Fermi that this was so and lent him a book on the subject. Amidei then went on to say, in a letter to Segré, that after a few days Fermi had read the introduction and the first three lessons, and after two months he had mastered the text, demonstrated all the theorems, and quickly solved the more than 200 problems at the end of the book.

At the age of 17 Fermi finished his secondary school studies and took the entrance examination for admission to the Scuola Normale Superiore at Pisa. His examination paper must have amazed his examiners. Rather than giving the usual high-school level solutions to the problems, he applied the most advanced mathematical techniques, such as partial differential equations and Fourier analysis, to problems in sound.

Fermi was one of those remarkable people who excel in everything they attempt. He was an outstanding, if not the number one, student in all his subjects at Pisa. He was particularly good in languages and had no trouble in learning them by working with a dictionary. He spoke and wrote German perfectly. After beginning his university career as a mathematician, Fermi changed to physics and began studying the most advanced treatises available. Whatever he read he quickly assimilated and made part of his own technique.

Fermi's quality of mind and his deep physical insight between the ages of 18 and 20 are clearly indicated by the books he read and his own notebooks. He devoted himself only to the most important material and authors; the pertinent points of each subject were carefully developed in neatly written pages.

After obtaining his doctorate in 1922 from the Scuola Normale, Fermi returned to Rome where he lived with his parents until he received a scholarship in 1923 to study with Max Born at Göttingen, then one of the major centers of quantum-mechanical research activity. He spent another year at Leyden with Paul Ehrenfest and then took a temporary post at the University of Florence; while there he discovered the Fermi statistics. In 1926 the first chair in theoretical physics was established at the University of Rome; Fermi became its first occupant. Here be began one of the most fruitful periods in all scientific history. He accomplished much research and also trained physicists of great ability.

When Fermi was awarded the Nobel Prize in 1938, he did not return to Italy. The collaboration between Mussolini and Hitler had made the expression of free thought impossible; indeed, it threatened the very safety of his family. He came to Columbia University as professor of physics in 1930 and remained there until 1942, when he went to the University of Chicago to direct the experiments that led to the first chain-reacting pile and to the atomic bomb.

After the war he accepted a professorship at the Institute of Nuclear Studies at the University of Chicago and remained there until his death from cancer on November 29, 1954. During these later years he applied himself to the physics of high-energy particles and cosmic rays. As in all his other ventures into physics, he sought to emphasize the physical content of a theory rather than the formalism.

Fermi was a rare human being who illuminated the lives of all who came into contact with him. He had the remarkable quality of making everyone he spoke to feel that there was no real intellectual difference between them and what he could accomplish could be done by anyone else with a little effort. From this humility stemmed Fermi's entire philosophy of teaching, writing, and lecturing. He was convinced that many of the difficulties of the student arose because of a formalistic rather than a physical presentation of the material by the teacher. Fermi's books, articles, and papers are models of excellence and a delight to the reader.

Just as he had a zest for physics and took great joy in its pursuit, he approached all of his life in that spirit. He had a great amount of physical stamina and was fond of sports, excelling in tennis, swimming, mountain climbing, and skiing.

Numerous honors and prizes were awarded to him for his work. He was the first recipient of the $50,000 prize that bears his name and is awarded by the United States government for achievement in atomic research.

Artificial Nuclear Disintegration: John Douglas Cockcroft (1897–1967) and Ernest Thomas Sinton Walton (1903–)

The fact that atoms are not immutable and may change was shown in 1902 by Rutherford and Soddy. But their results, while founded on incontrovertible chemical evidence, gave no clue as to how such changes came about. At that time the structure of the atom was unknown. Rutherford solved the puzzle of atomic structure in 1911 by proposing a nucleus and planetary electrons, an arrangement amply confirmed by the experiments of Geiger and Marsden. The subsequent investigations of Moseley showed that the system of increasingly heavier natural elements was based on the successive addition of units of positive charge to the nucleus, balanced electrically by the addition of extra-nuclear electrons. These developments made it clear, as Soddy pointed out in 1913, that radioactivity was a nuclear phenomenon.

As a next step along this path of reasoning it would have been logical to ask if nuclear changes took place naturally in radioactivity and whether it was practical to produce similar changes artificially, thus transmuting elements at will. If such thoughts arose they do not seem to have been voiced; the attention of physicists was elsewhere, deeply engrossed in the multitude of problem posed by the Bohr theory of the extra-nuclear electrons. But the next great problem was surely the understanding of the nucleus. The first step was taken by Rutherford when he effected, in 1919, the first artificial transmutation by bombarding the nitrogen nucleus with high-energy alpha particles.

But another decade was necessary for the growth that would lead to an effective technology: the accurate determination of nuclear masses, the development of wave mechanics, and, finally, the production of machines to produce highly energetic charged particles. By the end of the third decade of the 20th century, Gamow, and independently Gurney and Condon, had shown that the wave properties of particles could account for the radioactivity of such atoms as uranium. But if nuclear theory is to be compared with actual nuclear structure, a nuclear spectrum is required; only from such a spectrum can energy levels be determined and compared with predictions of the theory. Atomic spectra can be obtained by applying excitation energies of the order of a few volts, but the production of nuclear spectra requires excitation energies of the order of a million volts or larger. Actually a two-fold problem was posed by nuclear research: how to build machines to produce charged particles (or photons) of the required energies; and how to determine the kinds of particles best suited to be effective nucleus-penetrating projectiles.

In principle, we can use the same kinds of particles to excite both nuclei and the outer electrons of an atom. But in practice things are not so simple. Although the atomic electrons are easily disturbed by radiation, the excitation of nuclei with energetic photons (gamma rays) is a very tricky matter. Atomic energy levels lie very close together and are excited by a large frequency range of photons (from the X-ray end of the spectrum to the infrared). But the energy levels in a nucleus are spaced at fairly large intervals. For each nucleus only a few gamma ray frequencies will work. By 1930 the only gamma rays available were those emitted by radioactive nuclei like uranium. These were generally not of the correct frequencies to excite the nuclei that were being studied. In any case, the intensities of the gamma rays then available, and the range of their frequencies, were too small to be of much use. Nevertheless, various attempts were being made to obtain artificial radioactivity with the means at hand.

Since gamma rays were relatively ineffectual, physicists (in line with the researches and suggestions of Rutherford, Chadwick, Ellis and others) began intensive work with particle collisions. These investigators, particularly Rutherford, had used alpha particles from radioactive sources to bombard nuclei of various kinds, and some nuclear transformations had been produced. In the earliest experiments, Rutherford, in 1919, had bombarded light atoms, such as nitrogen and oxygen, with alpha particles and had disrupted the nitrogen nucleus. However, natural radioactive sources of particles were much too limited to be very useful for a systematic study of nuclei.

Since the neutron had not yet been discovered when particle accelerators were first produced, charged particles had to be used to probe nuclei. It was therefore necessary to devise various schemes to accelerate such particles to energies that would penetrate the nucleus. Transformations induced by particle bombardment of the nucleus involve two conditions. The nucleus and the bombarding projectile (alpha particle, proton, and so on) are electrically charged. The colliding particles have a wave character. Consequently one must apply quantum mechanics (the Schrödinger wave equation) to the interaction of charged particles. Since the nucleus and the alpha particles are both positively charged and repel each other, the problem is essentially that of the penetration of potential barriers by particles. This is how Gamow, and Condon and Gurney, analyzed the problems of radioactive emission in alpha-particle decay, the inverse of the penetration problem.

The quantum-mechanical theory of the penetration of potential barriers shows that for a given nucleus the depth of penetration of the bombarding particle increases with increasing energy but decreases with increasing mass and charge of the projectile. In other words, for a given energy it is advantageous to use as bullets the lightest particles with the smallest charge, all other things being equal. This fact immediately points to electrons or protons. However, electrons cannot form even quasi-stable systems within nuclei and therefore cannot excite them. In a sense, trying to excite nuclei with electrons (unless the electrons have enormously high energies and hence very short de Broglie wavelengths) is like trying to excite atomic electrons with radio waves. This predicament left the choice between protons and alpha particles. Since the theory of the penetration of potential barriers indicates that a million-volt proton has as much penetrating power as a 16-million-volt alpha particle, the proton was finally chosen as the projectile to use in probing the nucleus.

The only question that could not be answered in that pre-neutron period was whether the proton or alpha particle would be more effective in exciting the nucleus once the potential barrier was penetrated. Two phases are involved in nuclear excitation—the penetration of the barrier and the capture of the particle by the nucleus. Only if there is a good chance that a particle will combine, at least briefly, with the nucleus will the nucleus be excited or possibly disrupted. But this question can be answered only on the basis of a theory from which the energy levels of nuclei can be determined. Since such a theory did not exist, it was felt that, all other things being equal, the proton was to be preferred.

The production of particles of sufficient energy to excite nuclear reactions was attempted almost simultaneously by John Cockcroft and Ernest Walton in England, and by Robert Van de Graaff, and Ernest Lawrence and Milton Livingston in the United States. The methods adopted by these investigators for attaining the necessary accelerating voltages differed. Cockcroft and Walton decided that the most promising method was to rectify high-voltage, low-temperature frequency alternating current; Van de Graaff believed that the development of a high-voltage electrostatic generator offered the most advantages; Lawrence and Livingston favored the application of a series of incremental pulsed voltages applied to ions confined in a spiral path by a magnetic field. This device, the cyclotron, has proved to be the most useful for producing the highest energy particles.

The first system that successfully produced artifical nuclear reactions was that of Cockcroft and Walton. Of the three accelerators it was the least novel. The necessary high voltage was obtained by means of a conventional a.c. transformer. One of its secondary terminals was connected to the plate of a kenetron and the other to one plate of a suitable capacitor. The kenetron is a two element (filament and plate) electron tube capable of rectifying alternating current at high

Sir John Douglas Cockcroft (1897–1967)

potential. In the simplest arrangement the circuit was completed by connecting the filament of the kenetron to the other plate of the capacitor. The two electrodes of the accelerating tube were then connected in parallel with the capacitor. One transformer supplied the high voltage a.c. rectified by the kenetron; two other transformers were used in connection with the hydrogen discharge tube.

The current required for the operation of the accelerator is only a few microamperes so that the potential supplied to the accelerating tube is kept sensibly constant by a small capacitor. The protons from the discharge are used as projectiles to induce nuclear reactions. These protons enter the target chamber where they are incident on the material whose nucleus is to be studied.

The first substance studied with this apparatus was lithium. A round target of the element, 5 cm in diameter, was placed at an angle of $45°$ to the beam of the accelerated protons. An opening in the tube opposite the target and at right angles to the proton beam was closed by a zinc sulphide screen, suitable for giving scintillations on the impact of alpha particles. Mica absorbers placed in front of the screen prevented scattered protons from striking it. When 125 kilovolt protons were incident on the target, scintillations began to appear. Several tests proved that the scintillations were produced by alpha particles. Cockcroft and Walton therefore concluded that the bombardment had resulted in the formation of unstable beryllium 8 which then split into two parts. The reaction may be written as follows: $_1He^1 + _3Li^7 \rightarrow _4Be^8 \rightarrow _2He^4 + _2He^4$. Next, the evidence for the pair production of alpha particles was investigated with results supporting the above conclusions. A series of heavier elements up to uranium was then subjected to proton bombardment, in some instances up to 600 kV. The preliminary results suggested that, in most instances, the ejected particles were alpha particles.

It is interesting that the experiment of Cockroft and Walton on the artificial disintegration of lithium was the first of its kind to excite the imagination of the public and to give the layperson a feeling that physicists might be working on something of world-shaking importance. There were numerous interviews, articles in the papers, pictures of the apparatus, and inquiries from newspapers all over the

Ernest Thomas Sinton Walton (1903–)

world. One telegram to Rutherford from the Associated Press of America requested assistance in preparing an article for the American public and stated that the 1400 papers it served in North and South America "would be deeply grateful if you would grant us an explanatory interview on your recent experiments in splitting the atom. We have in mind nothing sensational." No answer to this telegram has ever been found.

The significance of this experiment must be emphasized. Cockcroft and Walton were the first to construct an ion accelerator of sufficient energy to produce nuclear disintegrations. The way was thus opened to study one of the most important but previously almost inaccessible fields of physics. The results showed that nuclei could be disrupted by particles of lower energy than previously supposed. From the balancing of the disintegration equation for Li^7, Einstein's mass-energy equivalence was proved without doubt. For this pioneering work Cockcroft and Walton received the Nobel Prize in physics in 1951.

Sir John Douglas Cockcroft was born in Todmorden, England, on May 27, 1897. His father was a cotton manufacturer, but Cockcroft showed no interest in the family business; instead, he was strongly attracted to science and mathematics. After a traditional British education, he matriculated at Manchester University and studied mathematics under Horace Lamb. With the advent of World War I, Cockcroft joined the British Armed Forces in 1915 and served in the Royal Field Artillery. At the end of the war he returned to Manchester to study electrical engineering at the College of Technology, after which he worked as an engineer for the Metropolitan-Vickers Electrical Company. Soon his deep interest in mathematics and the physical sciences drew him back to the university. Like Dirac, he left engineering to study mathematics at St. John's College, Cambridge. There he took the mathematical tripos in 1924 and embarked upon physics as a career. Like many brilliant young experimentalists of the period, he went to the Cavendish Laboratory and worked under Lord Rutherford.

Cockcroft did not go at once into experimental nuclear physics. At that time it appeared that the only way of penetrating the nucleus was with the alpha particles emitted by such radioactive nuclei as uranium and radium; this procedure was not too promising for large-scale investigations. Only a few nuclei could be transmuted. Moreover, Rutherford himself had already done the basic work in this field, so that little more could be done along these lines. Cockcroft therefore chose another research field and began working with the great Russian physicist P. Kapitza on the production of intense magnetic fields and the generation of low temperatures.

In 1928 Cockcroft turned to nuclear physics with an entirely new idea for penetrating nuclei with artificially acclerated protons rather than natural alpha particles. Experimental studies had shown that the energies of alpha particles emitted by radioactive nuclei were, in general, much smaller than one would expect if these particles were propelled from the nucleus by the full Coulomb repulsion. This puzzling fact was not understood until the theoretical investigations of Gamow, and of Condon and Gurney, showed that alpha particles, because of their wave nature, do indeed penetrate the Coulomb potential barrier at relatively low energies.

In 1928 when Cockcroft was still formulating his ideas, Gamow visited the Cavendish Laboratory and Cockcroft outlined his plan to him. Supported by Gamow, he sent a memorandum to Rutherford proposing that boron and lithium be bombarded by accelerated protons. In this memorandum he showed that boron could be penetrated by a proton of only 300 kilovolts of energy and that the conditions for lithium were even more favorable. Rutherford agreed to the proposal, and Cockcroft was joined in his project by Ernest Walton, who was then developing one of the first linear accelerators, as well as one of the earliest betatrons. Their collaboration in 1932 resulted in the first proton-induced artificial nuclear disintegrations.

Cockcroft continued his experimental work on the artificial transmutation of elements and in 1933 produced a wide variety of such phenomena, using both protons and deuterons as his projectiles. At the same time he produced artificial radioactivity of various nuclei by proton bombardment. His experimental abilities were quickly recognized and he was appointed director of the Royal Society Mond Laboratory in Cambridge. At that time he was already a fellow of St. John's College, having been elected in 1929; he then became, in turn, university demonstrator and lecturer. In 1939 he was appointed Jacksonian professor of natural philosophy.

When World War II broke out, Cockcroft accepted the post of assistant director of scientific research in the Ministry of Supply and devoted his skills to the development of a coast-to-coast radar defense system. In the autumn of 1940 he came to the United States as a member of the Tizard Mission and then returned to England to become head of the Air Defense Research and Development Establishment. In 1944 Cockcroft went to Canada as head of the Canadian Atomic Energy Project and became director of the Montreal and Chalk River Laboratories. He remained in Canada for two years and then, in 1946, returned to England as director of the Atomic Energy Research Establishment at Harwell. In 1954 Cockcroft was appointed a research member of the United Kingdom Atomic Energy Authority and remained with this agency as a full-time member until 1959, when he was elected master of Churchill College, Cambridge. He then continued with the Atomic Energy Authority on a part-time basis. He was later appointed chancellor of the Australian National University at Canberra. He received many honors for his scientific work and was president of the Institute of Physics, of the British Physical Society, and of the British Association for the Advancement of Science. He was a fellow of the

Royal Society and received honorary doctorates from many universities. He died on September 18, 1967, at Cambridge.

Cockcroft's partner, Ernest Thomas Sinton Walton, followed a direct path from his early schooling to a career in science. He was born on October 3, 1903, in Dungarvan, Waterford County, on the south coast of Ireland, the son of a Methodist minister from County Tipperary, who, because of his calling, had to move every few years. As a consequence Ernest attended various schools. His aptitude in mathematics and science was evident at a very early age and he was encouraged to continue these studies. He was therefore sent as a boarder in 1915 to the Methodist College in Belfast, where he did brilliant work in mathematics and physics. In 1922 he was awarded a scholarship and entered Trinity College, Dublin, where he read honors courses in mathematics and experimental physics. He graduated with highest honors in these subjects in 1926 and received his M.S. degree in 1927.

On receiving a research scholarship in 1927 from the Royal Commissioner for the Exhibition of 1851, he went to Cambridge University to work at the Cavendish Laboratory under Lord Rutherford. Although Walton's first research papers dealt with hydrodynamics, he shifted to particle accelerators when he began working with Rutherford. A senior research award from the Department of Scientific and Industrial Research in 1930 permitted him to continue with his graduate studies and research. In 1931 he received his Ph.D. degree.

During this period Walton met Cockcroft, who was already working on using relatively low-energy protons to penetrate nuclei by taking advantage of the wave character of the proton, which permits it to pass through the Coulomb barrier. Realizing that protons of only a few hundred kilovolts, rather than millions of volts, would do the job and that this was well within the technological capabilities of that time, Walton worked on improving the high-voltage X-ray and cathode-ray tubes that were then available. Finally, in collaboration with Cockcroft, he constructed a linear accelerator that produced protons of the right energy, which were then used to disintegrate lithium and to transmute boron. Walton shared the Nobel Prize with Cockcroft in 1951 for this pioneer work.

From 1932 to 1934 Walton was Clerk Maxwell scholar at Cambridge. He then returned to Trinity College, Dublin, as fellow. In 1946 he was appointed the Erasmus Smith professor of natural and experimental philosophy, and in 1960 he was elected senior fellow. In addition to his academic work, Walton participated in other educational, civic, and religious activities. He was connected with the Dublin Institute for Advanced Studies, the Institute for Industrial Research and Standards, the Royal City of Dublin Hospital, the Royal Irish Academy, the Royal Dublin Society, and other institutions. In addition to the Nobel Prize, he received the Hughes medal in 1938, jointly with Cockcroft. In 1959 the Queen's University in Belfast awarded him an honorary Doctor of Science degree. Walton published numerous papers on hydrodynamics, nuclear physics, and microwaves in the journals of various scientific societies.

The Electrostatic Generator:
Robert Jemison Van de Graaff (1901–1967)

In the previous chapter, emphasis was placed on the need to develop machines that could produce a wide range of particle energies in order to match the widely spaced energy levels of the nucleus. A successful start on this problem was made by

Cockcroft and Walton, who used an extension of conventional alternating-current-rectifier techniques to produce a steady, high voltage. But to produce extremely high, steady voltage over the necessary range raised inherent difficulties not easily overcome. The limitations of the high-voltage-rectifier method were realized by Robert Van de Graaff, who set about solving the problem of high voltage by a path that differed from that of Cockcroft and Walton. Realizing the advantages of a steady high potential for ion acceleration, Van de Graaff tried to find a simple means of achieving it.

The simplest solution seemed to demand a return to the earliest methods for generating high potentials, an electrostatic method in which charges were continuously carried to a hollow sphere to raise its potential. The sphere could be charged to voltages limited only by the corona breakdown at the surface of the sphere. To carry the charge, Van de Graaff hit on the idea of using a continuous belt on which charge could be sprayed by the brush discharge between a metal surface and a group of charged points. A belt moving over a pulley near the ground and another inside the sphere carried the charge from the charging position into the interior of the sphere where it was drawn off by a second series of sharp points and conveyed to the surface of the sphere.

Robert Jemison Van de Graaff was born in Tuscaloosa, Alabama, on December 20, 1901. He obtained his B.S. degree in 1922 from the University of Alabama and his M.S. degree in 1923. From 1924 to 1925 he was a student at the Sorbonne in Paris. In 1925, he won a Rhodes Scholarship and, for the next three years, continued his study of physics at Oxford, where he was awarded the Doctor of Philosophy degree in 1928. It was during this time that he conceived the idea of developing high-voltages by means of the continuous charging of a high potential spherical electrode. In 1929 a small model of what we now call the Van de Graaff Generator was built "to demonstrate the soundness of the principles involved." It performed as expected, generating a maximum potential of 80,000 volts.

From 1929 to 1931, Van de Graaff held a National Research Council Fellowship at Princeton University. This period was devoted to the further development

Robert Jemison Van de Graaff (1901–1967)

of his electrostatic method for generating high voltages. A larger generator was constructed and its operation tested in vacuum to determine the value of vacuum insulation. To test the usefulness of the generator as a particle accelerator and for scientific research in general, a 1.5 million-volt generator was completed in 1931. Thereafter, larger, higher-voltage generators were made at the Massachusetts Institute of Technology, where Van de Graaff was appointed research associate during the years 1931 to 1934. In 1934, he was appointed Associate Professor of Physics, an appointment he held until 1960. During World War II, he served as director of a project sponsored by the government's Office of Scientific Research and Development to develop radiographic equipment for surveying the internal mechanical structure of metals and welded seams. In 1926 he became a director of the High Voltage Engineering Company of Cambridge, Massachusetts, and, the following year, a member of its executive committee, positions which he held for many years. Before his death on January 16, 1967, Van de Graaff was honored by the award of the Elliot Cresson medal of the Franklin Institute and the Duddell medal of the Physical Society of Great Britain.

The Van de Graaff generator was used throughout the world, not only as an ion accelerator for studies in low-energy nuclear physics but also for such purposes as X-ray radiography, radiation therapy, food sterilization, and, on a limited scale, as an ion injector for the very high-energy particle accelerators.

The Cyclotron: Ernest Orlando Lawrence (1901–1958)

Before Lawrence and Livingston developed their method of accelerating ions, both Cockcroft and Walton and Van de Graaff had reported methods for obtaining high-speed ions. Both of these methods required the generation of high voltages, which, as Lawrence and Livingston pointed out, led to serious technical difficulties that had not yet been overcome. Lawrence therefore introduced a new procedure to accelerate ions to very high speeds in a series of steps, each of which would involve only a relatively small voltage. If this were done by moving the ion in a straight line, with its speed increased at regular intervals by a stepped accelerating voltage, the length of the device would be unmanageable. Lawrence therefore proposed a spiral path for the ion, which would cause it to move back and forth across a voltage increment in the plane of the spiral. To arrange this so that the electric field always accelerates the ion, one must first have a magnetic field at right angles to the plane of the ion's path and then an alternating electric field that changes direction periodically in phase with the motion of the ion. Such a device is called a cyclotron.

To understand the nature of the design problem, consider an ion that starts moving at right angles to a region across which there is a voltage. Initially, the motion of the ion is parallel to the electric field. If a constant magnetic field is now introduced at right angles to the electric field, the ion will move in a circular orbit of fixed radius. If the speed of the ion were constant, the radius of its circular orbit would always be the same. But if the speed of the ion is increased, and the magnetic field remains the same, it then moves in a larger orbit. We now come to a crucial question. Can we change the electric field periodically in such a way that every time an ion passes through the region where the voltage jumps, the field accelerates the ion?

Ernest Orlando Lawrence (1901–1958)

If the ion were always moving with the same speed, the radius of its orbit would always be the same, and the ion would always cross the voltage region at equal time intervals. But the speed of the ion is constantly increasing, so that its orbit gets larger and larger. It is not at once apparent that the time intervals between passages across this region remain the same. But this is, indeed, the case because the circumference of the orbit and the speed of the ion increase proportionately so that the ion traverses the larger orbit in the same time. This means that, if the direction of the electric field is reversed at intervals equal to the time it takes the ion to go halfway around, all the ions will be acted upon by the field in exactly the same way at all times. In other words, the electric field can be regularly alternated in such a way as always to give the ion an additional push when it crosses the voltage region. This is the basic principle of the cyclotron as it was developed by Lawrence.

It is easy to see that with a device of this sort protons or other light ions can be accelerated to high speeds by having them cross the electric field frequently. The cyclotron is important becase it performs this operation in a relatively small space. As Lawrence and Livingston pointed out, one need only have a potential of 4,000 volts across the accelerating gap to obtain speeds corresponding to 1,200,000 volts for protons, if each proton is sent across the gap 150 times. By then, it is moving in so large an orbit that it is near the edge of the apparatus.

In their analysis of the cyclotron action, Lawrence and Livingston showed that the energy finally acquired by an ion in its maximum orbit (when it reached the periphery of the apparatus and was ready to be expelled) depended on the square of the magnetic field strength and on the square of the radius of the final orbit. The stronger the magnetic field, the more the orbit is curved, and the more frequently the ion is forced to cross the electric field to be accelerated before it circles out to the periphery. The larger the radius of the orbit, the faster the ion must be moving, hence possessing more energy.

Milton Stanley Livingston (1905–)

Although the principle of the cyclotron is fairly simple, the actual construction and operation presented many problems. Factors such as the arrangement of the ion source, the constancy and uniformity of the magnetic field, and the focusing action of the magnetic and electric fields required care to acheive optimum results. Although in their initial experiments Lawrence and Livingston obtained proton beams of 1,200,000 volts, they were already thinking in terms of 25,000,000 (and higher) volt protons. In their work we see the first indications of what the future would hold for high-energy accelerators.

In 1928, three years after Ernest Orlando Lawrence had received his doctorate in physics from Yale University and just after he had joined the faculty at the University of California, he carefully took an inventory of what research work he had already done and laid out a plan of action for the future. At that time, although atomic physics seemed to offer the most exciting opportunities for a young physicist, Lawrence was strongly attracted by nuclear physics, which he correctly evaluated as "the next great frontier of the experimental physicist." He was particularly stimulated by the pioneering experiments of Rutherford and saw that they could be carried out extensively only if it were possible to devise some means "of accelerating charged particles to high velocities—to energies measured in millions of electrons volts, a task which appeared formidable, indeed!"

It must be remembered that the neutron had not yet been discovered. Only charged particles (protons or alpha particles) could be used to probe the nucleus and had to be speeded up to very high energies to overcome the repulsion of the nucleus. Keeping this point in mind, Lawrence decided that the greatest promise lay in constructing ion accelerators, and he began to investigate the machines that were then available. Discarding the idea of improving such devices, since he felt that they were already in the hands of very competent people, he searched for new methods of producing high voltages. By chance, in 1929, he ran across an article by a German engineer, Wideröe, that contained the germ of the idea for the cyclotron. Although Lawrence could not read German, the diagrams in the article were enough to start him off in the right direction, and so the era of large, nuclear-smashing machines was born. Although the first cyclotron was a small device, about one foot across, there appeared to be no limit, in principle, to the ultimate size of such an instrument. Since that time, accelerators have steadily increased in size.

Lawrence, himself, was well equipped intellectually, and by training, to do the kind of imaginative experimental work that the cyclotron project required. He was born on August 8, 1901, at Canton, South Dakota, of Norwegian immigrants. His parents had good educational backgrounds and his father became a superintendent of schools. Lawrence attended elementary and high school in Canton and then went on to St. Olaf College. In 1919 he entered the University of South Dakota and received his B.A. degree in chemistry in 1922. From there he went to the University of Minnesota, receiving his M.A. in chemistry, to the University of Chicago, where he studied physics, and then to Yale University, where he received his Ph.D. in physics in 1925. This was a remarkable achievement in those days since few students were able to complete their doctorate in less than six years after graduating from college.

Receiving a National Research Fellowship, he spent the next two years at Yale and was then appointed an assistant professor—again achieving something of a record since he was only 26 years old. He remained on the Yale faculty for one year and then accepted an appointment as associate professor of physics at the University of California at Berkeley. Two years later he was promoted to full professor, the youngest at Berkeley, and in 1936, at the age of 35, was named director of the university's radiation laboratory. He had already invented a cyclotron in 1929 so that his reputation as a top physicist was well established.

Lawrence was deeply involved in nuclear physics when World War II began. It was natural for him to be assigned one of the major roles in the development of the atomic bomb, to which he made important contributions. Like most of those working on this project, he was firmly convinced of the need to bring about international control of the bomb or, at least, international agreement on the suspension of testing. He worked hard toward this end and was a member of the 1958 Geneva Conference.

Lawrence's interests in and contributions to physics were extremely broad, as indicated by his published papers, which averaged three and one half a year from 1924 to 1940—an almost unbelievable productivity. During this time he made better and larger models of the cyclotron, discovered many radioactive isotopes of known elements, applied the cyclotron to medical and biological problems, became consultant to the Institute of Cancer Research at Columbia University, invented a method of obtaining time intervals as short as three billionths of a second, and devised very precise methods for measuring the values of atomic constants.

He received numerous awards and honors such as the Elliott Cresson medal of the Franklin Institute, the Comstock Prize of the National Academy of Science, the Hughes Medal of the Royal Society, the 1939 Nobel Prize, the Duddell medal of the Physical Society of Great Britain, the Faraday medal, the Enrico Fermi Award, and the Medal for Merit. He was an officer of the Legion of Honor and held honorary doctorates from one British and 13 American universities.

Lawrence was a very vigorous man, interested in many intellectual and physical activities including literature, music, boating, tennis, and ice-skating. He was devoted to his wife, the former Mary Kimerly Blumer, whom he married in 1923, and their six children. He died on August 27, 1958, at Palo Alto, California.

The Discovery of Induced Radioactivity:
Jean Frédéric Joliot-Curie (1900–1958) and
Irène Curie Joliot-Curie (1897–1957)

Irène Curie, the elder of the two daughters of Pierre and Marie Curie, was born in Paris in 1897. She received her early education in a cooperative school arranged by Marie Curie for the children of her intimate friends. This unusual school, which existed for only a few years, had remarkable teachers: Paul Langevin, Jean Perrin, Mme. Perrin, and Mme. Curie. For her secondary education Irène was sent to a private school, the Collège Sévigné. During World War I, she assisted her mother with the operation of mobile X-ray units for the French army and at the Radium Institute with the training of X-ray technicians. With such a background it is not surprising that after the war Irène became an assistant in the Institute and began independent research in radioactivity. It was at the Institute in 1925 that she met Frédéric Joliot who, through the influence of Langevin, had been appointed as assistant to Mme. Curie.

Frédéric Joliot was also a Parisian, the youngest of four girls and two boys in a middle-class family. His father, the owner of a hardware store, was 57 when Frédéric was born and his mother, the daughter of one of the chief chefs of Napoleon III, was 45. Frédéric was given an education designed to prepare him for a favored position in society. He attended the Lycée Lakanal where his high spirits and amiable manners made him popular with his schoolmates and his outstanding prowess in football made him, as he once remarked, "a semiprofessional."

At the conclusion of his course at the Lycée, Joliot decided to be an engineer. To prepare for engineering instruction he attended the École Lavoisier for two years before entering the School of Physics and Chemistry of the City of Paris, the same school where Pierre and Marie Curie had made their discovery of radium. Here he was a student in the physics courses given by Paul Langevin. The acuteness of Langevin's mind, his grasp of physics, and his radical social ideas all made a deep and lasting impression on his young and able pupil.

Jean Frédéric Joliot-Curie (1900–1958)

In 1923 Joliot, then 23 years old, graduated with high honors from the École, having majored in physics and chemistry. Although anxious to devote his future to scientific research, he felt himself ill prepared and for the next year worked as an engineer in the Arbed steel mills in Luxembourg. This position was terminated by military service in which he was reunited with an engineering school friend, R. Biquard, at the artillery school at Poitiers. Biquard also shared Joliot's scientific ambitions and, at the end of a year's service, asked Langevin about the possibility of both men engaging in laboratory work at the École. Langevin engaged Biquard and obtained an appointment for Joliot as assistant to Madame Curie at the Radium Institute.

In 1926, after a year at the Institute, Joliot married Irène Curie. In order not to lose the magic of the Curie name, the couple combined both surnames into the family name Joliot-Curie. This hyphenation led, perhaps because of Frédéric's lively personality and some alliteration, to the nickname of the "Jolly-Curios." Following their marriage the couple engaged in joint research. Some of their early notable accomplishments were the demonstration of the materialization of gamma radiation into positron-electron pairs and the opposite phenomenon of annihilation radiation. In addition, they developed a very intense source of alpha particles from polonium, which they used to induce nuclear reactions in atoms of the light elements. Bothe and Becker carrying out the latter research in Germany noticed that the incidence of these alpha particles on beryllium produced a very penetrating radiation, which they assumed to be hard gamma rays. The Joliot-Curies, following up this experiment, which was reported in January 1932, found that if a block of paraffin were placed in the path of these supposed gamma rays, protons were ejected from the paraffin with velocities up to a tenth that of light and energies up to 5.3 MeV. To explain this result it was assumed that a Compton type head-on collision had occurred between the gamma-ray photon and the recoiling proton. But this explanation required the gamma-ray photons to have energies up to 52 MeV, considerably in excess of known gamma-ray energies from radioactive substances.

Irene Joliot-Curie (1897–1957)

In thinking over these difficulties James Chadwick at the Cavendish Laboratory, Cambridge, realized that the energy problem could be resolved if one supposed that neutral particles rather than gamma rays were emitted from beryllium. The idea of a neutral particle was not new. In fact, Rutherford had suggested the existence of such a particle some years earlier. Adopting this hypothesis, Chadwick was able to confirm it within a month. It was in this way that the neutron was discovered.

In continuing their investigation of nuclear reactions induced by alpha particles, the Joliot-Curies found that, in the cases of aluminum and boron, the emitted particles consisted of neutrons and positrons, the latter discovered by Carl D. Anderson at the California Institute of Technology in September of 1932. In common with other physicists, the Joliot-Curies believed at first that both particles were emitted simultaneously from the struck nucleus. However, to check this assumption they removed the aluminum target from the path of the alpha rays and found that, while the neutron emission stopped at once, the positron emission continued, but at a decreasing rate. It was clear that a new phenomenon of unusual importance had been discovered—induced radioactivity. It was also clear that the induced radioactivity appeared because an unstable nucleus had been created. The supposed reactions $_2He^4 + _{13}Al^{27} \rightarrow _{15}P^{30} + _0n^1$ and $_{15}P^{30} \rightarrow _{14}Si^{30} + _1e^0 + \nu$ involving the formation of unstable phosphorus 30, which decays to stable Si^{30} by positron emission, were soon verified by chemical methods. Thus, if the aluminum sample, after exposure to polonium alpha radiation, is dissolved in HCl and the solution rapidly evaporated, the dry residue shows no positron activity. As the reaction equations indicate, the activity should accompany the phosphorus that is produced. This passes off as the gas PH_3. If it is collected and tested, the positron activity is immediately found.

The Joliot-Curies published a short paper announcing the discovery of induced radioactivity in the *Comptes Rendus* on January 15, 1934. Their discovery set off similar researches in physics laboratories around the world, and it was soon found that neutrons were very effective in producing radioactive isotopes of many elements. The application of such isotopes as tracers in many processes has given added importance to this effect. For their discovery the Joliot-Curies were awarded the 1935 Nobel Prize in chemistry. Following the receipt of the Nobel Prize, increasing administrative demands were made on the couple and the research team of husband and wife was no longer able to continue. In 1936 Irène was appointed to fill the post that her mother had held as Professor and Director of the Curie Radium Laboratory at the Sorbonne. Frédéric was appointed as Professor at the Collège de France in 1937. He began at once to develop a laboratory there for nuclear research and was involved also in the development of the National Center for Scientific Research (CNRS).

In September 1938, Irène and a co-worker, P. Savitch, following up Hahn and Strassman's research on the beta activity resulting from the irradiation of uranium by slow neutrons, showed that one of the four activities was apparently due to lanthanum, an element of much smaller atomic number than uranium. This finding led Hahn and Strassman, who ascribed the activity to isotopes of uranium, to repeat their work. After a very careful investigation they concluded, in December 1938, that the active product was an isotope of barium and that its decay product was chemically identical with lanthanum. The presence of these elements remained a puzzle until the following month when Meitner and Frisch proposed as the explanation the fission of the uranium nucleus. Frédéric immediately began investigation of the fission process; almost simultaneously with Frisch he showed that the fission energy was about 200 MeV and that the process gave rise to elements between bromine (atomic number 35) and cerium (atomic number 58). With Halban and Kovarski he showed that about three neutrons are produced per fission and predicted the possibility of a chain reaction with the release of enormous energy.

The conquest of France by the Nazi armies in June 1940 ended any further work on uranium fission. Elsewhere such investigations went on with increasing vigor because of their possible military significance, but the results no longer appeared in the professional journals. The 200-liter supply of heavy water, which Joliot wished to use as a neutron moderator for uranium experiments, he sent to England just one step ahead of the Germans. It was carried by his colleagues Halban and Kovarski. Frédéric remained at the Collége de France and became active in organizing and directing units in the Resistance movement. Throughout the war he continued to lead anti-German activities, often at great personal peril, and was twice arrested by the Nazis.

Following the liberation of France in 1944, Frédéric assumed, as Director, the heavy administrative duties necessary to revitalize and extend the activities of the National Center for Scientific Research. He began efforts to create a French Atomic Energy Commission. When this body was established by Charles de Gaulle in 1946, Frédéric was appointed High Commissioner. In this capacity he was responsible for the construction of the first French nuclear reactor, which commenced operation in the Fort de Chantillon in December 1948. In 1946 he was elected a foreign member of the Royal Society and a year later awarded the Society's Hughes medal. Despite these signal achievements and awards, his commitment to political activity brought about his dismissal as High Commissioner in 1950 on the grounds of his membership in the French Communist party. Irène, also a member of the Commission, terminated her connection with that body the following year.

Frédéric returned to his posts in the CNRS and the Collége de France, but his health began to fail and he had to restrict his activity. In 1955 he fell seriously ill. While he was recuperating in March 1956, Irène died of leukemia induced by long exposure to radiation. Shortly thereafter Frédéric succeeded to her chair at the Faculté des Sciences in the Sorbonne and to the directorship of the Curie laboratory. But he himself never completely recovered and, while on vacation in Brittany in August 1958, suffered an accident from which he died several days later. A daughter, Hélène, and a son, Pierre, carried on the scientific tradition of the family.

The Prediction of the Meson: Hideki Yukawa (1907–1981)

We have already noted that before the discovery of the neutron in 1933 only two fundamental particles of matter were recognized by physicists, the proton and the electron. Strictly speaking, the photon also had some claim to being a fundamental particle. But since it is a corpuscle of radiation, it was felt that it should not be counted among these particles. The neutrino had been postulated as a necessary consequence of beta decay (emission of electrons) from radioactive nuclei. But, although Fermi had constructed a beautiful beta-decay theory using neutrinos, these elusive particles were still not fully accepted since they had not yet been detected experimentally. The neutron was thus the first of a series of new particles that were soon to flood the detecting devices of physicists and to demand their rightful places in whatever model of nature (or the structure of matter) physicists were constructing.

The very discovery of the neutron pointed clearly to the need for additional fundamental particles. Neutrons and protons, the basic constituents of the nucleus, exert enormous short-range forces on each other to keep the nucleus a tightly bound particle. But how are these short-range forces exerted? One way of discussing such forces is to introduce the concept of action at a distance, but such an approach cannot be treated by standard quantum-mechanical methods. Rather than action at a distance, we can introduce a force field, which is then quantized. Let us consider briefly how the field concept helps us understand how electrically charged particles interact with one another.

Although in classical physics it is customary to speak of the force between two electrons or between an electron and a proton in terms of action at a distance, given mathematically by Coulomb's law of force, the modern description of electromagnetic forces between charged particles is expressed in terms of the electromagnetic field. This automatically leads to a quantum-mechanical formulation of the interaction between charged particles because the electromagnetic radiation field can be quantized. Indeed, quantum mechanics had its origin in the quantum theory of radiation, beginning with Planck's radiation formula for black-body radiation and then more fully developed with the introduction of Einstein's concept of the photon. To picture radiation as consisting of photons is, of course, a quantization procedure. One can then introduce a quantum theory of the interaction of a charged particle with a surrounding electromagnetic field by picturing the charged particle as absorbing and emitting the photons of this radiation field. Such a quantum-mechanical theory was first developed in a series of elegant papers by Dirac and later, from a slightly different and somewhat simpler point of view, by Fermi. The photons are thus the quanta of the radiation field.

In terms of these quanta one describes the interaction between two charged particles (e.g., two electrons or an electron and a proton) as arising from the mutual emission and absorption of photons. Thus, one electron emits a photon, which is then absorbed by the other electron (or proton), and vice versa.

This tossing back and forth of photons between two charged particles gives rise to the electromagnetic forces between them. The phenomenon was first demonstrated mathematically in a famous paper by Bethe and Fermi, who showed that if two charged particles are at rest, their mutual absorption and emission of photons (treated quantum mechanically) leads directly to Coulomb's law of electrostatic force. Later it was shown by other physicists, such as Møller and Breit, that by the same procedure one can obtain the electromagnetic force between two particles when they are in motion.

Another interesting phenomenon can be discussed in terms of the quantized electromagnetic field. A single charged particle such as an isolated electron is surrounded by its own electromagnetic field and can therefore interact with this field. This idea was first introduced into physics from a classical point of view by H. A. Lorentz, who described this interaction by picturing the electron as exerting a force on itself. He calculated this force by a straightforward application of classical electromagnetic theory to a spherical electron. He considered each little piece of the sphere as exerting an electromagnetic force on every other little piece and obtained the total force of the electron on itself by adding up all of these constituent forces.

The quantum-mechanical description of the interaction of the electron with itself, that is, with its own electromagnetic field, proceeds differently. We picture the electron as emitting a photon momentarily (called a virtual photon) and then immediately reabsorbing it. The electron actually need not absorb this same photon. As long as it immediately absorbs another photon from its own radiation field, the electron interacts with its own radiation field and hence with itself. This virtual emission and absorption of photons, giving, as it does, a picture of the charged particle, indicates its so-called self-energy—the total energy concentrated in the particle, which would be released if the particle were completely destroyed.

This amount of energy should exactly equal mc^2 according to the theory of relativity, where m is the mass of the particle and c is the speed of light. It is also the work required to assemble the particle against the repulsion of its own total electric charge. In other words, it is the work that must be done to assemble the particle against the force of the particle on itself. Hence, one should be able to calculate the self-energy of an electron solely by analyzing its interaction with its own surrounding electromagnetic field.

This analysis is done quantum mechanically by computing the contribution to the interaction energy of the electron with its own field when it emits and immediately reabsorbs a virtual photon. The sum of all such virtual emissions and absorptions should give the total self-energy of the electron. Unfortunately, this gives an infinite value for the self-energy, because the electron can emit virtual photons of any frequency and hence very energetic photons (high-frequency photons). These photons contribute heavily to the self-energy of the electron and ultimately, since there is an infinitude of very high-frequency photons, give an infinite result. This is a severe deficiency of the theory, which, in a sense, has been swept under the scientific rug by special mathematical schemes that enable us to overlook, or at least to work with, these infinities.

We come now to the important contribution of Hideki Yukawa, who developed a quantized field theory of the forces between nucleons by quantizing the nuclear force field in complete analogy with the electromagnetic radiation field.

As soon as nuclear forces were discovered, physicists realized that these forces would have to be described in terms of a quantized force field. A serious difficulty arose, however, when the neutron was first discovered. Unlike the electromagnetic field that revealed its quanta (photons) very early in the game, the nature of the nuclear force field and the properties of its quanta remained a mystery.

One attempt to pierce the mystery and develop a quantum theory of nuclear forces was made by Fermi in his famous 1934 paper in which he presented a theory of beta radioactivity (the emission of electrons and nuclei by radioactive nuclei). He proposed that a neutron and a proton interacted with each other by their mutual emission and absorption of electrons and neutrinos. According to this picture, the electrons and neutrinos together constitute the quanta of the nuclear force field just the way the photons are the quanta of the electromagnetic radiation field.

Initially this view seemed reasonable, but it soon became clear that it could not be the true picture, since nuclear forces are extremely strong, whereas the interaction between nucleons and electrons and neutrinos is extremely weak. In other words, if a gas of nucleons is bombarded by streams of electrons and neutrinos little interaction takes place. Consequently, Fermi gave up the idea of introducing electrons and neutrinos as the quanta of the nuclear force field even though his theory gave an excellent description of beta radioactivity. Nevertheless, physicists everywhere were convinced that some kind of quanta would have to be introduced to describe the nuclear forces between nucleons, although they would not be electrons and neutrinos.

The first step in establishing a quantum field theory of nuclear interactions was taken by the Japanese physicist Hideki Yukawa in a well-known paper that appeared in 1934. Yukawa assumed that the Fermi process (the emission of an electron and a neutrino when a neutron changes into a proton) does not always occur; in fact, it is rather rare as compared to another process. Yukawa described this second process as one in which a much heavier particle than an electron-neutrino pair is emitted by the neutron and then absorbed by the proton. If such a process does occur and is much more probable than the Fermi process, it must generate strong interactions between the neutron and the proton and thus account for nuclear forces. Such a process would have no effect on the Fermi beta-decay process. Therefore, no conflict arises between the Yukawa theory and the experimentally verified Fermi theory.

With this idea as a starting point, Yukawa set out to develop a quantum field theory of nuclear forces and to derive some of the properties of the quanta of this field. He proceeded by analogy with the electromagnetic field that is also described by quanta (the Einstein photons). The photons of the electromagnetic field are described by Maxwell's equations, from which one can deduce that the motion of the photons is described by a wave equation. This wave equation, of course, is valid only for particles like photons that have no rest mass. Since the Coulomb force is much smaller than the nuclear forces between nucleons, and of a much longer range, the same kind of wave equation that describes photons cannot describe the quanta that are required for the strong nuclear interactions. Yukawa saw at once where the difficulty lay and how the correct answer could be obtained. He pointed out in his paper that the wave equation for photons leads to an interaction

Hideki Yukawa (1907–1981)

between two charged particles that varies inversely as the distance between them (the Coulomb potential) and thus gives rise to an inadequately small force at short distances. To obtain nuclear forces one needs a law of interaction that increases much faster as the distance gets smaller. Yukawa noted that this could be obtained from the Coulomb law by multiplying the latter by an exponential function of the distance. It could not be obtained by the kind of wave equation that describes photons; in fact, as Yukawa indicated in his paper, the wave equation for the quanta of the nuclear force field must differ from that of photons by an additional term. The rationale for the introduction of this term is that the quanta of the nuclear force field must be massive particles—their rest mass must be considerably larger than that of the electron—if they are to account for nuclear forces, whereas the electromagnetic photons have no rest mass. The added term in the wave equation for the heavy quanta signifies the mass of these particles.

Yukawa showed that the solution of this wave equation for heavy quanta gives nuclear interactions of the right size and of the type demanded by the experimental evidence; that is, they are exchange forces of the sort that Heisenberg first suggested to describe the interactions of neutrons and protons. According to Yukawa's theory, a neutron emits a heavy quantum, which is then absorbed by the proton. In the process, the neutron changes to a proton and the proton becomes a neutron. Immediately following this change, the proton (which is now a neutron) emits its own heavy quantum that is absorbed by the original neutron (now a proton), and so on. Thus the neutron and proton behave as though they were changing places by tossing a heavy quantum between them.

In this analysis Yukawa introduced a new kind of charge (nuclear charge) between nucleons. It plays the same role in the nuclear force field as the electrostatic charge on electrons and protons plays in the electromagnetic field. However, the nuclear and electric charges are in no way connected, and the former is numerically much larger than the latter. To obtain the physical properties of the heavy quanta Yukawa applied his theory to certain experimentally-observed phenomena (e.g., scattering of nucleons, beta decay, and so on). He showed that the rest mass of the heavy quanta must be about 200 times that of the electron. In addition, it is clear from the analysis that heavy quanta must be particles with zero spin (or perhaps spin one) instead of spin $1/2$ as in the case of electrons, neutrinos, protons, and neutrons. To see this point we note first that these heavy quanta obey the same kind of wave equation (except for the mass term) as do photons; they must therefore have the same statistical properties as the latter and hence an integer spin value. (Photons have a spin one and all particles with integer spin, that is, 0, 1, 2, and so on, must obey the same statistics, the so-called Einstein-Bose statistics, as do photons.) Second, since a neutron becomes a proton when it emits a heavy quantum, and the proton spin is the same as the neutron spin ($1/2$), the spin of heavy quantum must be zero (or one).

When Yukawa first presented his theory, there was no evidence for the existence of the heavy quanta he proposed because, as he pointed out, heavy quanta pass back and forth between two nucleons, but they remain undetectable unless the nucleons have kinetic energies with respect to each other that are at least equal to the mass of a heavy quantum times the square of the speed of light. In other words, there must be enough kinetic energy present to create those heavy quanta.

Although the physics laboratories at that time did not have accelerators large enough to accelerate nucleons to the required energy, cosmic ray nucleons do have such large energies. Shortly after Yukawa had published his paper, a number of physicists working with cosmic rays, in particular Carl Anderson and S. H. Neddermayer, discovered heavy particles with masses more than 200 times the mass of the electron in cosmic ray showers. At first it was thought that these particles (now called mu-mesons or muons) were the Yukawa heavy quanta, but experimental evidence soon showed that this was not the case. The mu-mesons interact much too weakly with nuclei to be the Yukawa quanta. We also know now that the mu-mesons, which have a mass about 208 times that of the electron, have the wrong spin (their spin is $1/2$ and hence they obey the same statistics as do electron, the Fermi-Dirac statistics). Today we believe that the mu-meson is an excited state of the electron. In fact, a mu-meson has a very short life (about one one-millionth of a second) and decays into an electron by emitting two neutrinos. Electrons, neutrinos, and mu-mesons are called leptons and form a single family of particles.

A few years after the mu-mesons were discovered, other mesons, slightly more massive than mu-mesons, were discovered by C. F. Powell and G. P. Occhiallini in cosmic rays. These particles, the famous pi-mesons, have all the properties required by the Yukawa theory. We now know that they are indeed the heavy quanta of the nuclear force field. Nucleons interact with each other by tossing pi-mesons (also called pions) back and forth. Each pion has a rest mass about 273 times that of the electron and a spin zero. There are three kinds of pi-mesons with different electrical properties; one is electrically neutral, one has a unit positive electric charge (like the proton), and the other has a unit negative electric charge like the electron.

Since nucleons can interact by tossing any of the three kinds of pi-mesons back and forth between them, it follows that the nuclear force between nucleons is charge-independent. In other words, the nuclear force between two protons (disregarding the Coulomb electric force of repulsion) is equal to that between two neutrons and equal to that between a neutron and a proton. This plays a very important role in the structure of nuclei. The protons interact with each other by tossing neutral pions between themselves and this is also true for the interaction between two neutrons. But a neutron and a proton interact by tossing either positive or negative pions back and forth. A pion is not a stable particle and decays into a muon in about 3×10^{-8} second by emitting a neutrino.

The essential difference between modern physics and classical physics in their treatments of forces between particles is in the concept of the field as a carrier of these forces. The idea of the classical field of force began with Newton, was then greatly expanded by Maxwell, and reached its highest and most fruitful development in the hands of Einstein. This development was classical in the sense that the field was pictured as continuous and not quantized. Planck's discovery of the quantum of energy and Einstein's photon concept showed that the electromagnetic radiation field must be quantized. Later, Bethe and Fermi showed that the Coulomb force between two charges can be derived as a consequence of their absorption and the emission of photons. However, until 1935, when Yukawa published his famous paper on the forces between nucleons (nuclear forces), it was thought that only particles with zero rest mass, such as photons, could be the quanta of a force field. Yukawa showed that this was not so and that particles (now called heavy quanta or mesons) could be the quanta of a force field.

That Yukawa should have arrived at this important discovery before any of the European or American physicists did is all the more remarkable when one realizes that Fermi had almost arrived at a similar solution with his theory of beta decay and that Yukawa had very little contact with European physicists. In a sense, he was a self-taught physicist who created his own school of physics and was responsible for the rapid rise of theoretical physics in Japan.

Hideki Yukawa was born in Tokyo on January 23, 1907, the third son of Takiyi Ogawa, who later became professor of geology at Kyoto University. The young Yukawa was raised and received his early education in Kyoto and then went on to Kyoto University, from which he graduated in 1929. After graduation, he continued with his studies in theoretical physics and became interested in problems relating to nuclear physics and the structure of elementary particles. From the first, Yukawa's approach to the solution of these problems showed a high degree of originality and clear evidence of a brilliant mind. His ability was quickly recognized and he was appointed lecturer at Kyoto University and lecturer and assistant professor at Osaka University. In 1938 he was awarded his doctorate and in the following year became professor of theoretical physics at Kyoto University. He remained at these academic posts until 1939. This was a very fruitful period for Yukawa; he published many papers on nuclear physics, including his famous paper, in 1935, "On the Interaction of Elementary Particles," which he had completed at the age of 27. In this paper he developed his heavy quantum field theory of nuclear forces and postulated the existence of the meson.

When mesons were subsequently discovered by Anderson and Neddermeyer and proved to be mu-mesons and not the pi-mesons (discovered later by Powell) required by the Yukawa theory, Yukawa was greatly encouraged and continued to develop the theory of the meson field.

In 1948, after World War II, Yukawa was invited to the Institute for Advanced Study at Princeton as visiting professor and then, in 1949, became visiting professor of physics at Columbia University. While at Columbia he was awarded the Nobel Prize in physics in 1949. When Yukawa returned to Japan he devoted himself to the development of a new type of field theory called the nonlocal field, from which he hoped to derive the various elementary particles. For a few years during the 1950s Yukawa was professor of physics at Columbia University, where he lectured on his nonlocal field theory. He then returned to Japan, where he resided until his death on September 8, 1981. He spent the latter years of his career as professor of theoretical physics at Kyoto University, as well as director of the Research Institute for Fundamental Physics at Kyoto.

Yukawa was a member of the Japan Academy, the Physical Society, and the Science Council of Japan, and was professor emeritus of Osaka University. He was also a foreign associate of the National Academy of Sciences of the United States. He was honored by learned societies and universities throughout the world. He was awarded an honorary degree by the University of Paris and was an honorary member of the Royal Society of Edinburgh and the Indian Academy of Sciences.

Yukawa did much to bring Japanese physics to its present state of eminence, not only by his research work and his work with graduate students, but also by his interest in undergraduate education in the physical sciences. He stimulated interest in physics by his books, among which were *Introduction to Quantum Mechanics* and *The Theory of Elementary Particles*. He also was the editor of the *Progress of Theoretical Physics* (Kyoto) for many years.

Many of the outstanding theoretical physicists in Japan of the late 20th century owed their positions in part to the training they received under Yukawa and to the inspiration they found in his research work. One of his outstanding students was Tomonaga, who independently of Schwinger and Feynman introduced the basic ideas that finally led to the present improved formulation of quantum electrodynamics; this theory enabled physicists to answer questions relating to the interaction of electrons and the electromagnetic field without running into infinities.

References

Enrico Fermi, "Nobel Prize Address, 1938," *Collected Papers*. Chicago: University of Chicago Press, 1962.

14 Newer Developments in Atomic and Nuclear Theory

Veil after veil will lift—but there must be veil upon veil left behind.

—Sir Edward Arnold, *The Light of Asia*

Discovery of Mesons in Cosmic Rays: Cecil Frank Powell (1903–1969)

Determining the nature of nuclear forces, that is, the forces acting between two protons or neutrons, was extremely important for understanding high-energy processes and the structure of nuclei. Certain elementary facts were already known before the discovery of the way in which nucleons actually interact. Nuclear forces operate only at extremely short ranges. They are practically zero when the distance between the two interacting nucleons exceeds one-trillionth of a centimeter, and they are very strong. Because of these properties, physicists reasoned that the nuclear force field (the medium through which the force is transmitted from one nucleon to the other) must consist of massive particles that are tossed back and forth between the two nucleons.

We have already seen that this idea was the basis of Yukawa's theory of nuclear forces. By analogy with the emission and absorption of photons by electrons and protons, Yukawa pictured nucleons as emitting and absorbing massive particles (the quanta of the nuclear force field). He calculated a mass of about 150 times that of the electron for these massive quanta and showed that two nucleons tossing such particles back and forth would exert just the kind of forces on one another that are demanded by the stability and other observed properties of nuclei.

After Yukawa had published his paper, physicists began to look for these heavy quanta, or mesons, as they were called. Since these massive intermediate particles can come into being only if energy equivalent to their own mass is available (according to Einstein's formula $E = mc^2$), physicists naturally turned to cosmic rays for such particles. Cosmic rays enter the earth's atmosphere with vast amounts of energy and should therefore, on occasion, create mesons from their own energy store. Intermediate particles, which seemed to have the desired Yukawa properties, were discovered between 1936 and 1938 by Anderson, Neddermeyer, and others. These particles, which exhibit all the properties of electrons except that they are about 200 times more massive, were called mu-mesons (they are now called muons).

At first physicists thought that these were the heavy quanta that would account for the strong, short-range nuclear forces because it had been observed that the mu-mesons decay into electrons and neutrinos, as demanded by the Yukawa theory.

373

However, the lifetimes of these particles were found to be about 20 times longer than the theory predicted. This was the first indication that the mu-mesons might not be the heavy nuclear force quanta. Moreover, only positive and negative mu-mesons (both of the same mass) were found, the former decaying into positrons and the latter into electrons. This was another drawback, since one would also require neutral mu-mesons (of the same mass as the charged mesons) to account for the nuclear forces between protons and between neutrons, which are the same as the nuclear forces between a neutron and a proton.

Finally (and this was the telling argument against the mu-meson), experimental physicists found that these particles interacted only very weakly with nuclei. If they were the quanta required by the Yukawa theory, they should have been absorbed very strongly by nuclei. Instead, they tended to move in orbits around nuclei for their entire lifetimes (about 2 millionths of a second) before decaying. During this time, even though they were close to the nucleus, they scarcely interacted with it except for the Coulomb electrostatic interaction. If the mu-mesons were the Yukawa quanta, they should have been captured by nuclei, which would have disintegrated in the process. Since disintegration did not occur, physicists rejected Yukawa's quantum role for the mu-mesons and began to seek other heavy particles in cosmic rays.

It soon became clear from further analyses of cosmic ray events that other mesons were present in cosmic rays and that these did, indeed, cause the disintegration of nuclei when they were brought to rest in various substances. Such disintegration can be observed because, in the process, slow protons, neutrons, alpha particles, and other nuclear remnants are emitted. Furthermore, it was found that these very mesons at times decayed spontaneously in about 3 one-hundred-millionths of a second, with the emission of slower secondary particles; these secondaries were observed to have all the properties of mu-mesons. From all of these findings, it was clear that there were two mesons in cosmic rays, one of which was called the primary, or pi-meson (pion), and the other the mu-meson. Today we refer only to the pi-meson (or pion) as a meson and classify the mu-meson with the electron and the neutrino in the group that is called the leptons.

The pion has the very properties required by a Yukawa heavy quantum: its mass is about 273 times that of the electron (the mass of the muon is about 207 electron-masses); it has a very short lifetime; it causes the disintegration of a nucleus when close enough to it. Moreover, direct cosmic ray evidence shows that pions are ejected from nuclei when the latter disintegrate after collisions with very energetic protons (the primary particles in cosmic rays). Then the negatively charged pions (the secondary particles in cosmic rays) disrupt other nuclei in their paths as they come to rest. This shows that they interact very strongly with nuclei. Neutral pi-mesons, as well as negatively and positively charged pi-mesons, have been discovered. Pions are thus the carriers of the force field between nucleons postulated and described by Yukawa. In 1947 Powell and Occhialini discovered pion tracks on special photographic plates exposed to cosmic rays. Powell received the Nobel Prize in 1950 for developing special photographic techniques for the study of cosmic rays and applying the techniques to the analysis of mesons found in such rays.

Cecil Frank Powell was born on December 5, 1903, at Tonbridge in Kent, England, where his father's family had long been gunsmiths. Powell showed his intellectual aptitude quite early by winning a scholarship at the age of 11 to the

Cecil Frank Powell (1903–1969)

Judd School in Tonbridge, where he excelled in mathematics and the sciences. On leaving the Judd School, Powell won an open scholarship to the Sidney Sussex College, Cambridge, where he continued his scientific studies. He graduated in 1924 at the head of his class, with highest honors in science. By this time he had decided to become a physicist. He did graduate work at the Cavendish Laboratory under C. T. R. Wilson and Lord Rutherford, receiving his Ph.D. in 1927. During this period Powell developed the interests that ultimately led to his discovery of the pi-meson and the Nobel Prize in physics. Stimulated by his work with Wilson, he worked to improve techniques for tracking particles that might be used in cosmic ray studies during balloon flights. Upon receiving his doctorate, Powell moved to the University of Bristol as research assistant to A. M. Tyndall, in the H. H. Wills Physical Laboratory, and was subsequently appointed lecturer and then reader in physics. For a time he became engrossed in the physics of earthquakes and spent the year 1936 as a seismologist with an expedition that was studying earthquakes in the West Indies.

When he returned to Bristol, Powell began work on a Cockcroft generator for accelerating protons and deuterons to very high speeds in order to study the scattering of neutrons by protons. Although he used the Wilson cloud chamber to detect the scattered particles, he was not satisfied with it and sought better techniques for discerning the charged particles and measuring the lengths of their tracks. Photographic emulsions had been used early in the 20th century to detect emanations from radioactive elements. Yet few nuclear physicists resorted to photographic techniques because they appeared to be less reliable than the cloud chamber method and more difficult to interpret. Moreover, the photographic plates required special sensitizing in order to react to swift particles, and plates of varying sensitivity had to be used for different particles. In 1935, Zhadanov, in Leningrad, and the Ilford Laboratories in England, independently produced emulsions that could detect fast protons without previous sensitization.

One advantage of a photographic emulsion over the cloud chamber is that the emulsion detects and tracks charged particles continuously; the cloud chamber does so only momentarily, while it is expanding. Therefore, a short-lived particle like a meson can be overlooked in the cloud chamber. Moreover, the length of the track in the cloud chamber is not related in any simple way to the energy of the incident particle, whereas the length of the track in the emulsion gives an accurate measure of its energy. With these things in mind, Powell applied photographic techniques more and more extensively to nuclear research. Between 1939 and 1945 he had perfected photographic emulsions to such an extent that their superiority over the cloud chamber was undisputable.

During this same period, Powell decided to use his photographic techniques in conjunction with balloon flights to study cosmic rays. In 1947, plates with a new emulsion that Powell had developed the year before were exposed to cosmic radiation on Pic du Midi, 2,800 meters above sea level. A study of the tracks on these plates led Powell and Occhiallini to announce the discovery of the pi-meson—the particle needed to explain nuclear forces as predicted by the theory of Yukawa. The chance of detecting such a particle by a cloud chamber is extremely slight because its rate of decay in flight is most rapid. However, in an emulsion, such a particle can be brought to rest before decay takes place, thus enabling the interactions with atomic nuclei to be observed.

In 1948, following his discovery of the pi-meson, Powell was appointed Melville Wills professor of physics at Bristol, where he carried on his research and teaching activities for the remainder of his career. From 1952 to 1957 he was director of various European expeditions for making high-altitude balloon flights in Sardinia and in the Po valley. Powell did research in many fields and wrote extensively on the discharge of electricity through gases, as well as on nuclear physics, cosmic rays, and photographic techniques. He was a fellow of the Royal Society and a foreign member of the Academy of Sciences of the U.S.S.R. In 1961 he was awarded the Hughes and the Royal medals. The universities of Dublin, Bordeaux, and Warsaw awarded him honorary doctorates. Powell died on August 9, 1969, in Bellano, Italy.

The Antiproton: Emilio G. Segrè (1905–) and Owen Chamberlain (1920–)

Dirac had predicted not only the antielectron (the positron) but also the antiproton. However, to produce an antiproton required vastly more energy than was available from the comparatively simple accelerators of the 1930s. Today, the leading physics laboratories of the larger nations, such as the United States and the U.S.S.R., bear testimony to the increasingly successful technology of atomism, which has produced more and more powerful particle accelerators. In 1946 the synchrocyclotron, built by the University of California, accelerated particles to energies of 200 to 400 million electron volts (MeV). Later, larger synchrocyclotrons in the United States and in the U.S.S.R. raised obtainable energies to 700 to 800 Mev. Earlier versions of the Lawrence cyclotron had been unable to push particles beyond about 20 Mev because at that energy the particles traveled so fast that the mass increase with velocity became appreciable, as predicted by Einstein's special theory of relativity. This effect made the particles lag and fall out of phase with the electrical stimulus they received. To compensate, the synchrocyclotron synchronizes the alternations of the electric field with the increase in mass of the particles.

In 1955, with a Bevatron accelerator, so called because it elevated protons to energies measured in terms of billions of electron volts (BeV), two California physicists, Emilio Segrè and Owen Chamberlain, succeeded in creating and identifying the antiproton. After bombarding copper with protons of 6.2 Bev hour after hour, they were able to identify some 250 antiprotons. It was not a simple matter to recognize them because, for every antiproton they produced, some 40,000 particles of other types also came into existence. But an elaborate system of detectors was designed so that only the antiproton could touch all the bases. Segrè and Owen found that the antiproton is as unenduring as the positron. It is quickly snatched up by a normal, positively charged nucleus, where the antiproton and one of the resident protons annihilate each other. If the proton and antiproton pass close by without colliding, an antineutron is produced from the antiproton through the neutralization of its charge and the proton becomes a neutron.

In 1926 the new Italian school of physics at the University of Rome consisted of one man, Enrico Fermi. Senator O. M. Corbino, professor of experimental physics and director of the Physics Institute at the university had dreamed of a school for many years. He had succeeded in having a chair of theoretical physics established at Rome and had staunchly and successfully supported Fermi's candidacy. But one man, even a Fermi, does not make a school. Consequently, Corbino, when he met Rasetti, a friend of Fermi's when they were high-school students together in Pisa, induced him to transfer from Florence to Rome. With two such members in his school Corbino was greatly encouraged. One morning in June 1927 he announced to his undergraduate class in electricity for engineers that he was looking for one or two outstanding students who were interested in transferring from engineering to physics. He recruited no engineering students, but his glowing account of the great excitement to be found in the pursuit of modern physics induced a student of mathematics, Edoardo Amaldi, the son of a professor of mathematics, to switch to physics. Corbino's class in electricity was designed not only for engineers but for all other science students not studying physics, so that mathematicians and chemists also attended the course. On the day of Corbino's appeal one of the chemistry students in attendance was Laura Capon, who already knew Fermi quite well and was to become his wife a year later.

A fourth member of Corbino's boys, as his recruits became known, was Emilio Segrè, who had started out as an engineering student at the University of Rome in 1922. He had not heard of Corbino's appeal but came into the group through his friendship with Rasetti. Segrè was born in Tivoli, Rome, on February 1, 1905, the son of the industrialist Giuseppe Segrè and Amelia Treves. While studying engineering at Rome, he became acquainted with G. Enriques, the son of an outstanding Italian mathematician. Through Enriques he met Rasetti, who aroused his interest in physics. Even before this meeting, Segrè had heard some lectures by Fermi in 1925 and was greatly impressed by the beauty of the new physics Fermi revealed. However, it was only after many mountain-climbing discussions with Rasetti that Segrè learned enough about the new theories to begin to think of physics as a career.

During the summer of 1927 he came to know Fermi personally and through conversations with him saw that physics was more to his liking than engineering. However, he was not yet entirely convinced. Segrè began reading physics books in earnest during this period. In September 1927 he accompanied Fermi and Rasetti to the international conference at Como where he had the unforgettable experience of meeting the great physicists he had read about such as Lorentz, Rutherford, Planck, and Bohr, as well as a group of much younger men, including Heisenberg, Pauli, and Fermi.

In November 1927 Segrè began studying physics under Fermi as a fourth-year student and in 1928 received his doctorate, the first to be awarded under Fermi's sponsorship. Segrè interrupted his academic career to serve in the Italian army for one year but returned to the University of Rome in 1929 to become an assistant to Professor Corbino. Although the theoretical part of the physics program under Fermi was progressing nicely, the experimental work was still weak. To strengthen this part of the program and to import more advanced experimental techniques, Rasetti and Segrè each spent some time abroad. Rasetti went to Robert Millikan's laboratory in Pasadena, California, where he worked on the Raman effect, and Segrè went to Hamburg as a Rockefeller Foundation fellow to study with Otto Stern. He also worked for a short period with Professor Pieter Zeeman in Amsterdam.

In 1932 Segrè returned to Italy to become assistant professor of physics at the University of Rome. He remained there until 1936, working in nuclear physics with Fermi, Rasetti, Amaldi, and D'Agostino, the fifth of Corbino's boys. In 1936 Segrè was appointed director of the Physics Laboratory at the University of Palermo, where he remained until 1938. The collaboration between Hitler and Mussolini and the racial laws adopted by the fascist regime made it impossible for him to remain in Italy. In 1938 he left to become a research associate at the University of California Radiation Laboratory in Berkeley and then a lecturer in the physics department there. From 1943 to 1946 he was a group leader at the Los Alamos division of the Manhattan Project. He returned to the University of California in 1946 as professor of physics and remained at this post for the remainder of his career.

During his career, Segrè contributed to various areas of physics, including atomic spectroscopy, the Zeeman effect, molecular beams, neutron physics, nuclear fission, and elementary particle physics. He also did important work in radio chem-

Emilio G. Segre (1905–)

istry and collaborated in the discovery of the elements technetium, astatine, and plutonium 239.

At various times in his academic career he taught at Columbia University, the University of Illinois, and at Rio de Janeiro. For his important contributions to physics, Segrè was awarded numerous honors in addition to the 1959 Nobel Prize in physics, which he shared with Owen Chamberlain. Segrè was a member of the National Academy of Sciences and of various foreign academies.

Owen Chamberlain was born on July 10, 1920, in San Francisco, California, and grew up in a science-oriented family. His father, a prominent radiologist who was greatly interested in physics, stimulated his son's natural talents in this field, so that when Owen entered Dartmouth College in 1937 he was firmly committed to becoming a research physicist. After receiving his bachelor's degree in 1941, he entered the University of California to do graduate work. His studies were interrupted by World War II, and he joined the Manhattan Project to work on the atomic bomb. He began this work at Berkeley, California, under Segrè and then went to Los Alamos where he investigated the nuclear capture and scattering cross-sections of intermediate energy neutrons. He also studied the spontaneous fission of heavy nuclei.

After the war Chamberlain continued his graduate studies under Enrico Fermi at the University of Chicago and was awarded his Ph.D. degree in 1949 for experimental work on the diffraction of slow neutrons by liquids. In 1948, after he had completed his experimental work under Fermi, he accepted a teaching post at the University of California in Berkeley and began a series of experiments in collaboration with Segrè. Although these experiments dealt primarily with the scattering of protons by protons, they prepared the way for the discovery of the antiproton in 1955. For this work he and Segrè won the Nobel Prize four years later. After that time Chamberlain investigated the interaction of antiprotons with hydrogen and deuterium. He also studied the production of antineutrons from antiprotons and the scattering of pions.

Owen Chamberlain (1920–)

Chamberlain was a fellow of the American Physical Society and a member of the National Academy of Sciences. In 1957 he was a Guggenheim fellow at the University of Rome, and in 1959 he was Loeb lecturer at Harvard University. In 1958 he was appointed professor of physics at the University of California at Berkeley where he remained for the rest of his career.

In our discussion of Dirac's relativistic theory of the electron, we pointed out that his equations have two sets of solutions. One gives the motion of an ordinary positive-energy electron and the other describes electrons in states of negative energy. These two sets of solutions arise because the theory of relativity leads to an equation which relates the square of the energy, and not the energy itself, to the momentum and mass of a particle. To obtain the energy one has to take the square root of this relativistic expression and, as we know, a square root can be either positive or negative. In nonquantum, relativistic mechanics the negative square root is neglected because electrons with negative energy are not observed and have no meaning. But we cannot neglect the negative-energy states of an electron in quantum mechanics because electrons in states of positive energy must jump down into these negative-energy states with the emission of radiation. According to the Dirac theory, all the observable (positive-energy) electrons would jump down into the states of negative energy if these were empty. To avoid this difficulty, Dirac proposed that all the negative-energy states—an infinite number in each volume element of space—are filled with negative-energy electrons. The Pauli exclusion principle prevents the ordinary electrons from jumping down to states of negative energy and thus destroying the universe.

This very imaginative theory was not taken seriously until C. D. Anderson discovered the positron, which has all the properties of a hole in Dirac's infinite sea of negative-energy electrons. We know today that if enough energy—about 1 million volts—is supplied to a Dirac negative-energy electron under the proper conditions, it can be lifted to a state of positive energy, leaving a hole and thus creating a pair. The electron itself—being in a state of positive energy—now behaves like an ordinary negatively charged particle with positive energy. The hole, being the absence of negative charge and negative energy, behaves like a particle of positive charge and positive energy.

After Anderson's experimental verification of Dirac's theory of holes, physicists surmised that protons should also exist in states of negative energy and that a hole in this infinite sea of negative-energy protons should be detectable as a negatively charged particle with a mass of the proton. This particle, which was discovered in 1955 by Segrè and Chamberlain, is now known as the antiproton. Physicists believed in the existence of an antiproton before its discovery because an ordinary proton is governed by the same relativistic equations as an electron. Hence, a proton is described by the same multiplicity of wave functions as an electron, so that negative-energy states of protons must also exist. From this reasoning, the existence of antiprotons follows immediately.

Although the theory of the antiproton is quite straightforward, and quite similar to that of the positron, the discovery or the creation of an antiproton is much more difficult. Whereas positrons can be created in the laboratory by bombardment of matter into charged particles or gamma rays of a few MeV, at least two BeV are necessary to create a proton-antiproton pair because the mass of the proton is about 2,000 times that of the electron. In actual practice, a 6 billion-volt proton is necessary as the bombarding particle to produce a proton-antiproton pair because not all its energy, as calculated in the laboratory frame of reference, goes into the creation of the pair. The reason is that the energy that is available for creating a pair must be computed in the frame of reference in which the center of mass of the bombarding particle and the target is at rest, whereas the 6 billion volts refers to the energy of the bombarding particle in the frame of reference in which the target is at rest (the laboratory itself). The pair-creating energy must be computed in this way because the energy associated with the motion of the center of mass of the system does not contribute to pair creation.

Since 6 billion-volt particle accelerators were not available immediately after the discovery of the positron, physicists first looked for antiprotons in cosmic rays where large energies occur. But such pair creations were so rare, that no definite conclusions about the existence of antiprotons could be drawn. The search remained in abeyance until the Bevatron at Berkeley, California, was constructed. At the suggestion of Professor Ernest O. Lawrence, this accelerator had been designed specifically with enough energy to create proton-antiproton pairs.

In principle, the experiment to create and detect antiprotons is very simple. One subjects a target to bombardment by 6 billion-volt protons (as measured in the laboratory) and then looks for negatively charged particles with the mass of a proton, emanating from the target. The analysis of the particles coming from the target is carried out by means of a magnetic field, which deviates the negatively and positively charged particles in opposite directions to one another. One now chooses a magnetic field of such strength that it causes all the antiprotons (negatively charged particles) with a definite momentum to move along a definite circular path into a detector placed somewhere along its path. At the same time another detector is placed at a different point along the path of the particle to obtain its velocity. Both the momentum of the negatively charged particle, as given by the radius of its path in the magnetic field and the strength of this field, and the speed of the particle, as given by the time it takes this particle to cover the known distance between the two detectors, can be measured. From the measured momentum and the speed, one can then calculate the mass of the particle using the well-known relativistic formula that relates the mass to the speed and momentum of the particle.

Although the theory of the experiment is very simple, it is quite complicated in practice. To begin with, the arrangement of the apparatus and the alignment of the detectors must be very precise in order to detect a negative particle with just the mass of the proton. In fact, in the first stages of their work, Chamberlain and Segrè failed to detect antiprotons because of an error in their alignment. Furthermore, other negatively charged particles, such as mesons and electrons, having the same momentum as the antiprotons could move along the path of the antiproton. Two of these might trigger the two detectors, giving the impression of a single antiproton. Fortunately, mesons must move about ten times faster than antiprotons to have the same momentum. They then are moving fast enough to emit a characteristic type of radiation, first detected by the Russian physicist P. A. Cerenkov. Since the much more slowly moving antiprotons do not emit Cerenkov radiation, a Cerenkov counter can be placed in the path of the deflected particles to differentiate between antiprotons and the less massive particles.

With such an experimental arrangement, and with great care taken in the alignment, Chamberlain and Segrè detected antiprotons even though there was only one in the 30,000 particles in the magnetically analyzed beam. To prove beyond any doubt that they had detected antiprotons, they then showed that these particles annihilated protons and neutrons. When this happened, about five pi-mesons (pions) were emitted.

After discovering the antiproton, Chamberlain and Segrè began looking for the antineutron. Such a particle is very difficult to detect in the primary beam of particles coming from the target because it is electrically neutral and therefore cannot be separated from the beam by means of a magnetic field. However, the antineutron can be detected by what is called a charge exchange reaction: an antiproton meets a proton and, rather than annihilating one another, the proton gives up its positive electric charge to the antiproton. The proton thus becomes a neutron and the antiproton becomes an antineutron. The antineutron is then annihilated after it has moved a short distance; the products of this exchange can be detected in a bubble chamber. There the following phenomena are observed: (1) the curved track of the incoming antiproton; (2) a sudden end of the track at a point where the charge exchange occurs and the antineutron is created; (3) a point further along where a star is formed by the tracks of secondary particles created when the antineutron is annihilated. Between the point where the antineutron is destroyed and the point of the star, nothing is visible in the bubble chamber because the antineutron is an electrically neutral particle.

The final proof of the existence of the antinucleon came with the analysis of the energy released in its annihilation. Here the situation is not as straightforward as in the annihilation of an electron-positron pair, which occurs with the emission of two photons. To prove that the electron and the positron have annihilated each other, one simply measures the energy of the two photons (gamma rays) and notes that it equals twice the mass of the electron times the square of the speed of light (Einstein's mass-energy relationship). But when a proton and an antiproton annihilate each other, photons are not emitted directly. Instead, different kinds of mesons are released. These decay at different stages so that measuring the total energy of all these decay products and showing that it equals twice the mass of the proton times the square of the speed of light becomes rather complicated. It has been done, however, and the evidence for the antinucleon is conclusive.

When an antinucleon and a nucleon annihilate each other, one observes an annihilation star, with many different particles moving off in various directions. Most of these particles are pi-mesons, which decay into mu-mesons (muons) and neutrinos in about 10^{-8} seconds. The muons, in turn, decay into electrons or positrons and neutrinos in a few microseconds; the positrons and electrons annihilate each other to become photons. Thus, in a very small fraction of a second, the nucleon-antinucleon pair decays into photons and neutrinos—particles with zero rest mass—which move off with the speed of light.

The existence of the antinucleon greatly strengthened the belief of physicists that antimatter exists as the normal state of things in a different part of our universe. This opinion was already expressed by Dirac in his Noble lecture in 1933.

If we accept the view of complete symmetry between positive and negative electric charge so far as concerns the fundamental laws of nature, we must regard it rather as an accident that the earth (and presumably the whole solar system) contains a preponderance of negative electrons and positive protons. It is quite possible that for some of the stars it is the other way about, these stars being built up mainly of positrons and negative protons. In fact, there may be half the stars of each kind. The two kinds of stars would both show exactly the same spectra, and there would be no way of distinguishing them by the present astronomical methods.

More recently, L. Lederman and his collaborators at Columbia University have created the antineutron in the laboratory. Therfore, antimatter is now more than a mere hypothesis. In principle, one should be able to detect the presence of antimatter stars in our universe by using the nonconservation of parity discovered by Lee and Yang. Under ordinary conditions, when a nuclear process occurs involving the emission of beta rays, either neutrinos or antineutrinos are emitted. In the carbon cycle inside ordinary stars antineutrinos are emitted when carbon or nitrogen captures a proton. If the star consisted of antimatter, a neutrino would be emitted instead of an antineutrino. According to the discovery of Lee and Yang, the neutrino and the antineutrino have different chiralities: one advances like a left-handed screw and the other like a right-handed screw. If the neutrinos coming from a star like the sun did not have the same chirality as those coming from the sun, we would know that the star consists of antimatter.

Nuclear Magnetic Moment: I. I. Rabi (1898–1988)

As the experimental and theoretical aspects of nuclear physics developed, the need for greater accuracy in the measurement of nuclear parameters became ever more imperative. The construction of nuclear models and the introduction of a reasonable picture of nuclear forces depend on a knowledge of these things. The nuclear magnetic moment is of particular interest, since the magnetic moment must be carefully taken into account in any model of the nucleus and in any theory of nuclear forces. I. I. Rabi, in his molecular-beam laboratory at Columbia University, developed the most precise and elegant method for measuring the size of the magnetic moment of a nucleus as well as its sign. (If the magnetic moment is parallel to the angular momentum of the nucleus, its sign is positive; if antiparallel, the sign is negative.)

I. I. Rabi (1898–1988)

Rabi's molecular-beam method grew out of the experiments of Stern and Gerlach for measuring the magnetic moments of atoms by passing the atoms through inhomogeneous magnetic fields. The Stern-Gerlach experiments were developed to test the theory of space quantization according to which a spinning atom, whose total angular momentum is a certain multiple of the fundamental unit of angular momentum $h/2\pi$, can orient itself only in a discrete number of directions relative to a magnetic field. In the classical picture, a spinning atom behaves like a small magnet and therefore precesses about the magnetic field at a frequency (the Larmor frequency) that depends on the strength of the magnetic field multiplied by the ratio of the magnetic moment of the atom to its angular momentum (the so-called gyromagnetic ratio). In this respect, the classical and the quantum-mechanical pictures are the same. But the classical picture includes no restriction on the angle at which the atomic magnet can precess about the magnetic field. According to the quantum mechanics, the atomic magnet can precess about the magnetic field only at a discrete number of angular positions related in a simple way to the total angular momentum. If the total angular momentum of the atom is J, in units of $h/2\pi$, there are just $2J + 1$ angular orientations that the atomic magnet can assume with respect to the external magnetic field. The atom precesses about the magnetic field in any one of these orientations, but in no other.

To test this picture of space quantization (discrete orientations with respect to the direction of the magnetic field), Stern and Gerlach proposed the following simple experiment. Allow a parallel beam of similar atoms—all with the same total angular momentum, but of random spatial orientation—to pass through an inhomogeneous magnetic field. Analyze the transmitted beam to see whether its structure is unaffected or split into a number of beams. If the composition of the beam is the same as it was prior to its passage through the magnetic field, then there is no such thing as space quantization. If, on the other hand, the transmitted atoms are separated into a discrete number of beams, space quantization is a reality.

To understand this experiment in terms of the quantum-mechanical picture, note that the atomic beam consists of atoms with their angular momentum vectors, and hence their magnetic moments, oriented randomly in space. The beam of atoms is directed at right angles to the direction of a magnetic field whose strength is inhomogeneous. Since the inclinations of the angular momentum vectors of the atoms in the beam are initially randomly distributed, these atoms, behaving like little magnets, precess about the magnetic field at all possible angles. But in an inhomogeneous magnetic field, a force is exerted on a magnet in the direction of the inhomogeneity. This force is proportional to the product of the inhomogeneity and the magnitude of the projection of the magnetic moment of the magnet in the direction of the inhomogeneity. It changes the direction of motion of an atom passing through the magnetic field. Hence, if there were no space quantization, the directions of motion of the atoms in the initial beam would be displaced by random amounts perpendicular to the direction of motion of the initial beam of atoms. The transmitted beam would merely spread out like a fan. But if space quantization is present, then there are only $2J + 1$ orientations along which the atoms in the initial beam become oriented in the region of the magnetic field. As the beam passes through the field, the atoms in each of these orientations suffer a different force arising from the inhomogeneity of the field. Hence the transmitted beam must consist of $2J + 1$ component beams, each one displaced by a different amount relative to the original beam. This phenomenon is precisely what Stern and Gerlach discovered; it was the starting point of Rabi's investigation of the spins and magnetic moments of nuclei.

Rabi began his work by first improving the experimental techniques that ultimately led to the molecular-beam method. In the original work of Stern and Gerlach the magnetic field was confined to a rather small volume of space (the region between the poles of two magnets), and the beam of atoms interacted with the field for only a short time. Hence, these fields had to be quite strong to bring about an appreciable displacement of the component beams. Rabi improved upon this method by producing the necessary magnetic field with electric currents in long wires parallel to the direction of motion of the atoms in the beam. Even though the magnetic fields themselves were weak, sufficient effect was obtained because the beam was in contact with the magnetic field over a longer path than in the Stern-Gerlach experiments. Rabi introduced another important feature—auxiliary magnetic fields that could be altered at will and directed in any desired way relative to the inhomogeneous splitting field or the analyzing magnetic field. These additional fields are particularly interesting when they are changed periodically, either by making them rotate or by making them oscillate at a predetermined frequency. The use of these varying magnetic fields finally led to the molecular-beam resonance method, which gave the magnitudes of the magnetic moments of nuclei, as well as the signs of these moments, with greater precision.

The theoretical basis of Rabi's resonance method of measuring magnetic moments was presented in a paper in which Rabi dealt with the behavior of a particle with a magnetic moment, such as a nucleus, which was placed in an inhomogeneous magnetic field to which had been added a weak oscillating or rotating magnetic field. The particle with the magnetic moment is thus in a strong inhomogeneous field plus a weak varying electromagnetic field of a definite frequency. If only the inhomogeneous field were present, the spin axis of the particle would precess about this field at a definite angle in accordance with the requirements of space quantization. But only a finite, discrete number of precessional states would be available to the particle. The imposition of the weak varying electromagnetic field causes the particle to change from one precessional state—that is, from one state of space quantization—to another, either by absorbing or emitting a photon of a frequency equal to the frequency of the varying electromagnetic field. This can happen only if the energy of such a photon is just equal to the difference in energy between the state in which the particle is precessing and one of the other possible precessional states. Since these energy differences are very small, the varying electromagnetic field is a microwave or even a radiofrequency field. If the frequency of this field is equal to or very close to the Larmor precession of the particle around the inhomogeneous field direction, it causes transitions to other permissible states of space quantization. The addition of the microwave field was Rabi's major contribution to this investigation.

If we are dealing with an electron in a rotating magnetic field and assume that the electron's spin is at first lined up parallel with the magnetic field, we find that after a certain time the spin of the electron has turned over and is antiparallel to the field. This problem had been approach by E. Majorana and P. Güttinger before Rabi considered it, but they analyzed only a special case. Rabi, however, investigated the probability of such transitions of spin in a magnetic field rotating with any orientation in space. The analysis showed that if the frequency of rotation of the magnetic field was close to or equal to the Larmor precession frequency of the particle, the chance for a transition was much greater than in other cases. Rabi's analysis also showed that the probability for a transition was different for positive and negative magnetic moments. This means that a careful analysis of the behavior of the magnetic moment of a particle in a varying magnetic field (the flipping over of the spin) must yield information about the magnitude and the sign of the magnetic moment.

In collaboration with his associates and students, Rabi applied precisely this type of analysis to the magnetic moments of three different nuclei in a famous series of experiments. In this type of experiment a beam of atoms or molecules is subjected to an inhomogeneous field, displacing the atoms from their original path. After the atoms pass through a slit, they are subjected to a second field whose inhomogeneity is opposite to that of the first field, but equal in strength. The atoms are thus deflected back to their original path and arrive at the detector. If only these two fields were acting on the beam of atoms, the number of atoms detected would be the same as if there were no fields present because the second field would compensate exactly the action of the first field. The third field, either rotating or oscillating, is now introduced in the neighborhood of the slit at right angles to the constant magnetic field. This field induces transitions (flipping over) of the magnetic moments of the atoms just before they enter the second constant inhomogeneous field. Because of these transitions, the second inhomogeneous field cannot compensate the first one completely and the number of atoms arriving at the detector is not the same as before. This effect is most pronounced if the frequency of the oscillating field is close to the frequency of the Larmor precession of the atoms in the beam. Since this frequency is proportional to the gyromagnetic ratio of the atom, the ratio (the so-called Landé g factor) can be measured. Thus, the magnetic moment can be found if the spin or the angular momentum is known. Therefore, if the frequency of the oscillating field is slowly varied, a sharp decrease (the resonance phenomenon) occurs in the number of atoms arriving at the detector when the frequency of the field equals the Larmor frequency. Each such resonance then gives a g value and, hence, a magnetic moment.

In 1930, a small research room was set aside at Columbia's new Pupin physics laboratory to study the spins and magnetic moments of atomic nuclei. The physicist placed in charge of this project was the newly appointed assistant professor in theoretical physics, Isidor I. Rabi. This choice was singularly appropriate because Rabi had worked on the magnetic properties of crystals while a graduate student at Columbia and had spent two years in European laboratories working with Bohr, Pauli, Heisenberg, and Stern, who greatly influenced the direction of Rabi's future research.

Unlike most other outstanding physicists, Rabi did not go directly into physics upon completing his undergraduate education but came to it after an interruption of three years, during which he pursued a business career. Rabi was brought to this country when he was one year old from Raymanov, Austria, where he was born on July 29, 1898. After completing his grammar school and high-school education in Manhattan and Brooklyn, he went to Cornell University to study chemistry. From the economic point of view chemistry was considered more lucrative than physics and, in any case, it touched the day-to-day lives of people more intimately than did physics, which seemed esoteric and unrelated to any direct way of earning a livelihood. On receiving his bachelor's degree in chemistry from Cornell in 1919,

Rabi left academic pursuits but three years later returned to Cornell—this time as a graduate student in physics.

When economic reasons prevented him from continuing at Cornell, he matriculated as a graduate student at Columbia University with the idea of supporting himself with part-time work in New York City. Fortunately, he was given a part-time teaching program in physics at the College of the City of New York and thus was able to complete his doctoral work at Columbia, receiving his Ph.D. degree in 1927. With the aid of fellowships, Rabi spent two years at various European universities and then returned to Columbia as a lecturer in theoretical physics in 1929.

When he started his remarkable career at Columbia, Rabi was interested in both theoretical and experimental work. Although he was primarily an experimentalist, he also had an excellent grasp of theory. He had, in fact, published some theoretical papers on the quantum mechanics of rotating systems before he began to experiment on the spins and magnetic moments of nuclei. Once he began his experimental work, he devoted all his time to it, except for his hours of teaching.

His first experiment was the measurement of the spin of the sodium nucleus. The apparatus he used was essentially the same as that used by Stern. He began to improve the Stern molecular-beam technique and, a few years after the completion of this first experiment, applied the resonance principle to nuclei that were moving through a magnetic field, obtaining fantastic precision in his measurements. After discovering the resonance method of measuring spins, he applied it to a series of nuclei. These experiments won him the Nobel Prize in physics in 1944.

When Rabi came to Columbia in 1929, he organized a seminar in theoretical physics, conducted jointly with Professor Gregory Breit of New York University. It attracted outstanding students of physics in the New York metropolitan area and Columbia became a recruiting ground for future theoretical physicists. At the same time, Rabi's molecular beam laboratory became more and more successful. It began to attract outstanding students of experimental physics, one of whom, Polykarp Kusch, won the Nobel Prize for physics in 1955.

Rabi's method of work at that time was quite different from what one ordinarily would expect of an experimentalist. Although he spent the necessary time in his laboratory to ensure the success of an experiment, he spent most of his time on theory, leaving the technical details of the experiments to his assistants.

In those years, Rabi's office door was always open, as it would be throughout his career. Any student or colleague who wished to talk with him was free to do so. Often one would find one or more of his graduate students talking to him while he whittled away at a piece of wood. Although the discussions were generally about physics, it was not unusual to find other subjects, such as literature, politics, or economics, the topics of vehement arguments, for Rabi never allowed his intense pursuit of physics to blind him to the excitement to be found in other intellectual activities.

In 1940 he took a leave from Columbia to become associate director of the Radiation Laboratory at the Massachusetts Institute of Technology, where he worked both on radar and the atomic bomb. In 1945, at the end of the war, he returned to Columbia as executive officer of the physics department. After retiring from this post, he continued in the department as Higgins professor of physics. In 1964 a special academic appointment as university professor was created for him, allowing him complete freedom in teaching. He held this post as an emeritus until his death in January 1988, a few months shy of his 90th birthday. He also gave lectures on science and history—an outgrowth of similar lectures given at Princeton as visiting professor of history.

In addition to the Nobel Prize, Rabi received many other honors. In 1939 he was awarded the prize of the American Association for the Advancement of Science, and, in 1942, the Eliott Cresson medal of the Franklin Institute. In 1948 he received both the medal for merit and the King's medal for service in the cause of freedom. He was also an Officer of the Legion of Honor and was awarded the honorary Doctor of Science degree by Princeton, Harvard, and Birmingham Universities. He was an associate editor of the *Physical Review* for two periods and, in 1950, was elected president of the American Physical Society. He was a member of the National Academy of Sciences, the American Philosophical Society, and the American Academy of the Arts and Sciences.

In 1959 he was appointed a member of the Board of Governors of the Weizmann Institute of Science in Israel. In addition, he was a foreign member of the Japanese and Brazilian Academies. He was a member of the General Advisory Committee to the Arms Control and Disarmament Agency, the United States National Committee for UNESCO, and the Science Advisory Committee of the International Atomic Energy Agency.

Hydrogen and the Elementary Particles: Willis E. Lamb, Jr. (1913–)

When Bohr first introduced his theory of the atom, he explained the origin of the spectral lines (the Balmer lines) of hydrogen by means of a single quantum number n (the principle quantum number), which takes on all integral values $(1, 2, 3, \ldots)$ and corresponds to the various permitted orbits. The principle quantum number determines the size of the permitted orbit and is a measure of the total energy of the electron in this orbit, just as in classical physics the semimajor axis of the orbit of a planet determines the planet's total mechanical energy. The original Bohr theory, however, was inadequate in two respects. On the one hand, it gave only circular orbits for the electron in a hydrogen atom; on the other, it did not account for certain fine details of the structure of the hydrogen spectral lines. Moreover, it could not account for the presence of certain lines and the absence of others (the selection rules). Indeed, on the basis of circular orbits alone, the only lines that should be present are those associated with transitions in which the principal quantum number changes by one unit only.

As noted previously, Arnold Sommerfeld was the first to point out that the introduction of elliptical orbits would automatically introduce a fine structure into the spectral lines of hydrogen if one took into account the relativistic variation of a particle's mass with its speed. As long as the electron moves in a circular orbit, its speed is constant and there is no variation of its mass. But if the electron moves in an elliptical orbit, its speed becomes greater as it moves closer to the proton and slower when it is farther away. Thus, the mass of the electron varies continuously. This continuous variation causes the whole orbit to precess so that the electron acquires additional energy. To define the ellipticity of the orbit of the electron, Sommerfeld introduced a second quantum number l, the azimuthal quantum number, which corresponds to the eccentricity of the elliptical orbit of a planet. The azimuthal quantum number is a measure of (is proportional to) the angular momentum of the electron in its orbit. Sommerfeld pointed out that for an orbit of principal quantum number n, there are just n integral values of the azimuthal quantum number, namely, $0, 1, 2, 3, \ldots, (n - 1)$.

These various azimuthal quantum numbers correspond to the values of the angular momentum of an electron that can be associated with its motion. The smallest value of the azimuthal quantum number (the most highly eccentric ellipse) corresponds to the smallest angular momentum (actually zero); the largest value of the azimuthal quantum number (the roundest ellipse) gives the largest angular momentum of the electron. These n different energy sublevels (different because of the relativistic variation of mass with speed) are labeled S, P, D, F, G, and so on. Thus, $2S$ represents the energy sublevel of an electron with principal quantum number 2 (the second Bohr energy level of the atom) and zero angular momentum, $2P$ represents an electron in the second energy level of principal quantum number 2, with one unit of angular momentum, and so on.

With the introduction of the azimuthal quantum number, physicists thought that all the fine details of the spectral lines of hydrogen and other atoms would be accounted for; but a closer examination of the spectral lines, particularly those of the alkali atoms, such as sodium, where there is only one outer electron responsible for the lines, revealed that each of the orbital angular momentum sublevels associated with a given principal quantum number, except the level S, was split into two levels. This was accounted for by assigning a half unit of spin angular momentum to the electron. Since this half unit of spin can only add to the orbital angular momentum, or subtract from it, there are two sublevels (doublets) associated with each value of the orbital angular momentum except the zero value. For example, one must replace the single level $2P$ by the two levels $2P_{1/2}$ and $2P_{3/2}$. Here the subscripts refer to the total angular momentum of the electron, owing to the effects of the particle's spin. To see how we obtain the two possible values $1/2$ and $3/2$ for the total angular momentum of the electron when it is in the $2P$ sublevel, note that it has just one unit of orbital angular momentum. If its spin axis is parallel to its orbital angular momentum, we must add its one half unit of spin angular momentum to its orbital angular momentum: $1 + 1/2 = 3/2$. If the spin axis is antiparallel (opposite) to the orbital angular momentum, we must subtract the two: $1 - 1/2 = 1/2$. These are the only possibilities; hence, we obtain doublets. In general, the level with the higher total angular momentum, in this case the $2P_{3/2}$ level, is a state of higher energy than that with the smaller total angular momentum.

Since these refinements in the Bohr theory had been introduced within the framework of the Bohr postulates, before quantum mechanics and the Schrödinger wave equation had been applied to the atom, there was some reason for believing that the Schrödinger wave equation would give the azimuthal sublevels and the spin doublet structure automatically. This, however, proved not to be the case. Although the solution of the Schrödinger equation for the hydrogen atom does give the azimuthal quantum numbers, in addition to the principal quantum numbers in the proper relationships with one another, it assigns the same energy to each azimuthal sublevel associated with a given principal quantum number. Consequently, the total number of distinct levels in the Schrödinger theory is still given by the principal quantum numbers. The azimuthal quantum number energy sublevels are degenerate in the sense that to each principal quantum number n the Schrödinger equation assigns n wave or vibrational modes (that is, n wave functions), each with the same energy. Besides this deficiency, the Schrödinger equation gives no spin effects whatsoever.

The different energy values that must be assigned to the azimuthal quantum number sublevels arise because of the variation of the electron's mass with velocity, a relativity effect. The electron spin is also a relativity effect. Hence, it follows that these effects cannot be deduced from the Schrödinger equation since this equation is not in accord with the relativity theory. But since the Dirac equation of the electron was designed specifically to accommodate the special theory of relativity, it corrects the flaws of the Schrödinger equation. When applied to the hydrogen atom, Dirac's equation gives not only the fine structure of the spectral lines associated with the azimuthal quantum numbers of the electron but also the spin of the electron and the spin doublet structure of the spectral lines.

Some of the most interesting features of the Dirac theory are related to the azimuthal energy levels associated with the principal quantum number $n = 2$. For $n = 2$, there are two azimuthal quantum numbers, 0 and 1. Thus, there are two azimuthal sublevels, the $2S$ level and the $2P$ level. If we now take the spin of the electron into account, we obtain three levels, one from the S level, the so-called singlet $2^2S_{1/2}$ and the doublet P state consisting of the two levels $2^2P_{1/2}$ and $2^2P_{3/2}$. The solution of the Dirac equation gives these terms and, in addition, the energy values that must be assigned to each of these levels. In terms of energy the $2^2P_{3/2}$ level lies slightly above the $2^2P_{1/2}$ level. An electron jumping from the upper of these levels to the lower would emit a photon of wavelength 2.74 cm, which is in the microwave region of the spectrum. But the $2^2S_{1/2}$ level and the $2^2P_{1/2}$ level coincide exactly according to the Dirac theory.

This theoretical prediction of an exact coincidence was a challenge to experimental spectroscopists. Even before World War II, when microwave techniques were still in their infancy, spectroscopic workers obtained observation data that disagreed with the theory. Houston, Williams, Pasternack, and others had found discrepancies. Pasternack, in particular, had argued that the $2^2S_{1/2}$ level must lie above the $2^2P_{1/2}$ level. However, the experimental techniques were not sufficiently accurate then to give a definitive answer.

At the end of World War II, however, microwave technology in the neighborhood of the 3 cm wavelength had advanced to such a stage that Willis E. Lamb, then professor of physics at Columbia University, could design a very ingenious and beautiful experiment, based on microwave techniques, to analyze the fine structure of the hydrogen lines for the principal quantum number $n = 2$. This experiment used the metastable character of the $2^2S_{1/2}$ level, which should be long-lived against a radiative transition to the ground state. This state is metastable, that is, has a much longer life than normal, because of a selection rule that forbids the spontaneous transition of an electron from one S level to another. This means that the electron cannot jump down from the $2S_{1/2}$ level to the ground state because the ground state is also an S level. Hence, once the electron has been forced into the $2^2S_{1/2}$ level by some excitation process, it will stay there for a long time—about one seventh of a second—as compared to the normal life time of about 1 hundred-millionth of a second.

Lamb and his student R. C. Retherford used this metastable state of the hydrogen atom in the following way. They pointed out that if the hydrogen atom is first excited to the $2^2S_{1/2}$ metastable state, it will stay there long enough to be recorded

Willis E. Lamb (1913–)

by some mechanism that can detect the excitation energy, which is released when the excited atom strikes it. One should therefore be able to detect a beam of such metastable atoms, which are prepared by bombarding unexcited hydrogen atoms with a stream of electrons having the proper kinetic energy. If the beam of excited metastable atoms is perturbed by an external electric field or by radiation of the proper frequency, the metastable atoms can be induced to go from the metastable S state to one of the P states, from which they can jump down to the ground state in a very short time. If this happens, fewer metastable atoms reach the detector and the current in the detector drops. Lamb and Retherford used this technique to show that transitions of the metastable atom to the $2^2P_{1/2}$ state could be induced by microwaves of the proper frequency. In this way, they analyzed the fine structure of the $2S$ and $2P$ levels and showed that there was a 1,000 megacycle-per-second separation between the $2^2S_{1/2}$ and the $2^2P_{1/2}$ levels. This discovery disagreed with the prediction of the Dirac theory.

The remarkable experimental result opened up an entirely new field of inquiry for physicists. The questions that arose immediately were whether the Dirac theory of the electron was correct and, if not, where did it fail and how was it to be improved? The answers to these questions were given by a group of theoretical and experimental physicists working along different lines, both independently, and in teams. We shall consider the various theoretical contributions to the solution of this problem and their effects on our knowledge of the properties of the electron in subsequent chapters.

Willis Eugene Lamb, Jr., by education, experience, and intellectual equipment, was one of the very few physicists who could have performed the very delicate microwave experiment in 1947. He proved that the observed fine structure of the energy levels of the hydrogen atom L shell (the levels of principal quantum number $n = 2$) did not agree with the predictions of the Dirac theory. To do this experiment successfully, one had to have a thorough understanding of the Dirac theory of the electron, an excellent working command of the quantum mechanics, a practical understanding of the Rabi molecular-beam technique, and a thorough working knowledge of microwave technology. Lamb, a theoretical physicist by training, had all of these prerequisites when he first began to consider this problem.

Lamb was born on July 12, 1913, in Los Angeles, California, and spent his very early years in Oakland. His family then moved back to Los Angeles, and Lamb continued his education there. After graduating from high school in 1930, he entered the University of California at Berkeley, where he majored in chemistry and received his B.S. degree in 1934. By the time he had completed his undergraduate studies, his interest had shifted from chemistry to theoretical physics, and he decided to do graduate work in physics under J. Robert Oppenheimer.

In 1930 Oppenheimer had shown that if one calculates the interaction of an electron with the quantized electromagnetic field, using Dirac's theory in which the electron is treated as a point charge, one obtains infinite and therefore meaningless results. According to this calculation, the interaction of the electron with its own radiation field should cause all the hydrogen energy levels to be shifted by an infinite amount, which, practically speaking, is nonsense. In 1934, a calculation by Victor Weisskopf showed that this difficulty arose from the peculiar interaction that exists between the electron and the vacuum in the Dirac theory.

Lamb himself came into contact with this problem during his graduate studies, when Oppenheimer assigned to him a thesis dealing with the field theories of nucleons. His analysis led to a very small discrepancy from Coulomb's law for the electrostatic field about a proton. Such a discrepancy would, of course, present departures from the Dirac theory of the fine structure of the hydrogen spectral lines, then being investigated by spectroscopists. On completing his thesis and receiving his Ph.D. in 1938, Lamb went to Columbia University as an instructor in physics, with the unresolved question of the Dirac theory's validity still in his mind. Spectroscopists had indicated a discrepancy between theory and observations at that time, but the results of their experiments were inconclusive. Lamb therefore sought other methods of investigation. The elegant and highly precise molecular-beam technique, developed by Rabi at Columbia, provided him with the basis for his own experiment.

At that time the experiment that Lamb was to devise later could not have been carried out because the special microwave techniques it required were not yet known. During World War II, however, he worked in the Columbia Radiation Laboratory where he became acquainted with the theory of microwaves and vacuum tube construction techniques. Since his own project dealt with the absorption of centimeter-wavelength microwaves, he acquired the skills needed for his Nobel Prize-winning experiment, which involved the absorption of such waves by excited hydrogen atoms.

Lamb's interest in using microwave absorption to check the Dirac theory was aroused in the summer of 1946 while teaching a summer session course on atomic physics at Columbia University. In Herzberg's textbook, he ran across a reference to some unsuccessful experiments conducted from 1932 to 1935 that were intended to detect the absorption of short-wavelength radio waves by the excited hydrogen atoms in a discharge tube. At first he contemplated repeating these same experiments, using the greatly improved microwave detection techniques. But a careful theoretical analysis of the experiment revealed that the statistical distribution of the excited hydrogen atoms under the stated conditions was such as to defeat the purpose of the experiment. He realized that he could do the same thing by using the Rabi molecular-beam technique in which individual atoms could be excited to the particular state desired, irradiated by microwaves of the proper wavelength, and finally detected. In this way the absorption could be measured directly by studying the excited beam of atoms after they had passed through the microwave field.

Much preliminary theoretical work had to be done on every phase of this problem before the experiment could actually be performed; the preparation took Lamb a full year. In 1947 he persuaded one of his students, R. C. Retherford, to perform the experiment. It was completely successful and proved that the fine structure of the hydrogen lines, as predicted by the Dirac theory, did not agree with the observational data. This remarkable experiment led to the mass renormalization theories of Bethe, Schwinger, Feynman, and Tomonaga and indicated how the Dirac theory must be corrected to conform to the observed results. The discrepancy between the uncorrected Dirac theory predictions and the observed results is now called the Lamb shift.

During World War II, Lamb was promoted from instructor to associate at Columbia; in 1945 he became assistant professor and in 1947, associate professor. In 1948, after the publication of his experimental results, he was appointed full professor. He remained at the Columbia Radiation Laboratory until 1951, giving graduate physics courses and stimulating theoretical and experimental work in connection with his discovery of the Lamb shift.

In 1951 Lamb left Columbia to become professor of physics at Stanford University in California, where he remained until 1956. During this period he spent one year (1953–1954) as Morris Loeb lecturer at Harvard University. In 1956 he went to Oxford University, England, as fellow of New College and Wykeham professor of physics. He remained at Oxford until 1962, when he left to become the first Henry Ford II professor of physics at Yale University, where he engaged in teaching and research for the remainder of his academic career.

The experimental discovery of the Lamb shift and the new problems arising from it were but one small aspect of his overall research work. Lamb was a theoretical physicist who published important papers in almost every field of physics. In nuclear physics he contributed to an understanding of the interaction of neutrons and matter, the field theories of nuclear forces and nuclear structure, the beta-decay process, and the range of fission fragments in nuclear fission. In particle physics he investigated cosmic ray showers and pair production. He also contributed to solid state physics and the physics of crystals. His 1939 paper on the resonance capture of slow neutrons by crystals played an important role in the development of the Mössbauer effect. Lamb also did significant work in molecular physics, magnetron oscillators, diamagnetism, nuclear resonance, the theory of the deuteron and the helium nucleus, and microwave spectroscopy.

In addition to the Nobel Prize, Lamb received the Rumford Award of the American Academy of Arts and Sciences and the Research Corporation Award. In 1954 the University of Pennsylvania awarded him an honorary Doctor of Science degree. He is a member of the National Academy of Sciences and a fellow of the American Physical Society.

The Magnetic Moment of the Electron: Polykarp Kusch (1911–)

Lamb and Retherford's remarkable findings for the fine structure of the sublevels of the second hydrogen quantum state led many physicists to suspect that the

Dirac theory did not properly take into account the interaction of the electron with its own radiation field. Another discrepancy detected experimentally was the value of the magnetic moment of the electron. The Dirac theory predicts that the electron must have a half unit of angular momentum arising from its spin. The negative ratio of the electron's magnetic moment to its own angular momentum, called the g factor, is generally the quantity that is determined experimentally. The g factor of the free Dirac electron is 2 since, when the Dirac electron is moving freely in space, its only angular momentum is that of its spin, $1/2$. However, for an electron bound inside the atom we must consider both spin and orbital effects in determining angular momentum. We therefore have two g values: g_S, associated with spin, and g_L, associated with orbital angular momentum. (There is, of course, a g value for total angular momentum of the bound electron, combining the effects of spin and orbit; this value is labeled g_J.) Since the Dirac theory did not give the proper fine structure of the hydrogen levels, physicists also expected to find that the experimentally determined g_S of a free electron differed from 2, the Dirac value. Polykarp Kusch, using the molecular-beam techniques developed by Rabi, was able to demonstrate this discrepancy in experiments that measured the magnetic moment of the electron.

Measuring the magnetic moment of the electron is complicated because it can be done only while the electron is bound inside an atom. This introduces additional effects since the electron has an orbital magnetic moment (arising from its motion around the nucleus) in addition to its intrinsic spin magnetic moment. The external magnetic fields used in Rabi's molecular-beam methods interact with both of these magnetic moments simultaneously. In the early investigations this difficulty was eliminated by utilizing atoms that had a zero value for the electronic orbital angular momentum; hence, the orbital magnetic moment of the electron was zero. This is true for alkali atoms, such as sodium and potassium. However, in very precise measurements another complication arises from the nuclear magnetic moment, which is responsible for the hyperfine structure of the spectral lines of an atom. The nuclear magnetic moment combines with the intrinsic and the orbital magnetic moments of the electron to give a total magnetic moment that interacts with the field.

We have seen in our discussion of Rabi's molecular-beam resonance methods that it is possible to measure the nuclear magnetic moment by a resonance procedure that uses an oscillating magnetic field applied at right angles to a beam of atoms passing through an inhomogeneous magnetic field. Measurable disturbances in the beam are then induced if the frequency of the oscillating magnetic field is equal or close to the frequency of the Larmor precession of the nuclear magnetic moment of the atom. From the value of the Larmor precession, which can be deduced from the frequency of the oscillating field, the nuclear g factor and, hence, the nuclear magnetic moment can be calculated. This procedure was first applied to atoms or molecules with zero total spin and zero orbital angular momentum, so that only the nuclear magnetic moment interacted with the magnetic field. Techniques were improved, however; it became possible to use molecular-beam methods to study the interaction of the nuclear magnetic moment with the total electronic magnetic moment, in other words, to analyze the magnetic hyperfine structure of the atom.

Kusch and Millman applied these resonance techniques to the measurement of the magnetic moment of the proton, expressed in units of the electronic magnetic moment, by comparing the resonance effects for hydrogen atoms and molecules with those for alkali atoms. In this way one could simultaneously obtain information about the *g* factor for the spin magnetic moment of the proton. A comparison of these values, on the assumption that the *g* value for the electron spin is 2 (that the magnetic moment of the electron is just 1 Bohr magneton), showed a discrepancy between the value for the magnetic moment of the proton measured in this way and that measured by direct resonance methods. Rabi and his coworkers found such a discrepancy in analyzing the hyperfine structure of hydrogen. It became clear that the intrinsic magnetic moment of the electron must differ from 1 Bohr magneton by about 1 percent. This discrepancy suggested the need for a very precise determination of the *g* factor for the spin magnetic moment of the electron; this attempt was made by Kusch and Foley.

Their experiment consisted in determining the ratio of the *g* value for the spin of the electron to the *g* value for the orbital angular momentum of the electron. This was done by considering two of the same atoms in two different magnetic states and determining the *g* value for the total angular momentum in each of these states. The ratio of these *g* values gives the ratio of the spin *g* value to the orbital angular momentum value. Since the orbital *g* value is 1, this procedure immediately provides the *g* value for the spin. The experimental data show that the magnetic moment of the electron is somewhat larger than 1 Bohr magneton. The value thus found is equal to that first calculated by Julian Schwinger by taking into account the mass renormalization of the electron. This corrected value of the magnetic moment of the electron is referred to as the anomalous magnetic moment of the electron.

It is a remarkable coincidence that Polykarp Kusch and Willis Lamb, working independently in the Columbia University molecular beam laboratories on two apparently unrelated physical phenomena, should have almost simultaneously discovered two different discrepancies with the Dirac theory of the electron, which were soon to be resolved by a single theoretical correction. That Kusch should have discovered the anomalous magnetic moment of the electron is quite understandable in the light of his extensive background in the field of molecular and

Polykarp Kusch (1911–)

atomic beam spectroscopy, dating back to 1937. He first began experimental work with molecular beams at that time at Columbia, shortly after he received his Ph.D. in physics.

Kusch was born on January 26, 1911, in Blackenburg, Germany, but was brought to the United States at one year of age by his parents. His father, a clergyman, took the family to the Midwest in 1912, where young Kusch received his elementary and high-school education. While in high school, his interests turned to chemistry. Like Rabi and Lamb, he decided on chemistry as a career when he entered college. But soon after starting his undergraduate work at the Case Institute of Technology in Cleveland, Ohio, he shifted to experimental physics. After receiving his B.S. degree in physics in 1931, he entered the University of Illinois to pursue his graduate studies. The research for which he was granted his doctorate dealt with optical molecular spectroscopy and was performed under the guidance of Professor F. Wheeler Loomis.

After spending the year 1936 at the University of Minnesota working with Professor John T. Tate in the field of mass spectroscopy, Kusch came to Columbia as an instructor in physics and began his association with Professor I. I. Rabi at the latter's molecular-beam laboratory. There Kusch participated with Rabi's other collaborators in the series of experiments that won Rabi the Nobel Prize and prepared Kusch for his own important work. The general principles of radio frequency spectroscopy by Rabi's molecular-beam resonance methods were first described in a paper published in the *Physical Review* in 1939 by Rabi, Kusch, and others. In another group of experiments in which Kusch participated, the Rabi method was used to study the magnetic hyperfine structure of the spectral lines of various isotopes of the alkali atoms. A paper describing these experiments was published in 1940 by Rabi and his group, with Kusch collaborating.

In 1940, working with Millman, Kusch performed the first in a series of experiments that were to lead him step by step to the measurement of the anomalous magnetic moment of the electron and to the Nobel Prize in 1955. After this work, described in a 1941 paper in the *Physical Review*, was completed, Kusch, like most other physicists, participated in military projects. He first worked at the Westinghouse Electric Corporation, then at the Bell Telephone Laboratories, and finally at the Columbia University Radiation Laboratory. This experience gained for him a good working knowledge of microwave techniques and vacuum-tube technology and their application to problems in experimental physics.

After the war, Kusch returned to the Columbia physics department and quickly advanced through the various professorial ranks to become a full professor in 1949. During this period, he completed the research projects that the war had interrupted. In a series of papers, published jointly with his colleague H. Foley, he investigated the question of the anomalous magnetic moment of the electron. He showed experimentally that the intrinsic magnetic moment of the electron is about 0.1 percent larger than that predicted by the Dirac theory. This result was in complete agreement with the theoretical calculations of Schwinger, based on new procedures introduced by Schwinger in quantum electrodynamics.

Kusch and his collaborators subsequently increased the accuracy of their experiments and showed that the observed magnetic moment of the electron agreed with the improved theory to about one part in a billion. This discovery was extremely important since it demonstrated the high degree of accuracy of the improved quantum electrodynamics in analyzing the interaction of an electron and an electromagnetic field. After completing this work, Kusch became interested in problems in chemical physics, to which he applied the molecular-beam techniques that had proved so fruitful in the study of the atom and the nucleus. Although he continued to do research in the latter years of his career, Kusch devoted steadily greater amounts of his time to educational problems and to the problems arising from the impact of science on society.

High-Energy Physics: Hans Bethe (1906–), Julian Schwinger (1918–), Richard Feynman (1918–1988)

The subject of modern physics is unquestionably linked with men like Einstein, Planck, and Bohr, who developed its basic theories. But an examination of the literature shows important contributions by others who did not discover a new theory but who clarified, expanded, and analyzed the fundamental ideas. Hans Bethe is one of these great physicists whose work is of such importance that one can deal with almost no phase of the subject without referring to him. From the time quantum mechanics was discovered, while he was still a student, he has been publishing original papers that have contributed much to our understanding of modern physics.

Bethe was born on July 2, 1906, in Strasbourg, Germany, and studied physics at the University of Frankfurt from 1924 to 1926 and at the University of Munich, where he received his Ph.D. under Arnold Sommerfeld in 1928. The papers he began to publish at this stage of his career, young as he was, clearly indicated that he was to become one of the dominant figures of modern physics.

After receiving his doctorate, Bethe returned to Frankfurt in 1928, where he taught physics for a year. From there he went to the University of Stuttgart, where he spent another year teaching, and then was invited to the University of Munich as lecturer. He remained at Munich until 1933 when the Nazis came to power. He then left Germany, ultimately to become a citizen of the United States.

Hans Bethe (1906–)

During his three years in Munich, Bethe published numerous papers on the quantum theory of the atom. One of the most important, written jointly with Enrico Fermi while Bethe was a fellow at Rome, deals with quantum electrodynamics and is one of the earliest applications of this important subject to the study of the interaction of charged particles. This period in Bethe's career was also marked by his emergence as a great expositor of modern physical theories. His book-length article on the quantum mechanics of the atom in the original edition of the *Handbuch der Physik* was a model of clarity and an example of scientific writing at its best. For years this treatise was the constant companion of the graduate student of theoretical physics. While in Munich Bethe wrote another famous article in collaboration with Sommerfeld. It dealt with the electron theory of metals and was the first systematic treatment of the application of the Fermi-Dirac statistics to the study of metals. These two articles established Bethe as an important physicist.

On leaving Germany, Bethe first went to England and spent a year at the University of Manchester. From there he went to the University of Bristol where he remained for a year as a fellow. These two years in England were very fruitful. Bethe published papers in numerous fields of physics with special emphasis on atomic collision processes, the radiation from accelerated electrons, nuclear physics, and solid state physics. In collaboration with R. Peierls he gave the first theoretical analysis of the two-body nuclear system (the deuteron), showing that quantum mechanics could be applied to the study of the nucleus as successfully as to the study of the atom.

In 1935 Bethe came to the United States to accept an assistant professorship of physics at Cornell University. In 1937 he was promoted to a full professorship and held this post for the remainder of his career. From 1935 until the beginning of World War II, Bethe devoted himself to research (primarily in nuclear physics and its applications to astrophysics), teaching, and writing. His three well-known articles on nuclear physics, which first appeared in the *Reviews of Modern Physics*, quickly became standard treatises, essential to a comprehension of nuclear physics. Like his articles on the atom, these treatises showed Bethe's remarkable qualities as a teacher.

During this period Bethe contributed important papers to solid state physics, meson physics, and astrophysics. His work in astrophysics was trail-blazing. He laid down the general principles of energy production in stars and developed the first models of the solar interior, using thermonuclear energy sources. This work, for which he won the Nobel Prize, served as a guide to future research in this field which has expanded greatly since the war.

During World War II, Bethe became a staff member of the Radiation Laboratory at MIT and did research there during the year 1942 to 1943. He then worked on the atomic bomb project at the Los Alamos Scientific Laboratory as chief of the theoretical division from 1943 to 1946, when he returned to Cornell to resume his research, teaching, writing and lecturing. After the war his research dealt with quantum electrodynamics, meson theory, and shock wave theory. His 1947 paper on the Lamb shift was a crucial step in the subsequent development of the new formulation of quantum electrodynamics. It showed—though only by approximate methods—that the physical content of quantum electrodynamics agrees with the observations and that its apparent infinities are a result of the incorrect treatment of the mass of the electron.

After Bethe came to Cornell he acquired a national reputation as a great teacher and lecturer and was in constant demand by other universities. In 1941 he came to Columbia University as visiting professor and gave a course of lectures on nuclear physics. In 1948 he returned to Columbia, again as visiting professor, and lectured on the new methods in quantum electrodynamics developed by Schwinger, Feynman, and Dyson. In the year 1955 to 1956 he was visiting professor at Cambridge University, England.

Bethe received many honors and awards in addition to the Nobel Prize for his scientific work. On two occasions he received awards from the New York Academy of Sciences for his work in astrophysics. In 1946 he was awarded the United States Medal of Merit and in 1948 the Draper medal of the National Academy of Sciences. He received the much coveted Planck medal of Germany in 1955 and, more recently, the $50,000 Enrico Fermi prize for his contributions to nuclear physics in the field of atomic energy. He was awarded honorary Doctor of Science degrees by the Polytechnic Institute of Brooklyn, the University of Chicago, the University of Birmingham, England, and Harvard University.

Bethe was a member of the National Academy of Sciences, a fellow of the American Physical Society, of which he was elected President in 1954, and a fellow of the Royal Society of London. He was also a member of the American Astronomical Society and of the Philosophical Society.

In the 1960s and 1970s Bethe concerned himself with the application of atomic energy to peaceful uses. He was a dedicated advocate of the international control of atomic energy and argued vigorously for the abolition of the bomb and, as a minimum, for the discontinuance of the testing of nuclear weapons. As a member of the United States Delegation to Discussions on Discontinuance of Nuclear Weapons he was instrumental in bridging the gap between the Russian and American positions and thus helped materially in bringing about the present cessation of above-ground tests. In the 1980s he became one of the leading opponents of the Reagan Administration's efforts to develop a space-based intercontinental ballistic missile defense system, arguing that such a system, if successful, would both destabilize the present balance of terror between the United States and the Soviet Union and be unimaginably costly.

Although American theoretical physics began its remarkable growth in the late 1920s, its full flowering did not come until a decade later. This rapid development was spurred on by the scientific needs of the military during World War II. The period just preceding the war was particularly noteworthy because it marked the debut of two outstanding theoretical physicists, Julian Schwinger and Richard Feynman, both of whom were born in the same year, in the same city; they were educated in the same school system and made their most important contributions in the same field of physics.

Julian Schwinger was born on February 12, 1918 in New York City, where he received all of his education, including his Ph.D. He demonstrated his remarkable mathematical ability while still in elementary school; by the time he was ready for high school he had taught himself calculus and was able to solve differential equations. Even at that time his interests tended to theoretical physics; he began to read advanced treatises in various branches of physics, with special emphasis on electromagnetic theory and relativity. Like most of the intellectually gifted New York youngsters of that period, Schwinger planned to attend the College of the City of New York (CCNY), which offered a free higher education to all city residents

Julian Schwinger (1918–)

who could qualify. At that time, CCNY had under its jurisdiction a preparatory high school, Townsend Harris High School, which was available to outstanding elementary-school graduates. Schwinger, quite naturally, entered this school, which stressed the classics, languages, and mathematics.

The close association between the Townsend Harris faculty and the CCNY faculty soon brought Schwinger to the attention of the members of the college physics department. By the time he matriculated as a freshman, his reputation as a prodigy was well established. He entered CCNY at the age of 15 and, although he attended the usual freshman and sophomore courses, he could usually be found spending his spare time in the physics library or working out problems in relativity theory or quantum mechanics in his notebook.

In 1935, his sophomore year, Schwinger attended a seminar in theoretical physics at Columbia University, taught by I. I. Rabi and G. Breit. Schwinger soon attracted Rabi's attention by clarifying a very subtle and puzzling point in the famous paper by Einstein, Podolsky, and Rosen that had just been published. Rabi was so impressed that he suggested that Schwinger leave CCNY and come to Columbia on a scholarship. Consequently, in September 1935, Schwinger entered Columbia as a junior. He had already made an impression on Wolfgang Pauli, who referred to him as "the physicist in knee pants." Bethe, in a letter to Rabi, expressed the opinion that Schwinger knew 90 percent of all the physics that was then known and that he could pick up the other 10 percent whenever he desired to do so.

During the spring and summer of 1935, Schwinger submitted his first original papers for publication. After entering Columbia, he continued with his own theoretical research. By the time he received his baccalaureate degree in the fall of 1936, he had already written his Ph.D. thesis on the scattering of neutrons. After spending two more years at Columbia to complete the residence requirement for his doctorate, he received a National Research Fellowship and spent some time at the University of Wisconsin and at the University of California, where he worked with Oppenheimer. During this period he published a series of important papers on nuclear forces.

In 1941 Schwinger went to Purdue University as instructor in physics and soon after was promoted to assistant professor. He left Purdue in 1943 to become a staff member of the Radiation Laboratory at MIT, where he remained until 1945. During this period he developed powerful mathematical methods for treating problems in electromagnetic theory. These so-called variational methods proved to be very useful later in the handling of nuclear scattering problems and in field theory. At the end of World War II Schwinger accepted an associate professorship in physics at Harvard. In 1947, at the age of 29, he was promoted to a full professorship—the youngest full professor in Harvard's history.

All the theoretical work that Schwinger had done up to this point was, in a sense, preparatory to his discovery of the mass and charge normalization in quantum electrodynamics. His amazing grasp of the physics of quantum field theory and the ease with which he handled its complex mathematical formalism led him quite naturally to an understanding of the difficulties that beset the theory and the way to eliminate them by recasting the theory in a completely covariant form, that is, in agreement with relativity at all stages. With this revised and relativistically correct version of quantum electrodynamics, Schwinger derived the Lamb shift from the theory and showed that the calculated magnetic moment of the electron is not 1 Bohr magneton, as predicted by the Dirac theory, but somewhat larger and in complete agreement with Kusch's measurements.

Schwinger continued working in quantum electrodynamics and field theory and published a series of important papers covering all phases of the subject. These papers are distinguished by their mathematical elegance and the use of powerful analytical techniques. During the 1960s he devoted himself to the construction of a field theory of fundamental particles in analogy with the electrodynamic field theory of photons.

For his contributions to quantum electrodynamics, Schwinger received the prize of the National Academy of Sciences and the Einstein award in 1951. He was a member of the National Academy of Sciences and a fellow of the American Physical Society and of the American Academy. In 1961 he was appointed visiting professor of physics at the University of California. He was awarded the 1965 Nobel Prize in physics together with Richard Feynman and Shinichiro Tomonaga.

Richard Feynman was born in New York City on May 11, 1918. Like Schwinger he received his early education in the city's public schools. While still in high school, Feynman demonstrated his precocity by doing advanced mathematics while his classmates did the usual class work. His high school physics teacher recalled that he had always allowed Feynman to sit in the back of the room solving problems in advanced calculus while the other students were busy with high school physics.

On graduating from high school, Feynman entered MIT to begin his studies in mathematics and physics. He received his B.S. degree from MIT in 1939 and went on to Princeton University as a Proctor fellow to do graduate work with Professor John A. Wheeler. While at Princeton he became interested in electrodynamics and, particularly, in the problems of the emission and absorption of radiation by charged particles. This interest led him to the fundamental problem of the interaction between charged particles and whether such an interaction is best treated as action at a distance or the action of a field. His concern with such problems later led to his graphic descriptions of phenomena in quantum electrodynamics.

After receiving his Ph.D. from Princeton in 1942, Feynman worked on the atomic bomb project at Los Alamos until 1945. From Los Alamos he went to

Richard Feynman (1918–1988)

Cornell University, first as associate professor and then as full professor of physics. During this time he worked in quantum electrodynamics and discovered what are now called the Feynman graphs, a very elegant and simple graphic procedure for analyzing the different processes that occur when a charged particle interacts with another charged particle or with the electromagnetic field.

Feynman arrived at his graphic representation of quantum electrodynamic processes after he had undertaken a critical analysis of quantum mechanics and had reformulated it in terms of a least action principle—the Lagrangian formulation. This approach to quantum mechanics is described in a famous paper that appeared in 1948 in the *Reviews of Modern Physics*. In this formulation the emphasis was not on the detailed behavior of a system from moment to moment, but rather on its overall behavior, as though its entire past and future were exposed to one's view. This orientation naturally led to a diagrammatic formulation of the problem. It is a remarkable coincidence that Feynman and Schwinger, working from such different points of view, arrived at the same solutions to the problems of quantum electrodynamics and published their results at about the same time.

Feynman left Cornell University in 1951 to become Richard Chase Tolman professor of theoretical physics at the California Institute of Technology at Pasadena. He remained there until his death in 1988. For a number of years after his first historic papers on quantum electrodynamics, Feynman continued publishing papers in this field, clarifying many subtle points and improving the mathematical techniques. After it became clear that the new quantum electrodynamics had obtained the desired plateau of excellence, Feynman turned his attention to the application of his field of theoretical methods to nuclear forces and to high-energy particle physics. Working with Murray Gell-Mann he reformulated the interaction between fundamental particles in terms of a general type of interaction that can be applied universally. Towards the end of his career, Feynman also did extensive work in low temperature physics, with particular emphasis on the properties of liquid helium and superconductivity. In 1954 he received the Einstein award for his contributions to quantum electrodynamics and was a corecipient in 1965 of the Nobel Prize in physics. He was a member of the National Academy of Sciences and the American Physical Society.

In all of his work, whether research, teaching, or writing, Feynman was always concerned with emphasizing the physical aspects of a problem and freeing it as much as possible from complex mathematical formalism. This concern was evident not only in his original papers but also in his advanced treatises.

One of Bethe's most important papers concerned the so-called Lamb shift. The shift in the $2S$ energy level of the hydrogen atom found by Lamb and Retherford did not agree with the Dirac theory of the electron. Therefore, theoretical physicists began to probe existing theories to discover the source of the difficulty and to determine just how the Dirac theory had to be corrected to account for the observed energy levels. Such outstanding contemporary theoretical physicists as Tomonaga, Schwinger, and Feynman tackled the problem and eventually solved it. In particular, Schwinger and Feynman improved quantum electrodynamics to such a degree that it was possible to reconcile many of the differences that had arisen. Thus, we can account for the Lamb-Retherford energy-level shift as well as the anomalous magnetic moment of the electron measured experimentally by Kusch. But it was Bethe who brought into sharp focus the inability of the Dirac theory to treat adequately the interaction of the electron with its own radiation field.

According to the Dirac theory, the electron must be treated like a point charge. But such a finite charge concentrated in a point, and therefore having an infinite charge density, would have an infinite interaction energy with its own radiation field. Consequently, if one uses the Dirac theory to calculate the effect that the interaction of the electron with its own field has on the energy levels of the electron in the hydrogen atom, one obtains an infinite answer. This was the result that Oppenheimer obtained in 1930 when he applied the standard perturbation techniques to the Dirac equations. This infinite result arises because the Dirac equations describe only a bare electron, without a surrounding electromagnetic field. The actual radiation field surrounding the true electron behaves like an additional mass (the electromagnetic mass) so that the mass that appears in a Dirac equation is not the quantity that applies in the true description of the electron.

To see why the radiation field of the electron influences the hydrogen energy levels, or, for that matter, the energy levels of any atom, consider a very energetic photon in the radiation field of the electron. It is transiently absorbed by a negative-energy electron (one of an infinitude according to the Dirac theory) in the neighborhood of the first electron. Momentarily, then, a positron-electron pair is created (the negative-energy electron enters a state of positive energy and leaves a hole). This pair is equivalent to an electric dipole (the polarization of the vacuum, as it is called). By means of its electric field, the dipole disturbs the electron in the hydrogen atom and therefore alters the energy levels. If one applies the standard perturbation techniques of the Dirac electron to calculate the value of this disturbance, one obtains an infinite result. Thus, the Dirac theory says that all the energy levels of the atom are shifted by an infinite amount as a result of the interaction of the electron with its own radiation field. This statement, of course, is nonsense. The error in the theory arises because the electron is regarded as a point; there is therefore no limit to the energy of the photons with which the electron can interact. According to Dirac, pairs can be created by photons ranging in frequency all the way from those corresponding to $2mc^2$ to infinity. Essentially, this is equivalent to saying that the interaction of the electron with the radiation field leads to an infinite correction to its mass.

Schwinger pointed out, and he and Feynman demonstrated later, that it is necessary to treat the interaction of the electron and the radiation field by a precise relativistic procedure in order to avoid these infinite corrections and obtain the correct electromagnetic shift of the energy levels. Bethe, however, was the first to obtain a fairly accurate value by an approximate nonrelativistic method. His reasoning was as follows. The Dirac and Schrödinger equations of the electron describe a mechanical electron, that is, one without a surrounding electromagnetic field. Any calculations performed with either of these equations, such as determining the energy levels of the hydrogen atom, omit the interaction of the electron with its field. However, when the results obtained with the theory are compared with observation, the numerical value for the mass of the electron that is put into the equation is the one obtained observationally. But this observed value is not the bare mechanical mass. What we detect in any experiment is the total mass of the electron, a combination of the bare mechanical mass and the electromagnetic mass arising from the interaction of the electron with its electromagnetic field. If we try to calculate this electromagnetic contribution to the mass, we get an infinite answer. This result, however, should not prevent us from calculating the effect that the interaction of the electron with its own electromagnetic field has on the energy levels of the electron when it is bound to a proton. For this purpose we should note that the infinite contribution to the energy level shift that we get from the theory (by perturbation calculations) is due entirely to the infinite contribution to the mass. In other words, when we try to calculate the energy level shift, we cannot avoid getting infinite terms that come from the contribution of the radiation field to the mass of the electron, since the mass of the electron is present in the expression for the energy levels. Consider this idea in another way. Since the wave equation of the electron takes into account only the bare electron without its radiation field, a perturbation calculation must be performed to find out how the interaction of the electron with its field affects the energy levels. But this perturbation calculation gives not only the effect on the energy levels but also a correction to the mass of the electron (the electromagnetic mass), which unfortunately turns out to be infinite. Hence, the shift in the energy levels obtained in this way is infinite because the corrected mass of the electron is involved.

How can this problem be avoided? We note first that we need not calculate the contribution of the radiation field to the mass of the electron because we obtain this quantity automatically by inserting the observed mass of the electron into our equations. If we work with the observed electron mass, the contribution of the radiation field to the total mass of the electron is automatically taken into account. Any calculation containing this mass contribution is redundant and should be corrected by subtracting from it the electromagnetic mass contribution. But how is this to be done without at the same time subtracting what we are seeking, namely, the contribution of the radiation field to the energy levels? Here Bethe used an ingenious procedure. The contribution of the radiation field (the interaction of the electron with its own electromagnetic field) to the total mass of the electron is the same whether the electron is moving freely in space or whether it is bound to the proton to form the hydrogen atom. However, in the latter case the perturbation calculation also contains the terms we seek—the radiation corrections of the energy levels of the bound electron. Hence, if we subtract the expression for the interaction of the free electron with its field from the expression for the shift in the energy levels, we should be left with the radiation-field corrections to the hydrogen-energy levels.

This is the nature of the analysis that Bethe carried out, but he had to take into account one more point because his calculation was nonrelativistic. Part of it involved summing over all the frequencies of the photons that can occur in the radiation field of the electron. Since the nonrelativistic theory places no limit on these frequencies, the sum (after the subtraction procedure described above) again gives an infinite result. Bethe argued, however, that the correct relativistic theory using the same subtraction technique would give a natural upper limit to the frequencies of the photons that must be summed over. He further argued that this natural limit is energetically equal to the total mass (in energy units, mc^2) of the electron. With this assumption Bethe obtained an expression for the correction to the shift in the $2S$ energy level of the electron in the hydrogen atom that was in close agreement with the experimental results of Lamb and Retherford.

Bethe's paper was important because it showed clearly that the difficulty in the theory lay in the infinite correction to the self-energy of the electron and that a proper mathematical procedure could cope with this difficulty by subtracting the infinite terms from the physically meaningful results. Later, a complete and relativistically correct procedure was developed by Schwinger and, independently, by Feynman in a self-consistent way. This work led not only to the proper correction to the mass of the electron but also to a correction to the charge on the electron. In both cases these corrections are infinite but cause no trouble because they are introduced as renormalization of the mass and the charge of the electron. These quantities in the equations are then replaced by their experimental values.

Bethe gave a theoretical explanation of the Lamb-Retherford effect by using a simple mathematical subtraction technique to eliminate the infinities that burdened the Dirac theory. But his procedure was only approximately correct and left the theory with all its deficiencies. Following Bethe's work, a series of papers, presented almost simultaneously by a number of authors working independently of each other, gave a systematic insight into the infinity difficulties of the theory and showed how these were to be overcome. These papers ultimately established a self-consistent mathematical scheme for using quantum electrodynamics in conjunction with the Dirac theory of the electron to obtain correct results for phenomena involving the interaction of the electron with the electromagnetic field. The theory as it now stands, however, is by no means a complete picture of quantum electrodynamics since it can say nothing about the structure of the electron and still gives an infinite result for the self-energy of the electron. In its present form, the theory, as developed by Schwinger, Feynman, Tomonaga and others, uses all the formalism that is present in quantum mechanics but organizes it in such a way that one can see where the infinities are and thus avoid them by a renormalization process. One group of infinities is related to the self-energy of the electron. These infinities can be sidestepped by incorporating them into the rest mass of the electron and then replacing the mass of the electron that appears in the Dirac equation by the experimental value of this mass. This is called mass renormalization. The other group of infinities stems from the interaction of the electron with the sea of negative-energy states in the vacuum. This is called vacuum polarization because the electron, by means of its radiation field, creates momentary virtual pairs or dipoles in the vacuum (positrons and electrons), which then annihilate each other.

These vacuum polarization infinities can be rendered harmless by incorporating them into the charge on the electron and then using the experimental value of the charge in the equations. This procedure is called charge renormalization.

Although Tomonaga, Schwinger, and Feynman worked on these developments independently and almost simultaneously, their formulations differed considerably even though their conclusions were much alike. Since the procedures of Tomonaga and Schwinger were similar we shall consider Schwinger's work as representative of the methods developed independently by these two physicists. Feynman's approach to this problem, however, was so radically different, at least in appearance, that one must consider his discoveries separately. The Tomonaga-Schwinger procedure, formulated and applied to its fullest mathematical extent by Schwinger, is a careful step-by-step mathematical analysis of the difficulties inherent in the original Dirac theory of the electron and an analytical reformulation of the theory. Consequently, the equations describing the interaction of an electron and a radiation field or the interaction of two electrons are free of the inconsistencies that lead to the infinities, which cannot be properly subtracted from the meaningful results in the Dirac theory. The importance of Schwinger's teatment lies in its theoretical completeness and generality.

To see what Schwinger accomplished in reformulating the Dirac theory we must first comprehend the difficulties that were encountered before this treatment. A simple analysis of the application of the Dirac equation to the description of the state of a system (e.g., the motion of an electron) shows that, although the Dirac equation is relativistically correct in that it is invariant to a Lorentz transformation, time is involved in such a way that the equation does not give a correct relativistic description of the state of the system. This is so because the Dirac equation (like the Schrödinger equation), despite its relativistic form, assigns the same instant of time to different points of space. As a result, the commutation relations have a nonrelativistic form even in the Dirac theory and the function that defines the state of a system depends on the time quite differently from the way it depends on space. Essentially, this means that the wave function that describes the state of a system and the commutation relations have meaning only in a particular Lorentz frame of reference. Thus, the theory is not relativistically correct in the fullest sense of the phrase.

Tomonaga was the first to point out that the theory can be made relativistically correct by recasting it in a form in which all references to a specific time for all points of space are eliminated. This means that the equations of motion must be so stated that they refer to space-time points and to world lines rather than to different points in space at a given time. In this way no particular frame of reference is singled out in which to express the commutation relations or the function that describes the state of a system. Tomonaga published his fundamental article in Japan in 1943, but physicists in this country did not become aware of it until after the Lamb-Retherford experiment. Schwinger and Feynman had already completed most of their basic work when reports of Tomonaga's theory and English translations of his paper began to reach American physicists in the winter of 1947 to 1948. It is interesting that whereas Schwinger and Feynman had available the experimental results of Lamb and Retherford to inspire them, Tomonaga reached his goals on the basis of theory alone. Although Tomonaga was the first one to recast the equations into this fully relativistic form, Schwinger was the first to carry out a complete self-consistent formulation.

As Schwinger first pointed out, the problem of isolating the infinities of the theory so that they can be properly subtracted from the physically meaningful results (that is, incorporated into "unobservable renormalization factors") is related to formulating the theory in a competely covariant form (relativistically correct) at each stage of its development. In referring to the sidetracking of the infinities, Schwinger asked "whether quantum electrodynamics can account unambiguously for the recently observed deviations from the Dirac theory of the electron, without the introduction of fundamentally new concepts?" His first paper in a series devoted to this question was occupied with the formulation of a completely covariant electrodynamics. "Manifest [that is, obvious at each stage] covariance with respect to Lorentz and gauge transformations is essential in a convergent theory since the use of a particular reference system or gauge in the course of calculation can result in a loss of covariance in view of the ambiguities that may be the concomitant of infinities."

To carry out this program Schwinger subjected the Dirac equation and the quantum-electrodynamic field theory to a searching analysis and showed precisely where the difficulties were to be found. He then reformulated the theory in such a way that it was completely relativistic. The new formalism made it possible to isolate the various singularities (infinities) and to make them harmless by a renormalization procedure. The remarkable result of Schwinger's theory was that it led to an equation of motion for the electron in a radiation field in which the interaction energy alone between the field and the electron (or between two fields) determined the behavior of the system. Since this interaction energy is relativistically invariant, the theory formulated in this way contained none of the objections to which the original Dirac theory was subject. With this formulation Schwinger was able to take care of the infinities arising from the self-energy of the electron and also of those introduced into the theory by the vacuum polarization. By isolating these infinities and incorporating the first into the mass (mass renormalization) and the second into the charge (charge renormalization) he was able to calculate with great accuracy the measured anomalous magnetic moment of the electron and the Lamb-Retherford displacement of the S and P levels of hydrogen.

Although Schwinger's solution of the infinities difficulty was the most general and theoretically complete exposition of the subject, the mathematical formalism in which it was presented and the mathematical techniques needed to apply it to most problems are so difficult that it is seldom used. Instead the methods and techniques developed by Feynman are applied to most problems. Starting from a point of view completely different from that of Schwinger, Feynman discovered a formulation of the completely covariant theory that enables one to write down the answer to most problems (if they are not too involved) almost by inspection. The beauty of the Feynman technique is that it gives a complete view of all the processes that are taking place, making it possible to write down mathematical expressions for each of these processes in turn. This procedure is further facilitated by a diagram or graph (the famous Feynman graphs) which gives a complete geometrical representation of the process being studied. These space-time diagrams consist of directed straight and curly lines. The straight lines represent the world lines of electrons and the curly lines represent the world lines of photons. The diagrams show schematically the time-space sequence in which the process being investigated unfolds. The use of world lines (space-time diagrams) to represent the entire process being studied has another great advantage. It leads easily to a relativistically invariant treatment of the problem, since world lines are the same for all observers and hence relativistically invariant. The meeting of a straight and a curly line means either the emission (creation) by the electron of a photon or the absorption (destruction) of the photon. In these diagrams a directed solid line may point either toward the future (in the direction of increasing time) or toward the past (the direction of decreasing time). Thus, there is no restriction as to how the straight lines are to be drawn. This corresponds to the finding that the equations of physics are symmetrical in time and are not altered when the time in these equations is replaced by its negative. This is called time inversion or inflection. One of the advantages of the Feynman method is that it automatically takes into account time inversion, and thus introduces the positron on the same basis as the electron. As Feynman pointed out, the straight line directed backward in time represents the motion of a positron, so that a positron is represented as an electron moving from the future into the past. Thus, these diagrams represent the creation or the annihilation of a pair (an electron-positron pair) by the intersection at a point of a straight line directed toward the future and a straight line directed toward the past. In this way all possible interactions that contribute to a process can be represented graphically. Since each of these graphs can be expressed mathematically according to certain fairly simple rules developed by Feynman, a complete mathematical expression giving the probability for the process can be written. In principle, the solution to any problem, taking into account all the interactions that contribute to it, can be written almost by inspection. But in practice the evaluation of the mathematical terms that occur is so complex in higher order calculations that the Feynman procedure is limited to the first few orders of approximation.

Feynman's discovery of the graphs and the methods for evaluating them evolved from an approach to quantum mechanics that departed considerably from the traditional procedure introduced by Schrödinger. It was closer to the matrix mechanics of Heisenberg. The standard way of treating a problem in quantum mechanics is to set up the Hamiltonian (the total energy expressed in terms of momenta and coordinates) and then to transform it to an operator (operating on the wave function or the state vector, as it is called) by replacing each component of the momentum by a differential operator with respect to the corresponding coordinate. The operator for the Hamiltonian thus obtained, operating on the wave function, is then equated to the rate at which the wave function varies with the time. This is the Schrödinger wave equation. Essentially, it says that the unfolding (or evolution in time) of the state of a system (the wave function) is determined by the differential operator that represents the total energy of the system.

The Schrödinger differential equation is usually solved by a process in which the change in the wave function (the state vector) is considered step-by-step in short time intervals. One assumes that one knows the state of the system at some initial moment and then obtains it at a very short interval later by noting that the change of the system from its initial state is just equal to the product of the small time interval and the Hamiltonian (as an operator) applied to the initial state (to the initial wave function). In this way one can, at least in principle, describe the evolution of the state of the system by degrees from some initial moment to any desired later time. Carried out with proper regard for relativistic accuracy, this is the procedure of Schwinger.

The disadvantage of this procedure is that it involves the investigator in the analysis of details that are superfluous and, in fact, meaningless as far as the overall picture of the process is concerned. Moreover, it is difficult to keep track of all the various terms that may contribute to the final result in the particular problem and to make sure that these various terms are relativistically correct. Feynman's method eliminated these problems because he considered the event in its entirety and represented the evolution of the system from its initial state to its final state as one continuous space-time path that starts from the initial moment and continues in a series of connected, directed straight lines to the final moment. All intermediate processes, such as the virtual emission and absorption of photons and the creation and annihilation of pairs, or the interaction of the electron with external perturbing fields, are represented on his continuous graph by sudden changes in direction of the directed straight line (which may take one backward in time) or by the appearance of curly lines that emerge from the straight lines and move off or come back to the straight line again to form a loop (the creation and annihilation of a virtual photon). The interaction of the moving electron with an external field on this space-time diagram is shown by a kink (a sudden change in direction) of the directed straight line and may be accompanied by the sudden appearance of a curly line at the kink (the emission or absorption of a photon).

Any particular event is now represented by the sum of all possible Feynman diagrams that can be drawn from the initial space-time point to the final space-time point. Since there are very definite rules for constructing these diagrams and each line in any diagram (as well as each point where there is a kink or change in direction) has a mathematical counterpart, the complete mathematical scheme for describing any event can be written down. This is particularly effective when one is involved in higher order calculations that are completely unmanageable with the traditional procedure. In the Feynman method each kink or change in direction of the directed straight line represents an additional order in the calculation. The graphic scheme representing the second-order interaction of an electron and an external field of force contains two changes in direction of the directed straight line (the world line) describing the motion of the electron.

Although the Feynman and Schwinger methods appear quite different at first sight, they are really completely equivalent, as was first demonstrated rigorously by Freeman Dyson. It is said that Feynman, on hearing of Dyson's proof that his and Schwinger's procedures were mathematically equivalent, remarked, "Now I have been translated into hieroglyphics." This is an interesting revelation of Feynman's attitude toward the highly formalistic approach to physical problems. The essential difference between the Tomonaga-Schwinger formalism and the Feynman approach is that the former is a field theory whereas the latter deals directly with the space-time history of the particles involved; it is a carryover into quantum mechanics of action at a distance between particles. Feynman's method introduces a set of simple rules that enable one to calculate the physically observable quantities that, according to Feynman, are the only things that can and should be calculable from a theory. In this sense, Feynman's ideas are closer to Heisenberg's point of view than to that of Schrödinger. Feynman's insistence on dealing directly and only with observable quantities has led to the growth and great importance of the S-matrix point of view in the study of fundamental particles today. This is a natural extension of Heisenberg's original matrix mechanics.

It should be emphasized that the discoveries of Schwinger, Feynman, Tomonaga, and Dyson have added nothing new to our picture of nature. The fundamental difficulties associated with the infinite self-energy and the structure of the electron still remain. The work of these physicists has shown, however, that the Dirac theory of the electron is more powerful and general than previously had been thought to be the case. All that is required for the application of the Dirac theory to its fullest productivity is a consistent and relativistically correct way of taking into account all the interactions that contribute to any event. This is the essential importance of the renormalization physics that has grown out of the work of Schwinger.

The renormalization techniques give a physically reasonable answer to any question that may arise in the interaction of electrons with the electromagnetic field. They can also be applied to eliminate some of the infinity difficulties that occur in the interaction of nucleons via the meson fields. But here the procedure is only moderately fruitful and most of the difficulties remain even with renormalization. The coupling between nucleons and the mesonic fields is many times stronger than that between electrons and the electromagnetic field. Hence, the renormalization procedure, which is essentially a sum of an infinite number of terms in successively higher powers of the coupling constant between the particle and the field (the number that measures the strength between the particle and the field), does not converge but becomes infinitely large for the interaction between two nucleons. Therefore, this procedure is not trustworthy even if one considers only first- or second-order perturbations.

The Nuclear Shell: Johannes Daniel Jensen (1907–)

When Johannes Jensen gave his 1963 Nobel Prize address, he offered a lucid and entertaining account of the history of nuclear theory as well as of his unique contributions to physics. For those not acquainted with the shell model of the nucleus, a few words explaining some of the concepts used by Jensen are worthwhile. The idea of the shell model stems from the arrangement of electrons in the outer regions of an atom. The electrons of heavier atoms arrange themselves in closed concentric shells. An atom in which all the shells are closed (each containing as many electrons as possible) is one of the nobel gases (helium, neon, etc.) and is inert and extremely stable. The electronic shells in the atom are the K shell with 2 electrons, the L shell with 8 electrons, the M shell with 18 electrons, and so on. We may therefore call the numbers 2, 8, 18, and so forth, the magic numbers of atomic structure. These arrangements of electrons in the outer region of the atom are due to the Pauli exclusion principle and to the fact that the electrons have a half unit of spin. An additional factor in this structure is that each electron has its own orbital angular momentum and, to a very good approximation, moves independently of the other electrons. The empirical evidence for the electronic shell structure is the periodic table of elements.

Since protons and neutrons also have half units of spin and are governed by the Pauli exclusion principle, one is immediately led to the idea of a nuclear shell structure. But there is one argument against this concept. Nuclear forces between individual nucleons are so large that nucleons cannot be pictured moving in orbits independently of one another as is true for electrons. However, suppose that, to a first approximation, we could picture each nucleon (neutron or proton) as moving in its own orbit in a kind of nuclear force field, independently of the motions of the other nucleons. We could then assign an orbital angular momentum as well as spin to each nucleon so that each nucleon would have its own distinct quantum numbers. The Pauli exclusion principle could then be applied to the nucleus, and

Johannes Daniel Jensen (1907–)

the nucleons would arrange themselves in shells. A closed nuclear shell would then manifest itself as an extremely stable (a very large binding energy or mass defect) nucleus. Such nuclei would also be more abundant than others. Observations show that such nuclei occur.

We find that when the number of neutrons or protons inside a nucleus has one of the values 2, 8, 20, 28, 50, 82, 126, that nucleus is unusually abundant and extremely stable; these are called the magic numbers. Jensen and Marie Goeppert-Mayer's contribution was to show how these numbers can be explained in terms of a shell structure. Eugene Wigner developed the mathematical techniques, in the form of group theory, that were needed to analyze nuclear structure.

Johannes Jensen was born on June 28, 1907. He obtained his doctorate from Hamburg University in 1932 and was a member of the faculty of that university from 1936 to 1941. in 1941, he became a professor of physics at the Hannover Institute of Technology, spending eight years there until called to a professorship at Heidelberg in 1949, where he later became the director of the Institute of Theoretical Physics. Frequently a visitor to the United States, in 1955 he published with Professor Maria Goeppert-Mayer of the University of California a paper on nuclear shell structure. It was for his shell structure theory of the nucleus that he shared the 1963 Nobel Prize with Goeppert-Mayer and Professor Eugene Wigner of Princeton University.

Radiocarbon Dating: Willard Frank Libby (1908–1980)

The discovery of the transformation of the radioactive elements, one into another through the step-by-step disintegration process, soon led to information on the half-lives of these unstable substances. The long half-lives of uranium and thorium, 4.5 billion and 10 billion years respectively, suggested that these elements, together with some of their disintegration products, might serve as accurate clocks to reveal the time of their formation and of the earth itself. Over the intervening years this hope has been realized. It would seem, now, that our planet originated some 4.5 billion years ago.

Until recently no radioactive substance was known whose properties and half-life were such that it could be used to date substances in the organic, as opposed to the inorganic, world. The dating of ancient events and civilizations through organic artifacts, refuse, or through animal or even human remains, could supply much valuable information to the historical record if a precise means and the right radioactive element were available. Fortunately, after World War II, through the discovery of the radioactive isotope, C^{14}, with a half-life of 5,568 years, a highly useful dating technique was developed. The isolation of this isotope, the investigation of its properties, and the refinement of its use in dating are all owed to Willard F. Libby and his associates.

Recorded history extends back only some 5,000 years; in that interval there are many gaps because of the lack of written records. But human culture is older than written records by many thousands of years. The clock of C^{14} ticks on in the artifacts, refuse, animal remains, and even the burned-out campfires of those ancient times. From radiocarbon, we have learned that the cave drawings in the Lascaux caves of France can be assigned to people of skull and skeletal structures similar to modern Europeans and that they flourished about 15,000 years ago. An even earlier dating from the Arignacian period of some 27,000 years ago indicates that Homo sapiens was then already fully differentiated as a species.

The appearance of early humans in North America has generally been linked to the maximum advance of the ice sheet in the most recent glaciation, previously assumed to have occurred about 25,000 years ago. This most southerly advance of the ice sheet has now been placed by radiocarbon analysis of wood samples from the Two Creeks forest bed in Wisconsin in the more recent past of 11,000 years ago. When analyzed, evidence for early habitation on this continent gives dates of the order of 10,000 years, corroborating the appearance of humanity in North America with the maximum of the last glaciation. The value of this method to the construction of an accurate archaeology is thus apparent.

The radiocarbon dating method depends upon the fact that C^{14} is being continuously formed in the upper atmosphere by the interaction of the incoming primary cosmic radiation with air atoms. This reaction produces neutrons that, entering the nuclei of nitrogen atoms in the air, result in the formation of C^{14} atoms. Such atoms quickly combine with oxygen in the air to form molecules of carbon dioxide. These soon become homogeneously mixed with molecules of stable CO_2 by the churning of the atmosphere. Thus, all samples of atmospheric carbon dioxide will be found to be radioactive as well as all plants, since plants grow by incorporating this substance. In the same way, since all animals ultimately subsist on plants, all animals and all humans are radioactive.

Present evidence leads to the conclusion that there has been a constant rate of production of radiocarbon in the atmosphere for at least the past 10,000 years and that this production has been in equilibrium with its decay. This balance has been a characteristic over that time of all living carbonaceous material. During the lifetime of any plant or animal, the radiocarbon assimilated from food will be in exact balance with the radiocarbon disintegrating in the tissues. When death occurs, the balance immediately ceases, and the radiocarbon atoms become fewer and fewer as time goes on. The mean number of disintegrations per minute per

Willard Frank Libby (1908–1980)

gram of carbon atoms from living material is 15.3; using the half-life of C^{14} we know that the same carbon 5,568 years after the death of the material would show an average of 7.65 disintegrations.

Radioactive C^{14} atoms decay by emitting a beta particle (an electron). To determine the age of a once-living substance such as wood, the sample must be very carefully reduced to pure carbon, taking care that all other material is excluded. If the activity is determined from the pure carbon itself the material must be so disposed that all beta particles released by the disintegrating items will be counted. Alternatively, the carbon may be reacted into a gaseous form such as carbon dioxide or acetylene. Whether gas or solid, the carbon is introduced into a sensitive Geiger counter and its activity determined. The measured disintegrations per gram per minute are then substituted in the equation $I = 15.3 \exp[-0.693(t/5568)]$ from which the desired value of t, the age of the material (in years), is found.

Willard Frank Libby was born in Grand View, Colorado, on December 17, 1908. He completed his undergraduate studies at the University of California at Berkeley in 1931 and received his Ph.D. at the same institution in 1933. He was then appointed instructor in chemistry and, in 1938, assistant professor. In 1941 he was awarded a Guggenheim Fellowship which he held at Princeton University. He later transferred to Columbia University, Division of War Research, serving from 1941 to 1945 on the uranium isotope separation project. After the war he was appointed professor of chemistry at the Institute of Nuclear Studies and the University of Chicago. It was during this time at Chicago that he carried on the C^{14} research and developed its dating techniques. In 1954, Libby was appointed a member of the U.S. Atomic Energy Commission. He served for five years, resigning in 1959 to become professor of chemistry at the University of California at Los Angeles where he remained until his death on September 9, 1980. The recipient of many awards, medals, and prizes for his distinguished contributions to chemistry, he was presented the Nobel Prize in chemistry in 1960 for his development of the C^{14} dating techniques.

15 *Nuclear Reactions and Nuclear Energy*

The best defence against the atom bomb is not to be there when it goes off.

—*The British Army Journal*, 1949

Nuclear Theory: Werner Karl Heisenberg (1901–1976)

Once the neutron was discovered, nuclear physics developed with great speed both experimentally and theoretically. Since the experimental physicists, such as those in the Fermi school at Rome, were collecting vast amounts of data, it was necessary for the theoreticians to collate these data and construct a rational model of the nucleus that could explain the evidence and indicate the direction for new research. The theoretical problem in nuclear research differs in one very important respect from that encountered in the structure of the outer regions of the atom. In ordinary atomic physics one deals with a swarm of electrons surrounding a nucleus; each electron has a mass that is negligible compared with the mass of the nucleus. This means that one can treat the problem as a central force (the Coulomb force) problem and apply a fairly simple perturbation procedure. The mass of each electron is negligible compared to the mass of the nucleus. The positive electric charge is concentrated in the nucleus, whereas the negative charge is distributed over the various individual electrons. Thus, we may neglect the interactions among the individual electrons to a first approximation.

But this is not the case inside the nucleus where each nucleon (neutron or proton) is of equal status. There is no single concentration of nuclear matter from which we may depict the emanation of the nuclear forces that keep the nucleus together. Hence, a simple perturbation procedure is not applicable, and we must treat the nucleus as a whole without being able to separate it into two more or less distinct parts.

Another difficulty arises in nuclear physics because the exact nature of the nuclear forces is not known. We do not have such a straightforward situation as in the case of the Coulomb force between two charged particles. For that reason nuclear physics developed along semiempirical lines, with little commitment to any particular type of nuclear force. Fortunately, it soon became clear that many properties of the nucleus can be discussed and understood without making specific assumptions about the exact mathematical form of the law of interaction between two nucleons. The reason for this state of affairs is in the very nature of the nuclear forces; they are of such short range that it hardly matters how one specifies

414

them over such a small distance as long as sufficiently large magnitudes are chosen to agree with the data.

Even without penetrating into the nucleus to determine its detailed structure, we can derive certain of its properties from the empirical data and then use these properties to draw conclusions about the structure. Thus, we have very accurate experimental data concerning the masses of the various nuclei, on the assumption that all nuclei are constructed of protons and neutrons. This information permits us to analyze the stability of these nuclei. As we stated in our discussion of Fermi's work in nuclear physics, the binding energy of a nucleus is equal to the work that is needed to separate (to large distances) all the protons and neutrons from each other, that is, to disrupt the nucleus entirely. In other words, it is the energy that is released (in the form of gamma rays, electrons, neutrinos, and positrons) when all the neutrons and protons come together to form a nucleus. If the total energy of a group of nucleons when they come very close together (within nuclear distances) is negative, these particles form a stable nucleus. If this energy is not negative, no stable nucleus is formed. The more negative this energy is (potential energy is negative and kinetic energy is positive), the more stable will be the nucleus that is formed.

To calculate the binding energies of nuclei one need only have accurate values for their masses and compare them with the sum of the masses of the individual nucleons of which they are composed. If the mass of the nucleus is smaller than the sum of a comparable number of free protons and neutrons, the nucleus is stable. Otherwise it is not stable. This result follows from Einstein's famous discovery of the equivalence of mass and energy, $E = mc^2$.

With a fairly accurate picture of the binding energies of the various nuclei, particularly in relationship to their atomic weights, one can see how they increase as the atomic weight increases. Then one can make important deductions about the overall nature of nuclear forces and the way in which the nucleons are distributed inside the nucleus of an atom. From the fact that the binding energy per nucleon remains the same as we go to heavier nuclei, Heisenberg pointed out that one may assume that the nucleons inside a nucleus interact (to a very good approximation) only with their nearest neighbors. This phenomenon is generally referred to as the saturation of nuclear forces. Moreover, since the volume of the nucleus is proportional to the number of nucleons, one may further conclude that the nucleons are spread out uniformly throughout the nucleus. This reasoning leads to the famous liquid drop model of the nucleus, first introduced by Niels Bohr, which has proved so useful in explaining nuclear fission and nuclear transformations in general.

Energy Production in Stars: Hans Bethe (1906–)

Hans Bethe was an international figure in physics when he left Germany as a young man in the 1930s to teach and work in England, but his most brilliant work was done in the United States after he became professor of physics at Cornell University. In 1934, the neutron was discovered, and there began the feverish work in neutron and nuclear physics that dominated all the major laboratories of the world. Bethe at once entered this field with his usual comprehensive approach. His work, theoretical in nature, dealt with the analysis of nuclear forces and the construction of models of light nuclei. One of the first problems he considered (in collaboration with Peierls) was the interaction of a neutron and a proton. Since a neutron-proton pair is the simplest type of compound nucleus that can exist, it plays a generic role in nuclear physics similar to that played by the hydrogen atom in atomic physics. Its theoretical treatment was therefore very important. Many questions had to be answered before nuclear theory could give a coherent picture of the structure of the nucleus. It was essential to know whether one could use quantum mechanics to describe the behavior of nucleons inside the nucleus just as one uses quantum mechanics outside the nucleus. The neutron-proton complex was the ideal system for studying this problem.

Bethe and Peierls treated this system theoretically by using the usual quantum-mechanical rules and assuming that the neutron and proton interaction could be represented by a simple law. This law was set forth as follows: As long as the neutron and the proton are separated by more than a critical distance (of the order of 10^{-13} cm) they have no influence on each other; if they are closer, there is a constant but very large pull between them. This type of interaction is known as the rectangular well and permits a very simple treatment of the problem. In spite of its simplicity, however, this model gives very good results. As Bethe and others showed, changing the type of interaction (as long as one keeps it very short range and very strong) does not significantly affect the final results. Bethe and Peierls solved the two-body nuclear problem with quantum mechanics and showed that the two-body nuclear system (now known as the deuteron, an isotope of hydrogen) is stable. Many of the properties of the deuteron were derived by Bethe theoretically; his progress stimulated experimental work by others. In addition, by showing that quantum mechanics was applicable inside the nucleus Bethe opened a new field for theoretical physics.

Although Bethe was not directly interested in astronomy during the decade from 1930 to 1940, his work led to the discovery of the nuclear reactions that generate the radiation of stars similar to the sun. It had been known for a long time that the energy radiated from stars could not come from ordinary chemical reactions such as combustion, nor could it be accounted for in most cases by a mechanical process such as a slow but steady gravitational contraction. To begin with, the temperatures inside stars are much too high for the existence of stable molecules, so that molecular reactions such as burning cannot occur. The amount of combustible material would be so meager that a star could not continue radiating for more than a few thousand years. Second, the release of gravitational energy, at a sufficiently high rate to account for the luminosity of the stars, could have continued only for a few million years in the past—in the case of the sun, about 30 million years, and in the case of the very luminous stars, such as Capella, a few hundred thousand years. Therefore, it was necessary to look for some unusual source of energy. It was clear to every physicist and astronomer, years before the neutron had been discovered and before much was known about the nucleus, that the stellar release of radiant energy involved some kind of subatomic or nuclear process.

In his well-known book *The Internal Constitution of the Stars*, the first systematic treatment of modern astrophysics, the great British astronomer Sir Arthur Eddington stated

> It is now generally agreed that the main source of a star's energy is subatomic. There appears to be no escape from this conclusion; but since the hypothesis presents many

difficulties when we study the details, it is incumbent upon us to examine carefully all alternatives. . . .

In seeking a source of energy other than contraction [gravitational] the first question is whether the energy to be radiated in the future is hidden in the star or whether it is being picked up continuously from outside. Suggestions have been made that the impact of meteoric matter provides the heat or that there is some subtle radiation traversing space which the star picks up. Strong objections may be urged against these hypotheses individually; but it is unnecessary to consider them in detail because they have arisen through a misunderstanding of the nature of the problem. No source of energy is of any avail unless it liberates energy in the deep interior of the star.

It is not enough to provide for the external radiation of the star. We must provide for the maintenance of the high internal temperature without which the star would collapse. (Eddington, 288).

In his book Eddington considered various types of subatomic processes and concluded that although energy can be released by the breakdown of the nuclei of atoms as in radioactive decay, the only process that can properly account for the great quantities of energy released by the stars and for their long lives as energy radiators is one in which nuclei are built up from less massive constituents. In this analysis, one can use Einstein's principle that mass and energy are equivalent and that when mass is destroyed an equivalent amount of energy is released whose quantity equals the mass destroyed multiplied by the square of the speed of light. Eddington pointed out that if four protons combined or coalesced to form a helium nucleus, energy would be released because the mass of the helium nucleus is about 1 percent less than the mass of the four free protons.

With the discovery of the neutron it soon became clear to physicists and astronomers alike that the helium nucleus is, indeed, built up from four protons. But just how the process is initiated or maintained inside the stars was not clear until Bethe in the United States and Weizsäcker in Germany began a series of independent investigations. Both physicists considered a series of reactions starting with the coalescence of two protons to form a deuteron and then a subsequent set of reactions ending in the formation of helium, after two other protons had been captured. This series of reactions is now known as the proton-proton chain and has been thoroughly and accurately investigated. The first complete analysis of the proton-proton chain was made in 1938 by Bethe and C. Critchfield, one of his students. In their paper they laid down the general procedure for an analysis of this type of reaction. The experimental data for certain physical parameters at that time were still in considerable error, and Bethe concluded that the proton-proton chain could account for only a small amount of the energy emitted by a star like the sun (the prototype of what astronomers call main sequence stars). He therefore sought other types of nuclear reactions that could account for the total energy generation.

Bethe found a solution of this intriguing and important problem after attending a meeting of the American Physical Society in Washington, part of which was devoted to the problem of stellar energy. Returning home on the train from this meeting, he analyzed all the interactions of the light nuclei with protons that can occur at the high temperatures found in stars, starting with helium and advancing to, but excluding, carbon. He was immediately struck by the speed with which all such nuclei interact with protons and ultimately end up as ordinary helium. This could only mean that these light nuclei could not be the source of stellar energy because they are changed into helium so fast that stars could not exist for more than a few thousand years.

The situation with carbon proved quite different. Although the carbon nucleus, like lighter nuclei, combines with a proton at temperatures found in stellar interiors in a relatively short time, astronomically speaking, this is only the first step in a cycle of six nuclear reactions involving three other protons in which the carbon nucleus finally reappears and the four protons are coalesced into a helium nucleus. This is the famous carbon cycle discovered by Bethe, which he described in 1938 in a short note to the *Physical Review*.

At that time, the difficulty in accounting for stellar energy generation in terms of nuclear reactions lay in explaining how positively charged protons could be absorbed by the positively charged nuclei. Even in the case of light nuclei, such as lithium, beryllium, boron, and carbon, the charge on the nuclei is sufficiently large to repel free protons very strongly. This repulsion can, of course, be overcome if the protons hit the nuclei with sufficient kinetic energy. But even when the temperature inside the stars is many millions of degrees, the average speed of the free protons relative to the nuclei is too small for this process to happen. However, because of the statistical distribution of the proton velocities about the average, there are always some moving with very high speeds and, hence, with enough kinetic energy to penetrate into the nucleus. In stellar interiors like the sun such high-speed protons are few in number, and energy released in this way would be inadequate to account for stellar radiation, except for considerations introduced by quantum mechanics and the wave properties of particles such as protons.

It had already been shown by George Gamow in his quantum-mechanical explanation of the emission of alpha particles from radioactive nuclei, and also by Atkinson and Houtermans, that the wave properties of protons enabled them to penetrate into charged nuclei even when these protons did not have enough kinetic energy to overcome the nuclear Coulomb repulsion. Bethe used this wave picture to calculate the probability for protons to penetrate into the various nuclei that took part in the carbon cycle. In this way he obtained a formula for the rate at which the carbon cycle proceeded in stellar interiors and showed that at the temperatures to be expected near the centers of stars like the sun, the carbon cycle should be responsible for the generation of most of the energy. In fact, with the aid of the carbon cycle, Bethe and his collaborators obtained a theoretical model of the sun that corresponded fairly well with what one might expect the interior conditions to be.

This work was done before World War II when some of the nuclear data were not very accurate. With the advances of postwar nuclear techniques, it became clear that the formula both for the carbon cycle and for the proton-proton chain had to be revised. The proton-proton chain is now known to be far more effective in releasing energy in stars like the sun than the carbon cycle. However, both the carbon cycle and the proton-proton chain operate simultaneously in stellar interiors; the latter is the dominant process in stars that are as luminous or less luminous than the sun and the former plays the major role in the stars that are hotter and more luminous. In any case, Bethe's pioneering work with the proton-proton chain and the carbon cycle laid the foundation for the great post-war advances in our knowledge of the structure and the evolution of stars.

There is an interesting point in connection with this work that Bethe mentioned in the last paragraph of his 1938 note to *Physical Review*. He noted that all the reactions of the light nuclei with protons ultimately lead to the end product helium. Since helium itself does not react with protons, the relative abundance of heavier elements above helium cannot change inside the stars on the main sequence. From this reasoning he concluded that the so-called Aufbau hypothesis, advanced by many physicists and astronomers, must be discarded. According to this hypothesis, all the heavy elements are built up in successive stages from protons. Bethe reasoned at that time that there was no stable nucleus of atomic weight 5, so that there was no way to jump the gap from ordinary helium to lithium, and other heavier elements, by proton capture. We now know that when enough hydrogen (about 12 percent) has been transformed into helium in the center cores of stars like the sun, the helium core begins to contract to such an extent that the core temperatures rises very rapidly until a temperature close to 100 million degrees is reached. At this high temperature, a new type of nuclear reaction, first described and analyzed by E. Salpeter, takes place. Three helium nuclei (alpha particles) coalesce to form an ordinary carbon nucleus. From then on, as the temperature in the core of the star increases, nuclear reactions involving helium nuclei and heavier nuclei occur, and increasingly heavy nuclei are formed, thus vindicating the Aufbau hypothesis.

Fission: Lise Meitner (1878–1968), Otto R. Frisch (1904–1979) and Niels Bohr (1885–1962)

In 1919, when Lord Rutherford first used alpha particles to disrupt atomic nuclei, very few people, even physicists, believed that this result was more than a scientific curiosity and that nuclear physics would ever have any practical value. The very fact that a vast amount of energy (the binding energy of the nucleus) must be poured into a nucleus to disrupt it, or even to tear out one of its nucleons, indicated to physicists that no energy could ever be obtained by the disintegration of a nucleus. And yet, just twenty years after Rutherford first disrupted a nucleus by bombarding it with very energetic alpha particles, uranium fission was discovered by Hahn and Strassmann and the world was thrust into the era of atomic energy and atomic bombs.

How can energy be obtained from the disintegration of a nucleus, which is a stable structure and apparently must absorb energy in order to undergo fission? Let us consider the formation of the simplest compound nucleus (the deuteron or heavy hydrogen) consisting of a proton and a neutron. If these two particles are initially at rest with respect to each other and are not very close together (if they are separated by more than a trillionth of a centimeter), we can say that they are in a state of zero energy with respect to one another (zero kinetic energy because they are at rest, and zero potential energy because the interaction between them is zero). Of course, each particle has its mass-energy mc^2 according to Einstein's formula, but this does not play a primary role in nuclear formation.

Suppose now that we bring the neutron and proton closer and closer together until they are within the range of their mutual nuclear attraction. At this point they suddenly rush violently together because they are attracted to each other, losing energy while they do so. Because this attraction does work on each particle they acquire kinetic energy. But since they lose more potential energy by coming close together than the kinetic energy they acquire, their total store of energy is reduced. The total store of energy at any given moment is the total mass of the system times the square of the speed of light—Einstein's famous relationship $E = mc^2$. The mass of the deuteron is thus smaller than the sum of the masses of the proton and the neutron. This difference in mass appears as a burst of energy in the form of a gamma ray when the proton and neutron combine to form the deuteron.

We may represent this process by the following crude picture. Consider a deep but narrow crater with steeply slanting walls, surrounded by a flat, smooth terrain on which there is a black (the proton) and a red (the neutron) ball. As long as the balls are on the flat surface, they are not attracted to each other and they remain at rest. If the two balls are brought to opposite points on the crater's edge so that they start rolling down the slope of the crater, they will move toward each other very rapidly as they roll down the slope. Of course, this movement is due to the gravitational pull on each ball. But, to pursue our analogy with the neutron and the proton, we may imagine that the balls move toward each other because of an attraction that arises when they are separated only by the diameter of the crater. If there were no friction along the slope of the crater, the two balls would roll down to the bottom and then right back up to the top again, and we would not have a compound system. But because of the friction along the slope, the balls reach the bottom with less mechanical energy than they had at the top. Their energy at the top is all potential (which we may take to be zero). But as they roll toward the bottom, they lose potential and gain kinetic energy. However, their net gain in kinetic energy is not equal to their loss in potential energy because some of the initial potential energy is dissipated by the friction and flows into the ground as heat. This process is analogous to the emission of a gamma ray when the neutron and the proton combine to form the deuteron.

Now suppose that below a certain level in the crater the walls are perfectly smooth so that the spheres generate no heat as they roll from that level to the bottom. It is clear, then, that after a few large oscillations back and forth in the crater, the spheres will reach an equilibrium configuration, moving around forever with a fixed amount of mechanical energy in the frictionless part of the crater. Thus, the two spheres form a stable dynamical system inside the crater because they lost some of their mechanical energy as they rolled down the wall of the crater. We can disrupt this system only if we supply mechanical energy in some form to the spheres. Of course, we must supply at least as much mechanical energy as the spheres lost by friction.

If we pursue this picture further, we note that we can send as many red spheres into the crater as we wish (assuming that each sphere takes up only a small amount of space). This is equivalent to adding more and more neutrons to the deuteron and thus building up more and more massive isotopes of hydrogen. At first, there appears to be no objection to this idea. All neutrons pull upon each other as strongly as they pull upon protons; consequently each neutron is like another sphere rolling into the crater. But there are definite quantum rules that prevent this outcome. Since all neutrons are identical and have a half unit of spin (like the electron) they obey the Pauli exclusion principle. This means that only one neutron can occupy the lowest energy state in the deuteron (only one red ball can move around in the frictionless region). Another neutron can still enter into a threefold combination with the proton and the other neutron by aligning its spin opposite to that of the neutron already in the deuteron. The Pauli exclusion principle allows two identical particles to move in very nearly similar orbits if their spins are antiparallel.

The Pauli exclusion principle prevents our carrying the analogy much further. Although we can send any number of red spheres into the crater, no more than two neutrons can form a stable structure with one proton. A third neutron is prevented by the Pauli exclusion principle from getting close enough to the other two to form a stable nucleus. How, then, can we get stable nuclei with more than three nucleons in combination? Only by adding more protons. Since protons are different from neutrons, they are not prevented by the Pauli exclusion principle from getting close to the neutrons already in the nucleus. With more protons in the nucleus, more neutrons can be added. But adding a proton to a nucleus presents another problem. Protons repel each other electrostatically. Hence, we cannot simply add them to nuclei the way we might roll spheres along a level surface into a crater. The protons already in the nucleus repel any new ones trying to get in. However, we can still carry out the analogy for protons if we picture the crater as surrounded by a very steep hill, resembling a lunar crater, that rises above the level ground to a fairly high ridge and then falls off gently. Before we can get a sphere into such a crater, we must first push it up the hill. Once the ball is at the top of the hill, we can let it go and watch it fall into the crater. It is more difficult to get it out now than previously because we must bring it from the bottom of the crater up to the top of the hill again. (Note that in this imaginary crater the red sphere, the neutron, must be pictured as being able to go through the hill with no resistance. A neutron has no charge and hence suffers no electrostatic repulsion.)

From this analogy, it becomes evident that it is not so easy to build up heavier and heavier nuclei. The more neutrons we add, the more protons we must add. The hill, because of the increased nuclear charge, gets steeper with each additional proton. But if the hill got steeper with additional protons, with no change in the manner in which the nucleons arranged themselves once they fell into the nuclear crater, they would all fall deep down into the crater. Energy could never be obtained by disrupting such a nucleus since as much energy would have to be used to accomplish this result as would be obtained when the nucleus was disrupted.

But something else happens when we go to heavier and heavier nuclei. The Pauli principle applies to protons just as it does to neutrons. The black and red spheres (protons and neutrons) do not all fall to the bottom of the crater because it is filled up when two black and two red spheres (2 protons and 2 neutrons) are present to form the helium nucleus or alpha particle. The other spheres must now find their places closer to the top of the crater. As more and more spheres are added, they lie closer and closer to the top. We can now see how much more energy can be obtained from a fission process than is supplied to the nucleus to induce the fission. Again we use the analogy with the crater, but this time we place our crater (which we make quite shallow, rather than deep, to take into account the spheres already inside it) on top of a very high peak (like the crater of a volcano). To get a black sphere (proton) into this crater we must first roll it up the side of the crater (do work on it, or give it energy). It then drops into the crater by itself. But since the crater is very shallow, the sphere remains close to the top of the peak. It takes only a small amount of energy to bring it back to the top and let it roll down to the outside again.

Thus, by supplying only a small amount of energy to the spheres in the shallow crater on top of the peak, we can bring them to the edge. In rolling down the side, they then acquire much more energy than is needed to lift them out of the shallow crater. To carry through the volcano analogy, if the lava in the volcano begins to boil violently some of the lava can overflow and move down the slopes of the crater, acquiring much more energy than it had while boiling.

We must apply this type of reasoning to understand the source of the vast amount of energy that is released in uranium fission, even though the fission has been induced by a neutron with only a small amount of energy. This is best analyzed by picturing the uranium nucleus as a liquid drop. Suppose now that we observe a drop of water suspended from a faucet high above the ground (the height being equivalent to the hill surrounding our crater). The drop is held to the faucet by the surface tension forces in the liquid. By blowing gently on the drop or tapping the faucet gently, we can dislodge the drop (it tears away). Once dislodged it acquires a great deal of energy while falling from its height above the ground. In the same way, the gentle disturbance that a slowly moving neutron communicates to a uranium nucleus separates the latter into halves, which were only loosely held together initially. However, the Coulomb repulsion between the two halves (like the pull of gravity on the drop of liquid) pushes them away from one another at very high speeds. Lise Meitner and Otto R. Frisch were among the first to analyze the experimental data correctly. After Hahn and Strassmann had demonstrated that one of the by-products of their bombardment of uranium with neutrons was barium (or its isotopes) Meitner and Frisch pointed out how such a splitting of uranium can occur. In a letter sent to *Nature* in 1939 they used the liquid drop model to show how fission can occur under the appropriate conditions.

Lise Meitner, who, together with Frisch, originated the idea of nuclear fission, was born in Vienna, Austria, on November 7, 1878. One of eight children (three girls and five boys), she was brought up in Vienna and obtained her doctorate from the University of Vienna in 1906. The following year, she studied in Berlin with Planck and soon began research in radioactivity in collaboration with Otto Hahn. From 1912 to 1915, she was assistant to Planck at the Institute for Theoretical Physics at the University of Berlin. In 1917, she was appointed head of the Physics Department in the Kaiser Wilhelm Institute for Chemistry. At that time, she and Hahn began studies of beta decay. Later, she and Von Beyer discovered homoge-

Lise Meitner (1878–1968)

neous groups in the beta emissions from radioactive elements. She continued her research in beta and gamma ray spectra until 1938, when Hitler's persecution of the Jews forced her to flee to Sweden. There she joined the staff of the Nobel Institute in Stockholm. At the end of 1938, she received a description from Hahn of his experiments on the interaction of neutrons with uranium. Meitner suspected from the results that the uranium nucleus was split approximately into equal parts; a subsequent discussion with Frisch led to their joint proposal of fission that appeared as a letter in Nature, entitled "A New Type of Nuclear Reaction." For this discovery she was presented with numerous awards and prizes. Lise Meitner was elected a foreign member of most of the scientific academies of Europe. She eventually emigrated to Great Britain and continued her research at Cambridge University. She remained at Cambridge until her death on October 27, 1968.

Otto Robert Frisch was born in Vienna, Austria, on October 1, 1904. His father, Dr. Justinian Frisch, was in the printing business; his mother, Auguste Meitner Frisch, was a pianist. Lise Meitner, with whom he collaborated, was his aunt. Frisch received his doctorate from the University of Vienna in 1926. He then carried on research in Berlin. He spent the years from 1930 to 1933 in Hamburg with Otto Stern working on molecular beams. He moved from there to London to work with Blackett in 1933 to 1934. The next five years he spent at Bohr's institute in Copenhagen. It was at the end of his stay there that he and Meitner advanced the idea of nuclear fission. In 1939, Frisch moved to England, first to Birmingham, then Liverpool and Oxford. In 1943, he went to the Los Alamos Laboratory in New Mexico as a member of the British team working on the Manhattan Project. On returning to England, he first went to the British Atomic Energy Research Establishment at Harwell and then to Cambridge, where in 1947 he succeeded Sir John Douglas Cockcroft as Jacksonian Professor of Natural Philosophy. The following year he was elected a Fellow of the Royal Society. He continued to teach and do research for the remainder of his academic career, dying on September 22, 1979.

The discovery of nuclear fission by Hahn and Strassmann, and Meitner's analysis of the energetic relations in uranium fission came at one of the most critical periods

Otto R. Frisch (1904–1979)

in the history of the world. World War II was about to erupt; the persecutions of scientists in Germany and, to a lesser extent, in Italy had brought many of the outstanding European physicists to the United States. When news of uranium fission was announced, Enrico Fermi was already a permanent member of the Columbia University physics department; most of his Rome group had left Italy to work in the Western democracies, many of them in the United States. In a sense this was also a critical period for physics. Neutron research seemed to have reached a plateau, and many of the top physicists were turning their attention to other fields. With the discovery of fission, however, a tremendous impetus was given to neutron and nuclear research. Such leaders as Fermi were soon deeply involved in nuclear-fission problems, aided by the development of the cyclotron and the new electronic techniques that were being rapidly introduced into physical research.

The most important problem that arose in connection with uranium fission dealt with the emission of secondary neutrons by the fission fragments. Meitner and Frisch had pointed out that Hahn and Strassman's discovery of a radioactive barium isotope in the fission fragments meant that a new type of nuclear reaction was taking place. Fermi realized that secondary neutrons would be emitted and that a chain reaction was possible. What was most puzzling was the absence of a natural chain reaction involving fission of the uranium in the earth and resulting in one vast atomic explosion. Several conditions prevent this catastrophe. Perhaps not enough secondary neutrons are emitted during each fission process to sustain a chain reaction (for each fission more than one secondary neutron must be emitted for a chain reaction to occur) or, rather than ordinary uranium 238, one of its rare isotopes undergoes fission—or both statements are true. It was demonstrated experimentally by Fermi and his collaborators that enough secondary neutrons are emitted for a chain reaction to occur. But it also became clear that under normal conditions, with low-energy neutrons, it is not uranium 238 but one of its less massive isotopes that undergoes fission.

The theories of nuclear forces and nuclear models had been developed sufficiently to enable physicists to analyze the fission process theoretically. In the forefront of this work was Niels Bohr. To account for certain properties of the nucleus, particularly what is known as the saturation of the density of the nucleus and the saturation of nuclear forces, Bohr introduced what is now known as the liquid-drop model of the nucleus.

Recalling the mass defect of the nucleus (that the total mass of the nucleus is smaller than the sum of the masses of its constituent neutrons and protons) and Einstein's theory of the equivalence of mass and energy ($E = mc^2$), it is apparent that the mass defect is in fact the measure of the amount of energy released in the formation of the nucleus. The mass defect multiplied by the square of the speed of light gives us the binding energy of the nucleus—the amount of energy needed to break down the nucleus into its component protons and neutrons. It was known that the binding energy per particle inside the nucleus, that is, the total binding energy divided by the total number of neutrons and protons inside the nucleus, is approximately the same for all nuclei. Thus, it appears that all nuclei are bound with about the same cohesiveness. In other words, there is a kind of saturation of the binding force in the sense that the neutrons and protons in the nucleus only interact with their immediate neighbors (at least to a very good approximation).

It was also known that the density of the nuclear material is fairly uniform from one nucleus to the other. This can be seen from the sizes of the nuclei. By bombarding nuclei with alpha particles and protons we can calculate the diameters and hence the volumes of the various nuclei, starting from the lightest and going to the heaviest. We find that the volume of a nucleus increases approximately as the number of particles in the nucleus increases. For example, the nucleus of uranium contains about 60 times as many particles as the helium nucleus; but the volume of the uranium nucleus is about 60 times larger than that of the helium nucleus. The density of the nuclear material in both nuclei is about the same.

The nuclear material is a kind of homogeneous substance spread out in a uniform matter in all nuclei. The substance in any nucleus consists of approximately equal numbers of neutrons and protons packed together at a constant density. In other words, the nuclear material is distributed in a nucleus just the way the material is distributed in a liquid drop. This liquid-drop model, first introduced by Bohr, can account very nicely for most of the observed properties of the nucleus. Liquid drops can be of various sizes but have the same binding energy per molecule since no matter how large a drop is, it takes the same amount of energy to rip out a molecule, say, by evaporation. Similarly, there can be nuclei of various sizes. The liquid-drop model leads to some other interesting characteristics that have been verified experimentally. The nuclear drop has surface-tension forces that keep the nucleus spherical in shape just as surface tension tends to keep a drop of water or a soap bubble round. Moreover, just as a liquid drop can be broken into two or more drops if it is set vibrating properly, so, too, can a nucleus. This is an important phenomenon in nuclear fission. Bohr used the liquid-drop model to determine which isotope of uranium undergoes fission when it absorbs a slow neutron. He attacked the problem by means of an analogy with what occurs when molecules are evaporated from a liquid drop. With this model Bohr pictured the fission as occurring in two steps.

In the first step, when the neutron is absorbed, a new compound nucleus is formed that has more internal energy than the original uranium nucleus. This energy is stored in the compound nucleus the way heat is stored in a drop. The second step, according to Bohr's analysis, depends on what happens to the energy introduced into the nucleus by the neutron. If all or most of the energy is concentrated on a single particle, such as a neutron, proton, or alpha particle, near the surface of the nucleus, the usual type of nuclear reaction occurs. The compound nucleus emits one of these particles and settles down to become a new stable nucleus. However, if the energy brought by the absorbed neutron is transformed into mechanical vibrations of the whole nucleus, it undergoes deformations that may shatter it into two equal or very nearly equal parts—and fission takes place.

Fission occurs when the short-range, strong nuclear forces that keep the neutrons and protons bound together are overcome by the long-range, repulsive electrostatic forces that are disruptive. As long as the nuclear matter is kept close together, the short-range forces prevail and the neutron causes the nucleus to vibrate so that it becomes elongated in one direction. The nuclear forces in that direction may be reduced sufficiently below the repulsive forces, which are scarcely affected by the elongation. Then fission occurs.

As Bohr pointed out in a 1939 paper in *Physical Review*, the energy required to bring about the necessary deformation of the nucleus decreases with increasing nuclear charge (the repulsive force becomes bigger and therefore helps the deformation get started). A point is reached for a given atomic number when the energy for deformation is comparable with the energy needed for the emission of a single particle. The probability for fission is then of the same order of magnitude as the probability for particle emission.

Bohr then considered how to determine which of the uranium or thorium isotopes undergo fission. Thermal neutrons move about as fast as molecules in the atmosphere under ordinary conditions; they are also called slow neutrons. For neutrons moving with energies of about 25 electron volts and hence considerably faster than thermal neutrons there is a strong increase in the absorption by uranium 238 (clearly a resonance phenomenon) without an accompanying increase of the fission process. It follows that the ordinary abundant isotope of uranium does not account for the observed fission. In other words, for both ordinary uranium and thorium the probability for emission of a single particle after fast neutron absorption is much greater than the probability for fission. Consequently, these nuclei are stable against fission.

But the situation is quite different for thermal or slow neutrons. Uranium displays more fission than thorium after slow neutron absorption. Uranium, then, would seem to be the active element in fission; but it must be some isotope of uranium other than 238. Bohr then pointed out that it is very probably the rare isotope 235. His reasoning was as follows. Since the excitation energy, and hence the energy available for deformation, increases with the energy of the absorbed neutron, the probability for fission should increase with the energy of the absorbed neutron. But since there is hardly any such increase of fission in uranium 238, another isotope must be involved. The situation for slow neutrons is then the following. Since uranium 238 is not involved we look to the isotope 235 and find that symmetry plays a role. When U^{235} absorbs a neutron, it becomes a nucleus of mass 236 (an increase in one unit of mass) and has an even number of neutrons and protons. Nuclear theory states that in an even-numbered nucleus the protons and neutrons are more tightly bound together than in an odd-numbered nucleus. Consequently, when 235 absorbs a neutron to become 236, the neutron gives up much more of its mass in the binding process than it does when 238 absorbs it to become the odd nucleus 239. Thus, the slow neutrons give up much more excitation energy to uranium 235, to uranium 236 in reality, than they do to uranium 238. This causes 235 (actually uranium 236) to split.

Because U^{235} is very scarce (only one gram is present in each 140 grams of uranium) fission occurs only rarely under ordinary conditions. This was an important factor in the development of the atomic bomb, since it was imperative to isolate a sufficient amount of 235 to ensure a chain reaction.

Chain-Reacting Pile: Enrico Fermi (1901–1954)

Shortly after Enrico Fermi accepted a professorship of physics at Columbia University in 1939, word reached this country via Bohr that Hahn and Strassmann in Germany had produced nuclear fission by bombarding the uranium nucleus with neutrons. This announcement caused great excitement in nuclear laboratories all over the world. Soon experimental nuclear physicists everywhere were busy splitting uranium nuclei, and theoretical physicists were equally busy trying to interpret and explain the results.

It was only natural that Fermi, by consensus the top nuclear physicist of the age, should take a major part in this activity. Soon he was involved in nuclear fission experiments at Columbia's Pupin physics laboratories. Working with him at the time was the Hungarian-born physicist Leo Szilard. Envisioning the implications of these experiments for the security of the world against the axis powers, Szilard initiated the steps that led to Einstein's famous letter to President Roosevelt. Both Fermi and Szilard were quick to see that if there were some neutrons among the fission products, they could, in turn, induce additional fissions, and a chain reaction might result.

A number of questions had to be answered in connection with this process and certain technical problems had to be solved. To begin with, which particular isotope of uranium undergoes fission most readily? Thanks to the theoretical analysis of Niels Bohr and John A. Wheeler, it was known that the isotope is U^{235}, and not the abundant isotope U^{238}, although the latter also suffers fission to a small extent. This is in agreement with the observations that fission—under ordinary conditions—is a very rare phenomenon, about as rare as the isotope U^{235} as compared to U^{238}.

Second, one had to know the speed of the neutrons emitted during fission, for it was known that U^{235} fission is caused by slow neutrons. Hence, if the neutrons emitted during fission were fast, they would first have to be slowed down if a chain reaction were to be achieved.

Third, one would have to determine how many neutrons were emitted in each fission process. This number must be bigger than one if a chain reaction is to result. If only one neutron were emitted per fission, there could be no geometric growth of the fission process, and the whole thing would die out very quickly. On the other hand, if each fission process gave birth to two neutrons, these could then split two nuclei, giving rise to four neutrons, and so on. This sequence is what we mean by a chain reaction, a geometric increase in the number of neutrons that can cause additional splitting of nuclei.

The emission of more than one neutron during each fission is in itself not sufficient to ensure a chain reaction, for these neutrons might be used up in absorption processes that compete with the fission process. The important point is that there must be more than one neutron per fission after the neutrons initially emitted in a fission are slowed down. This factor, called the reproductive factor, must be larger than one. A chain reaction can then occur since we have a situation similar to the growth of money in compound interest.

Finally, one would have to see whether the product neutrons (those emitted during fission) could be kept within the uranium to cause additonal fission. This depends on what is called the mean free path of the neutron. This is the average distance that a neutron travels before it is absorbed by a uranium nucleus. It is desirable for this distance to be short, for then a neutron cannot get very far (or escape from the uranium) before inducing fission. But even if the mean free path were long, one could still obtain a chain reaction by piling up (hence the name atomic pile) enough uranium so that the neutrons would ultimately have to be absorbed because of the great amount of uranium surrounding them on all sides.

These are the questions that Fermi set out to answer when Einstein's letter convinced President Roosevelt to assign a few thousand dollars to what ultimately was to become the Manhattan Project. In the early stages of this work at Columbia University, Fermi had already obtained some evidence that a chain reaction was possible. It was then decided to carry on the project at the University of Chicago.

It was there that the first chain-reacting pile was constructed and successfully operated. This momentous event was announced to the President in a code telegram stating that "the Italian navigator had reached shore safely and found the natives friendly."

A nuclear reactor or a pile, as the latter name implies, consists of slugs of pure uranium metal arranged in a space lattice embedded in a matrix of graphite. In present reactors the uranium slugs are generally situated at regular intervals in fuel rods. In addition, there are cylindrical openings in the pile for inserting control rods made of neutron-absorbing materials. The uranium is the fuel and the source of the neutron flux. The graphite is present to slow down the neutrons and thus prepare them to carry on the fission. The control rods, usually of boron steel, are used to control the rate of the chain reaction by absorbing neutrons and thus reducing the neutron flux.

In 1952 Fermi's official report of the first experimental production of a divergent chain reaction appeared in the *American Journal of Physics*. The reaction took place in a temporary laboratory, a squash court under the west stands of the stadium at the University of Chicago on December 2, 1942. This reactor was the prototype of all the power production reactors that have followed it. To produce a chain reaction or a self-sustaining pile a game of slowing down and catching neutrons must be played. The neutrons emitted during a fission process can have energies up to 1 MeV, and must be slowed down before they are lost in some fashion or other. This is the whole trick, but it is not a trick that can be mastered easily because there are so many hazards in the paths of the neutrons.

Not all neutrons emitted as fission products come out with the same energy because the total energy released in the fission process is divided among all the products, some of which are heavy nuclei. The amount of energy that any particular neutron gets (and hence its speed) is a matter of chance. In his paper Fermi stated that the fast neutrons are slowed down by "elastic collisions with the atoms of carbon and with inelastic collisions with the uranium atoms." By elastic collisions, the physicist meant that the total amount of kinetic energy after the collision is the same as before, but is distributed differently. When a neutron hits a carbon nucleus elastically, the neutron loses some of its kinetic energy and the kinetic energy of the carbon nucleus increases exactly by this amount. This phenomenon occurs when the neutron collides with a nucleus whose mass is not very much greater than that of the neutron.

When a neutron hits a very massive nucleus elastically, it just bounces off, losing none of its kinetic energy. Hence, atoms of small atomic weight like carbon are used in a pile for slowing down neutrons. When a fast neutron hits a uranium nucleus, it can lose some of its kinetic energy by exciting the nucleus itself (stirring up the protons and neutrons inside the nucleus to greater activity). This is called an inelastic collision, but it is only moderately effective in slowing down neutrons.

In his paper Fermi also spoke of thermal energies. The term simply means the kinetic energy of a neutron when it is moving with the average energy of molecules in a gas at room temperature. This is the final result of the slowing-down process; thermal neutrons are what we want for fission. As Fermi pointed out, it takes about 15 collisions of a neutron with carbon atoms to reduce the neutron kinetic energy by a factor of 10. This means that about 110 such collisions are required to bring a million-volt neutron down to thermal energy, which is 1/40 of a volt.

A number of things can happen to the neutron to end its life as an independent and, therefore, a fission-producing particle before it reaches thermal energy. Fermi discussed these neutron-capturing processes in terms of the cross section for the process. To understand this idea, let us suppose that bullets are being fired at a target. We can imagine that this target has only a certain effective area for the occurrence of some event, which is triggered only when the bullet strikes this effective area. We would then call this area the collision cross section for the event. We may now further imagine that each nucleus carries a target area with it for each type of event or process that can occur. Only if a neutron strikes within this area does the particular event take place that is associated with that area or cross section. The larger the cross-section for an event (a collision), the more probable the event. Cross sections, like all areas, are expressed as squares of lengths (square centimeters).

A few examples will illustrate this concept. Both cadmium and boron capture neutrons very rapidly; we say that they have a large neutron-capture cross section. This cross section is of the order of 10^{-24} square centimeters, which, though in itself a very small number, is very large for a nuclear cross section. In fact, a boron or a cadmium nucleus captures neutrons so easily that physicists referred to it as easy as "hitting a barn." Hence 10^{-24} square centimeters is referred to as a barn and taken as a unit of cross section.

The cross section for the capture of a neutron by a carbon nucleus is only 0.005 of a barn. In general, the absorption cross section for neutrons follows the inverse velocity law. This means that the smaller the velocity of the neutron, the larger is the capture cross section, and vice versa. We can see why this is so if we recall that the de Broglie wavelength of a particle gets bigger as its speed gets smaller. In other words, as the neutron slows down, it spreads out and hence can contact nuclei over a wider area.

Fermi then analyzed the various accidents that can happen to the neutron before it is slowed down. He dismissed the possibility of absorption by carbon because the cross section for this process is very small. There remained the absorption by the uranium itself. This can either result in the emission of a gamma ray, with the formation of a new isotope of uranium, or in fission. The first of these capture processes is called resonance capture and reduces the number of neutrons that are available for fission. Finally, some neutrons will be lost by escaping from the pile entirely. All of these losses must be considered in calculating the chain reaction.

If v fast neutrons are produced in a single fission, the number of neutrons available for further fission, and hence for reproduction of new neutrons, is smaller than v by a factor that is measured by or is proportional to the probability that these neutrons are absorbed in a fission process. If we start out with one neutron which produces one fission and hence v fast neutrons, the number of neutrons available for fission in the second generation will be Pv where P is the probability that a fast neutron is ultimately absorbed in the fission process. The product Pv is called the reproduction factor. It is clear that this product must be larger than one for a self-sustaining chain reaction. The problem then in building a chain-reacting atomic pile lies in making the factor P, "the probability that a fast neutron is ultimately absorbed by the fission process," as large as possible. This means reducing all the factors, such as resonance absorption, absorption by carbon, and so on, as much as possible and increasing the fission absorption as much as possible. This presents a difficulty since, in general, one cannot increase the favorable factors without at the same time increasing the unfavorable ones.

To slow down the fast neutrons as much as possible and thereby make them fission-producing neutrons, one should use large quantities of carbon. But this increases the chance that neutrons will be lost by carbon capture. On the other hand, if one uses large quantities of uranium to increase the probability of fission, one also increases the loss of neutrons by uranium resonance capture before the neutrons are slowed down. This difficulty can be partly overcome by concentrating the uranium in lumps and distributing these lumps in the carbon matrix like stones in earth.

To see how this change reduces losses by resonance absorption before the neutron is slowed down as compared to distributing uranium and carbon atoms homogeneously, consider a fast neutron that has an energy close to the energy for resonance capture. This is defined as capture that occurs only if the neutron is moving at a very definite speed. When the carbon and uranium are uniformly distributed the neutron will not have far to go before meeting a uranium nucleus and being absorbed. On the average, every second nucleus it meets will be a uranium nucleus. But if the uranium is distributed in lumps, the neutron will meet many carbon atoms before coming to a uranium lump; by that time it will have lost much of its energy and will not be near resonance. If, however, some neutrons with resonance energy do strike a lump of uranium, they will be absorbed by the uranium atoms on the surface of the lump and the interior nuclei will scarcely be affected.

By clever geometry and a proper distribution of the uranium atoms relative to the carbon atoms, Fermi and his coworkers achieved a chain reaction. To illustrate the reproduction factor under reasonable conditions Fermi devised a chart showing what happens in the second generation if two fast neutrons are produced per fission. He assumed that 3 percent of these fast neutrons are immediately recaptured in a fast fission process, 10 percent are resonance-absorbed, 10 percent are absorbed by carbon, and 77 percent end up as thermal neutrons, some of which give rise to fission and some of which are resonance absorbed. By adding up all the neutrons emitted in these processes, Fermi obtained the reproduction factor and showed that for this case a chain reaction is possible only if at least 1.22 of the original two neutrons become thermal neutrons and give rise to fission.

The construction of the first pile or nuclear reactor was effected under Fermi's direction by a group of able scientists working under the stress of wartime conditions to develop a means of producing plutonium and the superbomb that could be realized with its use. But it was also clear that the controlled release of fission energy would provide a new power source that in the years to come would compete with fossil fuels and possibly outdistance them as a prime source of power when the world's oil and gas deposits became depleted. About half a century has gone by since the initial operation of Fermi's reactor, but already power reactors are generating electricity in countries all around the world. Areas in which it was impracticable to operate conventional generating stations now have the advantages and convenience of electricity. For example, experimental power reactors have been used on the Antarctic continent and in northern Greenland. But this great advance also carries with it problems that will assume larger and larger proportions as time goes on. For purposes of this discussion and to avoid complexity, we shall limit ourselves only to the problems connected with power generation and with the uranium-graphite reactor.

The fission process itself produces a number of radioactive nuclei. This energy, resident in the fission products, amounts to about 5 percent of the total energy released. Although the initial decay of fission radioactivity is rapid, significant long-lived activities remain. The half-life of the latter is about 30 years. The fission products remain in the cylindrical slugs of uranium metal spaced in a lattice structure in the moderator matrix of the reactor. As the products accumulate they reduce the possibility of fission and the efficiency of the reactor decreases. Eventually, the slugs must be replaced as more and more of the uranium is burned up and the proportion of ash increases. But the metal removed from the reactor is highly radioactive. Until the short-lived activity has decayed sufficiently the slugs are stored under water, which serves as a radiation shield. They are then subjected to chemical treatment to reclaim the plutonium produced and the residual uranium. The liquid waste containing the radioactive ash must then be disposed of.

With few fission plants in operation, this problem, although difficult, is not large. But it has been estimated that if the world's power were generated with fission reactors, some 10^9 curies per year of activity would require disposal. This is such an enormous activity that if it were dispersed uniformly in the Atlantic Ocean it would significantly raise the activity of the water. In any event, to dump radioactive wastes in the ocean could cause damaging world-wide biological changes that could persist for years. Alternatively, the dumping of liquid wastes into the earth is accompanied by insupportable seepage hazards. The safest procedure appears to consist of fixing the waste in certain clay materials, in concrete blocks, or incorporating it in glass which is highly resistant to solution or leaching. Disposal of these solids in abandoned dry mines or in the desert appears feasible.

The disposal of radioactive waste products from fission reactors poses an increasingly urgent, world-wide biological and health problem that should be regulated by international agreement. To produce nuclear power without generating such a hazard is a highly desirable goal. Is there any possibility that it may be achieved? Fortunately, the possibility exists through the use of nuclear fusion instead of fission. Fission involves the splitting of heavy atoms located at the end of the periodic table; fusion involves the building of light atoms from hydrogen at the beginning of the periodic table. Interestingly enough, the production of huge amounts of power by nuclear fusion is a natural process and goes on throughout the universe. It is the process by which our sun generates heat and by which all the stars radiate. Nuclear fusion as a source of stellar radiation was first investigated in a theoretical paper by Atkinson and Houtermans in 1929. Very little was added to their work until about ten years later when Bethe discovered the specific nuclear reactions that take place to produce this energy. Unfortunately, to produce fusion artificially is a task of much greater magnitude than to produce fission.

In order to make two atoms stick together after a collision and produce a new nucleus it is necessary for their nuclei to approach each other closely enough so that the short-range nuclear forces can act to make them coalesce. But such a close approach requires relatively enormous energy in order to overcome the long-range Coulomb force of repulsion arising from the like charges of the nuclei. It appears that the simplest way to endow atoms with the requisite high speeds for fusion, consistent with allowing them repeated opportunities for collision, is to heat them as a gas. The thermal motion of the gas molecules thus provides the continual collisions. But gas temperatures in the millions of degrees are required for fusion and no container exists that will withstand such temperatures. There is, however, the possibility of creating an electrical discharge with the desired atoms—hydrogen, in this instance—and containing the moving ions by means of an impressed magnetic field. The ions therefore are essentially in a magnetic bottle. Many years of research have been expended on this method and considerable progress has been made, but a successful fusion reactor still appears far away.

References

Eddington, Arthur S., *The Internal Constitution of the Stars*. New York: Dover, 1930.

16 *High-Energy Physics*

This is the very womb and bed of enormity.

—Ben Jonson, *The Alchemist*

Parity and Its Ill Fortune: C. N. Yang (1922–) and T. D. Lee (1926–)

Among the many features of nature that suggest a universe governed by a well-ordered system of laws, the ubiquity of symmetries has been perhaps the most cogent. The structure of the cell, the morphology of higher organisms, the geometrical patterns of crystals, all lead to the conviction that the laws of nature must be intimately related to symmetry. Until the beginning of this century, symmetries observed in nature were used empirically to discover new and unusual properties of matter without looking for any deeper significance. The symmetry that Mendeleev discovered in the properties of the chemical elements led him to classify the elements into families, although no attempt was made to relate this symmetry to any fundamental law of nature.

In 1905 Einstein took the first important step in relating natural laws to symmetry in his formulation of the special theory of relativity. He based his derivation of the Lorentz transformation equations on the equivalence of space-time for all observers in inertial coordinate systems. Einstein's work was fundamental to the understanding of any relationship between symmetry and the laws of nature. Time and space, according to Einstein, must be treated together. Although there are certain purely symmetrical relationships that are valid for all observers, the most profound conclusions are to be derived from symmetry properties only if space and time are merged into a single entity. His work emphasized the need for a discussion of symmetry in terms of coordinate systems, with the expectation that the symmetry properties most significant for the fundamental laws of nature would be those revealed in moving systems. Finally, and of greatest importance, is Einstein's introduction of the concept of invariance and the delineation of its relationship to space-time symmetries.

The principle of invariance is related to symmetry in Einstein's treatment of the space-time interval, the basis for the special theory of relativity. Furthermore, symmetry, or invariance, is related to conservation principles. Before Einstein's introduction of the space-time concept, the spatial separation of events and the time intervals between them were treated separately in the belief that they were the same for all observers. Einstein insisted that one must treat the space-time interval as the invariant quantity when one transformed from one Galilean coordinate system (inertial frame of reference) to another. In other words, since space-time is symmetrical with respect to all observers, the space-time interval must be invariant. Thus, symmetry and invariance are equivalent. But the invariance of the space-time interval signifies that a certain quantity associated with a system (for example, a moving particle) must be the same for all observers, and hence conserved. Thus, symmetry and invariance are related to conservation principles. The quantity so conserved for a moving particle and corresponding to the invariance of the space-time interval is the square of the momentum of the particle minus the square of its energy divided by the square of the speed of light. This is essentially the rest mass of the particle. The invariance of the space-time interval, or the proper time of the moving particle, is equivalent to the conservation of rest mass. From this equivalence we may conclude that a particular type of symmetry in nature imples the existence of a conservation law. We shall see that this assumption has played a major role in the development of the physics of the atom, the nucleus, and elementary particles.

Einstein developed the general theory of relativity by using still another example of symmetry, namely, that the space-time interval must be the same for all observers, whether they are in inertial frames of reference or in accelerated coordinate systems. This is the substance of the principle of equivalence. It states that all the effects that can be observed within a small region of a gravitational field can be duplicated by an appropriately accelerated system. In other words, inertial forces cannot be distinguished from gravitational forces in small regions of space.

Although the remarkable relationship between symmetry (invariance) and conservation principles first became apparent in the theory of relativity, the most fruitful application of this relationship is found in quantum mechanics. Because quantum mechanics is essentially an algebra of mathematical operators, many of the properties of atoms, nuclei, and fundamental particles can be derived merely from a mathematical analysis of the symmetry principles of the quantum-mechanical operators that correspond to the physical observables of the motion or state of these objects. Thus, the symmetry of the operator that corresponds to the energy (the Hamiltonian) of the system is of particular importance. Because of these symmetries, its invariance to particular types of transformations must embody conservation principles. If the Hamiltonian is symmetrical with respect to the spatial coordinates (that is, if a translation of the origin of the coordinate system used to describe the motion of a particle does not alter the Hamiltonian), the momentum of the particle is conserved. In the same way, if the Hamiltonian is unaltered when the time is changed, the energy of the system is conserved. Finally, if the Hamiltonian is spherically symmetrical, so that a rotation of the coordinate system leaves it unaltered, the angular momentum of the system is conserved. Each of these conservation principles leads to a particular quantum number. Hence, the existence of quantum numbers is, in a sense, a consequence of the symmetries in nature.

We must now consider another aspect of the symmetry properties of a system in the quantum-mechanical scheme. The description of the system in quantum mechanics is given by its wave function, or, more generally, its state vector. The state of the system is described mathematically by a function of space and time (which may be treated as a vector in an abstract function space) that changes from moment to moment as the system interacts with external force fields. Each of the observable quantities associated with the system, such as its energy, momentum, or angular momentum, is treated as a mathematical operator influencing the state vector of the system. These operators determine how this state vector changes with passing time. Thus, the Hamiltonian energy operator, applied to the state vector, represents the change of the state vector with time; a rotation operator, applied to the state vector, represents a change in the orientation of the system, and so on. Therefore, the symmetry properties of the system may be investigated by considering the symmetry properties of the wave function or state vector of the system. The Hamiltonian of the system may have the kind of symmetry that implies that a certain physical quantity is invariant when the system is subjected to the physical operation corresponding to the mathematical operator. Then the wave function describing the state of the system must at the same time change to another state function in a definite way. This change is referred to as a transformation (or a rotation) of the state vector in the abstract mathematical function space in which the state vector is represented. (The mathematical function space is also called Hilbert space in honor of the great German mathematician David Hilbert who discovered the principle laws of abstract function spaces many years before quantum mechanics was discovered.) In other words, the symmetries of the system under consideration must be evident in the mathematical symmetries of the state vector.

We now consider an important type of symmetry not discussed in the previous paragraphs. The analysis of this symmetry as it relates to the decay of certain types of mesons (the theta- and tau-mesons) led to the important discovery that right- and left-handedness in nature are not indistinguishable. Not all phenomena in nature can be described with equal correctness as being either right-handed or left-handed. In other words, right- and left-handedness are intrinsic properties of certain systems. Nature does not mix up the right- and left-handedness in the construction of such systems.

A physicist working in the early 1950s or before would have told us that there was no experiment involving any of the known laws of nature that could provide an operational definition of left and right. An attempt to communicate the meaning of left and right to intelligent beings on a distant planet would have required some particular asymmetric structure—possibly a constellation of stars—that both we and they could observe in common. Nature, at the level at which forces operate, seemed to have no preferred directions or orientations. Left, right, north, south, or any other direction—all seemed to be conventions that had meaning only in the macroscopic world where people could observe temporary asymmetries. Nature itself was thought to be always and everywhere governed by symmetry. We shall see that this can now be demonstrated experimentally for certain systems.

Before discussing the discovery of the handedness of certain systems in nature, we consider the symmetry of some material system, such as an atom or a nucleus, when the coordinate system which describes its motion is changed from what is called a right-handed coordinate system to a left-handed one. A right-handed coordinate system is one with its three mutually perpendicular spatial axes (X, Y, Z) pointing in such a way that if the forefinger of the right hand points in the direction of the X axis, and the second finger of the right hand points in the direction of the Y axis, then the thumb of the right hand, held at right angles to the two fingers, points in the direction of the Z axis. This means that if a right-handed coordinate system is rotated so that the X axis moves toward the Y axis, then the Z axis must point in the direction that a right-handed screw would advance if it were rotated in the direction from X to Y (in a clockwise direction). A left-handed coordinate system is one in which X, Y, and Z point respectively in the directions of the forefinger, second finger, and thumb of the left hand if these are held in mutually perpendicular directions to one another.

What then is the nature of the operation that transforms a right-handed coordinate system to a left-handed one? We know what this means physically because we obtain such a transformation with a mirror. If we hold the forefinger, the second finger, and thumb of the right hand so that one of the fingers points directly toward a mirror, the image is exactly the same as the arrangement of the three fingers on the left hand. In other words, the passage or transformation from a left-handed to a right-handed system, and vice versa, is equivalent to the reflection of the coordinate system in a plane mirror. This type of transformation is called a spatial reflection or inversion. Now suppose that the Hamiltonian describing the state of a system remains unaltered when we pass from a right-handed to a left-handed coordinate system (in other words, the Hamiltonian is invariant to a spatial reflection of the coordinate system). What does this invariance mean physically? Since the Hamiltonian represents the evolution of the state of the system in time, this invariance means that if we watch the system in a mirror, the events as they are revealed are as valid a description of nature as the events that are actually happening in real space. We may express this important statement somewhat differently. If the Hamiltonian that describes the dynamic properties of a system is invariant to a reflection of the coordinate system, then the mirror image of any dynamic process that the system may undergo is also a possible and equally probable physical process that the system can experience. It is thus immaterial whether we look at the world directly or through a mirror. There is no way of distinguishing between left-handed and right-handed events since they are both equally probable; if suddenly right-handedness and left-handedness were interchanged, no observable difference would occur in the world. Of course, this state of affairs depends on the assumption that the Hamiltonian describing any part of the universe or the whole of it is symmetrical with respect to reflections.

For a long time this invariance was thought to be the case for all physical events. To see what this means from the quantum-mechanical point of view we must understand what a reflection of coordinates does to the wave function or state vector of a system. In addition, we must define what a reflection of the spatial coordinate system means mathematically. It is easy to see that we obtain such a transformation (a reflection or mirror image of a coordinate system) if we replace X, Y, and Z (the coordinates of all points) by $-X$, $-Y$, and $-Z$. In other words, if we replace all the coordinates of a system by their negative values, we obtain the description of the system in the mirror-image coordinate system.

From this description we learn at once what happens to the wave function, or state vector, of a system when we change the description from a right-handed to a left-handed coordinate system. We may treat this change as though it were a transformation of the state vector by the application of an operator (the reflection operator). If we apply the operator twice (that is, view the system through two mirrors), we obtain the original state vector since the second reflection cancels out the first.

We view this double operation quantum mechanically as follows. If R is the operator that describes the first reflection, then the two successive reflections are represented by R^2. The quantity R^2 must equal 1 since the effect of R^2 gives us the original state vector again. Hence, the numerical representation of the operator R itself must be either $+1$ or -1 since the square of either of these numbers is 1. In the reflected coordinate system, therefore, the wave function must either remain unchanged or change its sign. Events in nature may, according to this point of view, be divided into two groups: those for which the state vector remains the same when the handedness of the coordinate system is changed, and those for which the state vector changes its sign. We may refer to these as odd and even events, respectively, and we shall see that there appears to be a rule in nature about such events. Odd events were thought always to remain odd and even events always to remain even. This phenomenon, or law, called the conservation of parity, corresponds to the conservation of a physical quantity, the parity of a system. Parity is a property of the wave function or state vector. One speaks of odd parity if the wave function changes sign on reflection of the coordinate system and of even parity if the wave function does not change sign.

It can be shown that the parity of the wave function of a particle is determined by its orbital angular momentum. If the orbital angular-momentum quantum number (the azimuthal quantum number) of the particle is 1, the wave function of the particle is multiplied by $(-1)^1$ as the result of a reflection of the coordinate system (inversion). The parity of the particle is thus even or odd depending upon whether the orbital angular-momentum quantum number of the particle is even or odd. The importance of parity and its conservation in ordinary atomic processes was first pointed out by O. Laporte in 1924. He showed that all the energy levels in the atom can be divided into odd and even ones. When a photon is emitted by the transition of an electron in the atom, the electron must always go from an odd to an even level, or from an even to an odd level. In other words, during a transition the orbital angular momentum quantum number must change by plus or minus 1. This constitutes what is known as a selection rule since it limits the type of transition that can occur inside an atom and therefore reduces the total number of spectral lines that the atom can emit.

This selection rule, or the change in the sign of the wave function of the electron during a transition, is related to the principle of the conservation of parity. After the transition, if the emission of a photon is involved, there are two particles to be considered, the electron and the emitted photons (for a transition involving absorption we start with two particles and finish with one). A knowledge of the total parity of the system after emission must take into account both the parity of the electron and the parity of the photon. Every particle in quantum mechanics is assigned a parity. Since, as can be shown, the parity of the photon is always odd, the parity of the final state is the same as the parity of the initial state only if the parity of the electron changes from odd to even or vice versa during a transition. Note that the parity of any state is obtained by multiplying the individual parities of the particles. The product of an odd and even parity is odd and the product of two odd or two even parity particles is even.

Laporte's discovery of the conservation of parity for atomic systems led physicists to accept this principle as a general law of nature and to discard any Hamiltonian that did not have the kind of symmetry that would leave it invariant to a reflection of the coordinate system. No one had any reason to object to this invariance in the early years of quantum mechanics. It seemed eminently reasonable to insist that there be no distinction between the universe and its mirror image and to insist that the mathematical formulation of the laws of physics take this equivalence of right- and left-handedness into account. These assumptions seemed valid until the discovery of the famous theta-tau-meson paradox, which was first recognized in 1953. It was found that two K-mesons, which seemed to have identical masses and lifetimes, decayed in different schemes, with different parities. The theta-meson decayed into two pi-mesons, the tau-meson into three pi-mesons. Since the pi-meson had odd parity, this meant that the parent theta- and tau-mesons differed in parity (if parity was conserved in this decay process as it was in electromagnetic processes). The conservation of parity, applied to this decay process, thus led to the assumption that these particles were two different K-mesons, even though they were alike in all other respects.

T. D. Lee and C. N. Yang, on examining this situation, discovered that the principle of conservation of parity had never been experimentally verified for certain types of phenomena known as weak interactions. In the physics of elementary particles, one divides phenomena into two types (excluding electromagnetic and gravitational phenomena): those involving strong interactions between particles, and those involving weak interactions. Weak interactions involve the decay of mesons and, in general, any reactions in which neutrinos or antineutrinos are emitted. Lee and Yang therefore proposed that the conservation of parity be discarded in any reactions involving neutrinos and in weak interactions in general. The decay of the K-mesons is itself a weak interaction so that, according to Lee and Yang, this decay need not obey the principle of the conservation of parity. Hence the tau- and the theta-meson are the same kind of meson, which can decay either by conserving parity or not.

Since this suggestion was extremely revolutionary, it was necessary to test it in a way that nobody could question. Lee and Yang suggested that the simplest and most conclusive test could be carried out by examining some particular beta-decay process and its mirror image, since such processes involve neutrinos and should therefore not necessarily conserve parity. If parity is conserved, the actual process and its mirror image should occur with equal probability. But if parity is

Chen Ning Yang (1922–)

not conserved, the mirror-image process and the actual process should occur at different rates.

Such an experiment was carried out by C. S. Wu and her collaborators. They studied the electrons emitted in the beta decay of cobalt 60. If all the cobalt nuclei are lined up in a magnetic field so that their intrinsic spins are parallel, as many electrons should be emitted in a direction parallel to the cobalt spin as in the direction opposite to the cobalt spin—if parity is indeed conserved. This symmetry would be observed if the emission of electrons from the cobalt nuclei were viewed in a mirror, for the conservation of parity requires that the actual emission of electrons and the image of this process be the same. But if the emissions in the cobalt spin direction were different from those in the opposite direction, the mirror image of this process would show just the opposite situation so that parity would not be conserved. This was exactly what the experiments of Wu and others showed.

These results can be understood if, as suggested by Lee and Yang, the neutrino is pictured as a spiral structure with a left-handed helicity. According to this view, the direction of motion of the neutrino and the direction of its intrinsic spin are opposite to each other. The neutrino spins and advances like a left-handed screw. It follows that when the cobalt atoms undergo beta decay, they emit their neutrinos in the direction opposite to their lined-up spins, not in the direction of these spins. This occurs because the neutrino leaves with its spin parallel to that of the cobalt nucleus; hence, since it has the helicity of a left-handed screw, it must move off in the opposite direction. Since the theory of beta decay shows that there is a greater probability for electrons to be emitted in the same direction as the neutrinos in beta decay, more electrons are emitted with direction opposite to the cobalt spin than in the same direction. But this result disagrees with classical assumptions about mirror-image events; hence, it follows that parity is not conserved in this beta-decay process. Today we know that parity is not conserved in any interaction involving the emission of neutrinos or antineutrinos, the particles that are emitted, together with positrons, in beta-decay processes.

Tsung Dao Lee (1926–)

Although the principle of the conservation of parity does not apply to weak interactions, a more general conservation principle does apply if we consider the general asymmetry of charged particles in the universe. Our universe consists of massive positively charged particles (protons) and very light negatively charged particles (electrons). Thus, there is a lack of symmetry between charge and mass. However, the discovery of the positron (antielectron) and the antiproton shows us that a universe could just as well be constructed with massive negatively charged particles (antiprotons) and light positrons. If then we had a cobalt atom consisting of antiprotons, it would emit positrons which would leave the anti-cobalt nuclei in a preferential direction opposite to that of the electrons from ordinary cobalt. In other words, the mirror image of the ordinary cobalt beta-decay process is not really impossible; it exists in the world of antimatter. This means that if we transform from a right- to a left-handed coordinate system and, at the same time, replace all the particles by their antiparticle numbers (called charge conjugation), the Hamiltonian describing the system remains invariant. Thus, charge conjugation and reflection together form a higher type of symmetry in the universe.

The chirality of the neutrino requires that this particle (and the antineutrino) move with the speed of light and have no rest mass. In this respect it behaves like a photon. If a neutrino moved with less than the speed of light, it could have no definite chirality. It would always be possible to find a coordinate system moving faster than the neutrino in which its chirality would appear opposite to the chirality seen by an observer in a coordinate system moving with a speed less than that of the neutrino. In both cases, if the direction of motion of the neutrino is taken parallel to that of the two observers, the direction of spin of the neutrino is the same, while the directions of motion as seen by the two observers are different. To one observer the neutrino would look like a right-handed screw; to the other, like a left-handed one. Hence, if the neutrino is to have the same handedness for all observers, it must move with the speed of light so that no observer can overtake it and thus see it recede. Since the neutrino moves with the speed of light, it can have no rest mass, for its moving mass, and its energy, would be infinite by the Einstein relationship between the mass and the speed of a body.

Chen Ning (Franklin) Yang was born in Hofei, Anwhei, China, on September 22, 1922, and spent his early years on the campus of Tsinghua University, near Peiping, where his father, Ke Chuan Yang, was a professor of mathematics. He was the first of five children in a household devoted to cultural and intellectual pursuits.

On completing his secondary school education, Yang matriculated at the National Southwest Associated University in Kunming, China, and received his B.Sc. degree in 1942. By that time, the Tsinghua University had moved to Kunming because of the Sino-Japanese War; Yang received his M.Sc. degree there in 1944. At the end of the war in 1946 he went to the University of Chicago on a Tsinghua University Fellowship to do graduate work in nuclear physics under the direction of Edward Teller. He received his Ph.D. from the university in 1948 and remained there as an instructor for one year. In 1949 he went to the Institute for Advanced Study at Princeton, where he began the work on elementary particles and weak interactions that led to the discovery of the nonconservation of parity. In 1955 Yang became a professor at the Institute and held that post until 1965, when he accepted an endowed chair of physics at the Stoney Brook, Long Island, division of the State University of New York.

Like Lee, with whom he collaborated in the parity work, Yang devoted a great deal of his research efforts to the fields of statistical mechanics and symmetry principles. Although he was greatly influenced by Enrico Fermi while at the University of Chicago, his interest in symmetry principles went back to his undergraduate days when he wrote his baccalaureate thesis, "Group Theory and Molecular Spectra." His M.Sc. thesis, "Contributions to the Statistical Theory of Order-Disorder Transformations," stemmed from his interest in statistical mechanics. This early work, including his Ph.D. thesis, led to many subsidiary problems in which the principle of the conservation of parity had to be dealt with and analyzed. Quite naturally, Yang became involved in parity work. In one of his first papers in this field, he proved in a very ingenious manner that the parity of the pi-meson is odd (equal to -1) and that its intrinsic angular momentum (spin) is zero. This knowledge was very important in analyzing the so-called theta-tau puzzle, namely, that these two K-mesons, similar in all other observable features, decay by different modes, one breaking down into two pi-mesons and the other into three pi-mesons.

Yang wrote extensively, not only for the professional physicist but also for the layman and for scientists in other fields who were interested in a simplified review of modern developments in physics. As demonstrated by his 1957 Nobel Prize address, his writing and lecturing were marked by simplicity and charm. His ideas were so presented and developed that the physics rather than the formalism was emphasized and always easy to follow. In spite of his great achievements, he was modest, self-effacing, and easily approached, and in his discussions with others he was affable, patient, and understanding.

Yang was a fellow of the American Physical Society and of the Academia Sinica. He was also a member of the National Academy of Sciences. In 1957 he received the Albert Einstein Commemorative Award and in 1958 an honorary D.Sc. degree from Princeton.

Tsung Dao Lee, the son of Tsing Kong Lee, a Shanghai merchant, and Ming Chang Chang, was born on November 24, 1926. After graduating from the Kiangsi Middle School in Kanchow in 1943, he matriculated at the National Checkiang University in Kweichow Province. He was unable to complete his college education there because of the Japanese invasion and was forced to flee to Kunming, Yunnan, where he attended the National Southwest University. There he met Chen Ning Yang, who was to become his collaborator in the discovery of the nonconservation of parity and was to share the Nobel Prize with him in 1957.

The Chinese government recognized Lee's great talents in physics and awarded him a scholarship in 1946 that brought him to the University of Chicago, where he studied with Fermi. Being then greatly interested in the properties of very dense (degenerate) matter and in the structure of very dense stars (the white dwarfs), he wrote his Ph.D. disseration on the hydrogen content of white dwarf stars in 1950.

After spending some months as a research associate at the Yerkes Observatory at the University of Chicago, Lee went to the University of California at Berkeley as a lecturer and research associate. At the end of 1951, he went to the Institute for Advanced Study at Princeton as a fellow and remained there as a staff member until 1953. There he renewed his friendship with Yang and began working with him on various basic problems in elementary particle physics.

During his residence at the Institute, Lee published papers on various basic problems in statistical mechanics and in nuclear and elementary particle physics.

This soon led to his recognition as one of the most talented of the younger theoretical physicists. His approach to problems was always a nontraditional one, marked by great ingenuity and deep physical insight.

In 1953, Lee left Princeton to become assistant professor of physics at Columbia University. In three years he was promoted to a full professorship, thus becoming at the age of 29 the youngest full professor on the Columbia University faculty. This three-year period was marked by great scientific activity; Lee generally spent his weekends commuting to Princeton to work with Yang. This collaboration finally led to a joint paper, published in the *Physical Review* in 1956, which questioned the universal validity of the principle of the conservation of parity and proposed an experiment to test it. The confirmation of their analysis brought them the Nobel Prize the following year. Lee was the second youngest scientist ever to have received the Nobel Prize, the youngest having been Sir Lawrence Bragg, who received it in 1915 at the age of 25.

After receiving the Nobel Prize, Lee continued working on fundamental particle theory and field theories in collaboration with Yang and others. He remained at his academic post at Columbia for a number of years, lecturing on various phases of theoretical physics, and then went to the Institute for Advanced Study as a professor. After residing at Princeton for a few years, he returned to Columbia to become its first Enrico Fermi professor of physics.

Lee received many honors, among which were the Albert Einstein Commemorative Award of Yeshiva University in 1957 and the Science Award of the Newspaper Guild of New York. He was a member of the National Academy of Sciences, a fellow of the American Physical Society and the Academia Sinica. In 1958 he was awarded an honorary D.Sc. degree by Princeton University.

Nuclei and Nucleons: Robert Hofstadter (1915–)

On December 10, 1961, Robert Hofstadter was awarded the Nobel Prize in physics, a prize that he shared with Rudolf Mössbauer, for detailed investigations into the structure of matter. Hofstadter, who was professor of physics at Stanford University, conducted his research by extending the technique of previous electron-scattering studies to higher energies. The scattering technique had been familiar to physicists for nearly half a century, ever since the experiments of Ernest Rutherford at Manchester. But never had it yielded such precise knowledge. Hoftstadter discovered that nucleons (protons and neutrons) consisted of charged heavy mesonic clouds. In the proton, these clouds appeared to add to the effect of their charges; in the neutron the clouds seemed to cancel one another.

The proton and neutron, once assured a place as elementary particles in their own right, were thus revealed as complex bodies of constituent mesons. "What will happen from [this] point on?" asked Hofstadter at the close of his Nobel address. "One can only guess," he answered, ". . . but my personal conviction is that the search for ever-smaller and ever-more fundamental particles will go on as long as man retains the curiosity he has always demonstrated."

Traditionally, physicists have used two methods to investigate the structure of atoms and nuclei: the scattering and absorption by these systems of bombarding particles and the emission of radiation or material particles from such systems. Most of our knowledge about the electronic structure and the energy levels of the atom has come from an analysis of the photons (spectral lines) emitted by

Robert Hofstadter (1915–)

excited atoms. The electronic structure of the atom lends itself quite naturally to this type of analysis because the electrons in an atom can be excited to higher levels quite easily in ordinary flames, discharge tubes, and in sparks. The scattering of low-energy electrons, as in the Franck-Hertz experiments, can also be used for this purpose since the electronic configuration in an atom is quite distended rather than being highly concentrated. Therefore, the electronic atomic dimensions are of the same order of magnitude as the de Broglie wavelengths of low-speed electrons.

But our knowledge of the structure of the nucleus of the atom has come almost entirely from the scattering experiments since the nucleus is a small, very tightly bound structure which can be excited only with great difficulty. In other words, we have hardly any nuclear spectroscopy—certainly nothing comparable to atomic spectroscopy—so that we are forced to investigate the nucleus by means of scattering experiments. The earliest such experiments were performed by Rutherford with alpha particles. These experiments led to the discovery of the nucleus itself and to the first artificial transmutation of nuclei. However, using alpha particles to obtain a detailed picture of the structure of the nucleus itself, that is, of the distribution of the electric charge and magnetic moments, is not feasible. This is because of the large Coulomb repulsion suffered by an alpha particle as it approaches the nucleus. In addition, only a small range of alpha particle energies is available.

With the development of high-energy particle accelerators, high-speed protons became available. The scattering of these particles from light nuclei revealed many features of nuclear forces and nuclear structure. With the discovery of the neutron a new era began in nuclear theory. In the hands of physicists like Fermi the neutron became a powerful analytical tool.

Hofstadter's investigation of nuclei and nucleons (protons and neutrons) by particle scattering differed from previous methods because he used light particles (electrons) rather than massive particles. It may appear at first that the electron is the natural particle for probing nuclei since it is negatively charged and therefore is not electrically repelled by nuclei. This would, indeed, be the case if the electron were a classical particle. But the electron has wave properties which interfere with its ability to get close to a nucleus under ordinary conditions. The de Broglie wavelength of a particle varies inversely as the product of its mass and speed. Because of the small mass of the electron, its de Broglie wavelength at ordinary speeds is much larger than the diameter of a nucleus. In a sense, then, such an electron does not see the fine details of the nucleus any more than we can see the fine details of a molecule with radio waves. Because of its large wavelength, such an electron suffers a centrifugal repulsion that prevents it from coming close to the nucleus.

It was clear to all physicists that electrons could be used as nuclear probes if their speed was increased so that their de Broglie wavelengths would be reduced below the diameters of nuclei, that is, a few ten-trillionths of a centimeter. This requirement meant imparting energies of the order of one BeV to electrons. Although the technical difficulties in obtaining this result appeared forbidding, Hofstadter undertook this work in 1950 when he went to Stanford University, where a linear accelerator capable of supplying the required electron energies had been constructed.

To understand the nature of the experiment note that when an electron penetrates a nucleus it suffers electric and magnetic forces which deflect it from its original path. This deflection depends on the initial speed of the electron and on the structure of the nucleus. If one now separates the electrons after they have been scattered by a nucleus into groups according to velocity, one can use the quantum-mechanical laws of scattering to analyze the structure of the nucleus. In a sense this process is like using a very high resolving power electron microscope. The actual charge density in the nucleus is determined in terms of what is called a form factor, which depends on the size of the nucleus and the charge distribution inside it. A simple analogy with the motion of a comet in the gravitational field of the sun will illustrate the principles involved.

When a comet comes from a great distance into the neighborhood of the sun the sun's gravitational pull causes it to move in a very elongated orbit. An analysis of this orbit gives us the mass of the sun. We do not obtain the actual distribution of mass within the sun because it behaves like a gravitational point as far as the comet is concerned, and a point has no structure. But suppose now that the comet came very close to the sun or actually penetrated it. The details of the mass distribution would then play an important role in the motion of the comet and its orbit would give some idea of this distribution. If the comet passed right through the sun, but at a definite distance from the sun's center, the scattering action would be smaller than if the sun were concentrated in a single point. In the former case, only part of the sun's mass affects the comet's motion, whereas in the latter case the entire mass of the sun plays a role in the process.

A similar situation arises in the scattering of electrons by nuclei. If a nucleus were just a point charge, one could calculate its scattering effect on the electrons by a straightforward application of the Dirac theory and quantum electrodynamics. When combined with the Born approximation, this calculation leads to a fairly simple expression for the scattering cross section. This is the target area around a nucleus that the electron must hit to be deflected through a certain angle from its original direction of motion. But the nucleus is not a point charge and its finite size must be taken into account in analyzing the electron scattering. This can be done simply by multiplying the scattering cross section for a point charge by a form factor or structure factor. This factor depends only on the amount of momentum transferred by the scattered electron to the nucleus. Since the form factor is always smaller than one, the effect of the finite size of the nucleus is to reduce the scattering cross section below its value for a point charge.

The analysis of the charge distribution inside a nucleus by this scattering technique proceeds as follows. Using a fairly wide range of electron energies, one compares the scattering cross sections with the cross section for a point charge. The difference between this quantity and the observed cross section gives us the form factor for some particular value of the momentum transfer. If one knows the values of the form factor for a large enough range of momentum transfers, one can use a simple mathematical device to calculate the charge distribution inside the nucleus.

Hofstadter first began working with nuclei and then went on to the analysis of the structure of the proton and the neutron. His research with nuclei showed quite clearly that every nucleus has a core of electric charge surrounded by a thin skin in which the charge falls off to zero quite quickly. In this nuclear core the charge density is constant. The size of this core is different for different nuclei but is equal to about 1 ten-trillionth of a centimeter multiplied by the cube root of the number of neutrons and protons in the nucleus. For a nucleus with an atomic weight 64 the core has a radius of about 4 ten-trillionths of a centimeter. The thickness of the nuclear skin is the same for all nuclei and is equal to about $2^1/2$ ten-trillionths of a centimeter.

After this basic work in the analysis of nuclear structure Hofstadter applied these electron-scattering techniques to the study of the structure of nucleons. Until his research, there was no evidence that the proton or the neutron had a structure or a finite size, although it was already known that the proton did not obey the Dirac equation because its magnetic moment was not equal to that predicted by the Dirac theory. Hofstadter's experiments showed conclusively that the proton is not a point but has a fairly large size. Its radius is about 0.7 ten-trillionths of a centimeter.

In discussing the size and structure of the proton, we must consider two different form factors. The proton has a magnetic moment as well as a charge distribution and both play a role in the scattering of electrons. When an electron approaches a proton, the two particles interact electrostatically because of their charges, but they also interact magnetically because they both behave like small magnets. The scattering formula must therefore take into account the interaction between the magnetic moments of the proton and electron. This consideration introduces another form factor which shows that the magnetic moment of the proton, like its electric charge, is distributed over a finite radius and is not concentrated in a point. Hofstadter applied the same electron-scattering techniques to study the structure of the neutron and demonstrated that, like the proton, it is not a point but is distributed over a finite volume. Since the neutron has no charge distribution, its structure must be expressed in terms of its magnetic moment, which is distributed the same way as the magnetic moment of the proton.

On the basis of Hofstadter's work, physicists now believe that both the proton and the neutron consist of cores surrounded by clouds of mesons so arranged as to give the correct magnetic moment in both cases. According to this model of the nucleon, the value of the magnetic moment that is actually measured occurs because the nucleon oscillates between its two possible states. Thus the neutron, for a fraction of its lifetime, behaves as though it were a proton, a neutron and a pi-meson, and the proton oscillates between being a proton, a neutron, and a pi-meson.

Robert Hofstadter was born in New York City on February 5, 1915, to a family with a long tradition of learning and culture. His father was Louis Hofstadter and his mother the former Henrietta Koenigsberg. He received his elementary and secondary school education in the New York City public schools and attended the City College of New York for his higher education. He graduated from CCNY in 1935 with the B.S. degree, magna cum laude. He was awarded the Kenyon Prize for excellence in mathematics and in physics.

Later in 1935, on receiving the Coffin Fellowship from the General Electric Company, he entered Princeton University as as graduate student in physics and was awarded both the M.A. and the Ph.D. degrees in 1938. His doctoral thesis dealt with the infrared spectrum of organic molecules and with the general problem of the hydrogen bond in molecular structure. After receiving his doctorate, he continued doing research at Princeton as a Proctor Fellow. The work he pursued during this period—the photoconductivity in willemite crystals—aroused his interest in the use of crystals as electron detectors and in the general problem of scintillation counters. This was important for his later work because accurate electron detectors are essential in high-energy electron-scattering experiments. In 1939 Hofstadter received the Harrison Fellowship from the University of Pennsylvania, where he went to work on the large Van de Graaff machine being built there. This work stimulated his interest in nuclear physics, but the outbreak of the war interrupted his career in pure research, and he went on to industrial scientific work arising from military needs.

In the early years of the war, Hofstadter worked at the Bureau of Standards but then went on to the Norden Corporation where he remained until the end of the war. He returned to Princeton as assistant professor of physics and continued the work on crystal conduction counters that he had started before the war. He applied his detection techniques to Compton effect experiments and, in the process of improving his detectors, discovered that sodium iodide, activated by thallium, makes an excellent scintillation counter. The counters that he developed in this work, which were made from very well-formed crystals of sodium iodide doped with thallium, proved to be excellent energy measuring devices for gamma rays and energetic charged particles, such as electrons. In addition to being very efficient particle counters, they could also be used as spectrometers to measure the energies of particles. Thus, they provided just the kind of instrument that was needed for Hofstadter's later work.

In 1950 Hofstadter left Princeton to become an associate professor of physics at Stanford University, where a high energy accelerator was soon to be made available for the kind of electron-scattering experiments that he was interested in doing. Since the accelerator was still under construction when he arrived, he devoted most of his time to building the equipment that he would need for the scattering experiments. He also improved his scintillation counters and, in the process, developed new detectors for neutrons and X rays. When the accelerator was completed, Hofstadter dedicated himself completely to the high-energy electron-scattering experiments which finally led to his discovery of the charge and magnetic moment distributions in nucleons. Most of the results for which he won the Nobel Prize in 1961 were obtained in the years from 1954 to 1957. After 1957 he was concerned with improving the accuracy of his results and obtaining more precise form factors—the mathematical quantities that give the distributions of charge in the nucleons.

Hofstadter, who became a full professor at Stanford, was a member of the National Academy of Sciences, a Fellow of the American Physical Society, and a member of the Council of the American Physical Society. In 1959 he was named California Scientist of the Year and in 1958 to 1959 he was a Guggenheim Fellow.

Elementary Particles

Until recently physicists were able to explain the structures of complex systems in what appeared to be increasingly simple terms. The properties of matter were thus reduced to the behavior of molecules, which, in turn, were pictured as consisting of atoms, and so on; indeed, it was this search for simplicity that prompted many early scientists and philosophers to accept the Greek atomic concept of matter. What can be simpler than a universe consisting of aggregations of a single kind of indivisible matter? To solve the riddle of matter it was necessary merely to discover the nature of the forces between these primordial particles. Then the great variety in the properties of bulk matter would be accounted for by the countless number of ways these atoms could combine.

The discovery of two kinds of electric charge and Faraday's electrolytic experiments destroyed this naive atomic picture, but left most scientists with the hope that only two fundamental particles would be needed to account for the chemical and physical properties of matter. The discovery of the electron as the basic unit of negative electric charge and the proton as the basic unit of positive charge seemed to justify this hope. It appeared that one could now construct all kinds of matter by combining equal numbers of protons and electrons, and the electrostatic forces between these particles would keep things in equilibrium. The discovery of the photon complicated the picture somewhat. If the photon were to be treated as a particle in its own right, it would have to be in a category quite different from that of the electron and proton. However, it could play no role in the actual structure of matter since it could never be brought to rest to occupy some particular position in the structural model. The difference between the photon and the electron and proton may be presented most forcefully by noting that the latter both have nonzero rest masses, whereas the photon's rest mass must be zero. Looked at from this point of view, the photon offered no problem since it was not necessary to assign any role to it. One simply assumed that the photon was destroyed when it was absorbed by an electron or proton and came into existence when emitted by them. The photon was thus regarded as a transitory particle existing only from the moment that it was emitted to the moment that it was absorbed. In a sense, we might say that a photon is emitted only to be absorbed. On the other hand, from the point of view of modern quantum electrodynamics, the photon is the building block or fundamental particle of the electromagnetic field. The electromagnetic interaction between charged particles and the electromagnetic field occurs by the emission and absorption of photons.

With the photon assigned to this intermediary role—the carrier of the electromagnetic interaction between charged particles—it appeared that the structure of matter might be completely accounted for by just the two charged particles, the electron and the proton. The protons would contribute most of the mass of the atom and the electrons would supply the necessary negative charge to make the atom as a whole electrically neutral and to hold it together. Although this was a very appealing picture and seemed to supply all the essentials, it ran into serious trouble with the discovery of the Rutherford planetary atom. According to this proposed model of the atom (which is now firmly established) all the protons are concentrated into a tiny nucleus that occupies only 0.001 trillionth of the volume of the entire atom. This model presented an immediate problem because it indicated that, in addition to the electrostatic repulsion between protons, which would, if uncompensated, cause the nucleus to fly apart, there must be some kind of attractive forces inside the nucleus. At first a solution seemed available because protons, like all matter, exert attractive gravitational forces on each other. Therefore, gravitation appeared to provide the necessary forces in the nucleus. Moreover, since the gravitational force increased as the distances between protons got smaller and smaller, it appeared that the force inside the nucleus should be just large enough to keep it together. But simple calculations showed that this idea was untenable. In fact, the gravitational force is many orders of magnitude too small for this task, being the weakest of the known forces in nature. This state of affairs immediately implied that the proton is a much more complex particle than had been assumed initially. Not only do electrostatic and gravitational fields surround it, but other force fields must emanate from it to keep the nucleus together.

The simple picture of matter consisting of only electrons and protons was given another blow when the wave properties of the electron were discovered; their existence implied that electrons have too large a wave structure to permit containment inside the nucleus. Moreover, the predicted spin and statistical characteristics of nuclei would not agree with observation if nuclei contained electrons. On the other hand, in the 1920s it was thought that electrons must be inside nuclei to account for the net charge on the nucleus. Furthermore, the emission of beta rays (electrons) from certain types of radioactive nuclei seemed to require the presence of electrons.

Nothing much could be done about these difficulties as long as electrons and protons were the only fundamental particles known. But suddenly, almost as though physicists had stepped across a threshold, new particles began to appear in bewildering numbers. Whereas some of them were warmly welcomed by the theoretical physicists intent on constructing a model of the atom and the nucleus, most of them were at first accepted the way Job might have accepted his tribulations—things one had to live with but without any rhyme or reason.

The first new particle to appear on the scene after the discovery of the electron, proton, and photon did not show itself directly but was demanded by certain conservation principles. The existence of this particle, the neutrino, was postulated by Wolfgang Pauli as early as 1927 to account for the apparent discrepancy in the energy balance when beta rays were emitted by radioactive nuclei. Later, when the neutron was discovered, its decay into a proton and electron also required the emission of a neutrino (actually, as we shall see, an antineutrino), not only to maintain the energy balance, but also to balance the spins before and after the decay of the neutron.

Since a sensible picture of all the new particles can be constructed only with the aid of conservation principles, we must consider these briefly before discussing the new particles themselves. First of all, we have the conservation of momentum, which states simply that if there are no external forces acting on a system—a collection of interacting particles—the total momentum of the system must remain constant, regardless of how the constituent particles move about or interact with each other. If a particle like a neutron decays into a collection of other particles, the momentum of the particle before its decay must equal the total momentum (the sum of the individual momenta) of all the decay particles. This principle is very useful in helping the particle physicist identify and classify the decay products of any reaction. Closely associated with the conservation of momentum is the conservation of energy. Indeed, the theory of relativity tells us that the energy and momentum of a system are not conserved separately for all observers but a single quantity, called the energy-momentum 4-vector, is conserved. In the calculation of the momentum and energy of a system, the energy and the momentum of each photon must be included, and the energy corresponding to the mass of each particle must also be taken into account. The conservation of energy simply ensures that no process will occur if not enough energy is available for the process. If enough energy is available, a process will occur spontaneously, unless some other conservation principles prohibit it. Since the total energy of the system (the total measured mass times the square of the speed of light plus the energy of each photon) before the spontaneous process occurs is the same as after the process, what is the need for "enough energy"? Why should a process go in one direction rather than in the reverse direction if the total energy must be the same at each step of the process? To answer this question we must consider the masses alone. If the total mass of a system is larger than its total mass after the process, the process will occur spontaneously. In other words, the energy available for a spontaneous process is the difference between the system's total mass (the sum of the masses of all the particles in the system) before and after the process. Since the mass of the neutron is somewhat larger than the mass of a proton plus an electron, the neutron spontaneously decays into a proton and an electron. The total energy after the decay is the same as before, but not all of it appears in the form of mass. Some of the original energy appears as kinetic energy of the newly formed particles: the proton, electron, and neutrino.

In addition to conservation of momentum and energy, we also have conservation of angular momentum or rotational motion. The total rotational motion of an isolated system before and after a process must be equal. Rotational motion, in general, consists of two parts: one arising from the orbital motions of the particles in the system, and the other from the spin of each particle. The total rotational motion or angular momentum is obtained by summing the orbital and spinning motions for all the particles. In discussing the properties and behavior of elementary particles, we do not deal with orbital motions. The only thing that concerns us here is the spin of each particle. The conservation of angular momentum then tells us that, in any process involving the transformation of one group of elementary particles into another, the total spin (the spins of all particles added together) before and after the process must be the same.

Here we must be careful because spin is a vector (a directed quantity) and adding such quantities is unlike the usual process of addition. We overcome this difficulty by always considering the components of the spin in a particular direction and adding them together. The spins of particles occur in integer and half-integer multiples of a basic unit, Planck's constant h divided by 2π, and is written as $\hbar$. In terms of this unit, the spin of both the electron and the proton is $1/2$ and the spin of the photon is 1. Like the electron and the proton, the neutrino also has a spin of $1/2$.

Are there other conservation principles that must be taken into account in our analysis of the fundamental particles? There are a few others of importance, one of which deals with electric charge. The total charge in the universe is constant, at least as far as all evidence indicates; charge can neither be created nor destroyed. Therefore, the total charge of a system must be the same before and after the process occurs. If a new, positively charged particle suddenly appears during a process, a new negatively charged particle must appear at the same time to compensate for the positive charge. Charge occurs in integer multiples (positive and negative) of a basic unit, which is the charge on the proton. As far as is known, only three values of this unit of charge occur on fundamental particles: -1, 0, $+1$. Examples are the electron (-1), the neutrino (0), and the proton $(+1)$. The charge on the photon is also 0.

Another conservation principle deals with the total number of heavy particles or nucleons (protons and neutrons) in the universe. Since no experimental evidence has ever been adduced for the destruction or creation of a heavy particle, we must assume that the total number of heavy particles (protons plus neutrons or any other particles that finally become protons or neutrons) in the universe is conserved. This number must therefore be the same before and after any process. A neutron thus decays into a proton and two light particles (electron and neutrino) so that we start with one heavy particle (neutron) and end with another one (proton). Light particles like electrons, photons, and neutrinos are not conserved, as is clear from the beta-decay process (e.g., the decay of the neutron) and from processes involving the emission and absorption of photons. Note, however, that the total number of leptons (electrons, neutrinos, and muons and their antiparticle counterparts) is conserved. This is because a lepton always appears or disappears with an antilepton, so that the total number, counting antileptons as negative, remains constant.

There are two more conservation principles that have guided physicists and are still important in the classification of particles: the conservation of parity, which we have already discussed in the chapter on the work of Yang and Lee, and the conservation of strangeness, an entirely new kind of quantity that had to be introduced to understand the way certain heavy particles decay. It is proper that these two principles should be grouped together because they are not universally valid but are obeyed only in processes in which no neutrinos are emitted or absorbed. Later, we shall classify processes according to the strength of the interactions, that is, the forces involved. The weakest interactions are those involving the gravitational force. They play no role in particle processes. Next in increasing order of strength come the so-called weak interactions involving neutrino emission and absorption, as in beta-decay processes. These processes are also called Fermi interactions since Fermi developed a theory of beta decay based on such interactions. Then come electromagnetic interactions, involving the absorption and emission of photons, which are about a trillion times stronger than the weak interaction. Finally, we have the strong interactions, an example of which is the interaction between two nucleons, that is, the nuclear force. These interactions are about 137 times stronger than the electromagnetic interactions and hence about 100 trillion times stronger than the weak interactions. Conservation of parity and of strangeness is not valid for weak interactions but holds for electromagnetic and strong interactions.

At this point we discuss only parity; we shall return to the subject of strangeness later. By conservation of parity, we refer to the behavior of the wave properties of a particle. Since every particle is described by a wave amplitude, which depends on (is a function of) the coordinates of the particle, we can classify particles according to how their wave amplitudes or wave functions behave when the coordinates of the particles are replaced by their negative values. This simply means comparing the behavior of a particle in the real world with its behavior as seen through a mirror. If such a reflection leaves the wave amplitude unchanged, we say that the particle has even parity. But if the wave amplitude changes its sign when viewed through a mirror, we say that the parity is odd. Parity can thus have only two values: $+1$ (even parity) and -1 (odd parity). Conservation of parity means that the total parity of a system (obtained by multiplying together the parities of the individual particles in the system) is the same before and after a process occurs. As was first predicted by Lee and Yang in 1957, this principle is violated in weak interactions.

Before returning to our history of fundamental particles we must introduce one more principle that governs the classification of these particles, the Pauli exclusion principle, and the statistics of particles, which we have discussed in detail in earlier chapters of this book. We have seen that particles can be classified into two different groups according to the kind of statistics they obey. If particles are governed by the Pauli exclusion principle, they obey the Fermi-Dirac statistics and are called fermions; particles to which the Pauli principle does not apply obey the Bose-Einstein statistics and are called bosons. In general, particles with half a unit of spin are fermions and those with zero or one unit of spin are bosons. Electrons, nucleons (protons and neutrons), and neutrinos are fermions, whereas photons are bosons. A fermion can never change into a boson, and vice versa. However, a boson can break up into other bosons or into an even number of fermions plus additional bosons, and an even number of fermions (plus additional bosons when needed) can combine to form a boson. A fermion can only break up into an odd number of fermions plus bosons in any process.

Before leaving these general principles, which govern the overall behavior of particles, it is instructive to regard them from a different point of view that has a bearing on particle classification. As indicated in the chapter on symmetries (the work of Yang and Lee), a conservation principle is generally associated with some kind of symmetry in the dynamic properties of a system of particles. This means that there are quantities associated with the system that are invariant to certain changes and hence are related to quantum numbers. Thus, one should be able to describe particle properties by introducing the quantum numbers associated with the various symmetries and then arranging the particles into groups according to the numerical values of these numbers. We note that the conservation of momentum means that the dynamic structure of the system must be such that the structure is not altered when the system is shifted from one point to another in space. The conservation of energy implies that, in a system left to itself, the total dynamic properties do not change from moment to moment. The conservation of angular momentum means that the dynamic properties are not altered if the system is rotated in any way; hence, the system has symmetry about the axis of rotation. A quantum number is associated with each of these invariances. The rotational quantum number is of particular importance in analyzing atomic and molecular spectra.

The conservation of parity means the system is symmetrical with respect to a reflection; its dynamic properties are not altered when viewed through a mirror. The conservation of charge is associated with a quantity called gauge and implies that the dynamic symmetry of a system is such that its properties do not change when the gauge is changed. A change of gauge occurs when the wave amplitude of the system is multiplied by a factor. Conservation of charge means then that multiplying the wave function (or wave amplitude) by such a factor does not alter the state of the system. One can therefore introduce a quantum number, the so-called isotopic spin or isospin quantum number to represent charge just as we introduce a quantum number for angular momentum. In the same way, the conservation of strangeness and nucleon number is related to a symmetry that is of a mathematical rather than a physical nature and cannot be expressed in terms of spatial relationships like rotational symmetry or reflection symmetry. Nevertheless, these mathematical symmetries must be incorporated in our overall mathematical representation of particles and used in their classification. The conservation of strangeness and nucleon number means that quantum numbers must be assigned to represent these conserved quantities. As we shall see, the conservation principles of charge, of nucleon or heavy particle number, and of strangeness are related so that they are governed by a single type of symmetry in some kind of hypothetical space. The Pauli principle and the statistical properties of a system of particles are related to the invariance in the structure when two of the particles of the system are interchanged. This last type of symmetry is present because two identical particles, that is, electrons, cannot be distinguished.

Now that we have traced the relationship between conservation principles and symmetries, let us consider how these symmetries are investigated and are related to the dynamic properties of a system. This is done by a mathematical technique called group theory. The application of this technique to the analysis of the spectra of complex atoms and molecules was developed in its most productive form by Eugene Wigner, who was awarded the Nobel Prize in 1963 for his contributions to symmetry principles in physics. Where the dynamic symmetry properties of a system can be found by inspection, group theory does not have to be applied. But where the spatial symmetries are complex, as in the many electron atoms and in molecules, or are in a hypothetical space and are of a purely mathematical form with no apparent spatial counterpart, only group theory can lead to a complete analysis. This is another advantage to the group-theory approach. Mathematicians have been studying groups of all kinds for years, so that their properties are well known. One can therefore carry over these mathematical developments to physics, with the hope that the group properties uncovered by the mathematicians will reveal all the conservation principles and symmetries of the elementary particles in nature.

We can best illustrate the group-theory method by some simple examples. Suppose we have a system of particles and we shift it along a line by a small amount. This shift is called a translation and can be represented mathematically by changing the coordinate of each particle by this small amount. The wave function of the system is altered in some way by this process since the wave function depends on the coordinates. In quantum mechanics this change in the wave function is represented by a mathematical operator applied to the function—in this case, the momentum operator. Consider now all possible translations of the system along the given line. Each such translation is represented by an operator and all of these operators—an infinite number in this case—constitute a mathematical group. A collection or set of operators is called a group in two cases: (1) if the product of any two of the operators, that is, the application of one operator followed by another, is itself a member of the set; (2) if for each operator in the set the inverse operator is also present so that the application of an operator followed by the application of its inverse operator leaves the wave function and hence the state of the system unaltered.

To see how the analysis of a group provides insight into the physical structure of a system, consider a group of rotations about an axis. Because the dynamic structure of the system does not change when the system is rotated, the mathematical representation—in this case, a certain type of matrix—of the rotation group must have a structure such that when it is applied as an operator to the wave function of the system before rotation, the wave function (or wave amplitude) of the rotated system is obtained. For this to be true, the matrix representing the rotation, that is, the mathematical form of the group of rotations, must have a definite structure that can be derived from the mathematical properties of the group. The structural elements of the matrix representing the rotation group can be related to the various multiplets of the spectrum of the system. Group theory can thus be used to analyze spectra. It is hoped that symmetry groups can be used in the same way to classify particles.

We now summarize briefly the procedure that is used in classifying particles and arranging them into a structural scheme. One first finds all the conservation principles applicable to the system of particles and assigns quantum numbers to them. One then relates these quantum numbers to symmetries—spatial or purely mathematical—for which the group representations are sought. The mathematical structures of these groups then allow one to arrange the particles into multiplets in analogy with the spectral lines.

Let us return to our story of the discovery of the various particles and the ways in which they fit into the structural scheme of things. Disregarding for the moment the postulated neutrino, the first real break in the electron-proton picture of the universe came with the discovery of the positron, whose existence had already been predicted by the Dirac theory. Since this theory applies not only to the electron but also to protons, neutrons, and neutrinos, the existence of the positron, or, as we shall now call it, the antielectron, implies by analogy the existence of the antiproton, the antineutron, and the antineutrino. In fact, all particles must have their anticounterparts. The photon is its own antiparticle, but the antiparticles of spin $1/2$ particles are the mirror images of these particles. The introduction of antiparticles has in a sense saved the principle of conservation of parity, if this principle is enlarged to include charge. If the signs of all the charges in a system are changed, while at the same time the system is viewed through a mirror, its dynamic structure must remain unchanged. Even in weak interaction processes, reflections do not destroy the structural invariance, provided that each particle is replaced by its antiparticle. This phenomenon is called charge conjugation. We may say that our system must remain invariant to charge conjugation and reflection acting at the same time. This concept is generally referred to as CP (charge-parity) invariance. We thus see that the existence of antiparticles is demanded by the charge-parity symmetry in nature. This important property of antiparticles was not known when the positron was discovered but became apparent later.

The next great step away from the simple electron-proton dualism came with the discovery of the neutron. It proved its usefulness at once by fitting nicely into a consistent scheme for the nucleus and thus eliminating all the difficulties that would stem from a nucleus consisting of protons and electrons. In addition to simplifying nuclear structure, the neutron gave the first clue as to the nature of nuclear forces—the forces between nucleons (neutrons and protons) that keep the nucleus together. By analogy with the emission and absorption of photons, which account for electromagnetic forces between charged particles, Yukawa postulated the emission and absorption of heavy particles, now called mesons, to account for the strong forces between nucleons. The mesons would thus be the quanta of a nuclear force field just as the zero-rest-mass photons are the quanta of the electromagnetic field. These nuclear field quanta, now called pi-mesons or pions, were discovered about ten years after Yukawa's prediction and after physicists had been misled by the discovery of the mu-meson or muon. This is not a field quantum but a particle in its own right that seems to have no reason at all for existing. Today the muon is not referred to as a meson since the word meson is reserved for the quanta of the nuclear field and their excited states.

The pion has a mass of about 270 times the electron mass, a spin of zero units, obeys the Bose-Einstein statistics, and possesses three different electrical charges: positive, negative, and zero. Each of the charged pions decays into a muon and neutrino in about 3 one-hundred-millionths of a second. The neutral pion decays into two gamma rays in about a thousand-trillionth of a second. The positive pion is the antiparticle of the negative pion and the neutral pion is its own antiparticle. We shall presently see how the pion can be classified into a group-theoretical scheme, but for the moment let us examine the nucleon more carefully.

When the neutron was first discovered, it was treated as a fundamental particle in its own right—quite different from, but on a par with, the proton—except that it was not stable, being about 1.5 electron masses heavier than a proton and an electron, and hence decaying into a proton, electron, and antineutrino when left to itself. Then experimental evidence showed that the nuclear force between two protons was equal to that between two neutrons, or between a neutron and a proton. In other words, the strong interaction is charge-independent, so that, as far as it is concerned, the neutron and the proton appear to be identical particles. If there were no electromagnetic interaction, one could not distinguish between a proton and a neutron on the basis of their mutual nuclear interactions alone. We may therefore say that the proton and the neutron are a charge doublet, being two different energy states of a single particle, the nucleon.

This idea of a particle doublet led Heisenberg to introduce a new designation in analogy with the doublet spin states of an electron in a magnetic field. All electrons have a $1/2$ unit of spin, but this spin does not become apparent until a magnetic field is switched on. Two electrons with opposite spins are identical when no magnetic field is present. When the electrons are in a magnetic field their spins line up either parallel to the field or antiparallel to it. Those with parallel spins have a slightly different energy from those with antiparallel spins, so that one group of electrons can then be distinguished from the other. The electron is a magnetic doublet because of its spin. The reason the electron has only two magnetic states of energy is that its spin is $1/2$. If its spin were 0, it would have one magnetic state (singlet); if its spin were 1, it would have three magnetic states (triplet), and so on. In analogy with this idea, and because the proton and neutron cannot be distinguished when no electromagnetic interactions are present, Heisenberg suggested that the nucleon be assigned a new quantum number, $1/2$, called isotopic spin, with the components $+1/2$ or $-1/2$ (parallel or antiparallel) relative to some direction in an isotopic spin space, just the way the spin of the electron can have the components $+1/2$ or $-1/2$ along some chosen direction in actual space.

The following convention has been adopted for isotopic spin: a nucleon whose isotopic spin component is $+1/2$ is a proton; if its isotopic spin component is $-1/2$, it is a neutron. The charge independence of nuclear forces can now be expressed as the conservation of isotopic spin. When nucleons interact, the total isotopic spin is conserved if electromagnetic forces are disregarded.

Although isotopic spin was introduced at first as a purely mathematical device to express the charge independence of nuclear forces, physicists soon realized that it also presented a technique for expressing some of the symmetry properties of the new particles that were being so rapidly discovered. It also showed that group-theory methods could be applied to the classification of particle multiplets just as groups were used to classify spectral line multiplets. For many years mathematicians had been studying what are called unitary unimodular groups, which can be represented by matrices of a given number of rows and columns. Now the two-dimensional unitary unimodular group, called SU(2), can be represented by the three 2×2 Pauli spin matrices; hence, the ordinary spin multiplets of particles such as electrons can be derived from the structure of the SU(2) group. Heisenberg therefore suggested that a representation of the SU(2) groups also be used to describe the neutron-proton charge doublet. The three matrices would now represent the x, y, and z components of isospin, that is, of charge, but they would be governed by the same mathematical rules as the spin matrices.

This idea opened up a whole range of new possibilities. If we can have charge doublets, why not higher charge multiplets, such as a charge triplet. One would then describe such a triplet of particles by assigning to it the isotopic spin 1. This is in complete analogy with ordinary spin, where a particle having a spin 1 can have three different states of energy in a magnetic field. Its spin can orient itself in only three ways: parallel, perpendicular, and antiparallel to the field. If there were no magnetic field, these three states would be identical. In the same way, a charge triplet with isotopic spin 1 would represent three particles with $+1$, 0, and -1 units of electric charge, but otherwise identical. All the members of this charge triplet should behave exactly the same way in processes involving strong interactions but differently in electromagnetic interactions. The British physicist N. Kemmer was the first to point out that the three pions, π^-, π^0, and π^+, are precisely such a charge triplet and that they are indistinguishable in strong interactions. They differ only in an electromagnetic field and in their decay modes, which involve weak and electromagnetic interactions. The existence of a pion charge triplet is required by the conservation of isotopic spin, that is, the charge independence of nuclear forces. The nuclear force (disregarding the Coulomb repulsion) between two protons or between two neutrons can equal that between a proton and a neutron only if a neutral pion with very nearly the same mass as that of the two charged pions exists. Two protons or two neutrons can then interact via the virtual emissions and absorption of these neutral pions.

Kemmer's application of the isotopic spin concept to pions gave the first indication of how powerful symmetry principles might be in classifying elementary particles. Before we consider the most recently discovered particles and where they fit into the scheme of things, let us see what additional sense we can make of the particles already discussed. Apart from the muon and the neutron, there are two general groups of particles: those that are permanent like the electron and nucleon and those like the photons and mesons, which are the quanta of the electromagnetic and nuclear force fields, respectively. All the permanent particles (electrons and nucleons) have a half unit of spin, whereas the photons and mesons have unit or zero spin. This means that the statistics of permament particles are different from those of quanta. All known quanta (photons and mesons) are bosons, but the permanent particles are fermions.

Having separated particles into bosons and fermions, we can also separate the fermions into light and heavy particles. The light particles, like the electrons, are called leptons; the heavy particles are called baryons or hyperons. Thus far we have discussed only two baryons—the proton and neutron, which are really different states of a single charge doublet, the nucleon—but we shall presently consider many more baryons. In addition to the electron, we must count the neutrino and the muon as leptons. The muons are a compete mystery to physicists at the present time and seem to play no role in any structural scheme of matter. Except for their mass—about 208 electron masses—they are identical with electrons and positrons. They are fermions (spin $1/2$) and occur as particle μ^- and as antiparticle μ^+. Today physicists consider the muon a heavy electron, possibly an excited state of the ordinary electron or the positron. The neutrino is a permanent particle with zero charge, zero rest mass, and spin $1/2$ (a fermion). It is involved in all weak interaction processes. Several physicists have demonstrated experimentally that there are two different neutrinos: one associated with the electron, and the other with the muon. These two neutrinos are identical in all respects except that one always appears together with the electron in weak interactions, that is, beta-decay processes, and the other appears together with the muon in weak interactions, that is, the decay of the pion to a muon and a neutrino. Some physicists consider the neutrino as the zero charge state of the electron. This idea recommends itself if we consider the decay of a proton into a neutron, a positron, and a neutrino. Since an outgoing positron is equivalent to an incoming electron, we may describe the decay of a proton as an encounter between a proton and an electron in which an exchange of charge occurs. The electron gives up its charge to the proton, or vice versa, and in the process the proton becomes a neutron. At the same time, the electron reverts to its uncharged state—the neutrino.

Now that we have divided particles into fermions and bosons—the carriers of the field—and the fermions into leptons and baryons, we consider the baryons and the heavy mesons in more detail. The simple picture of the nucleon doublet and the pion triplet had to be discarded in 1950 when a whole range of new particles suddenly appeared in cosmic ray tracks. The most striking of these particles was what we now call the lambda or Λ-particle. This is a neutral particle associated with so-called V-shaped tracks. The two visible tracks forming the V shape can be analyzed and shown to arise either from the motion of a proton and a negative pion or from a negative and positive pion. Clearly these two particles cannot be created from nothing. Thus, this kind of V-shaped event means that some neutral particle, more massive than a proton and pion combined, must have decayed at the point where the tracks of the proton and pion meet. That this particle is neutral is indicated by the absence of any track leading to the V-shaped event. This Λ°-particle was shown to have a mass of 2182 electron masses and to decay either into a proton and negative pion or into a neutron and neutral pion. Since the Λ°-particle decays into a fermion and a boson, it must itself be a fermion with spin $1/2$.

We now know also that the Λ° is a neutral particle and hence a charge singlet with isotopic spin 0. However, when the Λ° was first discovered, its isotopic spin was not known. It was assumed to be $^1/_2$, even though no positively charged Λ° was found. This failure led to a certain difficulty which we shall discuss in a moment. Since the Λ° decays into a nucleon it must be included among the baryons. Its lifetime, about 3 ten-billionths of a second, led to difficulties related to its isotopic spin. In addition to the Λ° we also have the antilambda or $\overline{\Lambda}^\circ$. The bar over a symbol indicates the antiparticle.

At the same time that the V tracks arising from Λ particles were discovered, other kinds of V tracks were found. These were shown to arise from a π^+ and a π^- moving away from the point where the two tracks began. Again it seemed that some neutral particle must have decayed into a π^+ and π^- at this point since pions cannot come out of nothing. These neutral particles were called K°-mesons. Since they decay into bosons they must themselves be bosons with zero spin and are thus properly classified as mesons. Later, positive and negative K-mesons were found with masses close to that of the neutral K-meson—about 966 electron masses. This discovery prompted physicists to classify the K-meson as a charge triplet with isotopic spin 1 just like the pions. This led to difficulties of the same kind as those stemming from the Λ° particle. The lifetimes of the K-mesons range from about a ten-billionth of a second to about a millionth of a second.

We now consider the difficulties that were encountered when the Λ and K particles were first discovered and how they were resolved at about the same time by Murray Gell-Mann and K. Nishijima, working independently and starting from a suggestion by A. Pais. Problems with these particles first became evident from their lifetimes, which are much longer than one would expect in connection with their strong interactions and the way they must be produced. All processes involving strong interactions occur very rapidly—in about a trillionth of a trillionth of a second, the time it takes light to move across a nucleon—provided enough energy is available for the process. Now we know from the great abundance of V events found in cosmic rays that both Λ and K particles are very numerous and must therefore be produced very rapidly by strong interactions. One would, therefore, by the principle of reversibility, expect these particles to decay by strong interactions and thus have lifetimes of the order of 10^{-23} seconds. But they behave quite differently. Their lifetimes are about 100 trillion times longer than this quantity and must be the result of weak interactions probably related to beta decay. Even though the Λ° decays into what appears to be a strong reaction scheme—a proton plus a negative pion—the process is really a weak reaction. The puzzling question is why this is so. Because of this unexpected behavior, physicists referred to Λ and K particles as strange or queer.

The discrepancies between the way the strange particles are created and the way they decay were explained as follows. Consider an analogy with electrons and positrons. Suppose we allow a very energetic beam of gamma rays to impinge upon heavy nuclei. This beam would create electron-positron pairs in great abundance. The positrons would be destroyed quickly and we would only detect large numbers of left over electrons. If we did not know that positrons are created together with electrons we would be quite puzzled on finding that the electrons formed by the gamma rays do not decay but last practically forever. By the process of reversibility we would expect the electrons to decay as quickly into gamma rays as they are created by the gamma rays. But knowing that electrons are negatively charged particles and that charge must be conserved in all processes, we realize that every time an electron is created a positron is also created to keep the total charge zero, just as it was before the creation of the pair. We see then that an electron cannot decay into gamma rays by itself—it can only do so together with a positron. Thus, charge conservation keeps the electrons alive when no positrons are around.

It occurred to A. Pais, then at the Institute for Advanced Study at Princeton and now a professor of physics at Rockefeller University, that a similar process applies to the creation and decay of strange particles. Suppose that in addition to electric charge there is still another kind of charge in nature that is not found on electrons, nucleons, or pions; let us call this charge strangeness. Suppose, moreover, that this charge is conserved in all strong interactions but need not be conserved in weak interactions. If strangeness now occurs in integer units, both positive and negative, we can, following Pais, clear up the lifetime difficulty described above. We proceed in analogy with positron-electron pair creation, which conserves electric charge. We assume that strangeness is conserved in all strong interactions so that the two or more strange particles with opposite strangeness charges must be created simultaneously in a strong reaction. Pais did not call this pair production but associated production. This idea eliminates the lifetime difficulty. All we need do is assume Λ° and K° particles have opposite strangeness and that whenever a Λ° is produced in a strong reaction a K° must be produced. This process conserves strangeness. But once a K° and Λ° are created, they move away from each other and neither one of them has any way of decaying alone without violating the conservation of strangeness. This means that neither one can decay by a strong reaction in the very short time (about 10^{-23} seconds) associated with strong interactions. But after a long enough time (about a hundred-millionth of a second) they decay through a weak interaction for which strangeness conservation is not required.

We now come to the important contribution of Gell-Mann and Nishijima. They were the first to introduce a quantum number of strangeness and to show how it was to be calculated for a particle. They then used the concept of strangeness to organize elementary particles into a reasonable scheme and to analyze particle reactions. To see how strangeness is defined we separate the particles into baryons and mesons and consider baryons first. As we have already indicated, any particle that decays into, or can be formed from, one or more nucleons is a baryon. We assign to each such particle a baryon number B or nucleonic charge equal to the number of nucleons hidden in it or, put differently, to the number of protons that ultimately appear in its final decay. Thus, a nucleon (proton or neutron) has a nucleonic charge or baryon number $B = 1$ and a Λ° particle also has $B = 1$, and so on. All antibaryons have negative baryon numbers. Thus, antinucleons and antilambda particles have $B = -1$. To obtain the total baryon number in a system, we add all the Bs together (taking positive values for baryons and negative values for antibaryons). A system consisting of a proton, neutron, and Λ° has $B = 3$; a system with one proton and one antilambda has $B = 0$, and so on. The baryon number is conserved in all processes and is zero for all leptons and mesons.

We consider now the electric charge or isotopic spin of baryons. We have seen that we can group baryons into charge multiplets. Since the two nucleons form a charge doublet with isotopic spin $1/2$, it was assumed that all higher baryons would be charge doublets of the same kind. When the Λ° particle was first discovered, it was thought that its isotopic spin was $1/2$ and that a Λ^+ would be found sooner or later. This conclusion was based on an analogy with the neutron-proton charge doublet and the fact that Λ° has a baryon number $B = 1$. It was felt that Λ particles ought to be charge doublets like the nucleons since they were members of the baryon family. But one day, while discussing the heavy strange particles, Gell-Mann referrred to them inadvertently as particles of isotopic spin 1. He quickly corrected himself, saying he had meant $1/2$, not 1. Later, however, after thinking about this slip of the tongue, he had the brilliant notion that his first statement might have been the correct one and that Λ particles were strange precisely because they were not charge doublets. If they were not doublets they must then either be charge triplets with isotopic spin 1 or charge singlets with isotopic spin 0. Gell-Mann decided to pursue this idea. If, following Gell-Mann, we suppose that normal baryons, like the nucleon, are only those charge multiplets whose average charge (obtained by adding the charges of all the members of the multiplet and dividing by the multiplet number) equals $1/2$, then strange baryons are those whose average charge does not equal $1/2$. The strangeness would then be measured by how much the average charge of the strange charge multiplet differed from $1/2$.

Gell-Mann reasoned that one should be able to arrange the strange particles into an orderly scheme by assigning a strangeness quantum number S to each hyperon charge multiplet and imposing the condition that S be conserved in all strong interactions. Gell-Mann and Nishijima independently defined S as equal to twice the average charge of a charge multiplet minus its baryon number. One can show from this definition that the strangeness of an antibaryon multiplet is the negative of the corresponding baryon multiplet, since the baryon number B changes sign when we go from baryon to antibaryon.

Although the introduction of the concept of strangeness led to a simplification of particle classification, the situation was quite unsatisfactory from the point of view of a single basic principle from which the mass spectrum of the various families of particles could be deduced. Nor could scientists understand the occurrence of the various particles in terms of any single dynamic principle that would relate the massive baryons to the nucleon and the massive mesons to the pions. It was therefore necessary to uncover some universal principle or some all-encompassing symmetry from which all else could be derived. Two ideas were then current and are still being investigated. The first is the bootstrap theory which starts from the premise that none of the strongly interacting particles (baryons and mesons)—called hadrons as a group—are fundamental particles but that each one is a dynamic bound state of various combinations of all hadrons including itself. One then seeks to obtain a self-consistent way of describing all strongly interacting particles in terms of combinations of what are essentially force fields. This concept is closely related to Bohr's idea that the nucleus should not be treated as a collection of nucleons, but rather as a system of interacting force fields. The second idea is that a fundamental set of particles and fields does exist and all other particles can be considered as excited states or resonances of these basic particles. This view would go back to Democritus's atomistic concept and seek to explain all known particles as constructed of these basic particles.

Apart from the two schemes described above the first attempts to explain the baryon mass spectrum in terms of an overall symmetry—called global symmetry—were made independently by Gell-Mann and Schwinger in 1957. The idea behind global symmetry, which is not a very clearly defined concept, is that all baryons are identical particles as far as interactions via the pion field are concerned. However, they show mass differences when they interact via K-mesons. Thus, the π-meson field was treated as the source of very strong baryon interactions and hence highly (globally) symmetrical. But the K field was treated as only moderately strong and hence of an unsymmetrical nature. It was hoped that the baryon mass spectrum could be derived from the broken symmetry introduced by the K field, just as the electromagnetic field breaks the mass symmetry between a neutron and a proton, which otherwise are identical (charge independence of nuclear forces). This hope was dashed, however, when Abdus Salam demonstrated in 1959 that the observed scattering of pions by nucleons is different from their scattering by hyperons (baryons other than nucleons). This result disagreed with the global idea, which requires that pions interact equally with all baryons.

After the global symmetry theory was discarded, particle physicists continued their search for a basic underlying symmetry that could lead to a single classification scheme for all particles. Although one cannot say at present that such a basic symmetry has been found, some remarkable successes have been achieved in this field. The basic motive in all of this work is to establish a universal symmetry for all particles and then to show that the mass spectrum of these particles can be explained in terms of some kind of symmetry-breaking effect. A symmetry-breaking effect means that a weak force field upsets the symmetry imposed by the strong force. Since symmetries mean that certain quantities remain invariant when operators are applied to the dynamic structure of a system, and these operators can be arranged into mathematical groups, particle physicists reverted to group theories to find basic groups whose structures would give the desired results. This work has been variously designated as the Eightfold Way, the Octet Unitary Symmetry Model, and the SU(3) Model, and has been developed as an extension of the SU(2) symmetry model applied to nucleons.

Higher group symmetry was introduced as an extension of the SU(2) group. This type of group-theoretical approach led to a simple classification scheme. Since there were just eight known baryons when the search for symmetry was begun, it was argued that these should form a supermultiplet, an octet which would then break up into separate isotopic spin, or charge, multiplets as the result of a symmetry-breaking effect like the electromagnetic field. The SU(2) symmetry group, which gives the charge or isotopic spin multiplets, should be contained in a larger symmetry group. The baryons would then correspond naturally to an eight-dimensional irreducible representation of this enlarged symmetry group, and the familiar charge multiplets would appear when mass differences were introduced. Thus, by a simple generalization of charge independence (SU(2) symmetry), one would obtain the enlarged unitary symmetry that would give the desired result.

Since the SU(2) group is represented by all unitary 2×2 matrices—and there are just three such independent ones, which give the three components of isotopic spin—one is led to the SU(3) group as a natural extension of SU(2). The SU(3) group is represented by all unitary 3×3, that is, 3 rows and 3 columns, matrices. It can be shown that there are just eight such independent matrices and that these can be arranged to correspond to isotopic spin, to baryon number—or rather, hypercharge Y, which is defined as baryon number plus strangeness—and to strangeness. The first three of the eight matrices are just the components of isotopic spin, four others are related to strangeness, and the eighth is proportional to hypercharge Y. Thus, this unitary spin group has an eight-dimensional representation (hence the designation eightfold way), which corresponds to the baryon supermultiplet. When the symmetry of this group is broken, isotopic spin and hypercharge (baryon number and strangeness) are still conserved. But the four quantities associated with the four remaining matrices are not conserved and the baryons split into four groups: xi, lambda, sigma, and nucleons. The electromagnetic field partially destroys the conservation of isotopic spin so that these groups break up into charge multiplets. Finally, weak interaction breaks the symmetry associated with hypercharge or strangeness, and only conservation of baryon number and of electric charge remains as absolute.

The SU(3) symmetry group was extended to an SU(6) symmetry by combining ordinary spin with isotopic spin and strangeness, which together give the fundamental triplet of SU(3). Since angular spin can have two orientations in space, this immediately doubles the multiplicity of the symmetry and leads to an SU(6) group of operators. The results of this wider symmetry are in remarkably good agreement with observation. One of its most notable successes is that it gives the value $2/3$ for the ratio of the magnetic moment of the neutron to that of the proton. This is the first time that a theory has succeeded in explaining this ratio.

The various multiplets of the SU(6) symmetry group are easily represented by a particle graph in which the mass is plotted along the ordinate and the strangeness is plotted along the abscissa. We first consider the baryon spectrum. According to this representation the various particles in the graph are to be looked upon as resonances or excited states of the nucleon. This simply means that all of these short-lived states can be reached by supplying the appropriate amount of energy to the nucleon. But these excited states, in general, differ from the ground state (nucleon) not only in energy (mass) but also in ordinary spin, in charge, and in strangeness or hypercharge. These excited steps or resonances decay to the ground state in one or more steps with the emission of pions, K-mesons, leptons, or photons. The particular type of particle that is emitted in a transition from the excited to the ground state is determined by the difference in quantum numbers between the two states and the mass difference. If the two states have the same strangeness, pions are emitted. But if they differ in strangeness, K-mesons must be emitted since these are the carriers of strangeness.

Ultimately, the factor that determines the kind of boson that can be emitted when a system goes from one state to another is conservation of energy. A transition will not occur unless a boson with a mass equal to the energy difference between the two states can be created. To take the simplest and best known case, that of electronic transitions in excited atoms, the only bosons that can be emitted are photons—by electromagnetic coupling. The energy differences between the excited states of atoms are much too small for the emission of lepton pairs (electron-neutrino pairs) or for meson emission, even though the electron in its excited state is coupled to the meson and lepton fields in addition to the electromagnetic fields. In the transitions between the excited states of an atomic nucleus not only are photons emitted, but there is also enough energy for electron-neutrino pair emission by weak interaction coupling to the lepton field. But there is not enough energy for meson emission. Note that an electron-neutrino pair may be treated as a boson since it consists of two fermions. In transitions between the excited states of the nucleon (hyperons and resonances) enough energy is available for mesons. However, K emission can occur only between the highly excited hyperon states and the lower states because of the very large mass of the K-meson. Conservation of energy forbids the de-excitation of the lambda, sigma and xi states by the emission of *K*-mesons, even though this would be allowed if enough energy were available since these states differ in strangeness from the ground state (nucleon). Thus, if strangeness were absolutely conserved, the lambda, sigma, and xi states would never decay. But since there is enough energy for the emission of pions and lepton pairs through weak interactions, and since strangeness does not have to be conserved in such interactions, these states do decay by such emissions.

Thus far almost everything that we have described is not much more than very significant classification. We have said little about the few attempts that have been made to explain the baryon and meson spectrums in terms of a dynamic picture, that is, by means of some kind of internal structure of hyperons. The question is whether all of these hadrons (baryons and mesons) can be explained in terms of some basic entities that we might call elementary and which therefore have a permanence. Looked at from this point of view the baryons are essentially different from the mesons; baryons cannot disappear, whereas mesons can and do disintegrate. This fact indicates that there is a real difference between these two types of particles. This idea is supported by the work of Robert Hofstadter, which shows that the nucleons consist of a core surrounded by a pulsating cloud of mesons. In terms of such a picture any baryon may be described as some excited state of the nucleon combined with a few mesons. However, from another point of view, if we consider the properties of baryons and antibaryons, any meson may be pictured as a combination of baryons, antibaryons, and some mesons. By clever manipulation we can to some extent consider all hyperon states as combinations of each other. This idea is the basis of the bootstrap procedure that we have already discussed.

Since bootstrap physics leaves much to be desired and cannot point to very many successes, we must look for other avenues into the structure of hadrons. Here the concept of strangeness may be of great help because the supermultiplets of SU(6) are divided into submultiplets according to the value of the strangeness. It appears that strangeness is related to some kind of symmetry in the dynamics of hadron structure. Possibly this new symmetry is related to basic structural elements. First consider how the introduction of strangeness enlarges the symmetry properties of hadrons. If there were no strangeness, the basic symmetry of our particles would be the twofold isotopic spin or charge symmetry. Its multiplet structure would be described by the SU(2) symmetry group, which has a basic doublet structure. To see how a doublet symmetry group can be used to construct complex systems, consider the proton and the neutron, the two members of the isotopic doublet $I = 1/2$. By combining protons and neutrons in different ways we can obtain the known isotopes of the nuclei. Thus, a system of two nucleons gives us either a singlet with isotopic spin 0 (the isotopic spins of the two nucleons are antiparallel) or a triplet with isotopic spin 1 (the two individual isotopic spins are parallel). Three nucleons give us isotopic doublets (total isotopic spin $1/2$) and quartets (total isotopic spin $3/2$). Thus, we see from the example just given that the SU(2) multiplets in general are derived by combining two basic units, which we may take as the two possible projections along a hypothetical z axis of the isotopic spin $1/2$. The only possibilities for such a projection for $I = 1/2$ are $I_z = 1/2$ and $I_z = -1/2$.

Suppose now that to these two basic structural elements we add a third entity with $I_z = 0$ but with $S = -1$. If we assign the value $S = 0$ to the two entities with $I_z = +1/2$, we can get a basic triplet. Next, by combining these three basic elements (which have been named quarks) in various ways, we try to obtain the observed baryons. Various quark combinations have been shown to correspond to known hadrons. The two quarks with $I = {}^1/2$ and with $S = 0$ are represented as circles; the quark with $I = 0$ and $S = -1$ is represented as a square. We now picture each baryon as consisting of exactly three quarks, which may be chosen from among the three basic quarks in any way. Expressed pictorially, every baryon consists of circles and squares combined in groups of three. If we choose all circles, we must have $S = 0$, but we can have $I = {}^1/2$ or $I = {}^3/2$, in other words, charge doublets and triplets. By combining one round quark and two square quarks we get $S = -2$ (note that the strangeness quantum numbers are added algebraically) and $I = {}^1/2$, a charge doublet. Finally, by combining three square quarks we obtain $S = -3$ and $I = 0$, a singlet. These states of isotopic spin and strangeness are just those given by the multiplets of the SU(3) group. The introduction of strangeness has thus enlarged the symmetry group from SU(2) to SU(3) and has led to a basic triplet.

One can now enlarge the SU(3) to an SU(6) symmetry group by including ordinary spin. If we assume that each quark has an angular spin of $^1/2$, so that it can have just two orientations in space, we obtain six basic quark states, which leads to the SU(6) group. Note that if ordinary (angular) spin is not taken into account, the isotopic spin and strangeness lead to thirteen quark states: 2 for $I = {}^1/2$, 4 for $I = {}^3/2$, 1 for $I = 0$, and so on. If we now include ordinary spin we obtain twice this number of 26 states. These can be divided into two octets and one decuplet. The octets are associated with angular spin sigma $= {}^1/2$, $^3/2$, and the decuplet is associated with spin $^3/2$. There is also a singlet in this scheme with negative parity and spin $^1/2$ whose existence can be derived from a quantitative analysis of the SU(6) group.

One can derive the meson supermultiplets from quarks by picturing each meson as consisting of one quark and one antiquark. The strangeness of an antiquark is the opposite of that of the quark. The round antiquark has strangeness 0 and the square antiquark has strangeness $+1$. If all possible combinations of two such quarks are made from the $S = 0$ (round) and $S = -1$ (square) quarks and from their anticounterparts, we obtain the meson supermultiplets. This quark-antiquark combination is necessitated by certain properties of mesons. Each supermultiplet contains both mesons and their antiparticles with opposite strangeness; triplets with $S = 0$ consist of mesons and their own antiparticles; mesons with $I = {}^1/2$ have strangeness numerically, that is, regardless of sign, equal to 1. It is easy to see from these three properties that only a quark-antiquark combination can give mesons. If ordinary spin is included, the meson spectrum breaks up into two octets: one with spin sigma equal to 0 and the other with spin equal to 1. Just as in the baryon spectrum there is also a singlet state phi with negative parity and spin 1.

If the three types of quarks and antiquarks were identical in mass, all the members of a supermultiplet would have the same mass; of course, this situation does not exist. This state would correspond to complete symmetry. There must therefore be some weak force field between the quarks that breaks this symmetry and gives rise to mass differences among the members of a supermultiplet. We may represent this asymmetry by allowing the mass of the $S = -1$ quarks to be somewhat larger than the mass of the other two quarks. This means that the supermultiplet would break up into strangeness submultiplets of different masses. The more square (strangeness-carrying) quarks we include in a hadron, the more massive it is under these conditions. Therefore, the mass of a hadron should increase with increasing numerical value of the strangeness. Our baryon and meson spectrum graphs show that this is the case. The differences in the masses of the members belonging to the same submultiplet of given strangeness arise from the electric charge differences within the submultiplet. Put differently, the coupling to the electromagnetic field breaks the symmetry of the supermultiplet still further, thus leading to mass variations within each strangeness submultiplet.

When the basic quark triplet was introduced, no baryons of strangeness -3 were known. The possible quark combination □□□ indicates, however, that a baryon with $S = -3$ should exist if the triplet picture is correct. This baryon, the omega, was subsequently discovered, lending considerable support to the SU(6) theory.

Despite the many successes quark theorists have enjoyed in presenting an orderly and understandable picture of the hadron spectrum, there is no experimental evidence that the quarks are anything more than convenient mathematical fictions for expressing the known symmetries. We are certainly not justified at this point in concluding that a nucleon consists of three physical entities having the properties described above simply because the baryon spectrum can be explained in terms of such a model. In a sense, this would be equivalent to concluding that the photon is composed of an electron-positron pair because the unit spin and the zero charge of the photon can be derived from such a model. We are thus left with a tantalizing and incomplete picture in which the observations are arranged in a very orderly scheme based upon symmetry principles but give us no real understanding of what these symmetries mean for the dynamic structure of the baryon. We may, indeed, have only begun to scratch the surface of the baryon states that can be excited if enough energy is available. It may well be that only by going to very high energies that we can solve the problem of baryon structure.

The baryon spectrum itself seems to tell us the same. The existence in this spectrum of many different resonances lying relatively close together (the masses of all the hadrons lie within about 1 BeV of each other) means that these, taken together, are but the fine structure of a single ground state. Higher states at much greater energies are still to be found. This statement implies that there are perhaps two different kinds of strong forces at work, one extremely strong and the other moderately strong. The very strong force is responsible for the overall symmetry and it keeps the dynamical structure of the hadron invariant to the SU(6) group of operators. If this very strong force were the only force at work, we would observe no supermultiplets since the masses of all the baryons thus far observed would be equal. But the moderate strong force breaks the SU(6) symmetry and gives rise to the observed supermultiplets. This is similar to the fine structure of the spectral lines of an atom. If there were no electromagnetic interaction among the outer electrons of an atom, the rotational symmetry imposed on the atom by central force arising from the nucleus would give rise to well-separated spectral lines with no multiplet structure. But the electronic interactions destroy this symmetry and cause each of the lines to break up into multiplets. If we apply this same reasoning to baryons, we see that we are simply observing the effects of the moderately strong force. The excited states associated with the very strong force will only be observed when we use excitation energies many times larger (billions of electron volts) than the energy differences between the baryons now observed.

In our search for the hadron structure we must not forget that what is perhaps the most fundamental and simplest question of all remains unanswered: What is the structure of the electron and why does it have its observed properties? Although the quantum electrodynamics has given us a way of calculating the interaction of the electron with the electromagnetic field to an amazing accuracy, it has done so at the expense of denying us any insight into the origin of the mass or the electric charge of the electron. The values for these quantities are to be accepted as preordained; they are shrouded in the mystery of renormalization—a scheme that relieves us of the mathematical task of having to work with infinities, but burdens us with a deep sense of incompleteness. Added to the problem of the electron is the problem of the muon and the two neutrinos. Possibly, in analogy with the picture of the nucleon as a core surrounded by a cloud of mesons, we may picture the electron as a core surrounded by a cloud of photons. But how such a structure can maintain itself against very rapid decay is difficult to see, unless we introduce enormously large attractive force fields inside the electron itself.

We complete this essay on high-energy elementary particle physics with a brief description of its more recent developments, most of which are theoretical and, hence, still unconfirmed. With the introduction of the three kinds of quarks u, d, and s (called flavors) particle physicists immediately accounted for the splitting up of the baryons into two groups (multiplets), the octet and the decimet, by arranging the different flavors into subgroups such as udu (proton), dud (neutron), dus (Λ°) in the octet and uuu, and so forth, in the decimet. However, a difficulty concerning the statistics and the Pauli exclusion principle immediately arose with this quark model of the baryons. This model assumes that all three quarks in a baryon are in the same ground state, that is, that the wave function that defines the ground state of the three quarks is symmetrical in their spatial coordinates (the interchange of any two quarks leaves the sign of the wave function unchanged). However, since the quarks themselves are fermions (spin $1/2$ particles), their total wave function must be antisymmetrical. By the total wave function of the quarks we mean a function not only of the quark spatial coordinates but also of their spins and flavors. No problem with the statistics arises in this model for the octet because no member of the octet consists of three identical flavors (such as uuu, ddd, or sss). Therefore, no conflict with the Pauli principle occurs in the octet. The total wave function of any one of the octet members with two of its quarks with identical flavors can be made antisymmetrical by having the spins of the two identical flavors oppositely aligned so that interchanging the spins of the two identical quarks changes the sign of the wave function of the baryon as demanded by the Pauli principle.

The decimet, however, gives trouble because each of its members Δ^{++} (uuu), Δ^- (ddd), and Λ^- (sss) has three identical flavors and therefore is described by completely symmetrical wave functions; it is symmetrical in spin, flavor, and spatial coordinates, which is forbidden. To overcome this difficulty (or avoid it) Gell-Mann introduced the concept of quark colors. He suggested that each quark had an additional property (in addition to mass, spin, and electric charge) which he called color. This has nothing to do with the color sensation associated with light; it is just another property assigned to quarks to permit us to obtain antisymmetrical wave functions for quark triplets. A wave function of a quark triplet which is symmetrical in spin, charge, flavor, and spatial coordinates can be made antisymmetrical in color. Since baryons themselves exhibit no color, the colors of the three quarks in a baryon must be so chosen as to give the baryon zero color (white). This is achieved by having each quark flavor come in three colors—red, green, and blue, which give white when mixed together. An antiquark has a property called anti- or negative color. Therefore, mesons are colorless since a meson consists of a quark of definite color and an antiquark with the negative or anticolor. Color was thus perceived as something like electric charge.

At this point in the development of this very complex theory, particle physicists suggested that the very strong interaction between two quarks (the strong interaction) be described by a color field similar to the electromagnetic field that produces the electromagnetic interactions between electric charges. The analogy between the electromagnetic field and the color field was pushed further by the concept of gluons as the quanta of the color field in analogy with photons as the quanta of the electromagnetic field. The gluons, tossed between quarks, are supposed to produce the strong quark interaction, but they differ from photons in one very important property. Whereas photons do not interact with each other because they do not carry electric charge, gluons must carry color and therefore must interact with each other. But the colors of gluons are produced by color and anticolor combinations such as rr, rg, rb, gr, and so forth (where r, g, b stand for red, green, and blue, respectively). Combining the three colors in doublets in this way gives eight distinct gluons. Owing to their anticolor components, these gluons can change the colors of the quarks which emit or absorb them to other colors. Thus, if a blue quark emits a blue-antired (a $b\bar{r}$ gluon combination), the blue quark becomes a red quark and the red quark that absorbs this gluon becomes blue.

Just as quantum field theory led to quantum electrodynamics (QED) when applied to the electromagnetic field and its interactions with charged particles, so a quantum field theory called quantum chromodynamics (QCD) was developed to handle the color interaction between quarks. But QCD is many times more complex than QED owing to the need to introduce eight gluons instead of a single gluon. Moreover, since gluons interact with each other, QCD is a nonlinear field (a self-interacting field like the gravitational field). Fields such as the gluons and the photon fields are called gauge fields because the magnitudes (amplitudes) of these fields can be multiplied by a complex phase factor without changing the physical properties of the field. Such an operation is called a gauge transformation. The electromagnetic field is called an Abelian (named after the mathematician Abel) gauge field because photons do not interact with each other. The color field is called non-Abelian because gluons do interact with each other.

With all the complex mathematical machinery and all the sophisticated physical assumptions that have been introduced to handle QCD and obtain deductions from it, the results have been very meager and at the cost of replacing simplicity by complexity. Taking color and gluons into account we see that QCD as it now stands is burdened with some 30-odd arbitrary parameters. We can hardly speak of this theory as a simplication or unification of physics. This complexity has been aggravated by the introduction of three additional quarks called charm, bottom, and top, each in three colors, though these quarks play no essential physical role in the scheme of things.

QCD also leaves much to be desired in predicting the masses of the individual quarks and the magnetic moments of baryons. It gives individual quark masses which are nowhere near the empirical masses required to account for the measured masses of the nucleons. Nor can it give the magnetic moments of the proton and neutron. At best it gives the ratio of the two magnetic moments as $2/3$ without accounting for the fact that the proton magnetic moment is positive whereas that of the neutron is negative.

The scattering of high energy electrons by nucleons produced certain conclusions about the nature of the gluon gauge field inside baryons which complicate the entire picture even more. The information carried by the scattered electrons is that the three quarks they encounter inside a baryon behave as though they were free, with no forces between them when they approach each other. This phenomenon is called asymptotic freedom. On the other hand, the attractive gluon field seems to increase rapidly as the quarks try to recede from each other. This phenomenon is called infinite confinement. Thus, the gluon gauge field differs drastically from the gravitational field and the electrostatic field which decrease with distance.

Most of the difficulties and complexities of the QCD quark-gluon gauge model are eliminated if one differentiates between the masses of quarks when they are bound inside baryons and when they are free, as one of the authors, L. Motz, has proposed. If quarks, when free, have a mass of 10^{-5} grams (we identify quarks with particles called unitons which have the Planck mass), then the gravitational force is the dominant force in nature and inside baryons, which may be pictured as linear gravitational rotators. This uniton model of baryons gives the correct masses of the bound quarks and the correct magnetic moments of the baryons in the octet. The gravitational binding energy of the triplet of unitons (quarks) inside a baryon is so large (10^{19} GeV) that the unitons behave as though they were infinitely confined.

What about the weak interaction which governs the behavior of neutrinos? Particle physicists have tried to explain this phenomenon by a special gauge field. The theory of such a gauge field has been developed by Steven Weinberg, Sheldon Glashow, and Abdus Salam, for which they received the 1979 Nobel Prize in physics. Like all other gauge fields, the weak interaction gauge field requires a boson (a quantum). But in this case the boson is a massive (non-zero rest mass) particle of spin 1, so that it is called a vector boson. Weinberg showed that the rest mass of such a boson must be about 80 GeV and that three such bosons (two charged and one uncharged) must exist to account for all the observations of neutrino interactions. The two charge bosons are called W^+ and W^- and the neutral one is called Z°. In a special experiment performed at CERN with a beam of protons colliding with a beam of antiprotons at energies of 540 GeV, Carlos Rubbia and his collaborators obtained tracks of the products of proton-antiproton collisions which were interpreted as stemming from the decay of W^- bosons with a rest mass of about 84 GeV. Later, the discovery of W^+ bosons of about the same mass was announced and, still later, the Z°, with a mass of 92 GeV was proclaimed.

These discoveries were heralded as the unification of the weak and the electromagnetic interaction (the so-called electroweak force) even though the boson of the electromagnetic field, the photon, has zero rest mass and the intermediate vector bosons have very large rest masses. This asymmetry is accounted for by introducing the assumption that during the initial stages of the universe, immediately after the big bang, all forces were equal and all gauge bosons had zero rest mass. But, when the universe cooled slightly, this perfect symmetry was broken by a special field (or fields) which assigned masses to the intermediate bosons but left the photon massless.

Index